Google It

Newton Lee
Editor

Google It

Total Information Awareness

 Springer

Editor
Newton Lee
Newton Lee Laboratories, LLC, Institute for
  Education, Research and Scholarships
Woodbury University School of Media,
  Culture and Design
Burbank, CA
USA

ISBN 978-1-4939-8192-2   ISBN 978-1-4939-6415-4   (eBook)
DOI 10.1007/978-1-4939-6415-4

Printed on acid-free paper

This Springer imprint is published by Springer Nature
The registered company is Springer Science+Business Media LLC New York

To peace, love, and freedom

Acknowledgements

The research effort for *Google It* was partially supported by a Woodbury University faculty development grant and adjunct faculty micro grant.

Google It (Soundtrack) is a companion music album that features all the songs mentioned in the book. I would like to thank Roger Hodgson (Supertramp) for his permission to include "The Logical Song" in the album, and also to thank the amazing international musicians and administrators involved in the companion music album: René Baños, Jake Coco, Keith Cooper, Brody Dolyniuk, Stefanie Field, Linda Gianotti, Poney Gross, Chris Hathcock, Seyfulla Mustafayev, Patrick Printz, Athena Reich, Shakti Shivaya, Pete Stark, and Princess X.

Track listing:

1. "The Logical Song" written and performed by Roger Hodgson (Supertramp), USA
2. "Another Brick in the Wall, Pt 2" performed by Bizimkiler, copyright by ANS TV, Azerbaijan Republic
3. "Hotel California" performed by Vocal Sampling, Cuba
4. "In the Year 2525" performed by Pete Stark, Australia
5. "Time" performed by Brody Dolyniuk, USA
6. "Losing My Religion" performed by The Reticent (Chris Hathcock), USA
7. "With a Little Help from My Friends" performed by Adam Christopher, Tay Watts, Corey Gray, and Jake Coco, USA
8. "Free" performed by Princess X, USA

A Word From Roger Hodgson:

"The Logical Song was born from my questions about what really matters in life. Throughout childhood we are taught all these ways to be and yet we are rarely told anything about our true self. We are taught how to function outwardly, but not guided to who we are inwardly. We go from the innocence and wonder of childhood to the confusion of adolescence that often ends in the cynicism and disillusionment of adulthood. In The Logical Song, the burning question that came down to its rawest place was 'please tell me who I am.' I think in these modern times, the more complex life becomes, this eternal question becomes ever louder—which is why the song continues to strike such a deep chord in people around the world."

About the Book

From Google search to self-driving cars to human longevity, is Alphabet creating a neoteric Garden of Eden or Bentham's Panopticon? Will King Solomon's challenge supersede the Turing test for artificial intelligence? Can transhumanism mitigate existential threats to humankind? These are some of the overarching questions in this book, which explores the impact of information awareness on humanity starting from the Book of Genesis to the Royal Library of Alexandria in the 3rd century BC to the modern day of Google Search, IBM Watson, and Wolfram|Alpha.

The book also covers Search Engine Optimization, Google AdWords, Google Maps, Google Local Search, and what every business leader must know about digital transformation. "Search is curiosity, and that will never be done," said Google's first female engineer and Yahoo's sixth CEO Marissa Mayer.

The truth is out there; we just need to know how to Google it!

Contributors

Dennis Anderson, Ph.D., St. Francis College
Penelope Beery, Ed.D., Knowledge Building in Action
Dirk Bruere, the Transhumanist Party
Frank Buddenbrock, Google AdWords Certified Specialist
Lincoln Cannon, Mormon Transhumanist Association
John Cassel, Wolfram Research
Nicole Ciomek, Radiant PPC
Tina Courtney, Evolve Inc.
Andrew Donaldson, Bolder Super School
Robert Epstein, Ph.D., American Institute for Behavioral Research and Technology
Dennis Gamayunov, Ph.D., Lomonosov Moscow State University
Barret Havens, MLIS, Woodbury University

David J. Kelley, Microsoft MVP and Futurist
Christoph Lahtz, Ph.D., Medical Researcher and Transhumanist
Newton Lee, Institute for Education Research & Scholarship and Woodbury University School of Media Culture & Design
Sandra Lund-Diaz, M.Ed., Knowledge Building in Action
Trond Lyngbø, Search Planet AS
Aylin Manduric, University of Toronto and Center for the Study of the Presidency and Congress
Darren Manners, Sycomtech
Mireia Montane, Ph.D., Knowledge Building in Action
Robert Niewiadomski, Educator
Frances Eames Noland, University of Oxford
Emily Peed, Institute for Education, Research, and Scholarships
Jennifer Rosenfeld, Woodbury University
Cyrus Shahabi, Ph.D., University of Southern California
Tiana Sinclair, Futurist
Zach Tolan, the Polymathic Prodigy Institute
Natasha Vita-More, Ph.D., Humanity+ and University of Advancing Technology
Mikhail Voronov, Lomonosov Moscow State University
Lewis Watson, Marshall Fundamental School
Nyagoslav Zhekov, Whitespark Inc.

Contents

About the Editor

Newton Lee is CEO of Newton Lee Laboratories LLC, President of the Institute for Education, Research, and Scholarships (IFERS), Adjunct Professor at Woodbury University School of Media, Culture and Design (MCD), editor-in-chief of Association for Computing Machinery (ACM) Computers in Entertainment, and U.S. presidential campaign advisor for the Transhumanist Party.

Previously, Lee was a computer scientist at AT&T Bell Laboratories, senior producer and engineer at The Walt Disney Company, and research staff member at the Institute for Defense Analyses. He was the founder of Disney Online Technology Forum, creator of Bell Labs' first-ever commercial AI tool, and inventor of the world's first annotated multimedia OPAC for the U.S. National Agricultural Library.

Lee graduated Summa Cum Laude from Virginia Tech with a B.S. and M.S. degrees in Computer Science, and he earned a perfect GPA from Vincennes University with an A.S. degree in Electrical Engineering and an honorary doctorate in Computer Science. He has been honored with a Michigan Leading Edge Technologies Award, two community development awards from the California Junior Chamber of Commerce, and four volunteer project leadership awards from The Walt Disney Company.

Part I
The Gordian Knot

Hear him but reason in divinity,
And all-admiring with an inward wish
You would desire the king were made a prelate:
Hear him debate of commonwealth affairs,
You would say it hath been all in all his study:
List his discourse of war, and you shall hear
A fearful battle render'd you in music:
Turn him to any cause of policy,
The Gordian knot of it he will unloose,
Familiar as his garter: that, when he speaks,
The air, a charter'd libertine, is still.

—William Shakespeare, Henry V, Act 1 Scene 1

Chapter 1
To Google or Not to Google

Newton Lee

> *We are the nation that put cars in driveways and computers in offices; the nation of Edison and the Wright brothers; of Google and Facebook.*
>
> —Barack Obama

Prologue Google's mission is "to organize the world's information and to make it universally accessible and useful" [1]. As of June 2016, Google has crawled and indexed 60 trillion individual web pages [2], befitting the search engine name that was a play on the word "Googol" which means ten duotrigintillion, 10^{100}, or 1 followed by a hundred zeros [3]. The Internet is accelerating collective consciousness and revolutionizing economy, politics, and education, among others.

1.1 A Brief History of Time: From Research to Product

Google began in 1996 as a search engine called "BackRub" developed by Stanford University grad students Sergey Brin and Larry Page in an academic research project to estimate the importance of a website by checking its backlinks [4]. With Craig Silverstein added to the team, the research project gained steam under faculty guidance from Hector Garcia-Molina, Rajeev Motwani, Jeffrey D. Ullman, and Terry Winograd [5]. The research funding was provided by National Science Foundation (NSF), National Aeronautics and Space Administration (NASA), Defense Advanced Research Projects Agency (DARPA), and Interval Research Corporation headed by Microsoft co-founder Paul Allen and Stanford University consulting professor David Liddle.

N. Lee (✉)
Newton Lee Laboratories LLC, Institute for Education Research and Scholarships,
Woodbury University School of Media Culture and Design, Burbank, CA, USA
e-mail: newton@newtonlee.com

© Springer Science+Business Media New York 2016
N. Lee (ed.), *Google It*, DOI 10.1007/978-1-4939-6415-4_1

In their 1998 paper "The anatomy of a large-scale hypertextual Web search engine," Brin and Page presented a prototype of Google with 24 million web pages indexed on the university server http://google.stanford.edu/. They introduced PageRank (PR) as follows [6]:

> We assume page A has pages T1…Tn which point to it (i.e., are citations). The parameter d is a damping factor which can be set between 0 and 1. We usually set d to 0.85. Also C(A) is defined as the number of links going out of page A. The PageRank of a page A is given as follows:
>
> PR(A) = (1 − d) + d (PR(T1)/C(T1) +···+ PR(Tn)/C(Tn))
>
> Note that the PageRanks form a probability distribution over web pages, so the sum of all web pages' PageRanks will be one.

In September 1998, Google was incorporated in California near the height of the dot-com bubble [7]. Although many Internet companies went belly up when the bubble burst in the following years, financial support for Google was unfaltering. In a 2001 interview by BusinessWeek, Larry Page explained to technology reporter Olga Kharif, "I think part of the reason we're successful so far is that originally we didn't really want to start a business. We were doing research at Stanford University. Google sort of came out of that. And we didn't even intend to build a search engine originally. We were just interested in the Web and interested in data mining. And then we ended up with search technology that we realized was really good. And we built the search engine. Then we told our friends about it and our professors. Pretty soon, about 10,000 people a day were using it" [1].

A decade later in May 2011, Google had more than one billion unique monthly visitors [8]. President Barack Obama touted the importance of federal funding for innovative research and development in his 2011 State of the Union address: "Thirty years ago, we couldn't know that something called the Internet would lead to an economic revolution. What we can do—what America does better than anyone—is spark the creativity and imagination of our people. We are the nation that put cars in driveways and computers in offices; the nation of Edison and the Wright brothers; of Google and Facebook. In America, innovation doesn't just change our lives. It's how we make a living" [9].

By August 2014, Google stock has risen 1294 % since it went public in 2004. A Google search on "Google IPO" returns a Knowledge Graph with IPO price information from *Wall Street Journal* (see Fig. 1.1). A Knowledge Graph is a knowledge base used by Google to enhance its search engine's search results with semantic-search information gathered from a wide variety of sources (see Fig. 1.2).

In February 2016, Google surpassed Apple as the world's most valuable company with a market capitalization of $531 billion [10]. A historical timeline on Google is available online at:

http://www.google.com/intl/en/about/company/history/

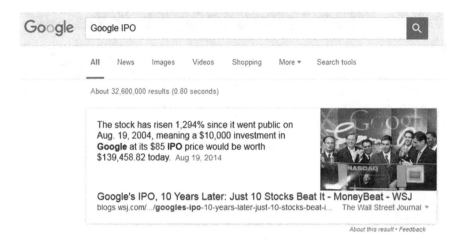

Fig. 1.1 Google search on "Google IPO" returns a Knowledge Graph with semantic-search information gathered from Wall Street Journal (August 19, 2014)

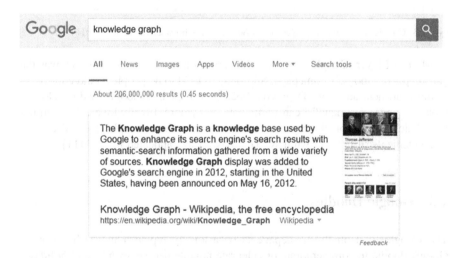

Fig. 1.2 Google search on "knowledge graph" displays a Wikipedia definition of Knowledge Graph as "a knowledge base used by Google to enhance its search engine's search results with semantic-search information gathered from a wide variety of sources"

1.2   Form Follows Function

The Google homepage (see Fig. 1.3) exemplifies the design principle of "form follows function"—a phrase coined by American architect Louis Sullivan who attributed the concept to Roman architect Marcus Vitruvius Pollio. Around 1490, Italian polymath Leonardo da Vinci drew the *Vitruvian Man* in pen and ink on paper, accompanied by notes based on the work of Vitruvius (see Fig. 1.4).

Fig. 1.3 Google homepage displayed on Google Chrome browser (as of June 2016)

Marissa Mayer, Google's first female engineer and Yahoo's sixth CEO, was the gatekeeper of Google's homepage. She said, "It used to be people would come over to my apartment and say, 'Does your apartment look like Google or does Google look like your apartment?' I can't articulate it anymore. I really love color. I'm not very knick-knacky or cluttery. My place has very clean, simple lines. There are some elements of fun and whimsy. That has always appealed to me" [11].

1.3 Google Doodles

The Google logo on the homepage is occasionally replaced for 24 hours by a Google doodle to commemorate or celebrate notable people, events, and holidays. The first Google doodle was in honor of the Burning Man Festival [12]:

> In 1998, before the company was even incorporated, the concept of the doodle was born when Google founders Larry and Sergey played with the corporate logo to indicate their attendance at the Burning Man festival in the Nevada desert. They placed a stick figure drawing behind the 2nd "o" in the word, Google, and the revised logo was intended as a comical message to Google users that the founders were "out of office." While the first doodle was relatively simple, the idea of decorating the company logo to celebrate notable events was born.

The first animated doodle appeared on January 4, 2010 showing an apple fall from a tree to pay tribute to Isaac Newton on his 367th birthday (see Fig. 1.5) [13]. The first interactive doodle game debuted on May 21, 2010 to celebrate Pac-Man's

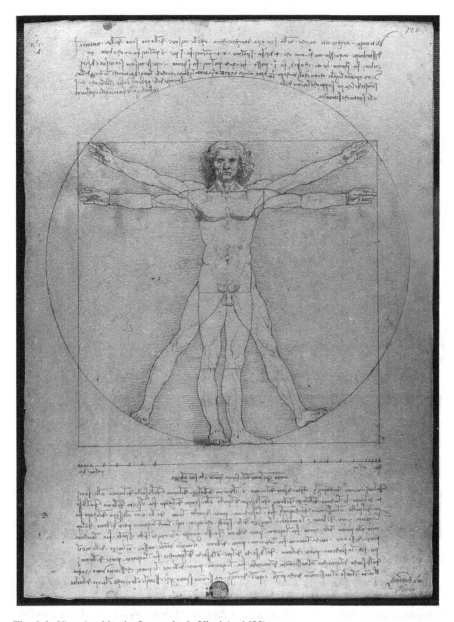

Fig. 1.4 *Vitruvian Man* by Leonardo da Vinci (c. 1490)

30th anniversary (see Fig. 1.6) [14]. Apart from its own curation, Google invites the general public to submit new doodle ideas to proposals@google.com. In addition, Google holds annual "Doodle 4 Google" competitions to "encourage eligible U.S. school students and their parents/guardians on their behalf to use their creativity to create their own interpretation of the Google logo" [15].

Fig. 1.5 First animated Google doodle on January 4, 2010 celebrating Isaac Newton's 367th birthday

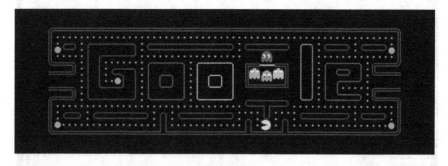

Fig. 1.6 First interactive doodle game on May 21, 2010 celebrating Pac-Man's 30th anniversary

Jennifer Hom, a long-time Google doodler, spoke to Emma Barnett of *The Telegraph* about the process of creating a doodle: "I draw mostly on the computer. I usually draw the thumbnail of the doodle on a sketchbook by hand and then take a photo of it on my phone and email it to myself so I can draw over the outline on Photoshop. Occasionally I will draw the entire doodle by hand. For example, when we celebrated the life of the Austrian artist, Gustav Klimt, [by camouflaging the word Google into his iconic painting 'The Kiss'] I did paint the idea out using gold leaf and everything" [16].

Of the Google doodles honoring notable people, female activist group SPARK Movement criticized Google for under representing women: "Google Doodles may seem light-hearted, especially when they're accompanied by quirky games and animation, but the reality is that these doodles have emerged as a new manifestation of who we value as a society—a sign of who 'matters.' Just like statues, stamps, and national holidays, you know that if someone is featured on Google's homepage, they've done something important." Ryan Germick, Doodle Team Lead, replied, "Women have historically been underrepresented in almost all fields: science, school curricula, business, politics—and, sadly, doodles. We've been working to fix the imbalance in our doodles. ... So far this year we've done doodles for as many women as men, a big shift from figures below 20 % in past years" [17].

1.4 I'm Feeling Lucky

As serious as a Google search might be, the "I'm Feeling Lucky" button adjacent to the "Google Search" button elicits the human side of Google. "It's possible to become too dry, too corporate, too much about making money," Marissa Mayer explained. "I think what's delightful about 'I'm Feeling Lucky' is that it reminds you there are real people here" [18].

In 2007, an estimated 1 % of all Google searches went through the "I'm Feeling Lucky" button. However, Google Instant has rendered the feeling lucky button practically unusable since 2010. Although you can't have your cake and eat it too, one can click on the "I'm Feeling Lucky" button without entering a keyword. As of June 2016, the lucky button would randomly take you to:

1. I'm Feeling Artistic—Art Project at Google Cultural Institute [19]
2. I'm Feeling Curious—a Google a day [20]
3. I'm Feeling Doodley—Google Doodles [21]
4. I'm Feeling Generous—Google One Today [22]
5. I'm Feeling Hungry—Local restaurants search results [23]
6. I'm Feeling Playful—Google Doodles [21]
7. I'm Feeling Puzzled—a Google a day [20]
8. I'm Feeling Stellar—Google Earth (Hubble Telescope) [24]
9. I'm Feeling Trendy—Google Trends [25]
10. I'm Feeling Wonderful—World Wonders at Google Cultural Institute [26].

Danny Sullivan, founding editor of *Search Engine Land*, visited Googleplex (see Fig. 1.7) in August 2007 and took notice of the wall art near one of the mini-kitchens (see Fig. 1.8). Designed by Joe Sriver, Fig. 1.9 is Larry Page on a box of Larryos with the tagline "Searching for delicious nutrition? Forget Google, eat. ... Larryos" and Fig. 1.10 is Sergey Brin on a box of Raisin Brin.

Fig. 1.7 Google search on "Googleplex" returns a Knowledge Graph showing a Wikipedia reference, address, phone number, reviews, and "People also search for."

Fig. 1.8 Google meets cereal brands (Courtesy of Danny Sullivan under Creative Commons 2.0)

Fig. 1.9 Larry Page meets
Cheerios in Larryos (Courtesy
of Danny Sullivan under
Creative Commons 2.0)

Fig. 1.10 Raisin Bran meets Sergey Brin in Raisin Brin (Courtesy of Danny Sullivan under Creative Commons 2.0)

In the 2013 comedy film *The Internship* shot on location at Googleplex, Nike Campbell (Owen Wilson) had a heart-to-heart conversation with Dana (Rose Byrne) [27]:

Nick Campbell: *You know, Google has single-handedly cut into my ability to bullshit.*

Dana: *Cramping your style?*

Nick Campbell: *Big time.*

Dana: *Make you a better person?*

Nick Campbell: *Yeah, true. 90 % Google, 10 % you.*

1.5 Have Fun and Keep Googling

"Google" as a transitive verb began in July 1998 when Larry Page wrote "Have fun and keep googling!" in his email to the Google-Friends subscribers (see Fig. 1.11) [28].

In October 2002, the verb "google" first appeared on American television in season 7, episode 4 of *Buffy the Vampire Slayer*. The title of the episode was coincidentally and appropriately named "Help." Willow (Alyson Hannigan), Buffy (Sarah Michelle Gellar), and Xander (Nicholas Brendon) were discussing their fellow student Cassie (Azura Skye) in their high school [29]:

Willow: *[to Buffy] Have you Googled her yet?*

Xander: *Willow! She's 17!*

Willow: *It's a search engine.*

Subject: New Features
From: L a r r y P a g e

Wed
8
Jul 98

First, I'd like to welcome a whole bunch of new subscribers to the
Google-Friends list. This list is an announcement list for the Google
search engine which is available at http://google.stanford.edu/.

We recently added some new features to Google. The most significant is a
summary for each result that highlights where your query matched. This
makes for a much more informative summary than most search engines provide.
You can actually see where your query matched without having to download
each page.

Also, you can fetch contents from our repository by clicking on the
"Cached" link near each result. This often is faster than going to the
original, but of course can give you outdated information since it is based
on our last crawl. This feature can be very useful if the original is not
available or the network is down, you will still have access to the page
(but not the images).

We also made some enhancements to our search code that produce more matches
for many queries, improving performance significantly, we have also done
some more tuning of the ranking functions.

After combining our web server and search engine for better performance, we
have been experiencing intermittent problems with our system being down for
short amounts of time fairly frequently. If you have trouble getting to
the system, try back in a minute or two, and it should be back up.

Expect to see a lot of changes in Google in the next few months. We plan
to have a much bigger index than our current 24 million pages soon. Thanks
to all the people who have sent us logos, HTML and suggestions. Keep them
coming!

Have fun and keep googling!

-Larry

Fig. 1.11 Larry Page's email to Google-Friends subscribers

In 2006, the Oxford English Dictionary and the eleventh edition of the Merriam-Webster Collegiate Dictionary added "Google" as a verb in their "definitive record of the English language," joining the ranks of pop-culture cachet like FedEx, TiVo, and Xerox [30].

In 2014, *The New York Times* portrayed the current generation "an era of Googled definitions" in its interview with the new chief of the Oxford English Dictionary (O.E.D.): "For the first time in 20 years, the venerable dictionary has a new chief editor, Michael Proffitt, who assumes the responsibility of retaining the vaunted traditions while ensuring relevance in an era of Googled definitions and text talk. ... Mr. Proffitt advocates links in digitized literature to O.E.D. entries; he wants more use by students, whose distinction between 'dictionary' and 'web search' is increasingly blurred" [31].

In 2016, Google assistant was introduced at the annual Google I/O developer's conference. "It's not enough to give them links," said Google CEO Sundar Pichai. "We really need to help them get things done in the real world. This is why we're evolving search to be more assistive" [32]. Google assistant enables conversational speech in human-computer interface, which is being used in voice-activated Google Home, messaging app Allo, and video calling application Duo [33].

Since the airing of "Help" in *Buffy the Vampire Slayer*, "Google" as a transitive verb has become popular in mainstream media and everyday usage. The following is one memorable dialogue between real estate magnate Donald Trump and U.S Senator Marco Rubio from Florida at the televised CNN-Telemundo Republican debate in Houston on February 25, 2016 [34]:

> *Trump*: *You haven't hired one person, you liar.*
>
> *Rubio*: *He hired workers from Poland. And he had to pay a million dollars or so in a judgment from.*
>
> *Trump*: *That's wrong. That's wrong. Totally wrong.*
>
> *Rubio*: *That's a fact. People can look it up. I'm sure people are Googling it right now. Look it up. "Trump Polish workers," you'll see a million dollars for hiring illegal workers on one of his projects. He did it.*
>
> *(Applause)*

Indeed, there was a 700 % spike in Google searches for "Polish workers" after the fiery exchange between Trump and Rubio. "Polish workers was on no one's radar but during those times that Rubio brought it up it piqued everyone's interest to know more and this is what they started searching for," said LaToya Drake, marketing manager and media outreach at Google [35].

1.6   Gatekeeper of Information

Google cofounder Larry Page said, "Basically, our goal is to organize the world's information and to make it universally accessible and useful. That's our mission" [1]. As of June 2016, Google has crawled and indexed 60 trillion individual web pages [2], befitting the search engine name that was a play on the word "Googol" which means ten duotrigintillion, 10^{100}, or 1 followed by a hundred zeros [3].

With the ever-changing search algorithm, Google ranks the results using more than 200 factors including PageRank, site quality, trustworthiness, and freshness. In a typical year, Google makes over 500 search improvements that are determined by:

1. Precision Evaluation—Human evaluators at Google run more than 40,000 precision evaluations per year to rate the usefulness of individual results for a given search input.
2. Side-by-Side Experiment—Evaluators compare two sets of search results from the old algorithm and the new algorithm before launching the search update. Google conducts over 9,000 side-by-side experiments annually.

3. Live Traffic Experiment—Google changes search algorithm for a small percentage of real Google users and analyzes their search behaviors. About 7,000 live traffic experiments are performed each year.

The Google search results are displayed in a multitude of formats:

1. Knowledge Graphs—Semantic-search information gathered from a wide variety of sources including a database of real world people, places, things, and the connections among them (see Fig. 1.12).
2. Snippets—Small previews of information such as a web page's title and short descriptive text for each search result.
3. News—Results from online newspapers and blogs.
4. Answers—Immediate answers and information for popular queries such as weather, sports scores, quick facts, and numeric computations including currency exchange.
5. Videos—Video-based results with thumbnails from YouTube, Vimeo, Dailymotion, and others.
6. Images—Image-based results with thumbnails.
7. Refinements—"Advanced Search," related searches, and other search tools to help Google users fine-tune their search results.

Google has become the de facto gatekeeper of information available on the Internet. Starting in December 2009, Google personalizes individual's search results based on the history of what they have clicked on. Consequently, Google displays more of what the users want to see and less of what they do not care about. To address the danger of creating information silos, Google keeps some search results similar between users. "We want diversity of results," said Google product manager Johanna Wright. "This is something we talk about a lot internally and believe in. We want there to be variety of sources and opinions in the Google results. We want them in personalized search to be skewed to the user, but we don't want that to mean the rest of the web is unavailable to them" [36].

Trond Lyngbø, founder of Search Planet AS, talks about search engine optimization (SEO) in the age of digital transformation (what every business leader must know) in Chapter 4 of this book. Frank Buddenbrock, Google AdWords certified specialist, demonstrates how to optimize your website to get to Google's first page in Chapter 5. Tina Courtney of Evolve Inc. offers 4 tips for writing outstanding SEO boosting content in Chapter 6. Nicole Ciomek, founder of Radiant PPC, teaches us Internet advertising and Google AdWords in Chapter 7. And Nyagoslav Zhekov, director of local search at Whitespark Inc., explains the use of Google Maps and Google Local Search for businesses in Chapter 8.

Fig. 1.12 Google search on "Burning Man Festival" returns a Knowledge Graph showing Burning Man 2016 information, location, dates, founders, nominations, social media profiles, related topics, and "People also search for."

1.7   Censorship of Information

In the fight against PROTECT IP Act (PIPA) and Stop Online Piracy Act (SOPA) in 2012, Google added a black "censored" bar atop its logo as well as a link "Tell Congress: Please don't censor the web!" to the Google Public Policy Blog that said, "You might notice many of your favorite websites look different today. Wikipedia is down. WordPress is dark. We're censoring our homepage logo and asking you to petition Congress. So what's the big deal? Right now in Washington D.C., Congress is considering two bills that would censor the web and impose burdensome regulations on American businesses. They're known as the PROTECT IP Act (PIPA) in the Senate and the Stop Online Piracy Act (SOPA) in the House. ... Fighting online piracy is extremely important. We are investing a lot of time and money in that fight. ... Because we think there's a good way forward that doesn't cause collateral damage to the web, we're joining Wikipedia, Twitter, Tumblr, Reddit, Mozilla and other Internet companies in speaking out against SOPA and PIPA" [37].

On the other side of the coin, Cary H. Sherman, CEO of the Recording Industry Association of America (RIAA) that represents music labels, criticized the blackout tactic in *The New York Times*: "Wikipedia, Google and others manufactured controversy by unfairly equating SOPA with censorship. ... The hyperbolic mistruths, presented on the home pages of some of the world's most popular Web sites, amounted to an abuse of trust and a misuse of power. ... The violation of neutrality is a patent hypocrisy. ... What the Google and Wikipedia blackout showed is that it's the platforms that exercise the real power. Get enough of them to espouse Silicon Valley's perspective, and tens of millions of Americans will get a one-sided view of whatever the issue may be, drowning out the other side" [38].

With great power comes great responsibility. Dr. Robert Epstein, senior research psychologist at the American Institute for Behavioral Research and Technology, warns about subtle new forms of Internet influence are putting democracy at risk worldwide in Chapter 9 of this book.

When the 2012 Google Transparency Report showed an alarming rise in government censorship around the world, Google's senior policy analyst Dorothy Chou wrote, "We've been asked to take down political speech. It's alarming not only because free expression is at risk, but because some of these requests come from countries you might not suspect—Western democracies not typically associated with censorship" [39]. The democratic countries include Australia, Austria, Belgium, Bulgaria, Canada, Czech Republic, Denmark, Finland, France, Germany, Greece, Ireland, Italy, Japan, Mexico, New Zealand, the Netherlands, Norway, Poland, Romania, Spain, Sweden, Switzerland, the United Kingdom, and the United States [40]. Figure 1.13 shows the number of content removal requests from courts and government agencies around the world between 2009 and 2015 [41].

Unlike Western democracies, communist China made very few content removal requests to Google. Instead, China opts for preemptive measures by severely censoring search results from Google. In March 2010 when Google ceased filtering

Removal requests by the numbers

See all data

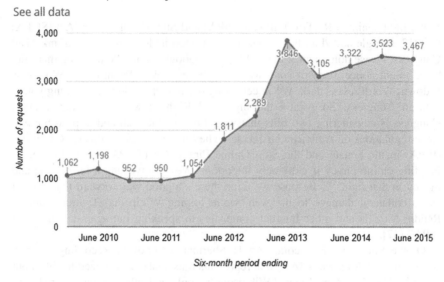

Six-month period ending

Fig. 1.13 Removal requests by the numbers (from courts and government agencies around the world to remove information from Google products)

its search results in China [42], the search giant had to scale back operations in China and redirect users from Google.cn to its uncensored Google.com.hk in Hong Kong [43].

1.8   Information Warfare and Connecting the Dots

At the 2014 DEF CON 22 conference in Las Vegas, American author Richard Thieme recalled a conversation with his friend from the National Security Agency (NSA) who spoke of the difficulty in knowing the truth: "You know enough to know what's not true, but you can't necessarily connect all of the dots to know what is true" [44]. Sometimes the answers are hiding in plain sight. A eureka moment came to Archimedes when he connected the dots between the ordinary routine of taking a bath and the scientific pursuit of determining the volume of an irregularly shaped object.

In the Academy Award-winning documentary *Citizenfour* (2014), former Central Intelligence Agency (CIA) employee and NSA whistleblower Edward Snowden told journalists Glenn Greenwald, Ewen MacAskill, and Laura Poitras [45, 46]:

> Any analyst, at any time, can target anyone; any selector, anywhere. Where those communications will be picked up depends on the range of the sensor networks, and the authorities that that analyst is empowered with. Not all analysts have the ability to target everything, but I, sitting at my desk, certainly had the authorities to wiretap anyone, from you or your accountant, to a federal judge, to even the President, if I had a personal email.

Snowden's revelation may come as a shock to the average American citizens, but certainly not to most foreign governments. Total information awareness has helped to stabilize relations among international powers, and to that end espionage is making the world a safer place. Interestingly and perhaps by pure coincidence, the English name "Snowden" (snow + hill) is uncannily related to the Russian name "Морозов" (Morozov) which means "frost" or "freeze." Павлик Морозов (Pavlik Morozov) was a 13-year-old patriot *or* traitor, depending on one's point of view.

According to an investigation by the House Oversight Committee in February 2016, the U.S. government may have used compromised software for up to 3 years. Cybersecurity researchers believe that foreign hackers may have repurposed an encryption backdoor created by the NSA to conduct their own cyber snooping [47]. In other words, foreign spies may have been eavesdropping on mobile phone calls, Skype chats, emails, and other means of communication by U.S. residents for years. An estimated 100,000 foreign spies are currently living and working in the U.S., according to Chris Simmons, a retired counterintelligence supervisor for the U.S. Defense Intelligence Agency.

The issue of counterintelligence came up during my meeting with the Federal Bureau of Investigation (FBI) on November 27, 2015. To my surprise, the FBI was not overly concerned with foreign spies on U.S. soil. In fact, the FBI may decide to leave a suspected mole undisturbed for years in order to feed him false critical information at an opportune moment or to use him to catch a bigger fish. "We have our spies in their country too," said an FBI agent with a contented smile. Total information awareness should be a two-way street. Otherwise, secrecy and manipulation make things worse by clouding the truth and impairing people's judgment. For example:

- **U.S. government:** More than 50 intelligence officers filed complaints with the Pentagon in 2015 that their reports on ISIS and al Qaeda in Syria were inappropriately altered by senior officials for political reasons. One defense official told *The Daily Beast* that "the cancer was within the senior level of the intelligence command" [48].
- **News media:** In a controversial cover story in the October 2015 issue of *The New York Times Magazine*, reporter Jonathan Mahler opined that it was "impossible to know what was true and what wasn't" about the saga of the hunt for Osama bin Laden and his death in Pakistan. The official bin Laden story, he said, was "floating somewhere between fact and mythology" [49]. It quickly sparked rebuttals from *Washington Post* national security reporter Greg Miller, CNN analyst Peter Bergen [50], and *Black Hawk Down* author Mark Bowden, who all accused Mahler of elevating unsubstantiated conspiracy theories [51].
- **Hollywood:** During the height of the Cold War in 1954, the CIA secretly funded the film version of George Orwell's *Animal Farm* as propaganda against communism and Joseph Stalin [52]. Leni Riefenstahl's award-winning *Triumph of the Will* in 1935 helped the rise of Nazism in Germany. And D. W. Griffith's *The Birth of a Nation* in 1915 was partly responsible for the resurrection of Ku Klux Klan in Georgia.

- **Education:** A world geography textbook published by McGraw-Hill called African slaves "workers" in the section titled "Patterns of Immigration": "The Atlantic Slave Trade between the 1500 s and 1800 s brought millions of workers from Africa to the southern United States to work on agricultural plantations." After a school kid's mother openly complained on Facebook, the publisher agreed to revise the text, stating that "we will update this caption to describe the arrival of African slaves in the U.S. as a forced migration and emphasize that their work was done as slave labor" [53].
- **Social media:** During the escalated Gaza-Israel conflict in November 2012, Israeli Defense Force (IDF) live tweeted its military campaign in the Gaza strip during the weeklong Operation Pillar of Defense [54]. In response, Hamas tweeted its own account of the war along with photographs of casualties [55]. Both sides hoped to use social media to win world sympathy and shift political opinion to their sides [56]. Aylin Manduric, international presidential fellow at the Center for the Study of the Presidency and Congress (CSPC) in Washington, D.C., examines the use of social media as a tool for information warfare in Chapter 10 of this book.
- **Television:** Amid the crisis in the Ukraine, Russia and the European Union in 2015 unleashed a new bout of information warfare to sway public opinions. Russia increased its spending on RT (Russia Today) television network whereas the BBC planned to launch a new Russian satellite TV and video service [57].
- **The Internet:** In the aftermath of losing the bitter fight on Stop Online Piracy Act (SOPA) and Protect Intellectual Property Act (PIPA), CEO Cary H. Sherman of the Recording Industry Association of America (RIAA) wrote an opinion piece in *The New York Times*: "Misinformation may be a dirty trick, but it works. Consider, for example, the claim that SOPA and PIPA were 'censorship,' a loaded and inflammatory term designed to evoke images of crackdowns on pro-democracy Web sites by China or Iran. … Wikipedia, Google and others manufactured controversy by unfairly equating SOPA with censorship. They also argued misleadingly that the bills would have required Web sites to 'monitor' what their users upload…" [38].

In fact, the Internet is a haven for information warfare because search engines do not differentiate between authoritative and crowd-sourced knowledge, truth and fabrication, information and disinformation, or free speech and hate speech [58]. Google bomb and Googlewashing are common practices of manipulating search results for the purpose of making a political statement. The best known Google bombs are "miserable failure" leading to President George W. Bush and "Rick Santorum" to a sexually explicit definition. Accused fraudster Wayne Simmons fooled the U.S. Department of Defense into issuing him a security clearance as Deputy Assistant Secretary of Defense Allison Barber remarked that "there is quite a bit of info under 'Wayne Simmons and CIA' on a Google search" [59]. Google's Jigsaw program https://jigsaw.google.com/ uses the Redirect Method to target aspiring ISIS recruits. Yasmin Green, Jigsaw's head of research and development explained, "The Redirect Method is at its heart a targeted advertising campaign:

Let's take these individuals who are vulnerable to ISIS' recruitment messaging and instead show them information that refutes it."

The Wolfram|Alpha search engine addresses the problem of misinformation and disinformation on the Internet by using the curation work of library professionals and domain experts to answer user queries. "Wolfram is far more computational," said Amit Singhal, former Head of Google Search. In Chapter 11 of this book, John B. Cassel from Wolfram Research describes in details the Wolfram|Alpha computational knowledge "search" engine. In Chapter 12, Barret Havens and Jennifer Rosenfeld at Woodbury University discuss seamless access to libraries from Alexandria through the digital age. In Chapter 13, Frances Eames Noland of the University of Oxford talks about privileged and corporatization of information. And in Chapter 14, futurist Tiana Sinclair discourses on communication and language in the age of digital transformation.

1.9 Information Silos and Prison of Ideas

At TEDGlobal 2009, Nigerian novelist Chimamanda Ngozi Adichie said, "The single story creates stereotypes, and the problem with stereotypes is not that they are untrue, but that they are incomplete. They make one story become the only story. ... Our lives and our cultures are composed of many overlapping stories, and if we hear only a single story about another person or country, we risk a critical misunderstanding" [60].

Antonin Scalia was the longest-serving U.S. Supreme Court Justice appointed by President Ronald Reagan in 1986. In a 2013 interview with *New York Magazine*, Scalia said that he ditched *The Washington Post* and *The New York Times* because they just "went too far for me. I couldn't handle it anymore. It was the treatment of

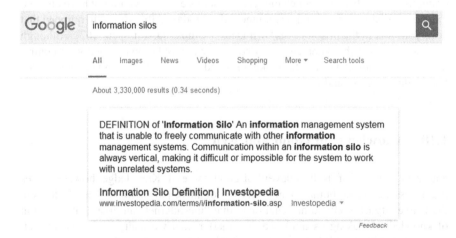

Fig. 1.14 Google Search on "information silo" returns a Knowledge Graph definition from Investopedia. In our metaphor, the information management system represents our brain, and the unrelated systems are people outside our own circle of families, friends, or society

almost any conservative issue. It was slanted and often nasty. And, you know, why should I get upset every morning? I don't think I'm the only one" [61]. Scalia was certainly not the only one. Too many people prefer to stay inside their own comfort zones with a one-sided liberal or conservative sentiment, creating their own information silos (see Fig. 1.14). When people refuse to see how things look from another point of view, their silo mentality has fueled arguments in families, disputes with neighbors, bigotry between races, and conflicts among nations.

In spite of the polar opposite opinions on same-sex marriage, all-male admissions policy, and other issues dividing conservatives and liberals, Justice Scalia had a long and close friendship with Justice Ruth Bader Ginsburg at the U.S. Supreme Court [62]. In December 2012, Jesuit priest and peace activist John Dear went to Kabul to meet the Afghan Peace Volunteers, a diverse community of students ages 15 to 27 who practice peace and nonviolence [63]. "I used to detest other ethnic groups," one of the youths told Dear, "but now I'm trying to overcome hate and prejudice. You international friends give me hope and strength to do this." Another youth added, "I used to put people in categories and couldn't drink tea with anyone. Now I'm learning that we are all part of one human family. Now I can drink tea with anyone" [64].

The tug of war between creationists and evolutionists is most evident in biology textbooks for high school students. In Russian schools where Darwinism prevails, a schoolgirl named Maria Schreiber from St. Petersburg asked the court in 2006 to replace an evolutionist biology textbook with an "Orthodox" version [65]. In a widely publicized live debate moderated by CNN journalist Tom Foreman in February 2014, Christian author Ken Ham and educator Bill Nye ("the Science Guy") presented their opposing answers to the question "Is creation a viable model of origins in today's modern, scientific era?" [66] Although no one has perfect answers and impartial views, we should allow debates to shape and reshape our opinions. If scientists had not challenged the status quo, we would not have enjoyed modern medicine and technological innovations today. Albert Einstein quipped that "everyone sits in the prison of his own ideas; he must burst it open" [67]. Although we may not have the complete knowledge or we may be bombarded with contradictory information, we can still make informed decisions based on wisdom and the knowledge of good and evil.

1.10 Knowledge of Good and Evil

Knowledge in and of itself is devoid of good and evil. Knowledge, however, does not necessarily make human beings wiser. Cognitive psychologist Tom Stafford at the University of Sheffield has cautioned that "the Internet can give us the illusion of knowledge, making us think we are smarter than we really are" [68]. In fact, knowledge without wisdom can be outright dangerous. King Solomon lamented that "for with much wisdom comes much sorrow; the more knowledge, the more grief" (Ecclesiastes 1:18). In 1965, Encyclopaedia Britannica ran an ad in *The*

Atlantic Monthly: "A little knowledge is a dangerous thing. So is a lot. The more you know, the more you need to know—as Albert Einstein, for one, might have told you. Great knowledge has a way of bringing with it great responsibility" [69].

First discovered by Friedrich Miescher in 1869, DNA (Deoxyribonucleic acid) is a molecule that carries most of the genetic instructions used in the development, functioning, and reproduction of all known living organisms [70]. In 2010 when new genomics data showed that all humans except for sub-Saharan Africans carry a small percentage of Neanderthal DNA [71], some racists started spinning the idea of racial superiority due to the presence of Neanderthal DNA in modern Europeans and Asians [72].

Einstein's mass-energy equivalence equation $E = mc^2$ was used to create weapons of mass destruction before its peacetime application in nuclear power plants [73]. In August 1939, Albert Einstein wrote to President Franklin D. Roosevelt to warn him about the possibility of the Nazis building an atomic bomb; and the Manhattan Project was subsequently born in 1941. "Woe is me," said Einstein upon hearing the news of the Hiroshima bombing (see Fig. 1.15). "Had I known that the Germans would not succeed in developing an atomic bomb, I would have done nothing" [74].

Fig. 1.15 Mushroom cloud from the atomic bombing of Nagasaki, Japan on August 9, 1945 (Courtesy of National Archives and Records Administration)

Einstein's regret is reminiscent of Apple CEO Tim Cook's reason for refusing to comply with the court order to bypass the iOS security on a terrorist's iPhone: "Specifically, the FBI wants us to make a new version of the iPhone operating system, circumventing several important security features, and install it on an iPhone recovered during the investigation. In the wrong hands, this software—which does not exist today—would have the potential to unlock any iPhone in someone's physical possession" [75]. Corporate spying and international espionage are always a grave concern, but who is to say that no one else besides Apple can hack the iOS, and no country besides the United States can create an atomic bomb? Indeed, the FBI cracked the terrorist's iPhone without Apple's help in less than 2 months [76], and nine countries—Russia, the United States, France, China, the United Kingdom, Pakistan, India, Israel and North Korea—possess nuclear weapons today [77].

It seems that evil oftentimes accompanies good in every scientific discovery or engineering marvel. For instance, nuclear power for electrical energy and weapons of mass destruction, GPS for navigating automobiles and guided missiles, and genetically-modified viruses for vaccines and bioweapons, just to name a few. Christian evangelist Billy Graham spoke about technology and faith at TED in February 1998: "You've seen people take beneficial technological advances, such as the Internet ... and twist them into something corrupting. You've seen brilliant people devise computer viruses that bring down whole systems. The Oklahoma City bombing was simple technology, horribly used. The problem is not technology. The problem is the person or persons using it" [78].

To encourage the use of technology for good instead of evil, dynamite inventor Alfred Nobel wrote his last will in 1895 to the establishment of the Nobel Prize to honor "men and women from all corners of the globe for outstanding achievements in physics, chemistry, physiology or medicine, literature, and for work in peace" [79].

Charlie Chaplin gave an impassioned speech in *The Great Dictator* (1940): "The aeroplane and the radio have brought us closer together. The very nature of these inventions cries out for the goodness in man, cries out for universal brotherhood, for the unity of us all. ... Let us fight for a world of reason, a world where science and progress will lead to all men's happiness" [80].

Bulletin of the Atomic Scientists issued a warning in its January 2016 newsletter: "The challenge remains whether societies can develop and apply powerful technologies for our welfare without also bringing about our own destruction through misapplication, madness, or accident" [81].

Circumstances can magnify both good and evil. War and destruction have taken a toll on many brilliant scientists. J. Robert Oppenheimer, head of the Manhattan Project, expressed his fear that the bomb might become "a weapon of genocide" [82]. In a 1965 television broadcast, Oppenheimer said in tears and agony (see Fig. 1.16), "We knew the world would not be the same. A few people laughed. A few people cried. Most people were silent. I remembered the line from the Hindu scripture, the Bhagavad Gita. Vishnu is trying to persuade the Prince that he should do his duty, and to impress him, takes on his multi-armed form and says, 'Now I am

Fig. 1.16 J. Robert Oppenheimer, head of the Manhattan Project, delivered a line from the Hindu scripture, the Bhagavad Gita, in his 1965 televised speech

become Death, the destroyer of worlds.' I suppose we all thought that, one way or another" [83].

1.11   Google's Mantra: Don't Be Evil

Google is the first and only company in history that has a "Don't be evil" manifesto written in their IPO letter (S-1 Registration Statement) [55, 84]:

> Don't be evil. We believe strongly that in the long term, we will be better served—as shareholders and in all other ways—by a company that does good things for the world even if we forgo some short term gains. This is an important aspect of our culture and is broadly shared within the company.

Whatever "evil" is or is not, Google's IPO statement seems to define evil as the failure to do "good things for the world."

The common theme between Google and world religions is faith, that is, complete trust or confidence. In the IPO letter, Page and Brin proudly proclaimed that "Google users trust our systems":

> Google users trust our systems to help them with important decisions: medical, financial and many others. Our search results are the best we know how to produce. They are

unbiased and objective, and we do not accept payment for them or for inclusion or more frequent updating. We also display advertising, which we work hard to make relevant, and we label it clearly. This is similar to a well-run newspaper, where the advertisements are clear and the articles are not influenced by the advertisers' payments. We believe it is important for everyone to have access to the best information and research, not only to the information people pay for you to see.

In a hilarious 2013 interview by NPR host Peter Sagal, Google's first CEO Eric Schmidt admitted that he thought the Google slogan "Don't be evil" was stupid but then he was surprised by how well it had worked [85]:

Well, it was invented by Larry and Sergey. And the idea was that we don't quite know what evil is, but if we have a rule that says don't be evil, then employees can say, I think that's evil. Now, when I showed up, I thought this was the stupidest rule ever, because there's no book about evil except maybe, you know, the Bible or something.

So what happens is, I'm sitting in this meeting, and we're having this debate about an advertising product. And one of the engineers pounds his fists on the table and says, that's evil. And then the whole conversation stops, everyone goes into conniptions, and eventually we stopped the project. So it did work.

In his 2014 book *How Google Works* coauthored with Jonathan Rosenberg, Schmidt wrote, "Yes, it genuinely expresses a company value and aspiration that is deeply felt by employees. But 'Don't be evil' is mainly another way to empower employees. … Googlers do regularly check their moral compass when making decisions" [86].

"In a certain sense, Google is being held to a higher standard," said Jon Fox of the California Public Interest Research Group (CALPIRG). "When Facebook does really nasty things, people are like, 'Oh well, it's Facebook, what can you expect from them?' But as Google is maturing, they are running up against that problem more and more of not doing evil" [87].

Over the years, some observers have questioned Google's practice of choosing business imperatives over social values amid a string of privacy concerns and violations. Eric Schmidt had vigorously defended the search giant by asserting that "there has to be a trade-off between privacy concerns and functionality" without belying the corporate motto "Don't be evil" [88].

In August 2005, CNET reporter Elinor Mills published some personal information of Eric Schmidt through Google searches. She wrote, "Like so many other Google users, his virtual life has been meticulously recorded. The fear, of course, is that hackers, zealous government investigators, or even a Google insider who falls short of the company's ethics standards could abuse that information. Google, some worry, is amassing a tempting record of personal information, and the onus is on the Mountain View, Calif., company to keep that information under wraps" [89].

In retaliation for publicizing Schmidt's personal information, Google blacklisted all CNET reporters for a year [90], which was an irony because in an interview by CNBC's "Inside the Mind of Google" in December 2009, Schmidt famously said that "If you have something that you don't want anyone to know, maybe you shouldn't be doing it in the first place" [91].

Fig. 1.17 Google Street View Car—Courtesy of Enrique Bosquet http://bosquetphotography.com/

In a January 2010 Town Hall meeting at Apple headquarters, Steve Jobs blatantly told his employees that he thought Google's "Don't be evil" mantra "bullshit" or "a load of crap" [92].

In May 2010, Google made a stunning admission that for over 3 years, its camera-toting Street View cars had inadvertently collected snippets of private information that people send over unencrypted WiFi networks [93] (see Fig. 1.17).

In October 2010, Google also admitted to accidentally collecting and storing entire e-mails, URLs, and passwords from unsecured WiFi networks with its Street View cars in more than 30 countries, including the United States, Canada, Mexico, some of Europe, and parts of Asia [94].

At the 2012 Black Hat security conference in Las Vegas, Jennifer Granick, director of civil liberties at the Stanford Law School Center for Internet and Society, asked the audience of security professionals who they trusted less, Google or the government? The majority raised their hands for Google. "I fear Google more than I pretty much fear the government," said panelist Jeff Moss, founder of Black Hat and DEF CON. "Google, I'm contractually agreeing to give them all my data" [95]. Indeed, one can download his or her entire Google search history from https://history.google.com/history/ including search strings, timestamps, geolocation coordinates, and accessed URLs.

In the October 2013 issue of *The Atlantic*, Georgia Tech professor Ian Bogost opined that "as both users of its products and citizens of the world it increasingly influences and alters, we would be wise to see Google's concern for evil as a pragmatic matter rather than an ethical one. … through its motto Google has effectively *redefined* evil as a matter of unserviceability in general, and unserviceability among corporatized information services in particular. … The company doesn't need to exercise any moral judgment other than whatever it will have done. The biggest risk—the greatest evil—lies in failing to engineer an effective implementation of its own vision. *Don't be evil* is the Silicon Valley version of *Be true to yourself*. It is both tautology and narcissism" [96].

In August 2014, technology analyst Rob Enderle complained that Google "didn't understand the difference between good and evil. I think they should change their slogan to 'evil are us.' It seems like every time you turn around they are doing something that is at best questionable and at worst anti-people" [97].

In December 2015, the Electronic Frontier Foundation (EFF) filed a complaint with the Federal Trade Commission (FTC) that for millions of K-12 students using Chromebooks, "Google is engaged in collecting, maintaining, using, and sharing student personal information in violation of the 'K-12 School Service Provider Pledge to Safeguard Student Privacy' (Student Privacy Pledge), of which it is a signatory" [98]. However, Jonathan Rochelle, Director of Google Apps for Education (GAFE), replied that GAFE, Chrome Sync, and other Google products and services comply with both the law and the Student Privacy Pledge [99].

1.12 Google for Education

Realizing the importance of computer technology in early education, Apple launched a program called Kids Can't Wait (KCW) in 1983 to donate an Apple IIe computer in each of the 9,250 elementary and secondary schools in California [100]. Fast forward to the January 2016 State of the Union address, President Barack Obama said, "In the coming years, we should build on that progress, by providing Pre-K for all and offering every student the hands-on computer science and math classes that make them job-ready on day one. We should recruit and support more great teachers for our kids" [101].

U.S. Chief Technology Officer Megan Smith elaborated, "Computer Science for All is the President's bold new initiative to empower all American students from kindergarten through high school to learn computer science and be equipped with the computational thinking skills they need to be creators in the digital economy, not just consumers, and to be active citizens in our technology-driven world. Our economy is rapidly shifting, and both educators and business leaders are increasingly recognizing that computer science (CS) is a 'new basic' skill necessary for economic opportunity and social mobility" [102].

Like Apple in its early days, Google is a champion of education. As of 2016, there are 50 million users of Google Apps for Education. 10 million students and teachers are using Google Classroom in 190 countries. Google also offers internships, student scholarships, and a host of educational programs including [103]:

1. AdCamp
2. BOLD (Building Opportunities for Leadership and Development) Immersion
3. Camp Google for kids
4. Certified Innovator program
5. Code Jam competitions
6. Computer Science for High School (CS4HS) grants
7. Computer Science Summer Institute (CSSI) program
8. CS First after school program
9. Doodle 4 Google
10. Exploring Computational Thinking (ECT) program
11. Google APAC MBA Summit
12. Google Code-in (GCI) contests
13. Google Online Marketing Challenge (GOMC)
14. Google Policy Fellowship
15. Google Science Fair
16. Google Student Veteran Summit
17. Google Summer of Code (GSoC) online program
18. Hash Code team programming competition
19. Made with Code for girls
20. RISE Awards
21. Student Ambassador Program

1.13 The Closing of the American Mind

Computer science education, however, is not the be-all and end-all. It should not come at the expense of other school subjects. In February 2016, Florida Senate approved a bill allowing high school students to count computer coding as a foreign language course. The NAACP's Florida Conference and Miami-Dade branch, the Florida chapter of the League of United Latin American Citizens (LULAC), and the Spanish American League Against Discrimination (SALAD) disapproved of the legislation: "Our children need skills in both technology and in foreign languages to compete in today's global economy. However, to define coding and computer science as a foreign language is a misleading and mischievous misnomer that deceives our students, jeopardizes their eligibility to admission to universities, and will result in many losing out on the foreign language skills they desperately need even for entry-level jobs in South Florida" [104].

It is all too easy to not see the forest for the trees. I once played a question card game with my college friends. I chose a card in random and the question was: "What would be your first order of business if you were elected President of the United States?" I answered, "Improve the educational system." A few years later in 1987, University of Chicago Professor Allan Bloom published the seminal book *The Closing of the American Mind* in which he described how "higher education has failed democracy and impoverished the souls of today's students" [105].

In 2011, PayPal cofounder Peter Thiel paid 24 kids $100,000 each to drop out of college to become entrepreneurs [106]. Larry Page and Sergey Brin suspended their Ph.D. studies to commercialize Google [107] [108]. Bill Gates and Mark Zuckerberg left Harvard University in their sophomore year to start Microsoft and Facebook respectively [109, 110]. What gives?

Formal education is supposed to nurture students into their full potential, but something is amiss. Since the Industrial Revolution in the 18th century, the division of labor has brought forth specialization in the workforce and university curriculums. For instance, in 1749, Academy of Philadelphia (predecessor to the University of Pennsylvania) was organized into three schools: the English School, the Mathematics School, and the Latin School [111]. By 2016, the Ivy League university has 12 schools and more than 100 majors of study [112].

Standardized tests and rote learning are churning out human workforce. Meanwhile, IBM Watson has won *Jeopardy!* in 2011 [113], and robots are displacing as many as 5 million human workers by 2020 [114]. Highly skilled workers are not immune either. In 2015, Google, Adobe, and MIT researchers at the Computer Science and Artificial Intelligence Laboratory (CSAIL) have created "Helium"—a computer program that modifies codes faster and better than expert computer engineers for complex software such as Photoshop [115]. What takes human coders months to program, Helium can do the job in a matter of hours or even minutes. Similarly, computers can outperform human physicians in diagnosing patients and recommending treatments [116].

"I have been in medical education for 40 years and we're still a very memory-based curriculum," said Columbia University professor Herbert Chase. "The power of Watson-like tools will cause us to reconsider what it is we want students to do" [113]. Long before IBM Watson's wake up call, Pink Floyd's 1979 song "Another Brick in the Wall, Pt. 2" has nailed down the problem: "We don't need no education. We don't need no thought control. No dark sarcasm in the classroom. Teachers leave them kids alone. ... All in all you're just another brick in the wall" [117].

We do need education, just not the one-size-fits-all education. Albert Einstein did not talk until he was four years old. Chinese-American author Yiyun Li wrote about early education for her son who was slow to start speaking: "I had been worrying more about Vincent not graduating from the programme than his real speech development. Is this something that all parents have to face in the modern world—that our children have to meet more and more standards, otherwise either we, the parents, the children themselves, or perhaps both, are considered by professionals to be failing?" [118]

Why have very few of child prodigies achieved adult eminence after graduating from universities? [119] Ted Kaczynski, commonly known as the Unabomber, is an infamous tragedy. Kaczynski was a child prodigy who entered Harvard University at the age of 16, earned a Ph.D. from the University of Michigan, and became an assistant professor at the University of California, Berkeley [120]. Despite his academic success, he became disillusioned with modern society and technology. The cookie-cutter education system has failed both genius kids and special-needs children. Status quo stifles creativity. A case in point: A National Geographic logic puzzle featured in Brain Games has found that 80 % of children under 10 gave the correct answer in less than 10 seconds whereas the majority of adults were left clueless [121].

In February 2016, cybersecurity expert John McAfee wrote in an op-ed article that "a room full of Stanford computer science graduates cannot compete with a true hacker without even a high-school education" [122]. That may be true, but cybercriminals are generally less educated than ethical hackers. For instance, Spanish police with support of INTERPOL arrested a 16-year-old girl for alleged cyber attacks [123], and security firm AVG linked a piece of malware to an 11-year-old boy in Canada [124].

"The Logical Song" by Roger Hodgson (Supertramp) in 1979 asked some pointed questions that still resonate with many young people today: "When I was young, it seemed that life was so wonderful... But then they sent me away to teach me how to be sensible... Please tell me what we've learned. I know it sounds absurd, but please tell me who I am. Now watch what you say or they'll be calling you a radical, liberal, fanatical, criminal" [125].

Luckily, we can turn to Albert Einstein for answers. Speaking at TED conference in February 1998, Rev. Billy Graham said, "Albert Einstein—I was just talking to someone, when I was speaking at Princeton, and I met Mr. Einstein. He didn't have a doctor's degree, because he said nobody was qualified to give him one" [78]. Joking aside, Einstein's view on college education is apparent in his autobiography and letters to American inventor Thomas Edison and African-American philosopher Robert Thornton:

1. Disagreeing with Thomas Edison's idea that education should be directed toward learning facts, Einstein wrote to Edison in May 1921, "It is not so very important for a person to learn facts. For that he does not really need a college. He can learn them from books. The value of an education in a liberal arts college is not the learning of many facts, but the training of the mind to think something that cannot be learned from textbooks" [126].
2. Supporting Robert Thornton in his efforts to introduce "as much of the philosophy of science as possible" into the modern physics curriculum at the University of Puerto Rico, Einstein wrote to Thornton in December 1944, "I fully agree with you about the significance and educational value of methodology as well as history and philosophy of science. So many people today—and even professional scientists—seem to me like somebody who has seen thousands of trees but has never seen a forest. A knowledge of the historic and philosophical background gives that kind of independence from prejudices of his generation from which most scientists are suffering. This independence created by philosophical insight is—in my opinion—the mark of distinction between a mere artisan or specialist and a real seeker after truth" [127].
3. Einstein wrote in his autobiography, "All religions, arts and sciences are branches of the same tree. All these aspirations are directed toward ennobling man's life, lifting it from the sphere of mere physical existence and leading the

individual towards freedom. It is no mere chance that our older universities developed from clerical schools. Both churches and universities—insofar as they live up to their true function—serve the ennoblement of the individual. They seek to fulfill this great task by spreading moral and cultural understanding, renouncing the use of brute force. The essential unity of ecclesiastical and secular institutions was lost during the 19th century, to the point of senseless hostility. Yet there was never any doubt as to the striving for culture. No one doubted the sacredness of the goal. It was the approach that was disputed" [128].

1.14   The Opening of the American Mind

A popular YouTube video titled "Top 10 Worst Teachers" garnered over 3 million views in a matter of 6 months. One of the top comments was posted by "TheRealSugarBitzSkelly (後輩)," a 14-year-old girl who is an aspiring author, artist, and animator. She wrote, "I hate school. Teachers are so mean. 6th/7th/8th grade girls are the worst sluts ever. Boys are just stupid.. Well, so are girls.. And, my education is horrible! I would like to go to a school with lots of nice kids and teachers, and get the best education" [129]. In one of the replies, a "Bob Larry" responded, "gotta say it.. most of what you learn in school won't apply to your life later. All it does is open up opportunities that need the different subjects but not others. for instance i'm going into programming. history, science, and a lot of the stuff they teach in english courses doesn't really apply to what I need to know to learn the career path i'm following."

Teachers, not students, are the ones who are failing. As Albert Einstein said that "all religions, arts and sciences are branches of the same tree," the detachment of philosophy—the forefather of all knowledge and academic disciplines—from mathematics, sciences, and technology is the fundamental reason for failure in modern-day K-12 and higher education.

Education needs a major overhaul. The closing of the American mind is not only an American problem but a global issue. The world needs more good teachers who can inspire. As William Arthur Ward said, "The mediocre teacher tells. The good teacher explains. The superior teacher demonstrates. The great teacher inspires" [130]. Duke University Prof. Kalman Bland said at a faculty roundtable about Allan Bloom: "He sees the university as an institution in society, and the function of the university in society as going against the grain. That's the good part of the book—showing that the university does fit into the social context, and that it defines itself in relationship to the needs and values of that context. And the book asks us to take a close look at whether or not we're serving the powers that be or whether we're being the gadflies—the Socratic model of shaking our students up and liberating them from their popular biases" [131].

We also need better methodologies and technologies to assist in teaching and learning. In Chapter 15 of this book, Sandra Lund-Diaz, Mireia Montane, and Penelope Beery from Knowledge Building in Action offer the key to knowledge-building pedagogy success in supporting paradigm shifts for student growth and the 4Cs (Critical thinking, Collaboration, Communication, Creativity) of future education. In Chapter 16, Zach Tolan of the Polymathic Prodigy Institute introduces educational ergonomics and the future of the mind. In Chapter 17, Lewis Watson from Marshall Fundamental School provides an answer to the math problem. And in Chapter 18, Andrew Donaldson of the Bolder Super School describes XQ Super Schools and online achievement.

The STEAM (Science, Technology, Engineering, Arts, and Mathematics) program is a good example of integrating multiple disciplines in the classroom to teach students to think critically. The arts include drawing, filmmaking, music, and photography that have been greatly empowered by computing and electronics technology [132, 133]. The arts also encompass cooking, dance, literature, and other creative expressions. John Keating (Robin Williams) lectured his students in *Dead Poets Society* (1989): "We don't read and write poetry because it's cute. We read and write poetry because we are members of the human race. And the human race is filled with passion. And medicine, law, business, engineering, these are noble pursuits and necessary to sustain life. But poetry, beauty, romance, love, these are what we stay alive for" [134].

Pragmatism over idealism, the percentage of undergraduates majoring in fields like English or philosophy has fallen by more than 50 % since 1970 [135]. The proportion of Stanford students majoring in the Humanities has plummeted from over 20 % to only 7 % in 2015 [136]. In an attempt to "breach the silos of students' lives," Stanford University created two new joint-major programs known as CS+X that allows students to study English and computer science or music and computer science [137]. Under the "One University" policy, University of Pennsylvania allows undergraduates access to courses at most of Penn's undergraduate and graduate schools [138]. Still, some people have opted for homeschooling or unorthodox educational institutions:

1. Academy Award winning director Laura Poitras attended Sudbury Valley School where there were no grades, no classrooms, and no division of students by age [139]. The school's cofounders Mimsy Sadofsky and Daniel Greenberg explained, "Students of all ages determine what they will do, as well as when, how, and where they will do it. ... The fundamental premises of the school are simple: that all people are curious by nature; that the most efficient, long-lasting, and profound learning takes place when started and pursued by the learner; that all people are creative if they are allowed to develop their unique talents; that age-mixing among students promotes growth in all members of the group; and that freedom is essential to the development of personal responsibility" [140].
2. Cofounded by Peter Diamandis (XPRIZE Foundation) and Ray Kurzweil (Google), Singularity University at the NASA Research Park offers educational and incubator programs based on "interdisciplinary, international and

inter-cultural principles" in order to "educate, inspire and empower leaders to apply exponential technologies to address humanity's grand challenges" [141]. The corporate founders are Genentech, Autodesk, Google, Cisco, Ewing Marion Kauffman Foundation, Nokia, and ePlanet Capital. "If I were a student, this is where I'd want to be," said Larry Page [142].

To realize the urgent need for education reform, American journalist George Packer succinctly summed up the perils of the current dysfunctional education system in the November 2011 issue of *The New Yorker* [143]:

> Thiel believes that education is the next bubble in the U.S. economy. He has compared university administrators to subprime-mortgage brokers, and called debt-saddled graduates the last indentured workers in the developed world, unable to free themselves even through bankruptcy. Nowhere is the blind complacency of the establishment more evident than in its bovine attitude toward academic degrees: as long as my child goes to the right schools, upward mobility will continue. A university education has become a very expensive insurance policy—proof, Thiel argues, that true innovation has stalled. In the midst of economic stagnation, education has become a status game, "purely positional and extremely decoupled" from the question of its benefit to the individual and society.

1.15   Internet Revolution and Collective Consciousness

Since the invention of the printing press by Johannes Gutenberg in the 15th century, mass media has ushered in a new era of collective consciousness—a set of shared beliefs, ideas, and moral attitudes that operate as a unifying force within society. Notwithstanding the danger of assimilation akin to the Borg in *Star Trek* (see

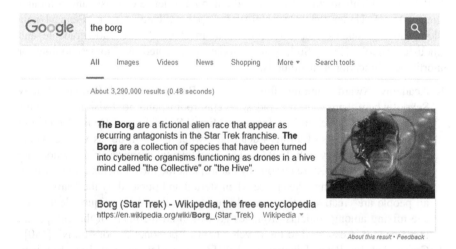

Fig. 1.18 Google search on "The Borg" displays a Wikipedia definition of the fictional alien race in the *Star Trek* franchise as "a collection of species that have been turned into cybernetic organisms functioning as drones in a hive mind."

Fig. 1.18), the Internet is accelerating collective consciousness and revolutionizing economy, politics, and education, among others:

1. New Industrial Revolution—Micha Kaufman, cofounder and CEO of Fiverr. com, published an article in the *Forbes* magazine titled "The Internet Revolution is the New Industrial Revolution." He wrote, "As we engage in a century where everyone is not only a global citizen, but a valuable 'Brand in Waiting,' we begin to understand that the Internet Revolution IS in fact the Industrial Revolution of our time. It's a sweeping social disruption that brings with it not only new inventions and scientific advances, but perhaps most importantly revolutionizes both the methods of work and we the workers ourselves" [144].
2. New Political Revolution—Social media has played a vital role in Arab Spring uprisings including the 2011 Egyptian revolution [145]. Activists organized through Facebook and Twitter the nationwide protests on January 28, 2011 to call for an end to President Hosni Mubarak's government [146]. Mubarak reacted by shutting down 88 % of the Egyptian Internet and 9 out of 10 Internet Service Providers (ISPs) [147]. Google responded to the Internet blockade by working with Twitter and SayNow to unveil a web-free speak-to-tweet service (Speak2Tweet), allowing anyone to send and receive tweets by calling a phone number [148]. In November 2012, more than 90 % of the Internet access in Syria was shut down by the government in an attempt to limit the dissemination of images and videos taken by the opposition activists [149]. Google again offered the Speak2Tweet service in Syria [150]. Prof. Yousri Marzouki and Olivier Oullier at Aix-Marseille University called the phenomenon "Virtual Collective Consciousness" [151].
3. New Educational Revolution—At the TED2013 conference, educational researcher Sugata Mitra talked about how poor children in an Indian slum were able to teach themselves English, along with advanced concepts in biology, chemistry, and mathematics, simply by following their own curiosity and helping each other with the use of a single personal computer and access to the Internet [152]. Mitra's vision is to build Self-Organized Learning Environments (SOLEs) in the School of Clouds, and to create the future of learning with a curriculum of "big questions." For example, Mitra said, "The way you would put it to a nine-year-old is to say, 'If a meteorite was coming to hit the Earth, how would you figure out if it was going to or not?' And if he says, 'Well, what? how?' you say, 'There's a magic word. It's called the tangent of an angle,' and leave him alone. He'll figure it out. ... I've tried incredible, incredible questions—'When did the world begin? How will it end?'—to nine-year-olds. This one is about what happens to the air we breathe. This is done by children without the help of any teacher. The teacher only raises the question, and then stands back and admires the answer."

In the foreseeable future, a poor child in a remote corner of the world will be able to create a killer app, solve the P versus NP problem [153], formulate the Theory of

Everything [154], and find a cure to cancer and other diseases—all without formal education. The Internet is the teacher.

1.16 Alphabet Slogan: Do the Right Thing

Parents often tell their children, "Don't do this" or "Don't do that." For a mature grown-up, however, the annoying "Don't" becomes "Do" as in "Do the right thing." In the 2015 corporate restructuring, Google's newly-created parent company Alphabet does not dwell on "don't be evil" anymore but instead adopts a code of conduct that entreats employees to "do the right thing" [155]:

> Employees of Alphabet and its subsidiaries and controlled affiliates ("Alphabet") should do the right thing—follow the law, act honorably, and treat each other with respect.

> We rely on one another's good judgment to uphold a high standard of integrity for ourselves and our company. We expect all Board members and employees to be guided by both the letter and the spirit of this Code.

A Google spokesman told *The Wall Street Journal*, "Individual Alphabet companies may of course have their own codes to ensure they continue to promote compliance and great values. But if they start bringing cats to work, there's gonna be trouble with a capital T" [156].

Humorous aside, "Don't be evil" and "Do the right thing" basically mean the same thing to Google cofounder Sergey Brin. In a 2004 *Playboy* interview, Brin explained [84]:

> As for "Don't be evil," we have tried to define precisely what it means to be a force for good—always do the right, ethical thing. Ultimately, "Don't be evil" seems the easiest way to summarize it. It's not enough not to be evil. We also actively try to be good.

> We deal with all varieties of information. Somebody's always upset no matter what we do. We have to make a decision; otherwise there's a never-ending debate. Some issues are crystal clear. When they're less clear and opinions differ, sometimes we have to break a tie. For example, we don't accept ads for hard liquor, but we accept ads for wine. It's just a personal preference. We don't allow gun ads, and the gun lobby got upset about that. We don't try to put our sense of ethics into the search results, but we do when it comes to advertising.

> I think we do a good job of deciding. As I said, we believe that "Don't be evil" is only half of it. There's a "Be good" rule also. We have Google grants that give advertising to nonprofit organizations. A couple hundred nonprofits—ranging from the environment to health to education to preventing various kinds of abuse by governments—receive free advertising on Google.

1.17   Philanthropy

Google.org, Bill & Melinda Gates Foundation, and Ford Foundation are some of the largest and most well-known philanthropic organizations in the world. In December 2015, Mark Zuckerberg announced that he would donate 99 % of his Facebook shares worth $45 billion to the Chan Zuckerberg Initiative, LLC for charitable projects [157]. In fact, there are more than 1.5 million charitable organizations in the United States spending over $1.57 trillion in cash and 7.9 billion hours of service to "do the right thing" by lending a helping hand [158].

Back in 1949, the Ford Foundation thoroughly reviewed the ideas and goals of philanthropy, and came to the following conclusions on why and how to spend half a billion dollars a year for charity [159]:

> At one time the gifts of individuals and benevolent organizations were intended largely to relieve the suffering of "the weak, the poor and the unfortunate." With the establishment of the modern foundation a much greater concept came into being. The aim is no longer merely to treat symptoms … but rather to eradicate the causes of suffering. Nor is the modern foundation content to concern itself only with man's obvious physical needs; it seeks rather to help man achieve his entire well-being—to satisfy his mental, emotional, and spiritual needs as well … what he needs and wants, what incentives are necessary to his productive and socially useful life, what factors influence his development and behavior, how he learns and communicates with other persons, and, finally, what prevents him from living at peace with himself and his fellow men.

The Apollo 11 Moon landing on July 20, 1969 was made possible because of the unprecedented national focus, collaborative spirit, and financial support for one tremendously difficult challenge. Imagine what $1.57 trillion in cash and 7.9 billion hours of service could have done to solve some of the most pressing issues today. We may not achieve an immediate unalloyed success without a few bumps along the way, but the successful Moon landing was preceded by many failures.

1.18   Moon Landing and a Little Help from My Friends

President John F. Kennedy spoke at Rice University on September 12, 1962: "We choose to go to the moon in this decade … not because they are easy, but because they are hard, because that goal will serve to organize and measure the best of our energies and skills, because that challenge is one that we are willing to accept, one we are unwilling to postpone, and one which we intend to win" [160].

America at the time intended to win the Space Race against Russia who launched the first satellite (Sputnik) in October 1957 (see Fig. 1.19) and sent the first astronaut (Yuri Gagarin) to outer space in April 1961 (see Fig. 1.20).

Moon landing was a huge challenge that was solved by human perseverance and ingenuity, in spite of the mere 50 % chance of success according to American astronaut Neil Armstrong [161]. Imagine what else we can accomplish if America

Fig. 1.19 Google search on "first satellite" displays a Knowledge Graph of Sputnik I

and the whole world is determined to eradicate wars, diseases, pollutions, global warming, poverty, homelessness, world hunger, and other human sufferings.

Scientists and engineers need unfaltering support akin to what NASA received in the 60's from the U.S. government and the American public. An older generation may feel nostalgic about the 1967 Beatles song *With a Little Help from My Friends* or the Joe Cocker's version that he performed with Jimmy Page and others at Woodstock in 1969. The Beatles song was played as the wake-up music on Space Shuttle Mission STS-61 on December 5, 1993 [162].

A little help from federal funding can go a long way. DARPA initiated and funded the research and development of Advanced Research Projects Agency

Fig. 1.20 Google search on "first human in space" displays a Knowledge Graph of Yuri Gagarin

Network (ARPANET) that went online in 1969 [163]. The success of ARPANET gave rise to the global commercial Internet in the mid-1990s and the new generation of Fortune 500 companies today including Amazon.com, Google, eBay, and Facebook. Another good example is the talking, question-answering Siri application on Apple's iPhone [164]. Siri originated from a DARPA-funded project known as PAL (Personalized Assistant that Learns)—an adaptive artificial intelligence program for data retrieval and synthesis [165].

President Barack Obama said at a campaign fundraiser in April 2012: "I believe in investing in basic research and science because I understand that all these extraordinary companies ... many of them would have never been there; Google,

Facebook would not exist, had it not been for investments that we made as a country in basic science and research" [166].

In a testimony before the U.S. Senate Commerce Committee on May 11, 2016, Robert D. Atkinson of the Information Technology and Innovation Foundation (ITIF) said, "It is no longer enough to simply fund scientific and engineering research and hope it somehow produces commercial results. Federal R&D funding as a share of GDP is lower today than before the Russians launched Sputnik. This means the nation needs to be much more efficient about transferring discoveries into commercial applications. Otherwise, we risk slowing the pace of innovation even more. ... Improving the efficiency of the scientific and engineering research system can provide significant benefits at a lower budgetary impact than increasing funding without improving the efficiency. But continuing to underfund research while also not improving the efficiency of the system ... is a recipe for underperformance. And to be clear doing both is ideal: more federal funding for R&D and a better commercialization and tech transfer system" [167].

In the 1983 film *The Right Stuff* adapted from 1979 book of the same name by Tom Wolfe, seven Mercury astronauts discussed their spacecraft [168]:

> **Gordon Cooper (Dennis Quaid)**: *You boys know what makes this bird go up? Funding makes this bird go up.*

> **Gus Grissom (Fred Ward)**: *He's right. No bucks, no Buck Rogers.*

Fig. 1.21 Apollo 11 astronaut Buzz Aldrin works at the deployed Passive Seismic Experiment Package on July 20, 1969. To the left of the United States flag in the background is the lunar surface television camera. Photo taken by Neil Armstrong. (Courtesy of NASA)

1.19   Faith in God and Trust in Google

Given that the first manned Moon landing only had a 50 % chance of landing safely on the moon's surface, it was indeed an exemplary faith in technology and human spirit (see Fig. 1.21). American astronaut Neil Armstrong said in a 2012 video interview, "I thought we had a 90 % chance of getting back safely to Earth on that flight but only a 50–50 chance of making a landing on that first attempt. There are so many unknowns on that descent from lunar orbit down to the surface that had not been demonstrated yet by testing and there was a big chance that there was something in there we didn't understand properly and we had to abort and come back to Earth without landing" [161].

When Armstrong and Buzz Aldrin were about to land on the moon, they disagreed with the on-board computer's decision to put them down on the side of a large crater with steep slopes littered with huge boulders. "Not a good place to land at all," said Armstrong. "I took it over manually and flew it like a helicopter out to the west direction, took it to a smoother area without so many rocks and found a level area and was able to get it down there before we ran out of fuel. There was something like 20 seconds of fuel left." The rest is history as Armstrong uttered his famous line, "One small step for a man, one giant leap for mankind."

Faith is defined as complete trust or confidence in someone or something. The Bible, the Quran, and Google all require faith from the great mass of the world population in order to thrive:

The Bible: "Truly I tell you, if you have faith as small as a mustard seed, you can say to this mountain, 'Move from here to there,' and it will move. Nothing will be impossible for you." (Matthew 17:20)

The Quran: "If Allah is your helper none can overcome you, and if He withdraws His help from you, who is there who can help you after Him? In Allah let believers put their trust." (Quran 3:160)

Google: "Google users trust our systems to help them with important decisions: medical, financial and many others." (Google founders' 2004 IPO Letter) [84, 169].

Sometimes Google is the only source of life-and-death information. In December 2013, 27-year-old Sanaz Nezami with severe head injuries was rushed to Marquette General Hospital in Michigan. "At the time the staff did not know anything about this young woman who came in with critical injuries," said nurse supervisor Gail Brandly [170]. The nurse Googled the patient's name and found a resume online with her picture and a phone number through which the hospital was able to reach her relatives in Iran [171]. In May 2016, Maritha Strydom in Brisbane, Australia had been following her daughter Maria's progress on climbing Mount Everest through a series of satellite pings from Maria's phone. "I was worried when the pings stopped, and we started calling but no one could give us any answers," Strydom told CNN in an interview. "So my other daughter, also lives in Brisbane, just before bedtime Googled and found in the Himalayan Times that my daughter had passed away" [172].

The World's Most Reputable Companies In 2015

Rank	Company	Home country
1	BMW	Germany
2	Google	United States of America
3	Daimler	Germany
4	Rolex	Switzerland
5	LEGO	Denmark
6	The Walt Disney Company	United States of America
7	Canon	Japan
8	Apple	United States of America
9	Sony	Japan
10	Intel	United States of America

Fig. 1.22 Top 10 of the world's most reputable companies in 2015 according to Reputation Institute's Global RepTrak 100

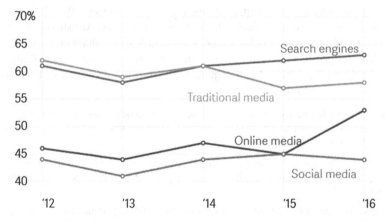

Fig. 1.23 More people trust search engines for their news than traditional media, online-only media, and social media in 2015 and 2016 according to the Edelman Trust Barometer

Indeed, Google ranked number 2 in the world's most reputable companies in 2015 according to Reputation Institute's Global RepTrak survey on innovation, governance, citizenship, and other factors [173] (see Fig. 1.22). More people trust search engines for their news than traditional media, online-only media, and social media [174] (see Fig. 1.23). In the 2016 Edelman Trust Barometer report, a survey of 33,000 people across 28 countries has affirmed that 63 % of respondents trust search engines for news and information, comparing to 58 % for traditional media and 53 % for online-only media [175].

Speaking at a small technology dinner at the Aspen Institute, executive chairman of Alphabet Inc. Eric Schmidt said, "There's a particular religion that we all represent, and it goes something like this: 'if you take a large number of people and you empower them with communication tools and opportunities to be creative, society gets better.' ... The combination of empowerment, innovation, and creativity will be our solution, but that is a religion in-of-itself" [176]. Chapter 2 of this book unravels the Gordian knot of religious, moral, and political entanglement.

Google Life Sciences, formerly a division of Google X, was renamed to Verily in December 2015 as a subsidiary of Alphabet [177]. The 13th century Middle English "verily" fell out of common use except in the King James Bible. "I can't think of another association for Verily but the Bible," said Greg Balla, creative director at the branding and naming agency Zenmark. "The challenge for them is to try to move away from the heavy-handed quality attached to Verily from association with the scripture—due to everything that's happening in our world right now. But if they can deliver on the promise, 'you can trust what we are doing,' then it fits perfectly" [177]. Chapter 3 of this book offers some scientific insights into Google vs. Death.

Verily, trust Google. The truth is out there; we just need to know how to Google it!

References

1. **Harbrecht, Douglas.** Google's Larry Page: Good Ideas Still Get Funded. *Bloomberg.* [Online] 2001, 12 March. http://www.bloomberg.com/news/articles/2001-03-12/googles-larry-page-good-ideas-still-get-funded.
2. **Google.** How Search Works: From algorithms to answers. *Google Inside Search.* [Online] http://www.google.com/insidesearch/howsearchworks/thestory/.
3. **Bialik, Carl.** There Could Be No Google Without Edward Kasner. *The Wall Street Journal.* [Online] June 14, 2004. http://www.wsj.com/articles/SB108575924921724042.
4. **Google.** Our history in depth. *Google Company.* [Online] http://www.google.com/intl/en/about/company/history/.
5. **Google.** About Google! *Internet Archive Wayback Machine.* [Online] September 27, 1998. https://web.archive.org/web/19990204033714/http://google.stanford.edu/about.html.
6. **Brin, Sergey and Page, Lawrence.** The anatomy of a large-scale hypertextual Web search engine. *Proceedings of the seventh international conference on World Wide 7.* [Online] April 1998. http://infolab.stanford.edu/ ~ backrub/google.html.
7. **Ironman.** Here's Why The Dot Com Bubble Began And Why It Popped. *Business Insider.* [Online] December 15, 2010. http://www.businessinsider.com/heres-why-the-dot-com-bubble-began-and-why-it-popped-2010-12.
8. **Efrati, Amir.** Google Notches One Billion Unique Visitors Per Month. *The Wall Street Journal.* [Online] June 21, 2011. http://blogs.wsj.com/digits/2011/06/21/google-notches-one-billion-unique-visitors-per-month/.
9. **Obama, Barack.** 2011 State of the Union Address. [Online] PBS, January 25, 2011. http://www.pbs.org/newshour/interactive/speeches/4/2011-state-union-address/.
10. **Clark, Jack.** Google Parent Overtakes Apple as World's Most Valuable Company. *Bloomberg.* [Online] February 2, 2016. http://www.bloomberg.com/news/articles/2016-02-02/google-parent-to-overtake-apple-as-world-s-most-valuable-company.
11. **Holson, Laura M.** Putting a Bolder Face on Google. *The New York Times.* [Online] February 28, 2009. http://www.nytimes.com/2009/03/01/business/01marissa.html.
12. **Google.** About Google Doodles. [Online] 2013. https://www.google.com/doodles/about.

13. Isaac Newton's birth marked by Google Doodle. *The Telegraph*. [Online] January 4, 2010. http://www.telegraph.co. uk/technology/google/6933008/Isaac-Newtons-birth-marked-by-Google-Doodle.html.

14. **Nelson, Randy.** Google celebrates Pac-Man's 30th anniversary with playable logo. *Engadget.* [Online] May 21, 2010. http://www.engadget.com/2010/05/21/google-celebrates-pac-mans-30th-with-playable-logo/.

15. **Google.** Contest Rules. *Doodle 4 Google Competition* . [Online] https://www.google. com/doodle4google/rules.html.

16. **Barnett, Emma.** Creating a women's Google Doodle was too frightening. *The Telegraph.* [Online] February 19, 2013. http://www.telegraph.co.uk/women/womens-business/9879577/ Creating-a-womens-Google-Doodle-was-too-frightening.html.

17. **Molloy, Antonia.** Are Google Doodles sexist and racist? Report says the graphics under-represent women and favour white men. *The Independent.* [Online] February 28, 2014. http://www.independent.co.uk/life-style/gadgets-and-tech/news/are-google-doodles-sexist-and-racist-report-says-the-graphics-under-represent-women-and-favour-white-9159500.html.

18. **Carlson, Nicholas.** Google Just Killed The "I'm Feeling Lucky Button". *Business Insider.* [Online] September 8, 2010. http://www.businessinsider.com/google-just-effectively-killed-the-im-feeling-lucky-button-2010-9.

19. **Google.** Art Project. *Google Cultural Institute.* [Online] https://www.google.com/ culturalinstitute/collection/mathematisch-physikalischer-salon-royal-cabinet-of-mathematical-and-physical-instruments?projectId=art-project.

20. **Google.** a Google a day. [Online] http://www.agoogleaday.com/.

21. **Google.** Google Doodles. [Online] https://www.google.com/doodles/.

22. **Google.** Google One Today. [Online] https://onetoday.google.com/home/projects? utm_source=ifg.

23. **Google.** Restaurant search results. [Online] https://www.google.com/search? gws_rd=ssl&q=restaurants.

24. **Google.** Hubble Telescope. *Google Earth.* [Online] https://www.google.com/earth/explore/ showcase/hubble20th.html#tab=ngc-6302.

25. **Google.** Google Trends. [Online] https://www.google.com/trends/hottrends.

26. World Wonders. *Google Cultural Institute.* [Online] https://www.google.com/ culturalinstitute/entity/%2Fm%2F06519j?hl=en&projectId=world-wonders.

27. **IMDb.** The Internship. *IMDb.* [Online] June 7, 2013. http://www.imdb.com/title/ tt2234155/trivia?tab=qt&ref_=tt_trv_qu.

28. **Page, Larry.** Google Search Engine: New Features. *Wayback Machine Internet Archive.* [Online] July 8, 1998. http://web.archive.org/web/19991009052012/http://www.egroups. com/group/google-friends/3.html.

29. **Kirshner, Rebecca Rand.** Buffy the Vampire Slayer: Help. *Wikia.* [Online] October 15, 2002. http://buffy.wikia.com/wiki/Help.

30. **Bylund, Anders.** To Google or Not to Google. *The Motley Fool.* [Online] July 5, 2006. http://www.fool.com/investing/dividends-income/2006/07/05/to-google-or-not-to-google. aspx.

31. **Rachman, Tom.** Language by the Book, but the Book Is Evolving: O.E.D.'s New Chief Editor Speaks of Its Future. *The New York Times.* [Online] January 21, 2014. http://www. nytimes.com/2014/01/22/books/oeds-new-chief-editor-speaks-of-its-future.html?_r=0.

32. **Sullivan, Danny.** Meet Google assistant: A new search platform, rather than a gadget or an app. *Search Engine Land.* [Online] May 18, 2016. http://searchengineland.com/google-assistant-249903.

33. **Pichai, Sundar.** I/O: Building the next evolution of Google . *Google Official Blog.* [Online] May 18, 2016. https://googleblog.blogspot.com/2016/05/io-building-next-evolution-of-google.html.

34. **Fix, Team.** The CNN-Telemundo Republican debate transcript, annotated. *The Washington Post.* [Online] February 25, 2016. https://www.washingtonpost.com/news/the-fix/wp/2016/ 02/25/the-cnntelemundo-republican-debate-transcript-annotated/.

35. **Chmurak, Elizabeth.** Polish Worker Speaks Out Following Trump-Rubio Spar. *Fox Business.* [Online] February 26, 2016. http://www.foxbusiness.com/politics/2016/02/26/polish-worker-speaks-out-following-trump-rubio-spar.html.
36. **Sullivan, Danny.** Google Now Personalizes Everyone's Search Results. *Search Engine Land.* [Online] December 4, 2009. http://searchengineland.com/google-now-personalizes-everyones-search-results-31195.
37. **Drummond, David.** Don't censor the web. [Online] Google Public Policy Blog, January 18, 2012. http://googlepublicpolicy.blogspot.com/2012/01/dont-censor-web.html.
38. **Sherman, Cary H.** What Wikipedia Won't Tell You. [Online] The New York Times, February 7, 2012. http://www.nytimes.com/2012/02/08/opinion/what-wikipedia-wont-tell-you.html.
39. **Chou, Dorothy.** More transparency into government requests. [Online] Google Official Blog, June 17, 2012. http://googleblog.blogspot.com/2012/06/more-transparency-into-government.html#!/2012/06/more-transparency-into-government.html.
40. **Google.** Download the data . *Google Transparency Report.* [Online] https://www.google.com/transparencyreport/removals/government/data/.
41. **Google.** Google Transparency Report. [Online] Google. [Cited: June 22, 2014.] http://www.google.com/transparencyreport/removals/government/.
42. **Helft, Miguel and Barboza, David.** Google Shuts China Site in Dispute Over Censorship. [Online] The New York Times, March 22, 2010. http://www.nytimes.com/2010/03/23/technology/23google.html.
43. **Drummond, David.** A new approach to China: an update. [Online] Google Official Blog, March 22, 2010. http://googleblog.blogspot.com/2010/03/new-approach-to-china-update.html.
44. **Thieme, Richard.** DEF CON 22 - Richard Thieme - The Only Way to Tell the Truth is in Fiction . [Online] DEFCONConference, December 21, 2014. https://www.youtube.com/watch?v=EdsJulQdUcg.
45. **Snowden, Edward, et al.** Interview On NSA Whistleblowing (Full Transcript). *Genius.* [Online] June 9, 2013. http://genius.com/Edward-snowden-interview-on-nsa-whistleblowing-full-transcript-annotated.
46. **Harding, Luke.** How Edward Snowden went from loyal NSA contractor to whistleblower . *The Guardian.* [Online] February 1, 2014. http://www.theguardian.com/world/2014/feb/01/edward-snowden-intelligence-leak-nsa-contractor-extract.
47. **Bennett, Cory and Williams, Katie Bo.** Government software may have let in foreign spies. *The Hill.* [Online] February 2, 2016. http://thehill.com/policy/cybersecurity/267826-government-software-may-have-let-in-foreign-spies.
48. **Harris, Shane.** Exclusive: 50 Spies Say ISIS Intelligence Was Cooked. *The Daily Beast.* [Online] September 9, 2015. http://www.thedailybeast.com/articles/2015/09/09/exclusive-50-spies-say-isis-intelligence-was-cooked.html.
49. **Mahler, Jonathan.** What Do We Really Know About Osama bin Laden's Death? *The New York Times Magazine.* [Online] October 15, 2015. http://www.nytimes.com/2015/10/18/magazine/what-do-we-really-know-about-osama-bin-ladens-death.html.
50. **Bergen, Peter.** The New York Times triples down on bizarre bin Laden story. *CNN.* [Online] October 24, 2015. http://www.cnn.com/2015/10/24/opinions/bergen-times-triples-down-bin-laden-story/index.html.
51. **Calderone, Michael.** NYT Magazine's Bin Laden Cover Story Sparked Controversy In The Newsroom. *The Huffington Post.* [Online] October 22, 2015. http://www.huffingtonpost.com/entry/new-york-times-bin-laden-story_56253edae4b08589ef485969.
52. **Chilton, Martin.** How the CIA brought Animal Farm to the screen. [Online] The Telegraph, November 5, 2014. http://www.telegraph.co.uk/culture/film/11209390/How-the-CIA-brought-Animal-Farm-to-the-screen.html.
53. **Wang, Yanan.** 'Workers' or slaves? Textbook maker backtracks after mother's online complaint. *The Washington Post.* [Online] October 5, 2015. https://www.washingtonpost.

com/news/morning-mix/wp/2015/10/05/immigrant-
workers-or-slaves-textbook-maker-backtracks-after-mothers-online-complaint/.

54. **Fung, Brian.** Military Strikes Go Viral: Israel Is Live-Tweeting Its Own Offensive Into
Gaza. [Online] The Atlantic, November 14, 2012. http://www.theatlantic.
com/international/archive/2012/11/military-strikes-go-viral-israel-is-live-
tweeting-its-own-offensive-into-gaza/265227/.

55. **Hachman, Mark.** IDF vs. Hamas War Extends to Social Media. [Online] PC Magazine,
November 16, 2012. http://www.pcmag.com/slideshow/story/305065/idf-vs-hamas-war-
extends-to-social-media.

56. **Sutter, John D.** Will Twitter war become the new norm? [Online] CNN, November 19,
2012. http://www.cnn.com/2012/11/15/tech/social-media/twitter-war-gaza-israel/index.html.

57. **Ennis, Stephen.** Russia in 'information war' with West to win hearts and minds. *BBC News.*
[Online] September 16, 2015. http://www.bbc.com/news/world-europe-34248178.

58. **Lee, Newton.** Facebook Nation: Total Information Awareness. [Online] Springer Science
+Business Media, October 17, 2014. http://www.amazon.com/Facebook-Nation-Total-
Information-Awareness/dp/1493917390/.

59. **Wiedeman, Reeves.** The Rise and Fall of a Fox News Fraud. *Rolling Stone.* [Online] January
26, 2016. http://www.rollingstone.com/politics/news/the-rise-and-fall-of-a-fox-news-fraud-
20160126.

60. **Adichie, Chimamanda.** Chimamanda Adichie: The danger of a single story. [Online] TED,
October   2009.   http://www.ted.
com/talks/chimamanda_adichie_the_danger_of_a_single_story.html.

61. **Senior, Jennifer.** In Conversation: Antonin Scalia. [Online] New York Magazine. http://
nymag.com/news/features/antonin-scalia-2013-10/index1.html.

62. **Vogue, Ariane de.** Scalia-Ginsburg friendship bridged opposing ideologies. *CNN.* [Online]
February 14, 2016. http://www.cnn.com/2016/02/14/politics/antonin-scalia-ruth-bader-
ginsburg-friends/index.html.

63. **Dear, John S.J.** Afghanistan journal, part one: Learning a nonviolent lifestyle in Kabul.
[Online] National Catholic Reporter, December 11, 2012. http://ncronline.org/blogs/
road-peace/afghanistan-journal-part-one-learning-nonviolent-lifestyle-kabul.

64. **Dear, John S.J.** Afghanistan journal, part two: bearing witness to peacemaking in a war-torn
country. [Online] National Catholic Reporter, December 18, 2012. http://ncronline.
org/blogs/road-peace/afghanistan-journal-part-two-bearing-
witness-peacemaking-war-torn-country.

65. **Blancke, Stefaan, Hjermitslev, Hans Henrik and Kjærgaard, Peter C.** Creationism in
Europe. *Google Books.* [Online] Johns Hopkins University Press, November 27, 2014.
https://books.google.com/books?id=-
gOhBQAAQBAJ&pg=PT183&lpg=PT183#v=onepage&q&f=false.

66. **Foreman, Tom, Nye, Bill and Ham, Ken.** Bill Nye Debates Ken Ham - HD (Official).
YouTube. [Online] February 4, 2014. https://www.youtube.com/watch?v=z6kgvhG3AkI.

67. **Einstein, Albert and Shaw, George Bernard.** Einstein on Cosmic Religion and Other
Opinions and Aphorisms. *Google Books.* [Online] April 23, 2009. https://books.google.
com/books/about/Einstein_on_Cosmic_Religion_and_Other_Op.html?
id=9YFCAwAAQBAJ.

68. **Stafford, Tom.** The web has deluded you, and don't pretend it hasn't. *BBC News.* [Online]
October 20, 2015. http://www.bbc.com/future/story/20151020-the-web-has-deluded-you-
and-dont-pretend-it-hasnt.

69. **Encyclopaedia Britannica.** The Atlantic, Volume 216. *Google Books.* [Online] 1965.
https://books.google.com/books?id=TuMmAQAAIAAJ.

70. **Dahm, R.** Friedrich Miescher and the discovery of DNA. *PubMed.gov.* [Online] National
Institutes of Health, February 15, 2005. http://www.ncbi.nlm.nih.gov/pubmed/15680349.

71. **Rincon, Paul.** Neanderthal genes 'survive in us' . *BBC News.* [Online] May 6, 2010. http://
news.bbc.co.uk/2/hi/science/nature/8660940.stm.

72. **Heguy, Adriana.** What makes Neanderthal DNA superior? *Quora.* [Online] August 11, 2015. https://www.quora.com/What-makes-Neanderthal-DNA-superior.

73. **Levy, Charles.** Mushroom cloud from the atomic bombing of Nagasaki, Japan on August 9, 1945. *National Archives and Records Administration.* [Online] August 9, 1945. http://www.archives.gov/research/military/ww2/photos/images/ww2-163.jpg.

74. **American Museum of National History.** The Manhattan Project. *American Museum of National History.* [Online] [Cited: December 6, 2015.] http://www.amnh.org/exhibitions/einstein/peace-and-war/the-manhattan-project.

75. **Cook, Tim.** A Message to Our Customers. *Apple.* [Online] February 16, 2016. http://www.apple.com/customer-letter/.

76. **Segall, Laurie, Pagliery, Jose and Wattles, Jackie.** FBI says it has cracked terrorist's iPhone without Apple's help. *CNNMoney.* [Online] March 29, 2016. http://money.cnn.com/2016/03/28/news/companies/fbi-apple-iphone-case-cracked/index.html.

77. **Friedman, Howard Steven.** 9 Countries In the Nuclear Weapons Club. *The World Post.* [Online] March 10, 2012. http://www.huffingtonpost.com/howard-steven-friedman/countries-with-nuclear-weapons_b_1189632.html.

78. **Graham, Billy.** On technology and faith . *TED.* [Online] February 1988. https://www.ted.com/talks/billy_graham_on_technology_faith_and_suffering/transcript?language=en.

79. **Nobel Media AB.** Alfred Nobel 1833-1896 Established the Nobel Prizes "for the Greatest Benefit to Mankind". *Nobel Media AB.* [Online] http://www.nobelprize.org/alfred_nobel/.

80. **Chaplin, Charlie.** The Great Dictator. *IMDb.* [Online] 1940. http://www.imdb.com/title/tt0032553/quotes.

81. **Bulletin of the Atomic Scientist.** Volume 72 Issue 1. *Bulletin of the Atomic Scientist.* [Online] January 2016. http://thebulletin.org/overview.

82. **Valiunas, Algis.** The Agony of Atomic Genius . *The New Atlantis.* [Online] November 14, 2006. http://www.thenewatlantis.com/publications/the-agony-of-atomic-genius.

83. **Oppenheimer, J. Robert.** J. Robert Oppenheimer: "I am become Death, the destroyer of worlds." . *YouTube.* [Online] 1965. https://www.youtube.com/watch?v=lb13ynu3Iac.

84. **Google Inc.** Amendment No. 9 To Form S-1 Registration Statement. *Securities And Exchange Commission.* [Online] August 18, 2004. http://www.sec.gov/Archives/edgar/data/1288776/000119312504142742/ds1a.htm.

85. **Sagal, Peter.** Google Chairman Eric Schmidt Plays Not My Job. *NPR.* [Online] May 11, 2013. http://www.npr.org/2013/05/11/182873683/google-chairman-eric-schmidt-plays-not-my-job.

86. **Mirani, Leo.** What Google really means by "Don't be evil". *Quartz.* [Online] October 21, 2014. http://qz.com/284548/what-google-really-means-by-dont-be-evil/.

87. **From Our Press Services .** Turning 'evil': Critics say Google's recent moves belie search giant's motto. *The Commercial Appeal.* [Online] March 15, 2012. http://www.commercialappeal.com/business/turning-evil-critics-say-googles-recent-moves-belie-search-giants-motto-ep-386504157-323900811.html.

88. **Westhoven, Jennifer.** CNET: We've been blackballed by Google. *CNNMoney.* [Online] August 5, 2005. http://money.cnn.com/2005/08/05/technology/google_cnet/.

89. **Mills, Elinor.** Google balances privacy, reach. *CNET.* [Online] August 3, 2005. http://www.cnet.com/news/google-balances-privacy-reach-1/.

90. **Westhoven, Jennifer.** CNET: We've been blackballed by Google. *CNNMoney.* [Online] August 5, 2005. http://money.cnn.com/2005/08/05/technology/google_cnet/.

91. **Newman, Jared.** Google's Schmidt Roasted for Privacy Comments. [Online] PCWorld, December 11, 2009. http://www.pcworld.com/article/184446/googles_schmidt_roasted_for_privacy_comments.html.

92. **Abell, John C.** Google's 'Don't Be Evil' Mantra Is 'Bullshit,' Adobe Is Lazy: Apple's Steve Jobs (Update 2). *Wired.* [Online] January 30, 2010. http://www.wired.com/2010/01/googles-dont-be-evil-mantra-is-bullshit-adobe-is-lazy-apples-steve-jobs/.

93. **Stone, Brad.** Google Says It Inadvertently Collected Personal Data. [Online] The New York Times, May 14, 2010. http://bits.blogs.nytimes.com/2010/05/14/google-admits-to-snooping-on-personal-data/.

94. **Landis, Marina.** Google admits to accidentally collecting e-mails, URLs, passwords. [Online] CNN, October 22, 2010. http://articles.cnn.com/2010-10-22/tech/google.privacy. controls_1_wifi-data-alan-eustace-google-s-street-view?_s=PM:TECH.

95. **Kelly, Heather.** Is the government doing enough to protect us online? [Online] CNN, July 31, 2012. http://www.cnn.com/2012/07/25/tech/regulating-cybersecurity/index.html.

96. **Bogost, Ian.** What Is 'Evil' to Google? Speculations on the company's contribution to moral philosophy. *The Atlantic.* [Online] October 15, 2013. http://www.theatlantic. com/technology/archive/2013/10/what-is-evil-to-google/280573/.

97. **Thompson, Cadie.** Does 'Don't be evil' still apply to Google? *CNBC.* [Online] August 19, 2014. http://www.cnbc.com/2014/08/19/does-dont-be-evil-still-apply-to-google.html.

98. **Williams, Rob.** EFF Files FTC Complaint Alleging Google Chromebooks Invade Student Privacy, Collect And Share App Usage, Browsing History. *Hot Hardware.* [Online] December 2, 2015. http://hothardware.com/news/eff-files-ftc-complaint-alleging-google-chromebooks-invade-student-privacy-collect-and-share-app-usage-browsing-history.

99. **Rochelle, Jonathan.** The facts about student data privacy in Google Apps for Education and Chromebooks . *Google for Education.* [Online] December 2, 2015. http://googleforeducation. blogspot.com/2015/12/the-facts-about-student-data-privacy-in.html.

100. **Uston, Ken.** 9,250 Apples for the teacher (free computers for California schools). *Creative Computing.* [Online] October 1983. http://www.atarimagazines.com/creative/ v9n10/178_9250_Apples_for_the_teac.php.

101. **Obama, Barack.** Remarks of President Barack Obama – State of the Union Address As Delivered. *The White House.* [Online] January 13, 2016. https://www.whitehouse. gov/the-press-office/2016/01/12/remarks-president-barack-obama-%E2%80% 93-prepared-delivery-state-union-address.

102. **Smith, Megan.** Computer Science For All. *The White House Blog.* [Online] January 30, 2016. https://www.whitehouse.gov/blog/2016/01/30/computer-science-all.

103. **Google.** Programs for educators and students. *Google for Education.* [Online] https:// www.google.com/edu/resources/programs/.

104. **Iszler, Madison.** Florida Senate approves making coding a foreign language. *USA Today.* [Online] March 1, 2016. http://www.usatoday.com/story/tech/news/2016/03/01/florida-senate-approves-making-coding-foreign-language/81150796/.

105. **Kimball, Roger.** The Groves of Ignorance. [Online] 1987, 5 April. http://www.nytimes. com/1987/04/05/books/the-groves-of-ignorance.html?pagewanted=all.

106. **Shontell, Alyson.** PayPal Cofounder Peter Thiel Is Paying 24 Kids $100,000 To Drop Out Of School. *Business Insider.* [Online] May 28, 2011. http://www.businessinsider. com/paypal-cofounder-peter-thiel-is-paying-24-kids-100000-to-drop-out-of-school-2011-5.

107. **Page, Lawrence.** Research at Google. *Google.* [Online] http://research.google.com/pubs/ LawrencePage.html.

108. **Brin, Sergey.** Research at Google. *Google.* [Online] http://research.google.com/pubs/ SergeyBrin.html.

109. **Malone, Scott.** Dropout Bill Gates returns to Harvard for degree. *Reuters.* [Online] June 7, 2007. http://www.reuters.com/article/us-microsoft-gates-idUSN0730259120070607.

110. **Klepper, David.** Mark Zuckerberg, Harvard dropout, returns to open arms. *The Christian Science Monitor.* [Online] November 9, 2011. http://www.csmonitor.com/ Technology/Latest-News-Wires/2011/1109/Mark-Zuckerberg-Harvard-dropout-returns-to-open-arms.

111. **University of Pennsylvania.** THE ACADEMY: Curriculum and Organization. *University of Pennsylvania University Archives and Records Center.* [Online] http://www.archives.upenn. edu/histy/features/1700s/acad_curric.html.

112. **University of Pennsylvania.** Degrees & Programs. *University of Pennsylvania.* [Online] http://www.upenn.edu/programs/academics-degrees-and-programs.

113. **Markoff, John.** Computer Wins on 'Jeopardy!': Trivial, It's Not. *The New York Times.* [Online] February 16, 2011. http://www.nytimes.com/2011/02/17/science/17jeopardy-watson.html?pagewanted=all.

114. **Hirschler, Ben and Willard, Anna.** Robots Will Replace 5 Million Workers By 2020: Report. *The Huffington Post.* [Online] January 18, 2016. http://www.huffingtonpost. com/entry/robot-job-replacement_us_569cf3b3e4b0778f46f9f9b3.
115. **Conner-Simons, Adam.** Computer program fixes old code faster than expert engineers. *MIT News.* [Online] July 9, 2015. http://news.mit.edu/2015/computer-program-fixes-old-code-faster-than-expert-engineers-0609.
116. **Shortliffe, Edward.** Computer-Based Medical Consultations: MYCIN. *Elsevier.* [Online] 1976. https://books.google.com/books?id=i9QXugPQw6oC.
117. **Pink Floyd.** Pink Floyd - Another Brick In The Wall, Part Two (Official Music Video) . *YouTube.* [Online] August 9, 2014. https://www.youtube.com/watch?v=HrxX9TBj2zY.
118. **Li, Yiyun.** 'Einstein didn't talk until he was four' . *The Guardian.* [Online] March 2, 2005. http://www.theguardian.com/lifeandstyle/2005/mar/02/familyandrelationships.features11.
119. **Wikipedians.** List of child prodigies. *Wikipedia.* [Online] https://en.wikipedia. org/wiki/List_of_child_prodigies.
120. **Altman, Michael E.** Unabomber's Secluded Plot of Land for Sale . *The Harvard Crimson.* [Online] December 6, 2010. http://www.thecrimson.com/flyby/article/2010/12/6/harvard-kaczynski-currently-plot/.
121. **Kligman, Julia.** The Key to Decoding This Puzzle Is to Think Like a Child. *National Geographic.* [Online] February 11, 2016. http://tvblogs.nationalgeographic. com/2016/02/11/the-key-to-decoding-this-puzzle-is-to-think-like-a-child/.
122. **McAfee, John.** JOHN MCAFEE: I'll decrypt the San Bernardino phone free of charge so Apple doesn't need to place a back door on its product. *Business Insider.* [Online] February 18, 2016. http://www.businessinsider.com/john-mcafee-ill-decrypt-san-bernardino-phone-for-free-2016-2.
123. **Whiteman, Hilary.** Interpol arrests suspected 'Anonymous' hackers. [Online] CNN, February 29, 2012. http://www.cnn.com/2012/02/29/world/europe/anonymous-arrests-hacking/index.html.
124. **Dunn, John E.** AVG finds 11 year-old creating malware to steal game passwords. [Online] TechWorld, February 8, 2013. http://news.techworld.com/security/3425185/avg-finds-11-year-old-creating-malware-steal-game-passwords/.
125. **Hodgson, Roger.** The Logical Song. *YouTube.* [Online] April 25, 2010. https://www. youtube.com/watch?v=iIyVDHlJgSE.
126. **Frank, Philipp, Rosen, George and Kusaka, Shuichi.** Einstein: His Life and Times. *Da Capo Press.* [Online] 2002. https://books.google.com/books?id=Qs724uDW-rIC.
127. **Howard, Don A.** Einstein's Philosophy of Science. *Stanford Encyclopedia of Philosophy.* [Online] February 11, 2004. http://plato.stanford.edu/entries/einstein-philscience/.
128. **Einstein, Albert.** Out of My Later Years. *Google Books.* [Online] 1956. https://books.google.com/books/about/Out_of_My_Later_Years.html?id=OBPAA3ZI4zcC.
129. **MostAmazingTop10.** Top 10 Worst Teachers. *YouTube.* [Online] December 19, 2015. https://www.youtube.com/watch?v=rC_gQNA0QW8&lc=z121dtsrkxmfy1xc422bx5kgrky2xdswj04.
130. **Fred, Herbert L.** The True Teacher. *Texas Heart Institute Journal.* [Online] 2010. http://www.ncbi.nlm.nih.gov/pmc/articles/PMC2879191/.
131. **Bliwise, Robert J.** Reopening The Closing of the American Mind. *Duke University Libraries.* [Online] Fall 2007. https://blogs.library.duke.edu/magazine/2007/10/28/reopening-the-closing-of-the-american-mind/.
132. **Lee, Newton.** Digital Da Vinci: Computers in Music. [Online] Springer Science+Business Media, April 12, 2014. http://www.amazon.com/Digital-Vinci-Computers-Newton-Lee/dp/149390535X.
133. **Lee, Newton.** Digital Da Vinci: Computers in the Arts and Sciences. [Online] Springer Science+Business Media, August 1, 2014. http://www.amazon.com/Digital-Vinci-Computers-Arts-Sciences/dp/1493909649.
134. **IMDb.** Dead Poets Society. *IMDb.* [Online] June 9, 1989. http://www.imdb. com/title/tt0097165/trivia?tab=qt&ref_=tt_trv_qu.

135. **Saul, Scott.** The Humanities in Crisis? Not at Most Schools. *The New York Times.* [Online] July 3, 2013. http://www.nytimes.com/2013/07/04/opinion/the-humanities-in-crisis-not-at-most-schools.html.

136. **Yang, Andrew.** What's eating Silicon Valley. *Quartz.* [Online] January 8, 2016. http://qz.com/586941/whats-eating-silicon-valley/.

137. **Flaherty, Colleen.** Not So Different: New Stanford programs aim to give computer science students a boost – by adding arts and humanities. *Inside Higher Ed.* [Online] March 7, 2014. https://www.insidehighered.com/news/2014/03/07/stanford-will-start-new-joint-computer-science-programs.

138. **Klemas, Amanda K.** History of Institutional Planning at the University of Pennsylvania: Martin Meyerson, President (1970-1981). *University of Pennsylvania.* [Online] 2004. http://www.archives.upenn.edu/histy/features/uplans/meyerson.html.

139. **Neyfakh, Leon.** Filmmaker focuses on Edward Snowden, his leaks. *Boston Globe.* [Online] October 27, 2014. http://www.bostonglobe.com/arts/movies/2014/10/26/the-woman-who-documented-edward-snowden/pP0uAAUjFKc4aiqQiTPbCM/story.html.

140. **Sadofsky, Mimsy and Greenberg, Daniel.** The Kingdom of Childhood. *The Sudbury Valley School.* [Online] January 1, 1994. https://books.google.com/books?id=XqvEu1aDFy4C.

141. **Singularity University.** The Founders. *Singularity University.* [Online] https://singularityu.org/community/founders/.

142. **Singularity University.** Frequently Asked Questions . *Singularity University.* [Online] http://singularityu.org/faq/.

143. **Packer, George.** No Death, No Taxes. *The New Yorker.* [Online] November 28, 2011. http://www.newyorker.com/magazine/2011/11/28/no-death-no-taxes.

144. **Kaufman, Micha.** The Internet Revolution is the New Industrial Revolution. *Forbes.* [Online] October 5, 2012. http://www.forbes.com/sites/michakaufman/2012/10/05/the-internet-revolution-is-the-new-industrial-revolution/.

145. **Alexander, Anne.** Internet role in Egypt's protests. [Online] BBC, February 9, 2011. http://www.bbc.co.uk/news/world-middle-east-12400319.

146. **Fathi, Yasmine.** In Egypt, nationwide protests planned for January 28. [Online] Ahram Online, January 27, 2011. http://english.ahram.org.eg/News/4953.aspx.

147. **Williams, Christopher.** How Egypt shut down the internet. [Online] The Telegraph, January 28, 2011. http://www.telegraph.co.uk/news/worldnews/africaandindianocean/egypt/8288163/How-Egypt-shut-down-the-internet.html.

148. **AFP.** Google unveils Web-free 'tweeting' in Egypt move. [Online] Google, January 31, 2011. http://www.google.com/hostednews/afp/article/ALeqM5h8de3cQ8o_S2zg9s72t7sxNToBqA?docId=CNG.ddc0305146893ec9e9e6796d743e6af7.c81.

149. **CNN Wire Staff.** Virtually all Internet service in Syria shut down, group says. [Online] CNN, November 29, 2012. http://www.cnn.com/2012/11/28/world/meast/syria-civil-war/index.html.

150. **Gross, Doug.** Syria caused Internet blackout, security firm says. [Online] CNN, December 3, 2012. http://www.cnn.com/2012/11/30/tech/web/syria-internet/index.html.

151. **Marzouki, Yousri and Oullier, Olivier.** Revolutionizing Revolutions: Virtual Collective Consciousness and the Arab Spring. *The Huffington Post.* [Online] September 16, 2012. http://www.huffingtonpost.com/yousri-marzouki/revolutionizing-revolutio_b_1679181.html.

152. **Mitra, Sugata.** Build a School in the Cloud. *TED2013.* [Online] February 2013. https://www.ted.com/talks/sugata_mitra_build_a_school_in_the_cloud.

153. **Wolfram Research.** P Versus NP Problem. *Wolfram MathWorld.* [Online] http://mathworld.wolfram.com/PVersusNPProblem.html.

154. **Wikipedians.** Theory of Everything. *Wikipedia.* [Online] https://en.wikipedia.org/wiki/Theory_of_everything#Modern_physics.

155. **Google.** Alphabet Code of Conduct. *Google Investor Relations.* [Online] October 2, 2015. https://investor.google.com/corporate/code-of-conduct.html.
156. **Barr, Alistair.** Google's 'Don't Be Evil' Becomes Alphabet's 'Do the Right Thing'. *The Wall Street Journal.* [Online] October 2, 2015. http://blogs.wsj.com/digits/2015/10/02/as-google-becomes-alphabet-dont-be-evil-vanishes/.
157. **Goel, Vindu and Wingfield, Nick.** Mark Zuckerberg Vows to Donate 99% of His Facebook Shares for Charity. *The New York Times.* [Online] December 1, 2015. http://www.nytimes.com/2015/12/02/technology/mark-zuckerberg-facebook-charity.html.
158. **National Philanthropic Trust.** Charitable Giving Statistics. *National Philanthropic Trust 2015 Annual Report.* [Online] 2015. http://www.nptrust.org/philanthropic-resources/charitable-giving-statistics/.
159. **MacFarquhar, Larissa.** What Money Can Buy: Darren Walker and the Ford Foundation set out to conquer inequality. *The New Yorker.* [Online] January 4, 2016. http://www.newyorker.com/magazine/2016/01/04/what-money-can-buy-profiles-larissa-macfarquhar.
160. **Kennedy, John F.** John F. Kennedy Moon Speech - Rice Stadium. *NASA Software Robotics and Simulation Division.* [Online] September 12, 1962. http://er.jsc.nasa.gov/seh/ricetalk.htm.
161. **Jha, Alok.** Neil Armstrong breaks his silence to give accountants moon exclusive. *The Guardian.* [Online] May 23, 2012. http://www.theguardian.com/science/2012/may/23/neil-armstrong-accountancy-website-moon-exclusive.
162. **Fries, Colin.** Chronology of Wakeup calls. *NASA History Division.* [Online] March 13, 2015. http://history.nasa.gov/wakeup%20calls.pdf.
163. **National Science Foundation.** NSF and the Birth of the Internet. [Online] The National Science Foundation. [Cited: August 7, 2014.] http://www.nsf.gov/news/special_reports/nsf-net/home.jsp.
164. **Ackerman, Spencer.** The iPhone 4S' Talking Assistant Is a Military Veteran. [Online] Wired, October 5, 2011. http://www.wired.com/dangerroom/2011/10/siri-darpa-iphone/.
165. **SRI International.** PAL (Personalized Assistant that Learns). [Online] SRI International. [Cited: May 28, 2012.] https://pal.sri.com/.
166. **Lucas, Fred.** Obama: 'Google, Facebook Would Not Exist' Without Government. [Online] The Washington Times, April 6, 2012. http://times247.com/articles/obama-google-facebook-would-not-exist-without-big-government.
167. **Atkinson, Robert D.** Testimony of Dr. Robert D. Atkinson, President, Information Technology and Innovation Foundation, Before the U.S. Senate Committee On Commerce, Science, & Transportation. *Information Technology & Innovation Foundation (ITIF).* [Online] May 11, 2016. http://www2.itif.org/2016-senate-competes-act-testimony.pdf.
168. **IMDb.** The Right Stuff. *IMDb.* [Online] February 17, 1984. http://www.imdb.com/title/tt0086197/trivia?tab=qt&ref_=tt_trv_qu.
169. **Page, Larry and Brin, Sergey.** 2004 Founders' IPO Letter. *Google Investor Relations.* [Online] Google, August 18, 2004. https://investor.google.com/corporate/2004/ipo-founders-letter.html.
170. **Amani, Elahe.** Silence did not make Sanaz Nezami strong: Facing lethal immigrant domestic violence. [Online] Women News Network, December 2013. http://womennewsnetwork.net/2014/02/18/silence-did-not-make-sanaz-nezami-strong-facing-immigrant-domestic-violence/.
171. **The Associated Press.** Family watches online as Iranian woman dies in U.S. [Online] New York Daily News, January 1, 2014. http://www.nydailynews.com/news/national/family-watches-online-iranian-woman-dies-u-s-article-1.1563678.

172. **Curnow, Robyn and Dewan, Angela.** 'A life wasted': Mother of Mount Everest victim wants answers. *CNN.* [Online] May 29, 2016. http://www.cnn.com/2016/05/25/world/everest-deaths-climb-maria-strydom/index.html.

173. **Reputation Institute.** See Who Made the 2015 Global RepTrak 100. *Reputation Institute.* [Online] [Cited: January 27, 2016.] http://www.reputationinstitute.com/thought-leadership/global-reptrak-100.

174. **Epstein, Adam.** People trust Google for their news more than the actual news. *Quartz.* [Online] January 18, 2016. http://qz.com/596956/people-trust-google-for-their-news-more-than-the-actual-news/.

175. **Groden, Claire.** More People Trust Google For Their News Than Traditional News Groups. *Fortune.* [Online] January 19, 2016. http://fortune.com/2016/01/19/edelman-public-trust/.

176. **Ferenstein, Greg.** An Attempt To Measure What Silicon Valley Really Thinks About Politics And The World (In 14 Graphs). *Medium.* [Online] November 6, 2015. https://medium.com/the-ferenstein-wire/what-silicon-valley-really-thinks-about-politics-an-attempted-measurement-d37ed96a9251#.sdspckax9.

177. **Piller, Charles.** Verily, I swear. Google Life Sciences debuts a new name. *STAT.* [Online] December 7, 2015. http://www.statnews.com/2015/12/07/verily-google-life-sciences-name/.

Chapter 2
Google My Religion: Unraveling the Gordian Knot of Religious, Moral, and Political Entanglement

Newton Lee

> *All religions, arts and sciences are branches of the same tree. All these aspirations are directed toward ennobling man's life, lifting it from the sphere of mere physical existence and leading the individual towards freedom.*
>
> —Albert Einstein

Prologue "One word: Google. The questions have always been at hand, but now the answers are within our grasp," said former Pentecostal preacher Jerry DeWitt who became an atheist after 25 years of pastoral services [1]. Notwithstanding the know-it-all reputation of Google Search, cognitive psychologist Tom Stafford has cautioned that "the Internet can give us the illusion of knowledge, making us think we are smarter than we really are" [2]. Is it possible to unravel the Gordian knot of religious, moral, and political entanglement?

2.1 Book of Genesis and Theory of Relativity

"In the beginning God created the heavens and the earth" is the first verse in the Book of Genesis in the Hebrew Bible (the Tanakh) and the Christian Old Testament (see Fig. 2.1) [3]. Similarly, the Quran proclaims that "Surely your Lord is none other than Allah, Who created the heavens and the earth in six days." (Surah Al-A'raf 7:54) [4].

Whether the 6-day creation story is a fable or the truth alluding to time dilation in Albert Einstein's theory of relativity (see Fig. 2.2), the Bible and the Quran—sharing the same root as an Abrahamic religion—are undoubtedly amongst the

N. Lee (✉)
Newton Lee Laboratories LLC, Institute for Education Research and Scholarships,
Woodbury University School of Media Culture and Design, Burbank, CA, USA
e-mail: newton@newtonlee.com

© Springer Science+Business Media New York 2016
N. Lee (ed.), *Google It*, DOI 10.1007/978-1-4939-6415-4_2

Fig. 2.1 "In principio creavit Deus caelum et terram" on the first page of Genesis in a Latin bible dated 1481 (Courtesy of Bodleian Library and licensed under the Creative Commons Attribution 4.0 International license)

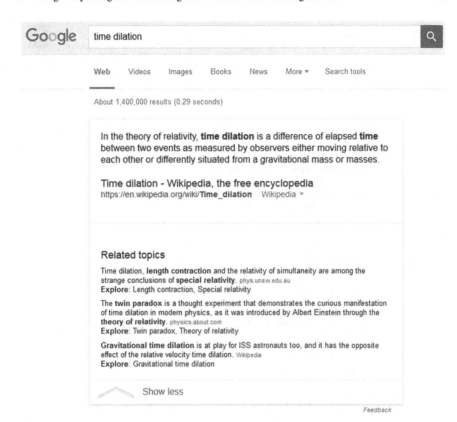

Fig. 2.2   Google Search on "time dilation" returns a Knowledge Graph definition from Wikipedia: "In the theory of relativity, time dilation is a difference of elapsed time between two events as measured by observers either moving relative to each other or differently situated from a gravitational mass or masses."

most profound and influential books ever written. The best known Abrahamic religions are Judaism, Christianity, Islam, and the Bahá'í Faith (see Fig. 2.3). Boston University professor emeritus Peter Berger sums up elegantly in his article on Abrahamic faiths: "There is also common ground, I think, between morally decent people of all faiths or no faith. That common ground is humanity. Jews, Christians and Muslims, the children of Abraham, believe that this humanity is part of the creation by the one God whom they worship" [5].

In April 2015, Pew Research Center reported that 31 and 23 % of the 7 billion world population are Christians and Muslims respectively [6]. In other words, the Bible and the Quran have touched the lives of almost 4 billion living souls worldwide among people of all ages, genders, educational levels, and socioeconomic status, regardless of whether they are devoted or hypocritical, conservative or liberal.

Fig. 2.3 Google Search on "Abrahamic religion" returns a Knowledge Graph definition from Wikipedia: "An Abrahamic religion is a religion whose people believe that Abraham and his descendants hold an important role in human spiritual development. The best known Abrahamic religions are Judaism, Christianity, Islam, and the Bahá'í Faith."

Thanks to the Internet and search engines, the full text of the Bible and the Quran are now at our fingertips. Long gone are the days when the Catholic priests discouraged the believers from reading the Bible on their own for fear that they would misinterpret the Scriptures. Former Google quantitative analyst Seth Stephens-Davidowitz disclosed that the number of Google searches questioning God's existence went up in the first half of this decade [7]. Albert Einstein acknowledged a very mysterious pantheistic God: "I see a pattern, but my imagination cannot picture the maker of that pattern. I see a clock, but I cannot envision the clockmaker. The human mind is unable to conceive of the four dimensions, so how can it conceive of a God, before whom a thousand years and a thousand dimensions are as one?" [8].

2.2 Information War and the Blame Game

Information war started in the very beginning of human history when the serpent half-deceived Eve by telling her, "You will not certainly die, for God knows that when you eat from it your eyes will be opened, and you will be like God, knowing good and evil" (Genesis 3:4–5). So Adam and Eve both ate the forbidden fruit, and they were overcome by shame and fear:

> Then the man and his wife heard the sound of the LORD God as he was walking in the garden in the cool of the day, and they hid from the LORD God among the trees of the garden. But the LORD God called to the man, "Where are you?" He answered, "I heard you in the garden, and I was afraid because I was naked; so I hid." And he said, "Who told you that you were naked? Have you eaten from the tree that I commanded you not to eat from?"
>
> (Genesis 3:8–11)

The omniscient and omnipresent God asked Adam, "Where are you?" without exercising his omnipotence. Adam and Eve were free, living in the Garden of Eden. Instead of owning up to one's mistakes, humanity set in motion the acrimonious blame game that is prevalent throughout human history:

> The man said, "The woman you put here with me—she gave me some fruit from the tree, and I ate it." Then the LORD God said to the woman, "What is this you have done?" The woman said, "The serpent deceived me, and I ate."
>
> (Genesis 3:12–13)

The man blamed the woman, and the woman accused the serpent. An ancient Chinese proverb says that 牛不飲水, 不能按牛頭低 (you cannot push a cow's head down to drink water). In other words, we all act according to our free will in spite of temptations and circumstances. It is easier to point fingers than to accept responsibilities. People complain about elected officials but they do not care to vote. They criticize some multinational corporations but their banks and 401K are profiting from the stocks of those companies. A modern-day serpent is anyone who disseminates misinformation and disinformation in their verisimilitude.

The Book of Job in the Hebrew Bible and the Christian Old Testament tells a story of a righteous man whose name is Job. One day, God allowed Satan to test Job by destroying all his properties and killing all ten of his children. Instead of being angry at God, Job said, "Naked I came from my mother's womb, and naked I will depart. The LORD gave and the LORD has taken away; may the name of the LORD be praised" (Job 1:21). The next day, God permitted Satan to afflict Job with painful sores from head to toe. Job's wife said to him, "Are you still maintaining your integrity? Curse God and die!" But he replied, "Shall we accept good from God, and not trouble?" (Job 2:10).

Job's ordeal calls into question the dichotomy of blessing and suffering, good and evil, and the role of God and Satan in humanity. Why would God bother to entertain Satan's suggestions about killing Job's ten innocent children? Could it be that the biblical author was trying to portray two sides of the same coin similar to yin and yang in Chinese philosophy? Must good and evil coexist in the universe like Jekyll and Hyde Captain Kirk in the 1966 *Star Trek* episode "The Enemy Within"? Is there a cosmic balance between creation and destruction as evident from 100 billion stars being born and dying each year [9]?

Despite the lack of complete answers, Job accused neither God nor Satan, not himself or other people—a rare quality for a millionaire who had lost everything, and a far cry from the rich who committed murder-suicide over financial ruin [10]. Job refused to play the blame game, acknowledging and accepting that life is not fair. "We cannot change the cards we are dealt, just how we play the hand," said Prof. Randy Pausch in his "Last Lecture" at Carnegie Mellon University [11].

2.3 Infinite Diversity in Infinite Combinations

Notwithstanding Thomas Jefferson's proclamation that "all men are created equal" in the 1776 U.S. Declaration of Independence (see Fig. 2.4), all men and women are *not* created equal in the literal sense unless we all share identical DNA. But even identical twins with the same DNA develop different fingerprints when the growing fetuses touch the amniotic sac in their mother's womb. The twins can become two very different people due to upbringing, social, economic, and environmental factors. Because these external factors can vary greatly from person to person, life cannot be fair. However, unfairness builds character and brings diversity to the otherwise homogeneous and isotropic existence. Who wants to live in a world where everybody looks identical, thinks alike, and acts the same? Nature has shown us that there are at least 9,956 species of birds on earth (see Fig. 2.5) and 30,000 species of fish in the oceans (see Fig. 2.6) according to Google search results. Diversity emanates beauty.

Even though all complex organisms are preprogrammed by their DNAs, they can exhibit individuality within the confines of nature. Mahatma Gandhi once said that "no two leaves are alike" [12]. Human DNA consists of about 3 billion bases, and more than 99 % of those bases are the same in all people [13]. Yet, each person is a unique manifestation of God. The commonality that unites all human beings is that we are created in the image of God: "So God created mankind in his own image, in the image of God he created them; male and female he created them." (Genesis 1:27). However, the image of God is not singular, but plural: "The LORD God said, 'The man has now become like one of us...'" (Genesis 3:22). One God, multiple manifestations—in both the spiritual realm and the physical world—just as Albert Einstein wrote about the wave-particle duality of light: "It seems as though we must use sometimes the one theory and sometimes the other, while at times we may use either. We are faced with a new kind of difficulty. We have two contradictory pictures of reality; separately neither of them fully explains the phenomena of light, but together they do" [14].

In January 2016, Spike Lee and Jada Pinkett Smith called for Oscar boycott because "for the 2nd consecutive year all 20 contenders under the actor category are white" [15]. American actress Stacey Dash dismissed the outrage over Oscars: "I think it's ludicrous. We have to make up our minds. Either we want to have

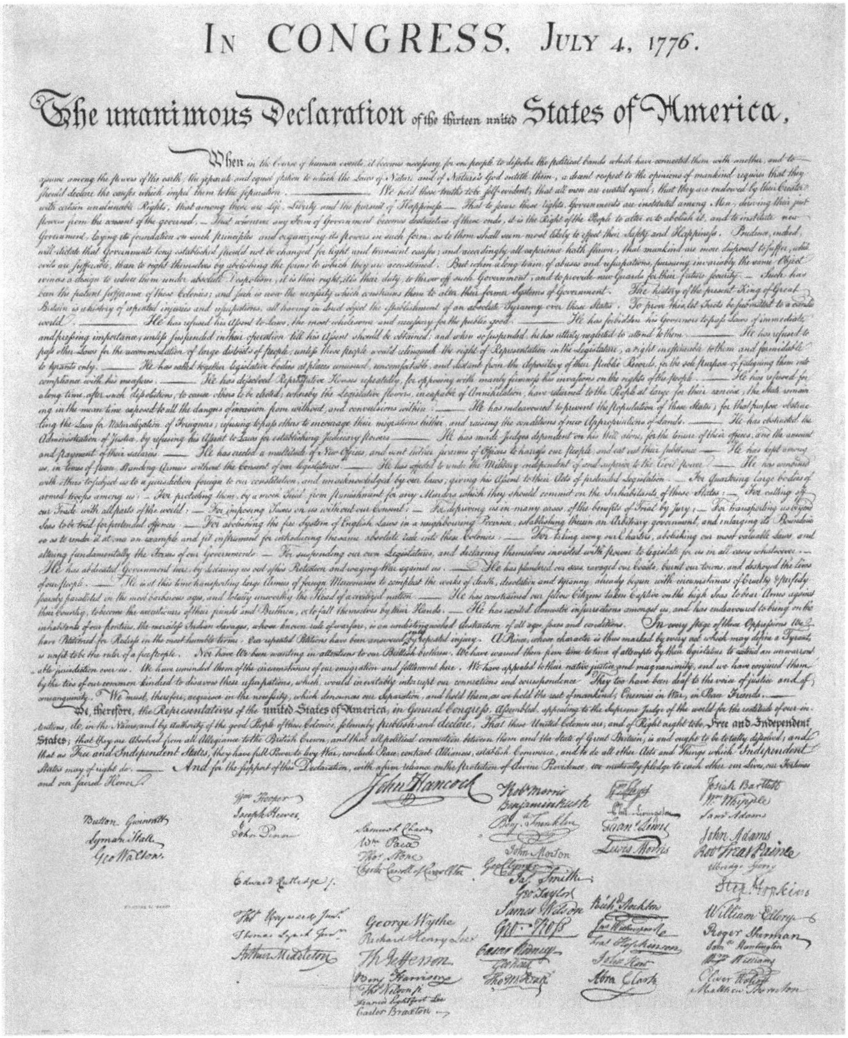
IN CONGRESS, JULY 4, 1776.

The unanimous Declaration of the thirteen united States of America,

Fig. 2.4 United States Declaration of Independence by Thomas Jefferson et al. (ratified by 56 signatories on July 4, 1776)

segregation or integration, and if we don't want segregation, then we have to get rid of channels like BETBlack Entertainment Television and the BET Awards and the Image Awards, where you're only awarded if you're black. If it were the other way around, we'd be up in arms. It's a double standard. There shouldn't be a Black History Month. We're Americans, period" [16]. A decade prior in December 2005, Oscar winner Morgan Freeman had also argued for abolishing Black History Month in an interview on CBS' *60 min*: "Black History is American history. … I'm going

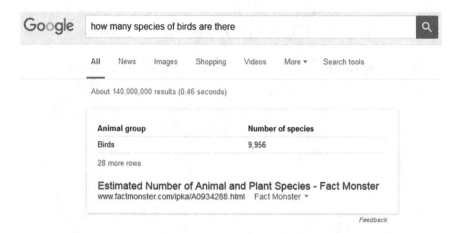

Fig. 2.5 Google search result for "how many species of birds are there"

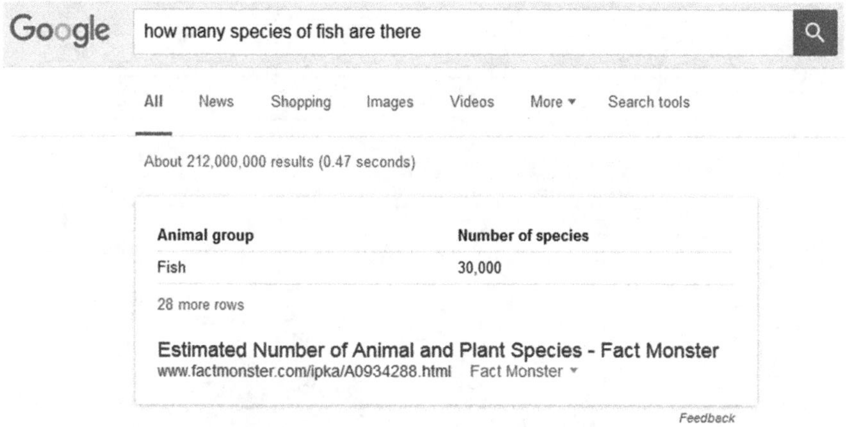

Fig. 2.6 Google search result for "how many species of fish are there"

to stop calling you a white man. And I'm going to ask you to stop calling me a black man. I know you as Mike Wallace, you know me as Morgan Freeman" [17].

On one hand, it is great to be inclusive without discrimination or prejudice. On the other hand, it is wonderful to celebrate diversity and uniqueness. Otherwise, where should we draw the line after getting rid of Black History Month? What about National Hispanic Heritage Month, Asian-Pacific American Heritage Month, Women's History Month, LGBT Pride Month, and many others that serve to raise awareness of diverse contributions to society? Instead of abolishing them, we should add a new Human Heritage Month to remind the world that we are human beings and not some savage animals.

Star Trek creator Gene Roddenberry said, "If man is to survive, he will have learned to take a delight in the essential differences between men and between cultures. He will learn that differences in ideas and attitudes are a delight, part of life's exciting variety, not something to fear" [18]. In the *Star Trek* episode "Is There in Truth No Beauty?" (1968), Roddenberry and writer Jean Lisette Aroeste introduced the notion of IDIC (Infinite Diversity in Infinite Combinations) [19]:

> **Dr. Miranda Jones (Diana Muldaur)**: *[regarding the Vulcan IDIC] I understand, Mr. Spock. The glory of creation is in its infinite diversity.*

> **Mr. Spock (Leonard Nimoy)**: *And the ways our differences combine, to create meaning and beauty.*

Mr. Spock is admired by Trekkies for his Vulcan logic and superior intelligence. In the 1987 seminal book *The Society of Mind*, MIT Prof. Emeritus Marvin Minsky described human intelligence as a result of diverse processes: "What magical trick makes us intelligent? The trick is that there is no trick. The power of intelligence stems from our vast diversity, not from any single, perfect principle" [20]. A few musical notes can morph into countless new songs, and a small set of vocabulary can create a congeries of poems. Infinite diversity in infinite combinations.

2.4 Questioning One's Religion

"We're long on search," said Marissa Mayer, Google's first female engineer and Yahoo's sixth CEO. "Search is curiosity, and that will never be done" [21].

Figure 2.7 shows the average monthly Google searches on major religious keywords in the year 2015 from January to December, topping 12 million queries in September. The individual keywords in descending number of searches were Jesus, Bible, Islam, Quran, angel, God, Pope Francis, Allah, Heaven, religion, Muslim, Christ, Hell, Satan, Devil, church, Christian, prayer, and Catholic. About 0.1 % of the 11.1 billion Google searches in December 2015 were religious queries listed above [22].

Figure 2.8 is a January 2016 snapshot of the Google search autocomplete predictions showing the most popular searches related to "Bible is." "Bible is fake" tops the list of Google searches, followed by the opposite "Bible is the word of God." Google search autocomplete predictions are good indictors of the current trends and states of mind for about 250 million unique visitors every month [23]. The autocomplete predictions are generated by a fully automated algorithm based on [24]:

(a) how often others have searched for a word; and
(b) the range of information available on the web.

Google search can drastically transform one's life forever, especially when it comes to religious beliefs and world views that aim to explain the meaning of life. "Seek, and ye shall find" (Matthew 7:7). Questioning one's religion can lead to a complete hundred and eighty degree turn. Morten Storm switched sides from being a radical Islamist in the al Qaeda organization to assisting the Danish intelligence

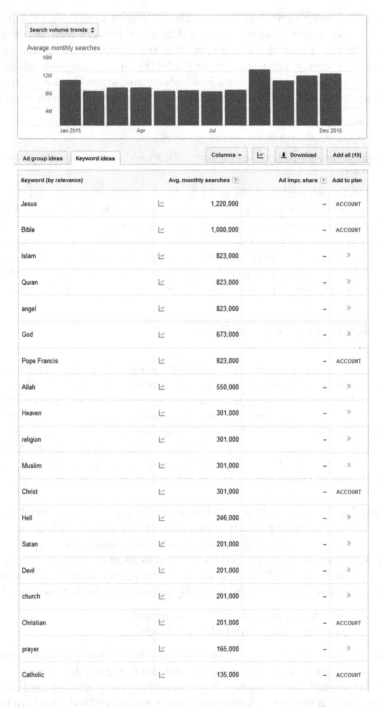

Fig. 2.7 Average monthly Google searches on major religious keywords from January to December 2015, topping 12 million queries in September

Fig. 2.8 Google search autocomplete predictions for "Bible is" (as of January 2016)

agency and the CIA fight the war on terror. Storm credited his change of heart on his laptop computer: "I hit the 'enter' and I saw plenty of websites talking about contradictions in the Quran. It took some time to research them, but once I concluded that they were genuinely contradictions, that's when it wiped totally away my faith. That's when I stopped being a Muslim in my heart – in my belief" [25]. By losing his religion, he has found God.

In the CNN documentary "Atheists: Inside the World of Non-Believers" aired in March 2015, former Pentecostal preacher Jerry DeWitt recalled how he became an atheist after 25 years of pastoral services: "One word: Google. The questions have always been at hand, but now the answers are within our grasp." David Silverman of American Atheists elaborated, "Religion is factually wrong. As a result, religion lives on ignorance of facts. The reason people are giving up on mythology is the Internet, and the access to information it represents. When religion can exist in a bubble, the lies it pushes cannot be challenged. But when there is a wealth of information at the fingertips of every believer, those lies can be refuted easily, from multiple sources and multiple perspectives." And Humanist Chaplain Greg Epstein at Harvard University added, "People are learning more about science." [1].

2.5 Religion and Science

Are religion and science at odds with each other? Not so according to Albert Einstein. In the November 1930 issue of the *New York Times Magazine*, Einstein wrote, "For science can only ascertain what is, but not what should be, and outside of its domain value judgments of all kinds remain necessary. Religion, on the other hand, deals only with evaluations of human thought and action: it cannot justifiably speak of facts and relationships between facts. According to this interpretation the well-known conflicts between religion and science in the past must all be ascribed to a misapprehension of the situation which has been described. … Science without religion is lame, religion without science is blind" [26].

Einstein believed that religion and science could coexist without being at odds with one another, in spite of the religious zealots and the gung-ho atheists being constantly at war with each other. *Salon* writer Peter Birkenhead ingeniously summarized the age-old heated debates on God: "The fundamentalists cultivate something like a sulky teenager's romanticized notion of love, and the atheists a grumpy old bugger's lack of belief in such nonsense" [27].

According to the second law of thermodynamics, the sum of the entropies of the participating bodies must increase. Yet, living organisms seem to exhibit a deliberate anti-entropic force that hints at "by design" rather than "by chance." Electrons, black holes, Higgs boson (aka god particle), and gravitational waves all exist, even though we cannot "see" them with our naked eyes. Ruling out the existence of God is as unscientific as believing in God. Religion and science are not enemies. Creationism and Darwinism are not necessarily contradictory; they each have an answer to the age-old catch-22 question "Which came first, the chicken or the egg?" (see Fig. 2.9) Only closed-minded people are the adversaries of truth.

Since Albert Einstein published his theory of photoelectric effect in 1905 and his general theory of relativity in 1915, it took some 100 years for scientists to take the first-ever photograph of light as both a particle and wave in February 2015 [28] and to detect gravitational waves as they ripple through spacetime in September 2015 [29]. It will take more time to scientifically prove the existence of God and to resolve the seeming contradictions in the Bible.

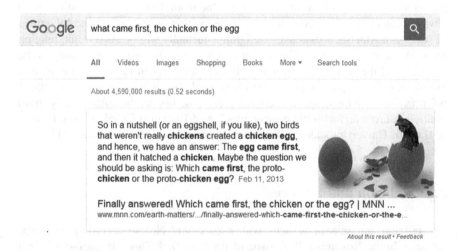

Fig. 2.9 Google Search on "Which came first, the chicken or the egg?" returns a Knowledge Graph with a new question "Which came first, the proto-chicken or the proto-chicken egg?"

University of Sheffield professor Tom Stafford wrote in a BBC News article that "the Internet is giving new fuel to the way we've always thought. It can be both a cause of overconfidence, when we mistake the boundary between what we know and what is available to us over the web, and it can be a cause of uncertainty, when we anticipate that we'll be fact-checked using the web on the claims we make" [2].

2.6 Religious Radicalization

In the historical drama *Agora* (2009) starring Rachel Weisz, philosophy and mathematics professor Hypatia of Alexandria was accused of witchcraft by Bishop Cyril who cited Apostle Paul's letter to the Corinthians that women must be in submission (1 Corinthians 14:34–35) while ignoring the fact that Jesus appeared first to a women—Mary Magdalene—after his resurrection and asked her to proclaim the good news to his disciples (John 20:11–18). Cyril deliberately misused the Scriptures in order to incite a mob to kidnap Hypatia and stone her to death (see *Mort De La Philosophe Hypatie.* by Louis Figuier in Fig. 2.10).

After the murder of Hypatia came widespread witch hunts in Europe, followed by the Crusades and a series of Inquisitions sanctioned or orchestrated by the Roman

Fig. 2.10 *Mort De La Philosophe Hypatie. A Alexandrie (Death of philosopher Hypatia, in Alexandria)* by Louis Figuier (1865)

Catholic Church to coerce the Jews and Muslims into a change of faith. The Christian Crusaders were essentially religious extremists and terrorists. Launched by Pope Urban II in November 1095, the series of Crusades against Muslims and pagans finally ended in 1291 with a death toll of 1 to 3 million people over a period of 197 years [30].

American University associate professor emeritus David Rodier said, "If people are intent on using religion to motivate terror or violence, they'll find an excuse there no matter what the actual text says. Religion, after all, speaks to our most basic and ultimate convictions, and if you are wanting to use violence, if you can find a religious justification, then you can find a very powerful motivation" [31].

While al Qaeda's long-term strategy is to "bleed America to the point of bankruptcy" [36], Islamic State of Iraq and the Levant (ISIL)—aka Islamic State of Iraq and Syria (ISIS)—claims religious, political, and military authority over all Muslims worldwide, inciting believers and pagans for an ominous apocalyptic war among 1.6 billion Muslims and 2.2 billion Christians [37]. Abraham must be rolling over in his grave.

2.7 Children in War and Peace

Innocent children are not immune to religious radicalization. Young adolescents and children are being manipulated by adults to join armed conflicts in war and terrorism as child soldiers [38]. The Children's Crusade is a purported event in 1212 about a failed Crusade by European Christians including children to expel Muslims from the Holy Land.

In modern times, as many as 1.5 million children were killed in the Holocaust during World War II [39]. Although the 1949 Geneva Conventions and additional protocols strictly prohibit the use of children under 15 in warfare, young children have been fighting on both sides of armed conflicts. Wasil Ahmad, an 11-year-old Afghan boy, commanded a police unit of 75 men for 43 days to fight the Taliban in 2015 [40]. Kids as young as eight have been used as bombers in Pakistan [41]. A 13-year-old Nigerian girl was coerced by her father to become a suicide bomber [42]. Children as young as 10 years old patrol the streets with AK-47s in Raqqa, ISIS' de facto capital city in Syria [43]. FBI counterterrorism chief Michael Steinbach told CNN that ISIS had been targeting and recruiting American children as young as 15 in the United States [44].

Hollywood has depicted children being used as leverage in counterterrorism. In the TV show 24 (season 2, episode 12), Jack Bauer (Kiefer Sutherland) extracted information from a terrorist detainee by forcing him to watch a live streaming video of the execution of his child. Bauer gave his order to the executioner, "Start with the oldest son. If it's not successful, move on to his younger son." Then Bauer turned to the detainee and said, "Tell me where the bomb is, or I'll kill your son. I know you think what you're doing is right. But it's my job to not let it happen. Please don't let me do this." Bauer finally managed to break the terrorist by staging a fake execution of his oldest son. "Tell me where the bomb is, and I'll spare the rest of your family." In *Homeland* (season 3, episode "Tin Man is Down"), Peter Quinn (Rupert Friend) assassinated a terrorist but accidentally killed the target's young son during the mission. Quinn subsequently used that incident (by pretending it was not an accident) to threaten a corrupted banker with the lives of his young children. The banker subsequently complied with Quinn's demand to hand over the terrorists' financial records to the CIA.

In sad reality, terrorists could care less about children, as Council on Foreign Relations analyst Ed Husain explained: "Al-Jazeera Arabic gives prominence to the popular Egyptian Muslim Brotherhood cleric Yusuf al-Qaradawi, who has repeatedly called suicide bombings against Israelis not terrorism, but 'martyrdom.' He argues that since Israelis all serve in the military, they are not civilians. Even children, he despicably argues, are not innocent. They would grow up to serve in the military" [45].

On the bright side, the tiny voices of children can make a big difference in a positive way. Malala Yousafzai is a Pakistani education activist and the youngest ever Nobel Prize laureate [46]. At age 11, Yousafzai fought for girls' right to education. At age 15, she suffered an attack on her life by Taliban gunmen on her way to school [47].

During the Cold War, 10-year-old American schoolgirl Samantha Reed Smith was a peace activist and child actress from Manchester, Maine. In 1982, Smith wrote a letter to Soviet Union General Secretary Yuri Andropov [48]:

Dear Mr. Andropov,

My name is Samantha Smith. I am ten years old. Congratulations on your new job. I have been worrying about Russia and the United States getting into a nuclear war. Are you going to vote to have a war or not? If you aren't please tell me how you are going to help to not have a war. This question you do not have to answer, but I would like to know why you want to conquer the world or at least our country. God made the world for us to live together in peace and not to fight.

Sincerely,
Samantha Smith

After a second follow-up letter addressed to Soviet Ambassador to the US Anatoly Dobrynin, Smith received a personal reply from Yuri Andropov and an invitation to visit the Soviet Union:

Dear Samantha,I received your letter, which is like many others that have reached me recently from your country and from other countries around the world.

It seems to me—I can tell by your letter—that you are a courageous and honest girl, resembling Becky, the friend of Tom Sawyer in the famous book of your compatriot Mark Twain. This book is well known and loved in our country by all boys and girls.

You write that you are anxious about whether there will be a nuclear war between our two countries. And you ask are we doing anything so that war will not break out.

Your question is the most important of those that every thinking man can pose. I will reply to you seriously and honestly.

Yes, Samantha, we in the Soviet Union are trying to do everything so that there will not be war on Earth. This is what every Soviet man wants. This is what the great founder of our state, Vladimir Lenin, taught us.

Soviet people well know what a terrible thing war is. Forty-two years ago, Nazi Germany which strove for supremacy over the whole world, attacked our country, burned and destroyed many thousands of our towns and villages, killed millions of Soviet men, women and children.

In that war, which ended with our victory, we were in alliance with the United States: together we fought for the liberation of many people from the Nazi invaders. I hope that you know about this from your history lessons in school. And today we want very much to live in peace, to trade and cooperate with all our neighbors on this earth—with those far away and those nearby. And certainly with such a great country as the United States of America.

In America and in our country there are nuclear weapons—terrible weapons that can kill millions of people in an instant. But we do not want them to be ever used. That's precisely why the Soviet Union solemnly declared throughout the entire world that never—never— will it use nuclear weapons first against any country. In general we propose to discontinue further production of them and to proceed to the abolition of all the stockpiles on earth.

It seems to me that this is a sufficient answer to your second question: "Why do you want to wage war against the whole world or at least the United States?" We want nothing of the kind. No one in our country—neither workers, peasants, writers nor doctors, neither grown-ups nor children, nor members of the government—want either a big or "little" war.

We want peace—there is something that we are occupied with: growing wheat, building and inventing, writing books and flying into space. We want peace for ourselves and for all peoples of the planet. For our children and for you, Samantha.

I invite you, if your parents will let you, to come to our country, the best time being this summer. You will find out about our country, meet with your contemporaries, visit an international children's camp—"Artek"—on the sea. And see for yourself: in the Soviet Union, everyone is for peace and friendship among peoples.

Thank you for your letter. I wish you all the best in your young life.

Y. Andropov

2.8   Terrorism and Abrahamic Religions

Most people believe that God is love. Figure 2.11 is a January 2016 snapshot of the Google search autocomplete predictions showing the most popular searches related to "God is."

A Google search on "terrorism" returns a Knowledge Graph definition as "the user of violence and intimidation in the pursuit of political aims" (see Fig. 2.12). The FBI defines terrorism as activities that appear to be intended [49]:

(a) to intimidate or coerce a civilian population;
(b) to influence the policy of a government by intimidation or coercion; or
(c) to affect the conduct of a government by mass destruction, assassination, or kidnapping.

Fig. 2.11 Google search autocomplete predictions for "God is" (as of January 2016)

While all eyes are on Islam after the 9/11 terrorist attacks, English film director and producer Ridley Scott raised an unanswered question in *Exodus: Gods and Kings* (2014) when Egyptian Pharaoh Ramses confronted Moses while holding his son's lifeless body. Ramses asked, "Is this your god? A killer of children? What kind of fanatics worship such a god?"

In handing down the death penalty to Boston Marathon bomber Dzhokhar Tsarnaev in June 2015, Judge George O'Toole said, "Surely someone who believes that God smiles on and rewards the deliberate killing and maiming of innocents believes in a cruel God. That is not, it cannot be, the God of Islam. Anyone who has been led to believe otherwise has been maliciously and willfully deceived" [50].

Arab-American comedian Dean Obeidallah wrote to CNN, "I'm an American-Muslim… I'm not going to tell you, 'Islam is a religion of peace.' Nor will I tell you that Islam is a religion of violence. What I will say is that Islam is a religion that, like Christianity and Judaism, is intended to bring you closer to God. And sadly we have seen people use the name of each of these Abrahamic faiths to wage and justify violence" [51].

Sadly indeed, from an extremist point of view, the Hebrew Bible and the Christian Old Testament only serve to reaffirm their radical religious conviction. In the context of wars and vengeance, religious militants and terrorists believe that:

1. God terrorizes and obliterates the enemies:

   (a) *I will send my terror ahead of you and throw into confusion every nation you encounter. I will make all your enemies turn their backs and run. I will send the hornet ahead of you to drive the Hivites, Canaanites and Hittites*

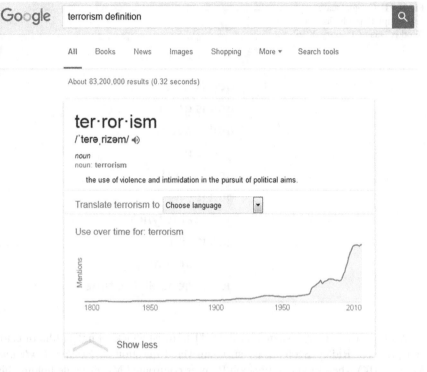

Fig. 2.12 Google search on "terrorism" returns a Knowledge Graph definition as "the user of violence and intimidation in the pursuit of political aims." The usage graph shows an exponential increase of mentions of "terrorism" towards the twenty-first century

> *out of your way. But I will not drive them out in a single year, because the land would become desolate and the wild animals too numerous for you. Little by little I will drive them out before you, until you have increased enough to take possession of the land. (Exodus 23:27)*
>
> (b) *The LORD sent you on a mission, and said, "Go and completely destroy those wicked people, the Amalekites; wage war against them until you have wiped them out." (1 Samuel 15:18)*

2. God is revengeful:

   (a) *If there is serious injury, you are to take life for life, eye for eye, tooth for tooth, hand for hand, foot for foot, burn for burn, wound for wound, bruise for bruise. (Exodus 21:23–25)*

   (b) *Show no pity: life for life, eye for eye, tooth for tooth, hand for hand, foot for foot. (Deuteronomy 19:21)*

   (c) *"Vengeance is Mine ... For the LORD will vindicate His people." (Deuteronomy 32:35–36)*

3. Those who worship other gods are punishable by death:

   (a) *While Israel was staying in Shittim, the men began to indulge in sexual immorality with Moabite women, who invited them to the sacrifices to their gods. The people ate the sacrificial meal and bowed down before these gods. So Israel yoked themselves to the Baal of Peor. And the LORD's anger burned against them. The LORD said to Moses, "Take all the leaders of these people, kill them and expose them in broad daylight before the LORD, so that the LORD's fierce anger may turn away from Israel." (Numbers 25:1–6)*

4. Those who disrespect God are punishable by death regardless of their intention:

   (a) *When they came to the threshing floor of Kidon, Uzzah reached out his hand to steady the ark, because the oxen stumbled. The Lord's anger burned against Uzzah, and he struck him down because he had put his hand on the ark. So he died there before God. ... David was afraid of God that day and asked, "How can I ever bring the ark of God to me?" (1 Chronicles 13:9–10, 12)*

5. God does not spare women and children:

   (a) *When the trumpets sounded, the army shouted, and at the sound of the trumpet, when the men gave a loud shout, the wall collapsed; so everyone charged straight in, and they took the city. They devoted the city to the LORD and destroyed with the sword every living thing in it—men and women, young and old, cattle, sheep and donkeys. (Joshua 6:20–21)*

   (b) *Moses was angry with the officers of the army—the commanders of thousands and commanders of hundreds—who returned from the battle. "Have you allowed all the women to live?" he asked them. "They were the ones who followed Balaam's advice and enticed the Israelites to be unfaithful to the Lord in the Peor incident, so that a plague struck the Lord's people. Now kill all the boys. And kill every woman who has slept with a man, but save for yourselves every girl who has never slept with a man." (Numbers 31:14–18)*

   (c) *From there Elisha went up to Bethel. As he was walking along the road, some boys came out of the town and jeered at him. "Get out of here, baldy!" they said. "Get out of here, baldy!" He turned around, looked at them and called down a curse on them in the name of the Lord. Then two bears came out of the woods and mauled forty-two of the boys. (2 Kings 2:23–24)*

Abrahamic religions—Islam, Christianity, and Judaism—are neither religions of peace nor religions of violence. They are storytellers of the past and forewarners of the future. Akin to Bible commentaries that aid in the study of the Scripture, the "Study Quran" is an English translation of the Quran with extensive commentaries from both Shiite and Sunni scholars who represent the two major denominations of Islam. "The commentaries don't try to delete or hide the verses that refer to violence," explained Editor-in-chief Seyyed Hossein Nasr. "We have to be faithful to the text. But they can explain that war and violence were always understood as a painful part of the human condition. ... The best way to counter extremism in modern Islam is a revival of classical Islam" [52].

Funded by King Abdullah II of Jordan and the El-Hibri Foundation which promotes religious tolerance, the Study Quran is meant to be a rebuttal to Islamic terrorists. However, since religious militants and terrorists believe that they are fighting a holy war and exacting God's vengeance on the infidels, the stories of violence in the Quran, the Hebrew Bible, and the Christian Old Testament only justify their combatant mindset and brutal tactics.

2.9   Ten Plagues of Egypt

The ten plagues of Egypt is the epitome of terrorism that eclipses all but the Holocaust when as many as 1.5 million children including Anne Frank perished in concentration camps [53]. According to the *Book of Exodus* and the *Quran*, the plagues of Egypt were ten calamities that God inflicted upon Egypt to persuade the Pharaoh to release the Israelites from slavery.

The plagues not only demobilized Egypt's mighty armed forces but also terrorized all Egyptians—military personnel and civilians alike—by inflicting emotional and physical sufferings that culminated in the death of all firstborn sons in Egypt.

History seems to hint that what goes around comes around. In chapter one of the *Book of Exodus*, a new king came to power in Egypt and began to oppress the Israelites with forced labor. The ruthless Pharaoh ordered the killing of newborn Hebrew sons:

> Then Pharaoh gave this order to all his people: "Every Hebrew boy that is born you must throw into the Nile, but let every girl live."
>
> (Exodus 1:22)

Moses was one of the Hebrew newborns abandoned in the Nile, but he was rescued and adopted by Pharaoh's daughter. Moses grew up in the Egyptian palace, but he eventually left to live among his own people. The king of Egypt passed away and his son became the new king who continued to oppress the Israelites. Moses was then chosen by God to free the Israelis from slavery:

> The Lord said to Moses, "When you return to Egypt, see that you perform before Pharaoh all the wonders I have given you the power to do. But I will harden his heart so that he will not let the people go. Then say to Pharaoh, 'This is what the Lord says: Israel is my firstborn son, and I told you, Let my son go, so he may worship me. But you refused to let him go; so I will kill your firstborn son.'"
>
> (Exodus 4:21–23)

> At midnight the LORD struck down all the firstborn in Egypt, from the firstborn of Pharaoh, who sat on the throne, to the firstborn of the prisoner, who was in the dungeon, and the firstborn of all the livestock as well. Pharaoh and all his officials and all the Egyptians got up during the night, and there was loud wailing in Egypt, for there was not a house without someone dead.
>
> (Exodus 12:29–30)

Passover is an important Jewish festival commemorating the liberation of the Israelites from slavery in Egypt. Islamic extremists may consider that the Hebrew Bible (the Tanakh) and the Christian Old Testament justify violent vengeance, killing of civilians, and slaughtering of children in the name of God. Some Islamic terrorists claim that their terrorist attacks are revenge for what Christians did in the Crusades [54]. For millenniums, the subconscious idea of terrorism has been ingrained in all Abrahamic religions. Pointing finger at one of them is to blame all of them—Judaism, Christianity, and Islam alike.

Considering the horror of the ten plagues and the long history of deep-seated hostility, the Egypt-Israel Peace Treaty in 1979 was no small feat brokered by U.S. President Jimmy Carter, Secretary of State Cyrus Vance, Egyptian President Anwar Sadat, and Israeli Prime Minister Menachem Begin [55]. The 1978 Nobel Peace Prize was awarded jointly to Sadat and Begin [56]. Jimmy Carter received the 2002 Nobel Peace Prize "for his decades of untiring effort to find peaceful solutions to international conflicts, to advance democracy and human rights, and to promote economic and social development" [57].

2.10   Xenophobia and Islamophobia

After the November 13th 2015 Paris attacks, U.S. Senator Ted Cruz from Texas proposed a "religious test" for Syrian refugees so that only Christians would be accepted into the United States [58]. In the wake of the December 2nd San Bernardino shooting, real estate magnate and 2016 presidential candidate Donald Trump called for surveillance against mosques and a complete ban on Muslims entering the United States [59]. Their knee-jerk rhetoric brings to mind the Chinese Exclusion Act of 1882 and the Japanese internment camps in World War II.

Is America in the twenty-first century no better than Communist China who denied entry visa to Miss World Canada Anastasia Lin for competing in Miss World 2015 pageant because of her religious ties to Falun Gong (see Fig. 2.13)? [60] China has reasons to consider religions as potential enemies of the state. For instance, Christians (including 17-year-old protest leader Joshua Wong) played a prominent role in the 2014 Umbrella Movement of sit-in street protests in Hong Kong [61].

Unlike Communist China however, the First Amendment to the United States Constitution guarantees religious freedom. Cruz and Trump are playing right into the hands of the terrorists and ISIS by dividing the country instead of uniting the American people [62]. Indeed, a 2016 terrorist recruiting video purportedly by al Qaeda-linked militant group Al-Shabaab used footages of Malcolm X and Donald Trump to highlight racism and religious discrimination in America [63]. Figure 2.14 is a January 2016 snapshot of the Google search autocomplete predictions showing the most popular searches related to "Islam is." "Islam is a religion of peace" tops the list of Google searches, followed by the exact opposite "Islam is not a religion of peace."

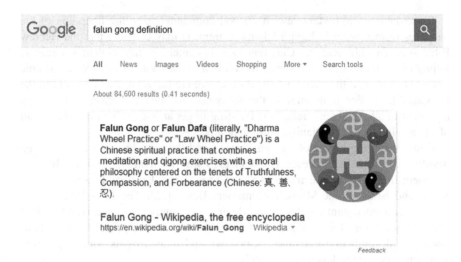

Fig. 2.13 Google search on "Falun Gong" returns a Chinese spiritual practice centered on truthfulness, compassion, and forbearance. The one large and four small Swastika symbols are considered sacred and auspicious in Hinduism, Buddhism and Jainism from around 200 B.C. [64]

Fig. 2.14 Google search autocomplete predictions for "Islam is" (as of January 2016)

Owing to safety concerns, all schools in Augusta County, Virginia were shut down for a day in December 2015, after a backlash over an Arabic calligraphy homework assignment from a standard workbook on world religions (see Fig. 2.15). The center of controversy was about copying by hand the calligraphy for *shahada*—the Islamic statement of faith—written in Arabic. An overreacting

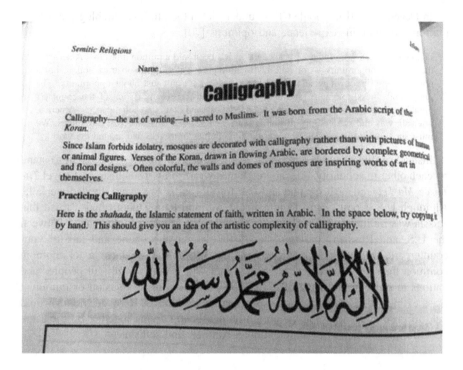

Fig. 2.15   An Arabic calligraphy homework assignment

mother, Kimberly Herndon, kept her 9th-grade son home from school and complained to the school officials, "There was no trying about it. The sheet she gave out was pure doctrine in its origin. I will not have my children sit under a woman [teacher] who indoctrinates them with the Islam religion when I am a Christian" [65]. Islamophobia has risen to a whole new level in America.

CNN commentator Mel Robbins wrote in response to the controversy, "For those Christians who assembled in fear and buried the school in an avalanche of fear and threats, it's also sad. As a Christian, I can only imagine how weak your own religious faith must be if you fear that a calligraphy assignment could change your child's faith" [66].

In response to the rise of anti-foreigner comments on Facebook, Mark Zuckerberg pledged over $1 million euros in launching an "Initiative for Civil Courage Online" in Europe to counter the racist and xenophobic Facebook posts [67]. Moreover, Facebook actively promotes "counter speech" by offering free advertising incentives to counter speakers in an effort to combat terrorism on social media [68]. In January 2014, a German group named Laut Gegen Nazis (Loud Against Nazis) launched a "Like Attack" by organizing more than 100,000 people to bombard neo-Nazi pages on Facebook with "likes" and nice comments to defuse racism, anti-Semitism, and anti-immigrant xenophobia [69].

In December 2015, Google CEO Sundar Pichai took to Medium blog-publishing platform to share his experience and opinions [70]:

I came to the US from India 22 years ago. … It's been said a million times that America is the 'land of opportunity'—for millions of immigrants, it's not an abstract notion, but a concrete description of what we find here. … And it's not just about opportunity. The open-mindedness, tolerance, and acceptance of new Americans is one of the country's greatest strengths and most defining characteristics. And that is no coincidence—America, after all, was and is a country of immigrants. … Let's not let fear defeat our values. We must support Muslim and other minority communities in the US and around the world.

Instead of dividing America along religious and ethnic lines, we ought to stand by each other in solidarity. Alexandre Dumas wrote in his 1844 historical novel *The Three Musketeers*: "All for one and one for all, united we stand divided we fall." Although U.S. Coast Guard Commandant Admiral Paul Zukunft had warned that Russia was militarizing the Arctic, he spoke of the robust communication between the U.S. and Russian coast guards: "You pick up the phone and talk to your counterpart. Operating in harsh environments, we find we have a lot more in common than we do differences" [71]. What will it take to rally all peoples and nations to unite in the name of humanity? An all-out alien invasion or imminent mass extinction? Perhaps a gentler proposal like a Human Heritage Month would help to raise awareness that we are all human beings living together on the same beautiful planet marred by undue human conflicts and selfishness.

2.11   Capital Punishment and Rehabilitation

Boston Marathon bomber, a 22-year-old Kyrgyzstani-American Dzhokhar Tsarnaev, hardly showed remorse in court for killing 8-year-old pacifist Martin Richard and two other marathon spectators [73]. Tsarnaev was sentenced to death, which might have helped to ease the pain of the victims' families. However, capital punishment is not the answer.

First, most Americans prefer the execution method to be as quick and painless as possible [74] given that the Eighth Amendment to the United States Constitution prohibits the government from imposing "cruel and unusual punishment." In his second grade class a year prior to his murder, 8-year-old Martin Richard handmade a sign that reads "No more hurting people. Peace" [75]. Since the innocent and peaceful boy was torn apart by hot shrapnel and slowly bled to death, why should the murderer meet his demise quickly and painlessly by electrocution or lethal injection?

Second and most importantly, capital punishment does not deter suicidal terrorists who are willing to die for their causes. Death penalty only makes them feel like martyrs. Prof. Jay Parini of Middlebury College wrote to CNN: "As a Christian, I can't but feel horrified at the news that a federal jury has imposed the death penalty on Dzhokar Tsarnaev. … This punishment only continues the cycle of violence, and it will not bring peace. In fact, the execution of Tsarnaev will transform him into a martyr, and millions around the world will find fresh reasons to dislike the United States" [76].

Rehabilitation is a better alternative to capital punishment. 21-year-old Abu Bakr Mansha was a convicted al Qaeda terrorist in the United Kingdom. Usman Raja, a renowned British cage-fighting coach, has managed to rehabilitate Mansha and de-radicalize terror convicts with a 100 % success rate using cage-fighting sessions and the teachings of Sheikh Aleey Qadir.

"Take away someone's hate and they feel liberated," explained Raja. "The key is to give them a sense of purpose" [77]. Now a transformed man, Mansha tries to prevent other young Muslims from following in his past footsteps of terror. "I could channel my energy straight away and build something for myself," said Mansha. "My transformation came over time" [77].

Roman Catholic "Dead Man Walking" nun Sister Helen Prejean met with Dzhokhar Tsarnaev five times since March 2015 and she concluded that "I had every reason to think he was taking it in and was genuinely sorry for what he did. The groundwork and the trust was there. And I knew. I felt it" [78].

What if Dzhokhar Tsarnaev were given a chance to be rehabilitated? What if a "born-again" Tsarnaev were to become a pacifist evangelizing peace among radicalized youths? Having been featured on the cover of *Rolling Stone magazine* and the *New York Times*, Tsarnaev could be an efficacious counterterrorism advocate.

2.12 Nine Familial Rehabilitation

During the May 2015 trial of Dzhokhar Tsarnaev, his aunts and other relatives from Russia flew to the United States to testify on his behalf. The otherwise emotionless defendant broke down and cried when he saw his aunt Patimat Suleimanova in her mid-60s desperately trying to defend him [79]. Notwithstanding the last-ditch effort, where was the family support when the young Tsarnaev needed guidance the most?

According to BBC News, children in east Jerusalem and at refugee camps in Lebanon celebrated news of the terrorist attacks as the Twin Towers collapsed on September 11, 2001 in Manhattan, New York City [80]. Instead of empathy and sympathy, what atrocities have the children learned from their parents?

Shortly after the historic Egypt-Israel Peace Treaty in 1979, Egyptian President Anwar Sadat was assassinated by Islamic jihadists in 1981. Khaled Al-Islambouli was one of the assassins who were convicted and executed in 1982 for treason. In a 2012 interview with Mrs. Qadriya by Iran's state-run Fars news agency, the 85-year-old mother of the assassin said, "I am very proud that my son killed Anwar Al-Sadat. [The government] called him a terrorist, a criminal, and a murderer, but they didn't say that was he was defending Islam. They didn't say anything about the oppressed people in Palestine, about Camp David, or how Sadat sold out the country to the Jews and violated the honor of the Islamic nation" [81].

In December 2015, Fox News asked Donald Trump how he would fight ISIS if he was elected President, and Trump replied, "We are fighting a very politically correct war. The other thing with the terrorists is you have to take out their families, when you get these terrorists, you have to take out their families. They care about their lives, don't kid yourself. When they say they don't care about their lives, you have to take out their families" [82]. His view resembles nine familial exterminations in ancient China as early as 1600 B.C. when death penalty for high treason applied not only to the criminal himself but also to his immediate and extended family members including his [83]:

1. spouse,
2. parents,
3. grandparents,
4. children above a certain age (usually 16),
5. grandchildren above a certain age (usually 16),
6. siblings,
7. siblings-in-law,
8. uncles, and
9. uncles' spouses.

Former U.S. Secretary of Defense William Cohen was among many officials who reproached Trump's egregious remark. Cohen told CNN, "The notion that we would attack and kill the families of terrorists is something that contravenes everything the United States stands for in this world" [84]. In March 2016, Trump reversed course on his vow to kill the families of terrorists [85].

Nine familial exterminations were absolutely inhumane and tyrannical. A more civilized derivative is nine familial rehabilitation which would require immediate and extended family members to be put on fair trials for the crimes of the perpetrator. In the case of Mrs. Qadriya, she should have been tried in court for encouraging her children to become assassins.

Family members serving prison time in the same correctional facility might just provide the much-needed family bonding and quality time that could have been absent otherwise. Nine familial rehabilitation can be a potent way to rehabilitate criminals, deter others from committing serious crimes, and encourage parents to do a better job in raising their children.

Toya Graham, a single mother of six in Baltimore, went to a riot and pulled her masked son away from a protest crowd, smacking him in the head and screaming at him. "I'm a no-tolerant mother," Toya Graham told CBS News. "That's my only son and at the end of the day I don't want him to be a Freddie Gray [who died in police custody]." Police Commissioner Anthony Batts thanked her in his remarks to the media, "I wish I had more parents that took charge of their kids out there tonight" [86].

When parents are unavailable or incapable, school teachers and community leaders ought to step into their shoes lest the vulnerable youth is either radicalized by terrorists or victimized by entrapment. A RAND research study discussed "the feelings of alienation—whether for social, economic, political, or psychological reasons—shared by terrorists, and the progression they follow from early, legal protest to acts of terrorism. ... In many cases it was a matter of chance whether the terrorist joined a left- or right-wing group" [87].

Entrapment artificially creates that chance, but it also generates deep resentment that makes rehabilitation extremely difficult if at all possible. A recent example in 2010 is the conviction of 19-year-old Somali-American college student Mohamed Mohamud for carrying out the Black Friday bombing of the Christmas tree lighting ceremony in Portland, Oregon, which was in fact a sting operation set up by the FBI. Instead of guiding the confused teenage student away from terrorism, an FBI undercover agent asked Mohamud with his power of suggestion, "So, what have you been doing to be a good Muslim? ... You can pray five times a day, train as a doctor and go overseas, donate money to the cause, become 'operational,' or become 'a martyr'" [88]. Mohamud chose to become "operational," and thus he was convicted by a jury and sentenced to 30 years in prison. Meanwhile, ISIS has lured more than 20,000 foreigners and an estimated 180 Americans into joining their fight since February 2015 [89]. The FBI agents should toil to entrap ISIS recruiters by posing as radical students rather than to enmesh troubled teenagers by pretending to be Islamic terrorists.

2.13 War on Drugs

Long before President George W. Bush declared the global war on terror in 2001, President Richard Nixon declared drug abuse "public enemy number one" in 1971. Today, the United States spends more than $51 billion annually on the war on drugs with no end in sight [90].

Actor and comedian Tim Allen, best known for his leading role in the sitcom *Home Improvement* and the voice of Buzz Lightyear in *Toy Story*, was given the highest honor as a Disney Legend for television, film, and animation-voice as well as a Star on the Hollywood Walk of Fame [91]. Before his showbiz success, however, Allen was arrested for drug trafficking in 1978 and was subsequently incarcerated for 28 months at Sandstone Federal Correctional Institution [92]. Drug trafficking can carry a mandatory death penalty in some countries like Singapore (up until 2011) [93] and China [94].

In a 2011 interview by ABC 20/20 anchor Elizabeth Vargas, Tim Allen talked about God and his rehabilitation [95]. He said that one day he got a call from his parole officer and the next day a call from Jeffrey Katzenberg, then chairman of The Walt Disney Studios, who asked him to become a part of the Disney family. It was nothing short of a miracle.

Surely one cannot compare Tim Allen smuggling cocaine versus Dzhokhar Tsarnaev planting bombs. True, but illicit drugs have killed several orders of magnitude more people than all acts of terrorism in America. Between 2001 and 2014, there were 3,030 Americans killed by terrorists on U.S. soil [96]. During that

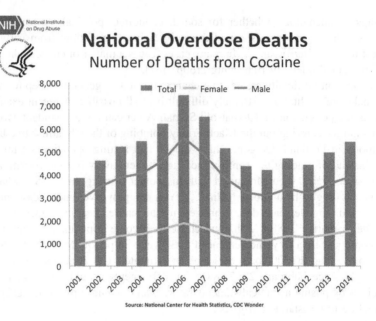

Fig. 2.16 National overdose deaths—number of deaths from cocaine. The bar chart shows the total number of U.S. overdose deaths involving cocaine from 2001 to 2014 (Courtesy of National Institute on Drug Abuse)

same 14-year period, 72,348 Americans died from cocaine overdose (see Fig. 2.16) and 52,830 lost their lives from heroin overdose (see Fig. 2.17) according to the statistics released by the National Institute on Drug Abuse at the Centers for Disease Control and Prevention [97]. Nationally, deaths from drug overdoses reached an all-time high in 2014 [98].

In a blatant confrontation of the epidemic, Los Angeles based street artist Plastic Jesus courageously placed a life-sized Oscar statue—bending over and snorting cocaine—on Hollywood Boulevard three days before the 2015 Academy Awards (see Fig. 2.18). "We only hear about drug addiction when a high-profile Hollywood celeb has a meltdown and goes into rehab or dies," Plastic Jesus told Variety [99]. For instance, actor Robert Downey Jr. spent years in substance abuse, arrests, rehab, and relapse before a full recovery from drugs and a return to his career, starring in blockbuster films such as *The Iron Man*, *The Avengers*, and *Sherlock Holmes*. Downey said in a 2004 interview, "Like Jung said about people using religion to avoid a religious experience, I have managed handily to avoid a religious experience. I don't know where I fall. Spiritual Green Party? There were times when I was into the whole Hare Krishna thing, which is pretty far out. Now I would call myself a Jew-Bu, a Jewish-Buddhist. But there were many times when Catholicism saved my butt" [100].

Apart from Hollywood, Silicon Valley is also at the center of the epidemic. In a high profile case in 2015, Google executive Forrest Hayes died of heroin overdose

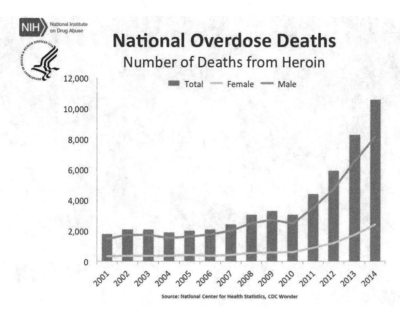

Fig. 2.17 National overdose deaths—number of deaths from heroin. The bar chart shows the total number of U.S. overdose deaths involving heroin from 2001 to 2014 (Courtesy of National Institute on Drug Abuse)

on his yacht off the California coast, and his prostitute companion Alix Catherine Tichelman was convicted of involuntary manslaughter [101]. "The billionaires I know, almost without exception, use hallucinogens on a regular basis," said Silicon Valley investor Tim Ferriss. "[They're] trying to be very disruptive and look at the problems in the world … and ask completely new questions" [102].

In a 2015 interview by CNN technology correspondent Laurie Segall, Apple's early employee Daniel Kottke reminisced about his LSD trips with Steve Jobs at Reed College: "We would take psychedelics and whole new vistas opened up. … Once Apple started, Steve was really focused with all of his energy on making Apple successful. And he didn't need psychedelics for that." Kottke added that Apple cofounder Steve Wozniak (who designed Apple I and II computers) was not interested in dropping acid: "Woz was in very close touch with the extent to which his mind is a miracle of nature. He's just fantastically interested in things … His mind was always working perfectly well and [he] didn't want to mess it up" [103].

Taking illicit drugs is like playing Russian roulette. In 2014, 16-year-old honor roll student Sam Motsay died from experimenting with a designer drug similar to the hallucinogen LSD for the first time [104]. Misusing prescription drugs is just as deadly. Prescription opioid overdose—the cause of death for musician Prince [105] —had killed more than 14,000 people in 2014 according to Centers for Disease Control and Prevention (CDC) [106].

An oft-used excuse from drug dealers is that their victims, unlike terrorist targets, willingly put themselves at risk. Mexican-American actress Kate Del Castillo,

Fig. 2.18 Life-sized Oscar statue—bending over and snorting cocaine—on Hollywood Boulevard in Los Angeles on February 19, 2015 (Courtesy of Plastic Jesus http://www.plasticjesus.net/)

who played crime boss Pilar Zuazo on the Showtime series *Weeds*, caught the attention of the notorious Mexican drug lord El Chapo Guzman after she posted online: "Mr. Chapo: Wouldn't it be cool if you started to traffic in goodness? With cures for diseases, with food for children in the street, with alcohol for nursing homes. … trafficking with corrupt politicians instead of with women and children that end up as slaves? With burning all the pimps that treat a woman like she's worth no more than a pack of cigarettes? With no supply there is no demand. Do it, sir, and you would be the hero of heroes. Let's traffic with love. You know how" [107].

Actress Castillo and actor Sean Penn met Guzman at a secret location in September 2015 to discuss an exclusive interview. In the January 2016 issue of *Rolling Stone* magazine, Penn wrote about his personal meeting with Guzman and he included the following Q&A [108]:

Penn: What is your opinion about who is to blame here, those who sell drugs, or the people who use drugs and create a demand for them? What is the relationship between production, sale and consumption?

Guzman: If there was no consumption, there would be no sales. It is true that consumption, day after day, becomes bigger and bigger. So it sells and sells.

Regardless of who is to blame, the war on drugs and the war on terrorism are equally futile if we treat only the symptoms but not the root causes [109]. It is high time we treated drug abuse and terrorism as diseases instead of wars—curing the patients rather than killing them. For example:

1. Dr. Carl Hart, a neuroscientist and associate professor of psychiatry and psychology at Columbia University, spoke at the TEDMED 2015 conference: "I grew up in the hood in Miami in a poor neighborhood. I came from a community in which drug use was prevalent. I kept a gun in my car. I engaged in petty crime. I used and sold drugs. But I stand before you today also—emphasis on *also*—a professor at Columbia University who studies drug addiction" [110].
2. Since 2010, Usman Raja—a renowned cage-fighting coach in the United Kingdom—has successfully rehabilitated released prisoners who were convicted terrorists into mainstream society [111]. Employing cage-fighting sessions and the teachings of Sheikh Aleey Qadir, Raja's mixed-martial arts (MMA) gym has de-radicalized terror convicts with a 100 % success rate. "Take away someone's hate and they feel liberated," explained Raja. "The key is to give them a sense of purpose" [77].

2.14 Forgiveness and Humility

In the Bible, Jesus won the war against evil not by fighting but by praying as he was being crucified: "Father, forgive them; for they do not know what they are doing" (Luke 23:34). Forgiveness, not vengeance, yields peace and security. As President Barack Obama said during his visit to Israel and the West Bank in March 2013, "Peace is the only path to true security ... because no wall is high enough, and no Iron Dome is strong enough or perfect enough, to stop every enemy from inflicting harm" [112].

Why is it so difficult for people to forgive? Perhaps it is because of our over-inflated ego. We may look down at a janitor, thinking that his life is not worth as much as ours. Now imagine a world without janitors: Every restaurant diner, mall shopper, movie patron, churchgoer, and office worker is required to clean the faucet and scrub the toilet after using a public restroom. How uncomfortable and inconvenient everyday life would become! We should be very thankful to janitors for doing their job well.

Whether we like it or not, nature forces chemical bonding among all living things. We smell food because it emits molecules that enter through our nostrils and bond with the odor receptors in our noses and lungs [113]. Through smell, babies bond with their mothers, and pet animals bond with their owners. As unpalatable as it is, we all have experienced public restrooms where unpleasant smelling molecules emitted from strangers' digestive tracts have bound to our olfactory receptors and entered our lungs.

Curt Stager, ecologist and climate scientist at Paul Smith's College, penned an amusing and eye-opening article titled "You Are Made of Waste: Searching for the ultimate example of recycling? Look in the mirror." [114]:

You may think of yourself as a highly refined and sophisticated creature—and you are. But you are also full of discarded, rejected, and recycled atomic elements. ... Look at one of your fingernails. Carbon makes up half of its mass, and roughly one in eight of those carbon atoms recently emerged from a chimney or a tailpipe. ... When you smile, the gleam of your teeth obscures a slight glow from radioactive waste. ... The oxygen in your lungs and bloodstream is a highly reactive waste product generated by vegetation and microbes. ... The next time you brush your hair, think of the nitrogenous waste that helped create it. All of your proteins, including hair keratin, contain formerly airborne nitrogen atoms. ... Every atom of iron in your blood, which helps your heart shuttle oxygen from your lungs to your cells, once helped destroy a massive star. ... The same blasts also released carbon, nitrogen, oxygen, and other elements of life, which later produced the sun, the Earth, and eventually—you.

The Book of Genesis certainly echoes Stager's main point about "recycled" human beings. Genesis 2:7 reads, "Then the Lord God formed a man from the dust of the ground and breathed into his nostrils the breath of life, and the man became a living being." In William Shakespeare's *Hamlet* (Act 2, Scene 2), the Prince of Demark said to Rosencrantz and Guildenstern [115]:

What a piece of work is a man! how noble in reason!
how infinite in faculty! in form and moving how
express and admirable! in action how like an angel!
in apprehension how like a god! the beauty of the
world! the paragon of animals! And yet, to me,
what is this quintessence of dust?

Jesus taught his disciples about humility in serving humanity: "Truly I tell you, whatever you did for one of the least of these brothers and sisters of mine, you did for me. ... Whatever you did not do for one of the least of these, you did not do for me" (Matthew 25:40, 45).

2.15 Masks of God and Moral Standard

Seeing is believing, but God said to Moses on Mount Horeb, "You cannot see my face, for no one may see me and live" (Exodus 33:20). Portraying deities as macho warrior gods was common among the Indo-European people. The masks of God were created by "the imperfections of man and the limits of reason"—a phrase borrowed from President Barack Obama's 2009 Nobel Peace Prize acceptance speech [116].

In *The Masks of God: Occidental Mythology*, American scholar Joseph Campbell wrote, "It is clear that, whether accurate or not as to biographical detail, the moving legend of the Crucified and Risen Christ was fit to bring a new warmth, immediacy, and humanity, to the old motifs of the beloved Tammuz, Adonis, and Osiris cycles."

Campbell recognized the significance of Jesus who told the world that "I am the way and the truth and the life. No one comes to the Father except through me" (John 14:6). Moreover, he said, "Do not think that I have come to abolish the Law or the Prophets; I have not come to abolish them but to fulfill them" (Matthew 5:17).

Jesus set a higher moral standard than the Law of Moses: "You have heard that it was said, 'You shall not commit adultery.' But I tell you that anyone who looks at a woman lustfully has already committed adultery with her in his heart. ... Anyone who divorces his wife, except for sexual immorality, makes her the victim of adultery, and anyone who marries a divorced woman commits adultery" (Matthew 5:27–28, 32).

Modern society tends to condemn prostitution but turn a blind eye to adultery. In June 2014, the FBI shut down RedBook escort service websites but not the Ashley Madison adultery site [117]. Shouldn't it be the other way around? Sex between two consenting adults is fine as long as it does not cause harm to one another or to any third party. In fact, three out of four women in the United States have lived "in sin" with a partner without being married by the age of 30 [118]. Regardless of cohabitation or marriage, adulterers oftentimes ruin the relationships with their partners, families, friends, and especially children if they have any [119]. In July 2015, hackers broke into AshleyMadison.com and leaked its database of 30 million customers online [120]. Resignations, divorces, and even suicides followed. Married with two adult children, 56-year-old Baptist pastor John Gibson in New Orleans killed himself after learning that his name was publicly exposed [121].

2.16 Prostitution and Gender Discrimination

The Bible surprisingly contains significant anti-stigmatization of prostitutes. For instance, when Israeli leader Joshua invaded the city of Jericho, only Rahab the prostitute and her family were spared from annihilation (Joshua 6:17). "You see that a person is considered righteous by what they do and not by faith alone. In the same way, was not even Rahab the prostitute considered righteous for what she did when she gave lodging to the spies and sent them off in a different direction?" (James 2:24–25).

Take another example, when the Israelis were in trouble with the Ammonites, their elders turned to Jephthah for help:

> Jephthah the Gileadite was a mighty warrior. His father was Gilead; his mother was a prostitute. ... Jephthah said to them [the elders], "Didn't you hate me and drive me from my father's house? Why do you come to me now, when you're in trouble?" The elders of Gilead said to him, "Nevertheless, we are turning to you now; come with us to fight the Ammonites, and you will be head over all of us who live in Gilead." ... Jephthah led Israel six years. (Judges 11:1, 11:7–8, 12:7)

Sex, food, water, and air are four basic survival elements without which humankind would cease to exist. 795 million people on earth are suffering from malnutrition [122], 663 million do not have access to safe drinking water [123], and 7 million people die annually from air pollution exposure [124]. Yet a whopping $186 billion dollars per year are squandered on prostitution worldwide [125]. One sex worker told Laurie Segall at CNN that she had earned nearly $1 million from affluent men in Silicon Valley [126]. A million U.S. dollars can feed a lot of hungry and thirsty people.

In January 2016, Missouri House of Representative Bart Korman introduced a bill that requires lobbyists to report sexual relations with state legislators as a "gift" in their disclosures: "For purposes of subdivision (2) of this subsection, the term 'gift' shall include sexual relations between a registered lobbyist and a member of the general assembly or his or her staff. Relations between married persons or between persons who entered into a relationship prior to the registration of the lobbyist, the election of the member to the general assembly, or the employment of the staff person shall not be reportable under this subdivision. The reporting of sexual relations for purposes of this subdivision shall not require a dollar valuation" [127].

Science fiction TV series *Firefly* created by Joss Whedon also challenged the moral question of high-society courtesans similar to Japanese geisha [128]. Although Hollywood features many heroines in popular films, they are mostly fantasy and sci-fi genres that do not translate into the real world. Even within the make-believe universe, Chloë Grace Moretz who played Hit-Girl in *Kick-Ass* (2010) and *Kick-Ass 2* (2013) told *Digital Spy* in a 2014 interview: "Whenever there's a female superhero, it's always a more sexual plotline rather than seeing an actual character on screen. I don't think that's cool. I think it's rather sad. I would love to change that" [129]. The semi-documentary blockbuster film *The Imitation Game* (2014) shows a solitary female codebreaker—Keira Knightley—in a male-dominated workplace when in fact 8,000 out of 12,000 codebreakers at Bletchley Park were women [130].

Changes are coming, albeit rather slowly. Every now and then there are movies such as *The Intern* (2015) where the lead female character played by Anne Hathaway is the founder of a successful Internet company while her husband is a stay-at-home dad [131]. It accurately reflects the steady increase of the number of househusbands since a 2002 survey at *Fortune* magazine's Most Powerful Women Summit [132]. In the 2008 remake of the 1951 film *The Day the Earth Stood Still*, Mrs. Helen Benson (Patricia Neal) as a secretary was rewritten to be Dr. Helen Benson (Jennifer Connelly) as an astrobiology professor at Princeton University.

Speaking of women in power, Ann Elizabeth Dunwoody was the first woman to become Commanding General of U.S. Army Materiel Command in 2008 [133]. Regina Dugan was appointed the first female director of Defense Advanced Research Projects Agency (DARPA) in 2009 [134]. Letitia "Tish" Long became the first woman in charge of a major U.S. intelligence agency—the National Geospatial-Intelligence Agency (NGA) in 2010 [135]. The White House named Megan Smith as the 3rd Chief Technology Officer of the United States in 2014 [136]. General Lori Robinson became the first female combatant commander to lead the North American Aerospace Defense Command (NORAD) and U.S. Northern Command (NORTHCOM) in 2016 [137]. At the time of writing, if former First Lady and U.S. Secretary of State Hillary Clinton becomes the first female President of the United States, she will join the prestigious list of some 77 female heads of state in world history.

Will all that breaking-the-glass-ceiling progress eradicate gender discrimination? Do people consider intelligent and powerful women a one of a kind or an anomaly which does not apply to the female population at large? Why are only 18 % of computer science graduates women [138] when the world's first computer programmer was English mathematician Ada Lovelace, best known for her algorithmic work on Charles Babbage's Analytical Engine?

It takes a lot more women than just some female geniuses, presidents, military combat generals, CEOs, and the new $20 bills with Harriet Tubman [139] to dismantle the entrenched mindset of chauvinism and patriarchy. Duke University student and porn star Belle Knox (Miriam Weeks) wrote in *xoJane*: "I, like all other sex workers, want to be treated with dignity and respect. … The virgin-whore dichotomy is an insidious standard that we have unfairly placed upon women. Women are supposed to be outwardly pure and modest, while at the same time being sexually alluring and available. If a woman does not have sex after a date, she will be labeled as a prude. If she does have sex, she will be referred to later as a ho or a slut" [140].

The long history of male supremacy has prompted British-Australian feminist Sheila Jeffreys to opine that marriage is a form of prostitution in some cases. "Prostitution and marriage have always been related," said Jeffreys. "The right of men to women's bodies for sexual use has not gone but remains an assumption at the basis of heterosexual relationships" [141]. The alternatives such as cohabitation, open marriage, and friends with benefits are not helping the situation either.

Online sexual harassment and the disproportionate underrepresentation of influential women in Wikipedia have motivated Emily Temple-Wood to cofound WikiProject Women Scientists in 2012 [142]. She has since written hundreds of Wikipedia articles about women in science [143] and has identified more than 4,400 notable female scientists who do not yet have a page on Wikipedia [144].

2.17 Pornography, Feminism, and Technology

In addition to gender discrimination, society's double standards legalize pornography as freedom of expression protected under the First Amendment while they penalize prostitution against personal rights and liberty. The truth is that pornography is essentially prostitution in front of cameras, and porn producers are no different from pimps who recruit women and make money from their sex acts.

Produced by Rashida Jones, Abigail Disney, and other notables, Netflix documentary *Hot Girls Wanted* (2015) exposes the dark side of the porn industry as naive young women are lured by money and fame [145]. Some viewers may see a parallel between that and Amy Berg's documentary *An Open Secret* (2014) which depicts underage abuse and sexual exploitation in the mainstream Hollywood entertainment industry [146]. There are good and bad people in every line of business, every stratum of society, every organization, and sadly every church.

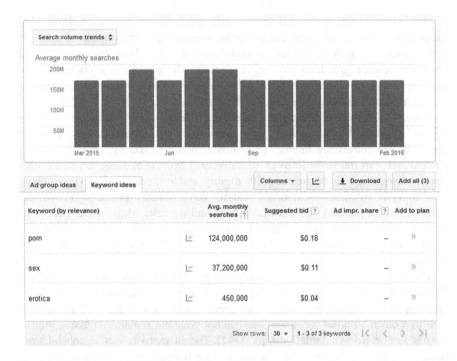

Fig. 2.19 Average monthly Google searches on the keywords porn, sex, and erotica from March 2015 to February 2016, averaging 162 million queries in a month

The question is: What do we do about it? As Morpheus (Laurence Fishburne) told Neo (Keanu Reeves) in *The Matrix* (1999), "You take the blue pill—the story ends, you wake up in your bed and believe whatever you want to believe. You take the red pill—you stay in Wonderland, and I show you how deep the rabbit hole goes. Remember: all I'm offering is the truth. Nothing more" [147]. The majority has chosen the blue pill and perpetuated the status quo.

Thanks to the Internet, pornography has become ubiquitous and ballooned to a $97 billion industry in 2015 [148]. Figure 2.19 shows the average monthly Google searches on the keywords porn, sex, and erotica from March 2015 to February 2016, averaging 162 million queries from month to month with little to no fluctuation. In a 2016 report by evangelical Christian polling firm the Barna Group, a survey of over 3,000 Americans uncovered that [149]:

- 57 % of pastors and 64 % of youth pastors have struggled with porn either currently or in the past,
- The attitudes of Americans towards pornography are shifting from "bad for society" to neutrality or "good for society," and
- Most teens and young adults view "not recycling" as more immoral than viewing porn.

In 2015, Danish sexology professor Christian Graugaard at Aalborg University suggested that pornography should be shown in schools as part of sex education classes [150]. Graugaard's philosophy is in sharp contrast to the 2015 story of the enraged mother Gail Horalek who complained to the school officials for assigning Anne Frank's *The Diary of a Young Girl* to her 7th-grade daughter, calling the passages on puberty in the unabridged version pornographic [151]. Horalek would probably have requested to ban the Christian Old Testament too if she had read the stories of adultery (e.g. 2 Samuel 12:11–12), prostitution (e.g. Ezekiel 23:1–21), rape (e.g. 2 Samuel 13:10–14), incest (e.g. Genesis 19:30–38), and other cardinal sins that rival any daytime soap opera on television. The Bible is the most brutally honest book that does not whitewash or sugarcoat history.

Prof. Graugaard admitted that some adolescents were incapable of differentiating between pornography and the reality of sexual relationships. At TED 2009, advertising consultant Cindy Gallop argued that "hardcore pornography had distorted the way a generation of young men think about sex" [152].

Feminism rejects chauvinism but accepts chivalry [153]. Harry Potter heroine Emma Watson has served as a U.N. Women's Global Goodwill Ambassador since 2014 [154]. In a February 2016 interview, American feminist Gloria Steinem and Emma Watson discussed the promotion of erotica as opposed to pornography that degrades or humiliates women. Steinem said, "We should at least have a word for sex that is mutual and pleasurable and not about domination, pain, violence, humiliation and so on. ... We were both worried about the envelopment of the earth in pornographic images. Young people especially. The right wing (right wing just in a general way) on one hand is suppressing sex education and allowing, or profiteering off pornography so young people look at porno and think that's it. That's what is supposed to be. I was hoping that having a word for erotica, for shared mutual pleasurable empathetic sex, real pleasurable sex would help us do something about pornography" [155].

Whether sex is called porn or erotica, the adult industry has been highly effective in driving innovation and mass adoption of new technologies: VCR, streaming video, online credit card transactions, haptics, and Snapchat, just to name a few [156]. The cover of Playboy's first-ever non-nude issue in March 2016 was made to look like a selfie on Snapchat [157]. At CES 2016 in Las Vegas, Mashable's product analyst Raymond Wong tried out VR porn on a Samsung Gear VR headset and was blown away by it [158]. Like social media, virtual reality could exacerbate the decrease of face-to-face communication and accelerate the decay of interpersonal relationships in the real world.

2.18 Unmasking of God and Enlightening Humankind

Jesus came to enlighten humankind about God, morality, and truth. The Sadducees and Pharisees had hoped that the Messiah would lead the Israelites to rebel against the ruling Roman Empire, just as Moses did against the Egyptian Pharaoh with the ten plagues and the parting of the Red Sea. To the dismay of the Sadducees and Pharisees, Jesus was a pacifist: "You have heard that it was said, 'Love your neighbor and hate your enemy.' But I tell you, love your enemies and pray for those who persecute you, that you may be children of your Father in heaven. He causes his sun to rise on the evil and the good, and sends rain on the righteous and the unrighteous. If you love those who love you, what reward will you get? ... And if you greet only your own people, what are you doing more than others? ..." (Matthew 5:43–47).

Jesus had unmasked God to reveal that:

1. God is forgiving:

   (a) *You have heard that it was said, "Eye for eye, and tooth for tooth." But I tell you, do not resist an evil person. If anyone slaps you on the right cheek, turn to them the other cheek also. (Matthew 5:38–39)*

   (b) *You have heard that it was said, "Love your neighbor and hate your enemy." But I tell you, love your enemies and pray for those who persecute you. (Matthew 5:43–44)*

2. God is a peacemaker:

   (a) *Blessed are the peacemakers, for they will be called children of God. (Matthew 5:9)*

   (b) *Put your sword back in its place, for all who draw the sword will die by the sword. Do you think I cannot call on my Father, and he will at once put at my disposal more than twelve legions of angels? (Matthew 26:52)*

3. God is nonjudgmental:

   (a) *The teachers of the law and the Pharisees brought in a woman caught in adultery. They made her stand before the group and said to Jesus, "Teacher, this woman was caught in the act of adultery. In the Law Moses commanded us to stone such women. Now what do you say?" They were using this question as a trap, in order to have a basis for accusing him. But Jesus bent down and started to write on the ground with his finger. When they kept on questioning him, he straightened up and said to them, "Let any one of you who is without sin be the first to throw a stone at her." (John 8:3–7)*

   (b) *A woman in that town who lived a sinful life learned that Jesus was eating at the Pharisee's house, so she came there with an alabaster jar of perfume. As she stood behind him at his feet weeping, she began to wet his feet with her tears. Then she wiped them with her hair, kissed them and poured perfume on them. When the Pharisee who had invited him saw this, he said to himself, "If this man were a prophet, he would know who is touching him and what kind of woman she is—that she is a sinner." ... Then Jesus turned toward the woman and said to Simon, "Do you see this woman? ... I tell you, her many sins have been forgiven—as her great love has shown. But whoever has been forgiven little loves little." (Luke 7:37–39, 44, 47)*

4. God is benevolent:

   (a) *Then the King will say to those on his right, "Come, you who are blessed by my Father; take your inheritance, the kingdom prepared for you since the creation of the world. For I was hungry and you gave me something to eat, I was thirsty and you gave me something to drink, I was a stranger and you invited me in, I needed clothes and you clothed me, I was sick and you looked after me, I was in prison and you came to visit me." Then the righteous will answer him, "Lord, when did we see you hungry and feed you, or thirsty and give you something to drink? When did we see you a stranger and invite you in, or needing clothes and clothe you? When did we see you sick or in prison and go to visit you?" The King will reply, "Truly I tell you, whatever you did for one of the least of these brothers and sisters of mine, you did for me." (Matthew 25:34–40)*

5. God loves children:

   (a) *At that time the disciples came to Jesus and asked, "Who, then, is the greatest in the kingdom of heaven?" He called a little child to him, and placed the child among them. And he said: "Truly I tell you, unless you change and become like little children, you will never enter the kingdom of heaven. Therefore, whoever takes the lowly position of this child is the greatest in the kingdom of heaven." (Matthew 18:1–4)*

   (b) *Then people brought little children to Jesus for him to place his hands on them and pray for them. But the disciples rebuked them. Jesus said, "Let the little children come to me, and do not hinder them, for the kingdom of heaven belongs to such as these." (Matthew 19:13–14)*

Following in Jesus' footsteps, Pope Francis expressed his nonjudgmental stand in *The Joy of Love (AMORIS LAETITIA)*: "A pastor cannot feel that it is enough simply to apply moral laws to those living in 'irregular' situations, as if they were stones to throw at people's lives. This would bespeak the closed heart of one used to hiding behind the Church's teachings, 'sitting on the chair of Moses and judging at times with superiority and superficiality'" [159]. The Pope denounced both favoritism and prejudice: "A gay person who is seeking God, who is of good will—well, who am I to judge him? The Catechism of the Catholic Church explains this very well. It says one must not marginalize these persons, they must be integrated into society. The problem isn't this [homosexual] orientation—we must be like brothers and sisters. The problem is something else, the problem is lobbying either for this orientation or a political lobby or a Masonic lobby" [160].

Indeed, Jesus had predicted religious infighting amongst Abrahamic religions and religious denominations: "Do not suppose that I have come to bring peace to the earth. I did not come to bring peace, but a sword. For I have come to turn 'a man against his father, a daughter against her mother, a daughter-in-law against her mother-in-law—a man's enemies will be the members of his own household.'" (Matthew 10:34–36) The Crusades, anti-Catholicism, anti-Protestantism, anti-Semitism, global Jihadism, and antichrists all have created havoc and suffering on earth that are more terrifying than any Stephen King novel. Abraham must be rolling in his grave many times over.

It is up to humankind to make peace on earth, not by force but based on free will. Albert Einstein said, "Whatever there is of God and goodness in the universe, it must work itself out and express itself through us. We cannot stand aside and let God do it" [8]. In February 2011, Muslims and Christians in Egypt set aside religious tensions and joined hands in anti-Mubarak protest in Tahrir Square, chanting "Arise O Egypt, arise. Arise Egyptians: Muslims, Christians and Jews" [161].

2.19   Separation of Church and State

If the first Christians—as described in the first chapters of the Book of Acts—were to rule the country, the United States of America would embrace socialism or perhaps even communism:

> All the believers were one in heart and mind. No one claimed that any of their possessions was their own, but they shared everything they had. With great power the apostles continued to testify to the resurrection of the Lord Jesus. And God's grace was so powerfully at work in them all that there were no needy persons among them. For from time to time those who owned land or houses sold them, brought the money from the sales and put it at the apostles' feet, and it was distributed to anyone who had need. (Acts 4:32)

America's Founding Fathers had a different idea. Article VI of the U.S. Constitution declares that "no religious Test shall ever be required as a Qualification to any Office or public Trust under the United States" [162] and the First Amendment to the U.S. Constitution states that "Congress shall make no law respecting an establishment of religion, or prohibiting the free exercise thereof" [163].

In the 1971 court case *Lemon v. Kurtzman*, the U.S. Supreme Court upheld the separation of church and state in siding with civil rights activist Alton T. Lemon's objection to state aid to religious schools. Chief Justice Warren E. Burger defined the "Lemon Test" as [164]:

> First, the statute must have a secular legislative purpose;

> Second, its principal or primary effect must be one that neither advances nor inhibits religion;

> Finally, the statute must not foster "an excessive government entanglement with religion."

However, the Chief Justice also cautioned that "judicial caveats against entanglement must recognize that the line of separation, far from being a 'wall,' is a blurred, indistinct and variable barrier depending on all the circumstances of a particular relationship" [165].

Indeed, according to the Pew Research Center report entitled "Faith and the 2016 Campaign," 51 % of adults said that they would be less likely to vote for a presidential candidate who does not believe in God, as compared to 37 % less likely for a candidate who had an extramarital affair and 20 % less likely for a former marijuana smoker [166]. Christians carry more clout in American politics. Separation of church and state is easier said than done. For example:

1. During the 2004 presidential campaign, George W. Bush met with the Amish (see Fig. 2.20) in Pennsylvania and Ohio—two of the swing states—to convince them to vote based on their conservative Christian outlook. As a result, some Amish voted for the first and only time in their lives [167].
2. U.S. House Representative Tulsi Gabbard from Hawaii is the first American Hindu elected to Congress in 2012. Opponents argued that "she shouldn't be allowed to serve because her religion doesn't align with the Constitution." Gabbard responded that "a pluralistic, secular government is the only way to ensure that all individuals have the freedom to follow the religious path of their choice" [168].
3. For the 2012 presidential election, a research study by Prof. Benjamin Knoll of Centre College showed that "about 1 out of every 20 Republicans decided to stay home instead of turning out to vote for their party's nominee [Mitt Romney] because they don't perceive Mormons as Christian" [169].
4. During the 2016 presidential race, Pope Francis suggested that Donald Trump was not a Christian and ABC's Jimmy Kimmel asked U.S. Senator Bernie Sanders from Vermont if he believed in God. Trump lashed out at the Pope, "If and when the Vatican is attacked, the pope would only wish and have prayed that Donald Trump would have been elected president" [170] whereas Sanders replied to Kimmel evasively, "Well, you know, I am who I am. And what I believe in, what my spirituality is about, is that we're all in this together—that I think it's not a good thing to believe, as human beings, that we can turn our backs on the suffering of other people. And this is not Judaism. This is what Pope Francis is talking about: That we cannot worship just billionaires and the making of more and more money. Life is more than that" [171].

2.20 Trinity of Religions, Arts, and Sciences

Whenever people talk about religions, morality, and politics, they often get into heated arguments. As a result, people tend to avoid these sensitive topics altogether. Partly due to the growing apathy, voter turnout in America has dropped to a 72-year low in 2015 [172] and church attendance has been in steady decline for years [173].

American comedian and television host Bill Maher told the audience on ABC's Jimmy Kimmel Live, "We have to stop saying 'Well, we should not insult a great religion.' First of all, there are no great religions. There are all stupid and dangerous. ... We should insult them, and we should be able to insult whatever we want" [174]. Pope Francis begs to differ. The Pope said, "You don't kill in God's

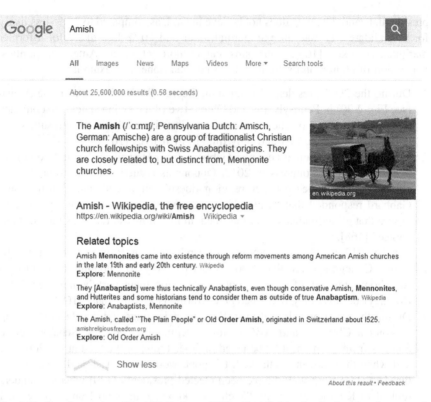

Fig. 2.20 Google search on "Amish" displays a Wikipedia definition for the Amish as "a group of traditionalist Christian church fellowships with Swiss Anabaptist origins" and related topics on Mennonites, Anabaptists, and "The Plain People."

name. ... [But] you cannot provoke, you cannot insult the faith of others. You cannot make fun of the faith of others" [175].

God exists with or without religions. A religion is a human expression and understanding of God. Mahatma Gandhi wrote in his weekly newspaper *Harijan* that "each religion has its own contribution to make to human evolution. I regard the great faiths of the world as so many branches of a tree, each distinct from the other though having the same source" [12]. Even the anti-religious and fascist Adolf Hitler invoked the name of God in many of his public speeches to motivate Nazi Germany. He said, "For one cannot assume that God exists to help people who are too cowardly and too lazy to help themselves and think that God exists only to make up for the weakness of mankind" [176].

Atheists do away with both God and religions altogether, and replace them with their own beliefs and expressions. Extropian Max More opines that the future belongs to posthumanity with "no more gods, no more faith, no more timid holding back" [177]. Nonetheless, there is an aphoristic saying that "there are no atheists in foxholes." During an edge-of-your-seat chase scene in the Academy Award winning film *Mad Max: Fury Road* (2015), the evil Immortan Joe (Hugh Keays-Byrne) and the fanatical War Boys were going after Max (Tom Hardy), Furiosa (Charlize Theron), and the Five Wives. One of the wives—The Dag (Abbey Lee Kershaw)—started to mutter a prayer, and Toast the Knowing (Zoë Kravitz) looked at her curiously:

Toast the Knowing: What are you doing?

The Dag: Praying.

Toast the Knowing: To who?

The Dag: Anyone that's listening.

God and religions may take on different names, forms, and meanings in different cultures, civilizations, and stages of human evolution. But as long as we are human, or posthuman for that matter, there is no escape from God and religions. Like the song *Hotel California* by the Eagles, "You can check out any time you like, but you can never leave!" Denying the existence of God the Creator is like an artificial intelligent machine doubting the existence of human inventors.

The religious, moral, and political entanglement is mirrored in physics where quantum entanglement—a word coined by Austrian physicist Erwin Schrödinger—describes "a physical phenomenon that occurs when pairs or groups of particles are generated or interact in ways such that the quantum state of each particle cannot be described independently—instead, a quantum state may be given for the system as a whole" [178].

Although Albert Einstein rejected quantum theory by calling the entanglement a "spooky action across distance" [179], he was correct in saying that "all religions, arts and sciences are branches of the same tree." Einstein explained that "all these aspirations are directed toward ennobling man's life, lifting it from the sphere of mere physical existence and leading the individual towards freedom. It is no mere chance that our older universities developed from clerical schools. Both churches and universities—insofar as they live up to their true function—serve the ennoblement of the individual. They seek to fulfill this great task by spreading moral and cultural understanding, renouncing the use of brute force" [180].

The ultimate wisdom comes from the trinity of religions, arts, and sciences (see *The Garden of Earthly Delights* triptych by Hieronymus Bosch in Figs. 2.21 and 2.22). Nature, for instance, shows us its artistic beauty that sciences explain the hows and religions contemplate the whys. Astrology, feng shui, and other pseudoscience are attempts to predict or account for life circumstances. Jesus offered humankind a profound enlightenment on the mystery of life, the universe, and everything (pun intended for paying homage to English writer Douglas Adams). Figure 2.23 is a January 2016 snapshot of the Google search autocomplete predictions showing the most popular searches related to "Jesus is." Overall, Jesus enjoys a better reception than God (see Fig. 2.11) and the Bible (see Fig. 2.8) in the top 10 list of Google searches among some 250 million unique visitors every month.

Fig. 2.21 Exterior (shutters) of *The Garden of Earthly Delights* triptych by Hieronymus Bosch (circa 1490–1510). The inscription reads "Ipse dixit, et facta sunt (He spoke and it was there)" and "Ipse mandāvit, et creāta sunt (He commanded and they were created)" from Psalm 33:9. Courtesy of Museo del Prado

Fig. 2.22 *The Garden of Earthly Delights*, oil on oak triptych by Hieronymus Bosch (circa 1490–1510). Courtesy of Museo del Prado

2.21   Death, Resurrection, and Future of Religions

In *Star Wars* (1977), Jedi master Obi-Wan Kenobi (Alec Guinness) sacrificed himself in the battle with Darth Vader (David Prowse) so that he could become a stronger "Force ghost" to help guide the young Luke Skywalker (Mark Hamill) (see Fig. 2.24) [181]. The assassination of Martin Luther King, Jr. on April 4, 1968 accelerated the American civil rights movement [182]. The crucifixion of Jesus brought forth the world's largest religion—Christianity—with about 2.2 billion followers in 2015 [6].

Jesus was raised from the dead on the third day when he appeared to his disciples:

> "Look at my hands and my feet. It is I myself! Touch me and see; a ghost does not have flesh and bones, as you see I have." When he had said this, he showed them his hands and feet. And while they still did not believe it because of joy and amazement, he asked them, "Do you have anything here to eat?" They gave him a piece of broiled fish, and he took it and ate it in their presence. (Luke 24:39–43)

Unlike other resurrection stories such as Zarephath (1 Kings 17:17–24), Shunamite (2 Kings 4:18–37), Jairus (Mark 5:35–43), and Lazarus (John 11:1–44), Apostle Paul spoke of immortality:

> But someone will ask, "How are the dead raised? With what kind of body will they come?"... When you sow, you do not plant the body that will be, but just a seed, perhaps of wheat or of something else. ... So will it be with the resurrection of the dead. The body that is sown is perishable, it is raised imperishable; ... Listen, I tell you a mystery: We will not all sleep, but we will all be changed – in a flash, in the twinkling of an eye, at the last trumpet. For the trumpet will sound, the dead will be raised imperishable, and we will be changed. For the perishable must clothe itself with the imperishable, and the mortal with

Fig. 2.23 Google search autocomplete predictions for "Jesus is" (as of January 2016)

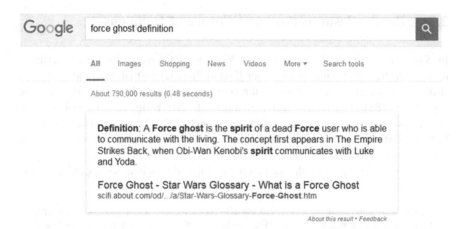

Fig. 2.24 Google Search on "Force ghost" returns a Knowledge Graph definition from about.com: "A Force ghost is the spirit of a dead Force user who is able to communicate with the living. The concept first appears in *The Empire Strikes Back*, when Obi-Wan Kenobi's spirit communicates with Luke and Yoda."

> immortality. When the perishable has been clothed with the imperishable, and the mortal with immortality, then the saying that is written will come true: "Death has been swallowed up in victory." (1 Corinthians 15:35, 37, 42, 51–54)

Paul's statement in 1 Corinthians 15:51—"We will not all sleep, but we will all be changed"—invigorates the belief of some transhumanists in achieving immortality during their lifetime. After all, no one in their current human form can see the face of God and live (Exodus 33:20). Lincoln Cannon, founder of the Mormon Transhumanist Association, talks about transfigurism as a future of religion as exemplified by religious transhumanists in Chap. 19 of this book.

2.22 And Now for Something Completely Different, or not

In the comical spirit of Monty Python, we could use a sense of humor while we unravel the Gordian knot of religious, moral, and political entanglement that forms the nexus between ideologies and actions in our everyday life (see Fig. 2.25):

1. Being a student preacher in high school, I co-wrote a gospel song with one of my best friends Chau Kai Ton for publication in our yearbook. My non-Christian friends oftentimes greeted me by saying, "Ah, here comes Jesus!" Three decades later, Chau and I collaborated again, this time for the academic book *Digital Da Vinci: Computers in Music* [183].
2. In an MTV interview with Dutch filmmaker Paul Verhoeven in 2010, the director called the original RoboCop "the American Jesus." He explained that "the point of 'RoboCop,' of course, is it is a Christ story. It is about a guy that gets crucified after 50 min, then is resurrected in the next 50 min and then is like

Fig. 2.25 *Alexander Cutting the Gordian Knot* by Jean-Simon Berthélemy (1767). Courtesy of The Athenaeum http://www.the-athenaeum.org/ [188]

the super-cop of the world, but is also a Jesus figure as he walks over water at the end. … I put something just underneath the water so [Peter Weller] could walk over the water and say this wonderful line… 'I am not arresting you anymore.' Meaning, 'I'm going to shoot you.' And that is of course the American Jesus" [184].

3. British singer-songwriter David Bowie said in an unaired CBS 60 min interview in 2003 that "searching for music is like searching for God. They are very similar. There's an effort to reclaim the unmentionable, the unsayable, the unseeable, the unspeakable, all those things come into being a composer, into writing music, into searching for notes and pieces of musical information that don't exist" [185].

4. If American rock band R.E.M. were to rewrite the lyrics to their 1991 hit song "Losing My Religion" [186] for Generation Z today, they could simply change one word, from "losing" to "Googling": ♪ *Oh life, it's bigger. It's bigger than you and you are not me. … Oh no, I've said too much. I haven't said enough. That's me in the corner. That's me in the spotlight Googling my religion. …* ♪

5. If we Google "prayers," the library of Catholic prayers is the number one organic search result (see Fig. 2.26). However, if we Google "Google prayers," the Church of Google becomes number one. Its webpage titled "Google Prayers:

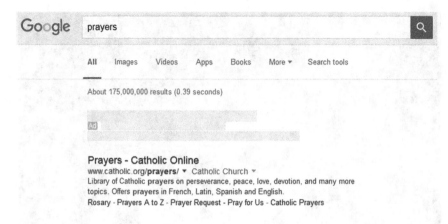

Fig. 2.26 Google search on "prayers" returns the library of Catholic prayers as the number one organic search result

Is Google God?—The Church of Google" contains a parody of the Lord's Prayer by Matt Bowen [187]:

Our Google, who art in cyberspace,
Hallowed be thy domain.
Thy search to come,
Thy results be done,
On 127.0.0.1 as it is in the Googleplex.
Give us this day our daily searches,
And forgive us our spam,
As we forgive those who spam against us.
And lead us not into temptation,
But deliver us from Microsoft.
For thine is the search engine,
And the power,
And the glory,
Forever and ever.
Amen.

References

1. **CNN Producers.** Atheists open up: What they want you to know. [Online] CNN, March 31, 2015. http://www.cnn.com/2015/03/28/opinions/atheists-q-and-a/index.html.
2. **Stafford, Tom.** The web has deluded you, and don't pretend it hasn't. *BBC News.* [Online] October 20, 2015. http://www.bbc.com/future/story/20151020-the-web-has-deluded-you-and-dont-pretend-it-hasnt.
3. **The Bible.** Genesis 1:1. *Bible Hub.* [Online] http://biblehub.com/genesis/1-1.htm.

4. **The Quran.** Surah Al-A'raf 7:54. *Towards Understanding the Quran.* [Online] http://www. islamicstudies.info/tafheem.php?sura=7&verse=54.

5. **Berger, Peter.** Do The Three Abrahamic Faiths Worship The Same God? *The American Interest.* [Online] December 14, 2011. http://www.the-american-interest.com/2011/12/14/do-the-three-abrahamic-faiths-worship-the-same-god/.

6. **Pew Research Center.** The Future of World Religions: Population Growth Projections, 2010-2050. *Demographic Study.* [Online] April 2, 2015. http://www.pewforum.org/2015/04/02/religious-projections-2010-2050/.

7. **Stephens-Davidowitz, Seth.** Googling for God. *The New York Times.* [Online] September 19, 2015. http://www.nytimes.com/2015/09/20/opinion/sunday/seth-stephens-davidowitz-googling-for-god.html.

8. **Einstein, Albert, Calaprice, Alice and Dyson, Freeman.** The Expanded Quotable Einstein (2nd Edition). [Online] Princeton University Press, May 30, 2000.

9. **Ask an Astronomer.** How many stars are born and die each day? (Beginner). *Cornell University Astronomy Department.* [Online] June 27, 2015. http://curious.astro.cornell.edu/about-us/83-the-universe/stars-and-star-clusters/star-formation-and-molecular-clouds/400-how-many-stars-are-born-and-die-each-day-beginner.

10. **Carter, Helen.** Millionaire facing ruin shot wife and daughter and then killed himself. *The Guardian.* [Online] April 3, 2009. http://www.theguardian.com/uk/2009/apr/04/8.

11. **Carnegie Mellon University.** Randy Pausch's Last Lecture. *Carnegie Mellon University.* [Online] September 18, 2007. http://www.cmu.edu/randyslecture/.

12. **Gandhi, Mahatma.** The essential unity of all religions. *Comprehensive Website by Gandhian Institutions - Bombay Sarvodaya Mandal & Gandhi Research Foundation.* [Online] http://www.mkgandhi.org/voiceoftruth/unityofallreligions.htm.

13. **U.S. National Library of Medicine.** What is DNA? *Genetics Home Reference.* [Online] March 14, 2016. https://ghr.nlm.nih.gov/handbook/basics/dna.

14. **Einstein, Albert and Infeld, Leopold.** The Evolution of Physics: The Growth of Ideas from Early Concepts to Relativity and Quanta. [Online] Simon & Schuster, 1942. https://books.google.com/books?id=rSg4AAAAIAAJ.

15. **Variety Staff.** Spike Lee, Jada Pinkett Smith Call for Oscar Boycott. *Variety.* [Online] January 18, 2016. http://variety.com/2016/film/awards/spike-lee-jada-pinkett-smith-oscar-boycott-1201682165/.

16. **Dash, Stacey.** I Was Right Today on Outnumbered: There Should Be No Black History Month. *patheos.* [Online] January 20, 2016. http://www.patheos.com/blogs/staceydash/2016/01/i-was-right-today-on-outnumbered-there-should-be-no-black-history-month/.

17. **Chapman, Michael W.** FLASHBACK: Morgan Freeman on Ending Racism: 'Stop Talking About It'. *CBS News.* [Online] August 22, 2014. http://cnsnews.com/news/article/michael-w-chapman/flashback-morgan-freeman-ending-racism-stop-talking-about-it.

18. **Roddenberry, Gene.** Quotable Quote. *goodreads.* [Online] [Cited: January 24, 2016.] http://www.goodreads.com/quotes/98079-if-man-is-to-survive-he-will-have-learned-to.

19. **IMDb.** Is There in Truth No Beauty? *IMDb.* [Online] October 18, 1968. http://www.imdb.com/title/tt0708433/quotes.

20. **Minsky, Marvin.** Society Of Mind. *Google Books.* [Online] Simon & Schuster Paperbacks, 1988. https://books.google.com/books?id=bLDLllfRpdkC&pg=PA308&lpg=PA308#v=onepage&q&f=false.

21. **Hargreaves, Steve.** Marissa Mayer: Yahoo isn't done with search. *CNNMoney.* [Online] February 11, 2014. http://money.cnn.com/2014/02/11/technology/yahoo-mayer/.

22. **comScore.** comScore Releases December 2015 U.S. Desktop Search Engine Rankings. *comScore.* [Online] January 20, 2016. http://www.comscore.com/Insights/Market-Rankings/comScore-Releases-December-2015-US-Desktop-Search-Engine-Rankings.

23. **statista.** Most popular multi-platform web properties in the United States in November 2015, based on number of unique visitors (in millions). *The Statistics Portal.* [Online] November 2015. http://www.statista.com/statistics/271412/most-visited-us-web-properties-based-on-number-of-visitors/.

24. **Google.** Autocomplete. *Google Search Help.* [Online] [Cited: January 20, 2016.] https://support.google.com/websearch/answer/106230.
25. **Robertson, Nic, et al.** Agent Storm: Inside al Qaeda for the CIA. [Online] CNN, January 12, 2015. http://www.cnn.com/2014/09/16/world/agent-storm/index.html.
26. **Einstein, Albert.** Religion and Science. [Online] New York Times Magazine, November 9, 1930. http://www.sacred-texts.com/aor/einstein/einsci.htm.
27. **Birkenhead, Peter.** Why do we let New Atheists and religious zealots dominate the conversation about religion?. [Online] Salon, April 25, 2015. http://www.salon.com/2015/04/25/why_do_we_let_new_atheists_and_religious_zealots_dominate_the_conversation_about_religion/.
28. **Papageorgiou, Nik.** The first ever photograph of light as both a particle and wave. *École Polytechnique Fédérale de Lausanne.* [Online] February 3, 2015. http://actu.epfl.ch/news/the-first-ever-photograph-of-light-as-both-a-parti/.
29. **Phys.org.** Gravitational waves detected 100 years after Einstein's prediction. *Phys.org.* [Online] February 11, 2016. http://phys.org/news/2016-02-gravitational-years-einstein.html.
30. **The Metropolitan Museum of Art.** The Crusades (1095–1291). [Online] The Metropolitan Museum of Art. [Cited: December 9, 2012.] http://www.metmuseum.org/toah/hd/crus/hd_crus.htm.
31. **Standring, Peter.** Koran a Book of Peace, Not War, Scholars Say. *National Geographic Today.* [Online] September 25, 2001. http://news.nationalgeographic.com/news/2001/09/0925_TVkoran.html.
32. **Wikipedians.** Charlie Hebdo shooting. *Wikipedia.* [Online] [Cited: December 8, 2015.] https://en.wikipedia.org/wiki/Charlie_Hebdo_shooting.
33. —. November 2015 Paris attacks. *Wikipedia.* [Online] [Cited: December 8, 2015.] https://en.wikipedia.org/wiki/November_2015_Paris_attacks.
34. —. 2015 San Bernardino shooting. *Wikipedia.* [Online] [Cited: December 8, 2015.] https://en.wikipedia.org/wiki/2015_San_Bernardino_shooting.
35. **Saifi, Sophia.** In Pakistan, Taliban's Easter bombing targets, kills scores of Christians. *CNN.* [Online] March 28, 2016. http://www.cnn.com/2016/03/27/asia/pakistan-lahore-deadly-blast/index.html.
36. **Bin Laden, Osama.** Full transcript of bin Ladin's speech. [Online] Aljazeera, November 1, 2004. http://www.aljazeera.com/archive/2004/11/200849163336457223.html.
37. **Wood, Graeme.** What ISIS Really Wants. *The Atlantic.* [Online] March 2015. http://www.theatlantic.com/magazine/archive/2015/03/what-isis-really-wants/384980/.
38. **Glazer, Ilsa M.** Armies of the Young: Child Soldiers in War and Terrorism (review). [Online] Anthropological Quarterly, Volume 79, Number 2, Spring 2006, pp. 373-384, 2006. http://dx.doi.org/10.1353%2Fanq.2006.0021.
39. **Holocaust Encyclopedia.** Children during the Holocaust. [Online] United States Holocaust Memorial Museum, August 18, 2015. http://www.ushmm.org/wlc/en/article.php?ModuleId=10005142.
40. **Popalzai, Masoud.** 11-year-old Afghan boy, hailed as hero for fighting Taliban, killed by militants. *CNN.* [Online] February 5, 2016. http://www.cnn.com/2016/02/04/asia/afghanistan-boy-hero-taliban-killed/.
41. **Mohsin, Saima and Khan, Shaan.** Police: Kids young as 8 used as bombers in Pakistan. [Online] CNN, March 14, 2013. http://www.cnn.com/2013/03/14/world/asia/pakistan-child-bombers/index.html.
42. **Abubakr, Aminu and Almasy, Steve.** Girl, 13: Boko Haram tried to force me to become a suicide bomber. [Online] CNN, January 4, 2015. http://www.cnn.com/2014/12/26/world/africa/nigeria-teenage-girl-suicide-bombing/index.html.
43. **Lee, Ian.** Refugee who fled ISIS 'capital': Kids patrolled streets with AK-47 s. *CNN.* [Online] December 7, 2015. http://www.cnn.com/2015/12/07/middleeast/raqqa-isis-refugee-testimony/index.html.

44. **Brown, Pamela and Bruer, Wesley.** FBI official: ISIS is recruiting U.S. teens. [Online] CNN, February 3, 2015. http://www.cnn.com/2015/02/03/politics/fbi-isis-counterterrorism-michael-steinbach/index.html.

45. **Husain, Ed.** Israel, face new reality: Talk to Hamas. [Online] CNN, November 21, 2012. http://www.cnn.com/2012/11/20/opinion/husain-hamas-israel/index.html.

46. **Laurence, Jon.** Malala Yousafzai: the youngest ever Nobel Peace Prize winner - in 60 seconds. *The Telegraph.* [Online] July 12, 2015. http://www.telegraph.co.uk/news/worldnews/asia/pakistan/11153860/Malala-Yousafzai-the-youngest-ever-Nobel-Peace-Prize-winner-in-60-seconds.html.

47. **Nobel Media AB.** Malala Yousafzai - Facts. *Nobelprize.org.* [Online] 2014. http://www.nobelprize.org/nobel_prizes/peace/laureates/2014/yousafzai-facts.html.

48. **Samantha Smith Foundation.** Samantha's Letter. *samanthasmith.info.* [Online] http://www.samanthasmith.info/index.php/history/letter.

49. **Federal Bureau of Investigation (FBI).** Definitions of Terrorism in the U.S. Code. *Federal Bureau of Investigation (FBI).* [Online] [Cited: December 10, 2015.] https://www.fbi.gov/about-us/investigate/terrorism/terrorism-definition.

50. **O'Neill, Ann, Cooper, Aaron and Sanchez, Ray.** Boston bomber apologizes, gets death sentence. [Online] CNN, June 24, 2015. http://www.cnn.com/2015/06/24/us/tsarnaev-boston-marathon-bombing-death-sentencing/index.html.

51. **Obeidallah, Dean.** I'm Muslim, and I hate terrorism. [Online] CNN, April 24, 2013. http://www.cnn.com/2013/04/24/opinion/obeidallah-muslims-hate-terrorism/index.html.

52. **Burke, Daniel.** Could this Quran curb extremism? *CNN.* [Online] December 4, 2015. http://www.cnn.com/2015/11/25/living/study-quran-extremism/.

53. **Holocaust Encyclopedia.** Children during the Holocaust. [Online] United States Holocaust Memorial Museum. http://www.ushmm.org/wlc/en/article.php?ModuleId=10005142.

54. **CBN.com.** What Were the Christian Crusades?. *The Christian Broadcasting Network.* [Online] [Cited: February 3, 2016.] http://www1.cbn.com/spirituallife/what-were-the-christian-crusades.

55. **Taylor, Alan.** On This Day 36 Years Ago: The Signing of the Egypt-Israel Peace Treaty. *The Atlantic.* [Online] March 26, 2015. http://www.theatlantic.com/photo/2015/03/on-this-day-36-years-ago-the-signing-of-the-egyptisrael-peace-treaty/388781/.

56. **Nobel Media AB 2014.** The Nobel Peace Prize 1978. *Nobelprize.org.* [Online] http://www.nobelprize.org/nobel_prizes/peace/laureates/1978/.

57. **—.** The Nobel Peace Prize 2002. *Nobelprize.org.* [Online] http://www.nobelprize.org/nobel_prizes/peace/laureates/2002/.

58. **Davidson, Amy.** Ted Cruz's Religious Test for Syrian Refugees. *The New Yorker.* [Online] November 16, 2015. http://www.newyorker.com/news/amy-davidson/ted-cruzs-religious-test-for-syrian-refugees.

59. **Johnson, Jenna and Weigel, David.** Donald Trump calls for 'total' ban on Muslims entering United States. *The Washington Post.* [Online] December 8, 2015. https://www.washingtonpost.com/politics/2015/12/07/e56266f6-9d2b-11e5-8728-1af6af208198_story.html.

60. **Reuters in Toronto.** Canada's Miss World entry claims China is trying to block her from final. *The Guardian.* [Online] November 11, 2015. http://www.theguardian.com/world/2015/nov/11/canada-miss-world-anastasia-lin-china-human-rights.

61. **Bell, Matthew.** Christians take a prominent role in Hong Kong protests. *PRI.* [Online] October 6, 2014. http://www.pri.org/stories/2014-10-06/christians-take-prominent-role-hong-kong-protests.

62. **Jones, Owen.** Islamophobia plays right into the hands of Isis. *The Guardian.* [Online] November 25, 2015. http://www.theguardian.com/commentisfree/2015/nov/25/islamophobia-isis-muslim-islamic-state-paris.

63. **Ap, Tiffany.** Al-Shabaab recruit video with Trump excerpt: U.S. is racist, anti-Muslim. *CNN.* [Online] January 2, 2016. http://www.cnn.com/2016/01/02/middleeast/al-shabaab-video-trump/index.html.

64. **Beer, Robert.** The Handbook of Tibetan Buddhist Symbols. *Serindia Publications, Inc.* [Online] 2003. https://books.google.com/books?id=-3804Ud9-4IC.

65. **Brumfield, Ben.** All schools shut down in Augusta County, Virginia, over Islam homework. *CNN.* [Online] December 18, 2015. http://www.cnn.com/2015/12/18/us/virginia-school-shut-islam-homework/index.html.

66. **Robbins, Mel.** Homework on Islam no threat to Christians. *CNN.* [Online] December 18, 2015. http://www.cnn.com/2015/12/18/opinions/robbins-augusta-county-islam/index.html.

67. **Carrel, Paul.** Sandberg: Hate speech 'has no place in our society' — Facebook cracks down on extremist posts in Europe. *Business Insider.* [Online] January 18, 2016. http://www.businessinsider.com/r-facebook-begins-europe-wide-campaign-against-extremist-posts-2016-1.

68. **Seetharaman, Deepa and Andrews, Natalie.** Facebook Adds New Tool to Fight Terror: Counter Speech. *The Wall Street Journal.* [Online] February 11, 2016. http://blogs.wsj.com/digits/2016/02/11/facebook-adds-new-tool-to-fight-terror-counter-speech/.

69. **Beltrone, Gabriel.** Germans Fight Neo-Nazis by Liking Their Facebook Page and Flooding It With Love. *AdWeek.* [Online] May 7, 2014. http://www.adweek.com/adfreak/germans-fight-neo-nazis-liking-their-facebook-page-and-flooding-it-love-157516.

70. **Pichai, Sundar.** Let's not let fear defeat our values. *medium.* [Online] December 11, 2015. https://medium.com/@sundar_pichai/let-s-not-let-fear-defeat-our-values-af2e5ca92371#.ltqul3jwr.

71. **Browne, Ryan.** Coast Guard seeks billion-dollar icebreaker as Russia makes Arctic push. *CNN.* [Online] January 19, 2016. http://www.cnn.com/2016/01/18/politics/icebreaker-russia-us-arctic/index.html.

72. **Strickland, Ashley.** Scott Kelly from space: Earth's atmosphere 'looks very, very fragile'. *CNN.* [Online] January 12, 2016. http://www.cnn.com/2016/02/11/health/scott-kelly-space-station-sanjay-gupta-interview/.

73. **O'Neill, Ann.** Boston Marathon bomber's trial leaves a lingering sadness. [Online] CNN, May 18, 2015. http://www.cnn.com/2015/05/17/us/tsarnaev-13th-juror-last-word/index.html.

74. **Swanson, Emily.** Americans Favor The Death Penalty, But Few Want The Executed To Suffer. [Online] The Huffington Post, January 25, 2014. http://www.huffingtonpost.com/2014/01/25/death-penalty-poll_n_4661940.html.

75. **Randall, Eric.** The Story Behind Martin Richard's Peace Sign. [Online] Boston Daily, April 30, 2013. http://www.bostonmagazine.com/news/blog/2013/04/30/the-story-behind-martin-richards-peace-sign/.

76. **Parini, Jay.** How Tsarnaev sentence traps us in cycle of violence. [Online] CNN, May 15, 2015. http://www.cnn.com/2015/05/15/opinions/parini-death-penalty-tsarnaev/.

77. **Robertson, Nic and Cruickshank, Paul.** Convicted terrorist calmed by cagefighting. [Online] CNN, July 28, 2012. http://edition.cnn.com/2012/07/22/world/europe/uk-caging-terror-mansha/index.html.

78. **O'Neill, Ann.** Nun: Tsarnaev 'genuinely sorry' about Boston bombing victims. [Online] CNN, May 11, 2015. http://www.cnn.com/2015/05/11/us/boston-bombing-tsarnaev-sentencing/.

79. **Seelye, Katharine Q.** Dzhokhar Tsarnaev Weeps as Relatives Try to Spare Him. [Online] The New York Times, May 4, 2015. http://www.nytimes.com/2015/05/05/us/russian-relatives-weep-as-jury-is-urged-to-spare-dzhokhar-tsarnaev-in-the-boston-marathon-bombing-trial.html.

80. **BBC.** In pictures: Atrocities' aftermath. [Online] BBC News, September 12, 2001. http://news.bbc.co.uk/2/hi/americas/1538664.stm.

81. **Ahram Online.** I'm proud my son Khaled killed Anwar Sadat: Mother. *Ahram Online.* [Online] February 19, 2012. http://english.ahram.org.eg/NewsContent/1/64/34912/Egypt/Politics-/Im-proud-my-son-Khaled-killed-Anwar-Sadat-Mother.aspx.

82. **LoBianco, Tom.** Donald Trump on terrorists: 'Take out their families'. *CNN.* [Online] December 3, 2015. http://www.cnn.com/2015/12/02/politics/donald-trump-terrorists-families/index.html.

83. **Leafe, David.** The Merciless Ming: The Chinese dynasty was unspeakably cruel and one of the most debauched in history - yet they produced the sublime art now on show at the British Museum. *Daily Mail.* [Online] September 17, 2014. http://www.dailymail.co.uk/news/article-2760019/The-Merciless-Ming-The-Chinese-dynasty-unspeakably-cruel-one-debauched-history-produced-sublime-art-British-Museum.html.

84. **Browne, Ryan and Gaouette, Nicole.** Donald Trump reverses position on torture, killing terrorists' families. *CNN.* [Online] March 4, 2016. http://www.cnn.com/2016/03/04/politics/donald-trump-reverses-on-torture/index.html.

85. **Haberman, Maggie.** Donald Trump Reverses Position on Torture and Killing Terrorists' Families. *The New York Times.* [Online] March 4, 2016. http://www.nytimes.com/politics/first-draft/2016/03/04/donald-trump-reverses-position-on-torture-and-killing-terrorists-families/?_r=0.

86. **Josh Levs, AnneClaire Stapleton and Steve Almasy.** Baltimore mom who smacked son at riot: I don't play. [Online] CNN, April 29, 2015. http://www.cnn.com/2015/04/28/us/baltimore-riot-mom-smacks-son/index.html.

87. **Hoffman, Bruce.** The Prevention of Terrorism and Rehabilitation of Terrorists: Some Preliminary Thoughts. *RAND Corporation.* [Online] 1985. http://www.rand.org/pubs/papers/P7059.html.

88. **Mora, Nicolas Medina and Hayes, Mike.** The Big (Imaginary) Black Friday Bombing. *BuzzFeed News.* [Online] November 15, 2015. http://www.buzzfeed.com/nicolasmedinamora/did-the-fbi-transform-this-teenager-into-a-terrorist#.oe8qL8Kow.

89. **Berlinger, Joshua.** The names: Who has been recruited to ISIS from the West. *CNN.* [Online] February 26, 2015. http://www.cnn.com/2015/02/25/world/isis-western-recruits/.

90. **Drug Policy Alliance.** Drug War Statistics. [Online] Drug Policy Alliance. [Cited: December 31, 2014.] http://www.drugpolicy.org/drug-war-statistics.

91. **Disney.** Disney Legend Time Allen. *D23.* [Online] [Cited: December 12, 2015.] https://d23.com/walt-disney-legend/tim-allen/.

92. **Biography.com.** Tim Allen Biography. [Online] [Cited: May 20, 2015.] http://www.biography.com/people/tim-allen-9542074.

93. **Yu, Katrina.** Singapore to amend death penalty for drug-trafficking. *Special Broadcasting Service.* [Online] February 24, 2015. http://www.sbs.com.au/news/article/2012/11/12/singapore-amend-death-penalty-drug-trafficking.

94. **Macfarlan, Tim.** Colombian model who has her own TV show faces death penalty in China for carrying drugs 'inside her laptop'. *Daily Mail.* [Online] July 29, 2015. http://www.dailymail.co.uk/news/article-3178690/Colombian-model-Juliana-Lopez-faces-death-penalty-China-carrying-drugs-inside-laptop.html.

95. **Gomstyn, Alice.** Funnyman Tim Allen: Serious About God. *ABC News.* [Online] November 19, 2011. http://abcnews.go.com/blogs/entertainment/2011/11/funnyman-tim-allen-serious-about-god/.

96. **Jones, Julia and Bower, Eve.** American deaths in terrorism vs. gun violence in one graph. *CNN.* [Online] October 2, 2015. http://www.cnn.com/2015/10/02/us/oregon-shooting-terrorism-gun-violence/.

97. **National Institutes of Health.** Overdose Death Rates. *National Institute on Drug Abuse.* [Online] December 2015. http://www.drugabuse.gov/related-topics/trends-statistics/overdose-death-rates.

98. **Jorgensen, Sarah.** Lethal strain of heroin strikes western Massachusetts. *CNN.* [Online] January 4, 2016. http://www.cnn.com/2016/01/04/health/heroin-lethal-strain-massachusetts/index.html.

99. Cocaine Snorting Oscar Statue Removed From Hollywood Boulevard. *Variety.* [Online] February 20, 2015. http://variety.com/2015/film/news/cocaine-snorting-oscar-statue-removed-from-hollywood-boulevard-1201438113/.

100. **Vries, Hilary De.** Robert Downey Jr.: The Album. *The New York Times.* [Online] November 21, 2004. http://www.nytimes.com/2004/11/21/arts/music/robert-downey-jr-the-album.html.

101. **Reuters.** Prostitute Pleads Guilty To Google Executive's Heroin Death. *The Huffington Post.* [Online] May 19, 2015. http://www.huffingtonpost.com/2015/05/19/alix-tichelman-guilty-google-heorin_n_7337496.html.

102. **Fink, Erica.** When Silicon Valley takes LSD. [Online] CNN, January 25, 2015. http://money.cnn.com/2015/01/25/technology/lsd-psychedelics-silicon-valley/index.html.

103. **Fink, Erica and Segall, Laurie.** I did LSD with Steve Jobs. *CNNMoney.* [Online] December 18, 2015. http://money.cnn.com/2015/01/25/technology/kottke-lsd-steve-jobs/.

104. **Wallace, Kelly.** The new 'Just say no to drugs'. *CNN.* [Online] January 15, 2016. http://www.cnn.com/2016/01/15/health/addiction-schools-education-prevention/index.html.

105. **Ellis, Ralph and Sidner, Sara.** Prince died of accidental overdose of opioid fentanyl, medical examiner says. *CNN.* [Online] June 3, 2016. http://www.cnn.com/2016/06/02/health/prince-death-opioid-overdose/index.html.

106. **CDC.** Injury Prevention & Control: Opioid Overdose. *Centers for Disease Control and Prevention.* [Online] March 16, 2016. http://www.cdc.gov/drugoverdose/.

107. **Shoichet, Catherine E.** A Mexican actress's tweet kick-started talks with 'El Chapo'. *CNN.* [Online] January 10, 2016. http://www.cnn.com/2016/01/10/world/mexico-actress-kate-del-castillo-el-chapo/index.html.

108. **Penn, Sean.** El Chapo Speaks. *Rolling Stone.* [Online] January 9, 2016. http://www.rollingstone.com/culture/features/el-chapo-speaks-20160109.

109. **Lee, Newton.** Counterterrorism and Cybersecurity: Total Information Awareness (2nd edition). [Online] Springer International, April 8, 2015. http://www.amazon.com/Counterterrorism-Cybersecurity-Total-Information-Awareness/dp/3319172433/.

110. **Hart, Carl.** Let's quit abusing drug users. *TEDMED 2015.* [Online] May 19, 2015. https://www.youtube.com/watch?v=C9HMifCoSko.

111. **Robertson, Nic and Cruickshank, Paul.** Cagefighter 'cures' terrorists. [Online] CNN, July 23, 2012. http://www.cnn.com/2012/07/20/world/europe/uk-caging-terror-main/index.html.

112. **Miller, Sara.** Obama: Peace is the only path to true security. [Online] The Jerusalem Post, March 21, 2013. http://www.jpost.com/Diplomacy-and-Politics/Obama-Peace-is-the-only-path-to-true-security-307323.

113. **Lutz, Diana.** Odor receptors discovered in lungs. *Washington University in St. Louis.* [Online] January 2, 2014. http://news.wustl.edu/news/Pages/26271.aspx.

114. **Stager, Curt.** You Are Made of Waste: Searching for the ultimate example of recycling? Look in the mirror. *Nautilus.* [Online] November 7, 2013. http://nautil.us/issue/7/waste/you-are-made-of-waste.

115. **Shakespeare, William.** The Tragedy of Hamlet, Prince of Denmark. *MIT.* [Online] 1603. http://shakespeare.mit.edu/hamlet/hamlet.2.2.html.

116. **Obama, Barack.** Remarks by the President at the Acceptance of the Nobel Peace Prize. *The White House.* [Online] December 10, 2009. https://www.whitehouse.gov/the-press-office/remarks-president-acceptance-nobel-peace-prize.

117. **Steuer, Eric.** The Rise and Fall of RedBook, the Site That Sex Workers Couldn't Live Without. *Wired.* [Online] February 24, 2015. http://www.wired.com/2015/02/redbook/.

118. **Lopatto, Elizabeth.** Unmarried Couples Living Together Is New U.S. Norm. *Bloomberg.* [Online] April 3, 2013. http://www.bloomberg.com/news/articles/2013-04-04/unmarried-couples-living-together-is-new-u-s-norm.

119. **Connell, HM.** Effect of family break-up and parent divorce on children. *Australian Paediatric Journal.* [Online] August 24, 1988. http://www.ncbi.nlm.nih.gov/pubmed/3064746.

120. **Lamont, Tom.** Life after the Ashley Madison affair. *The Guardian.* [Online] February 27, 2016. http://www.theguardian.com/technology/2016/feb/28/what-happened-after-ashley-madison-was-hacked.

121. **Segall, Laurie.** Pastor outed on Ashley Madison commits suicide. *CNNMoney.* [Online] September 8, 2015. http://money.cnn.com/2015/09/08/technology/ashley-madison-suicide/index.html.

122. **World Food Programme.** Hunger Statistics. *World Food Programme.* [Online] https://www.wfp.org/hunger/stats.
123. **World Health Organization.** Key facts from JMP 2015 report. *Water Sanitation Health.* [Online] 2015. http://www.who.int/water_sanitation_health/monitoring/jmp-2015-key-facts/en/.
124. **Jasarevic, Tarik, Thomas, Glenn and Osseiran, Nada.** 7 million premature deaths annually linked to air pollution. *World Health Organization.* [Online] March 25, 2014. http://www.who.int/mediacentre/news/releases/2014/air-pollution/en/.
125. **Havoscope.** Prostitution: Prices and Statistics of the Global Sex Trade. *Havoscope.* [Online] June 18, 2015. http://www.havocscope.com/prostitution-statistics/.
126. **Segall, Laurie and Fink, Erica.** Sex Valley: Tech's booming prostitution trade. *CNNMoney.* [Online] July 14, 2014. http://money.cnn.com/2014/07/11/technology/silicon-valley-prostitution/index.html.
127. **Korman, Bart.** House Bill No. 2059. *Missouri House of Representatives.* [Online] January 6, 2016. http://house.mo.gov/billtracking/bills161/billpdf/intro/HB2059I.PDF.
128. **Wiki Community.** Companion's Guild. *The Firefly and Serenity Database.* [Online] [Cited: February 16, 2016.] http://firefly.wikia.com/wiki/Companion.
129. **Reynolds, Simon and Tanswell, Adam.** Chloë Grace Moretz: 'I'm done with playing Hit-Girl'. *Digital Spy.* [Online] August 25, 2014. http://www.digitalspy.com/movies/interviews/a591530/chlo-grace-moretz-im-done-with-playing-hit-girl/.
130. **Smith, Michael.** Breaking the Enigma code was the easiest part of the Nazi puzzle. *The Telegraph.* [Online] November 15, 2014. http://www.telegraph.co.uk/history/world-war-two/11231608/Breaking-the-Enigma-code-was-the-easiest-part-of-the-Nazi-puzzle.html.
131. **IMDb.** The Intern. *IMDb.* [Online] September 25, 2015. http://www.imdb.com/title/tt2361509/.
132. **Fairchild, Caroline.** The rise of stay-at-home dads? For female execs, not so much. *Fortune Magazine.* [Online] November 20, 2014. http://fortune.com/2014/11/20/stay-at-home-dads-female-execs/.
133. **AMC Public Affairs.** Gen. Ann E. Dunwoody, U.S. Army Materiel Command commanding general. *The United States Army.* [Online] June 30, 2008. http://www.army.mil/article/10506/.
134. **Washington Post Live.** Regina Dugan. *The Washington Post.* [Online] August 21, 2012. https://www.washingtonpost.com/postlive/regina-dugan/2012/07/02/gJQAapMADY_story.html.
135. **Kelly, Suzanne and Benson, Pam.** 'Eye in the Sky': the case file on NGA Director Letitia Long. *CNN.* [Online] March 9, 2012. http://security.blogs.cnn.com/2012/03/09/eye-in-the-sky-the-case-file-on-nga-director-letitia-long/.
136. **Office of Science and Technology Policy.** Megan Smith. *The White House.* [Online] September 2014. https://www.whitehouse.gov/administration/eop/ostp/about/leadershipstaff/smith.
137. **Press Operations.** Statement by Secretary of Defense Ash Carter on Gen. Lori J. Robinson. *U.S. Department of Defense.* [Online] March 18, 2016. http://www.defense.gov/News/News-Releases/News-Release-View/Article/698279/statement-by-secretary-of-defense-ash-carter-on-gen-lori-j-robinson.
138. **Sankar, Pooja.** The pervasive bias against female computer science majors. *Fortune Magazine.* [Online] April 20, 2015. http://fortune.com/2015/04/20/the-pervasive-bias-against-female-computer-science-majors/.
139. **U.S. Department of the Treasury.** The New $20 Note. *Modern Money.* [Online] https://modernmoney.treasury.gov/new-notes.
140. **Knox, Belle.** I'm The Duke University Freshman Porn Star And For The First Time I'm Telling The Story In My Words. *xoJane.* [Online] February 21, 2014. http://www.xojane.com/sex/duke-university-freshman-porn-star.

141. **Bindel, Julie.** Marriage is a form of prostitution. *The Guardian.* [Online] November 11, 2008. http://www.theguardian.com/lifeandstyle/2008/nov/12/women-prostitution-marriage-sex-trade.

142. **Akst, Jef.** Student Fights Harassment with Wikipedia. *The Scientist.* [Online] March 10, 2016. http://www.the-scientist.com/?articles.view/articleNo/45541/title/Student-Fights-Harassment-with-Wikipedia/.

143. **Mosbergen, Dominique.** For Every Sexist Email She Gets, This College Student Will Write A Wikipedia Entry About A Woman Scientist. *HuffPost Women.* [Online] March 10, 2016. http://www.huffingtonpost.com/entry/emily-temple-wood-wikipedia_us_56e0f90ce4b065e2e3d4dc33.

144. **Paling, Emma.** Wikipedia's Hostility to Women. *The Atlantic.* [Online] October 21, 2015. http://www.theatlantic.com/technology/archive/2015/10/how-wikipedia-is-hostile-to-women/411619/.

145. **Snow, Aurora.** 'Hot Girls Wanted' Is Pornsploitation: The Porn Industry Fights Back. *The Daily Beast.* [Online] June 12, 2015. http://www.thedailybeast.com/articles/2015/06/13/hot-girls-wanted-is-pornsploitation.html.

146. **Kilday, Gregg.** Hollywood Sex Abuse Film Revealed: Explosive Claims, New Figures Named (Exclusive). *The Hollywood Reporter.* [Online] November 12, 2014. http://www.hollywoodreporter.com/news/hollywood-sex-abuse-film-revealed-748375.

147. **IMDb.** The Matrix. *IMDb.* [Online] March 31, 1999. http://www.imdb.com/title/tt0133093/.

148. **Morris, Chris.** Things Are Looking Up in America's Porn Industry. *NBC News.* [Online] January 20, 2015. http://www.nbcnews.com/business/business-news/things-are-looking-americas-porn-industry-n289431.

149. **Kinnaman, David.** The Porn Phenomenon. *The Barna Group.* [Online] January 19, 2016. https://barna.org/blog/culture-media/david-kinnaman/the-porn-phenomenon.

150. **Capon, Felicity.** Show Pornography in Schools, Urges Danish Sexology Professor. *Newsweek.* [Online] March 3, 2015. http://europe.newsweek.com/show-pornography-schools-urges-danish-sexology-professor-311090.

151. Anne Frank's diary isn't pornographic – it just reveals an uncomfortable truth. *The Guardian.* [Online] May 2, 2013. http://www.theguardian.com/commentisfree/2013/may/02/anne-franks-diary-pornographic-uncomfortable-truth.

152. **Trost, Matthew.** Cindy Gallop: Make love, not porn. *TEDBlog.* [Online] December 2, 2009. http://blog.ted.com/cindy_gallop_ma/.

153. **O'Sullivan, Erin.** Emma Watson Defines 'Feminism,' Sheds Light On Gender Inequality In Hollywood. *Access Hollywood.* [Online] March 8, 2015. http://www.accesshollywood.com/articles/emma-watson-defines-feminism-sheds-light-on-gender-inequality-in-hollywood-158350/.

154. **United Nations.** UN Women Goodwill Ambassador Emma Watson. *UN Women.* [Online] July 2014. http://www.unwomen.org/en/partnerships/goodwill-ambassadors/emma-watson.

155. **Blair, Olivia.** Emma Watson calls for feminist alternatives to pornography. *The Independent.* [Online] February 26, 2016. http://www.independent.co.uk/news/people/emma-watson-calls-for-feminist-alternatives-to-pornography-during-discussion-with-gloria-steinem-a6895146.html.

156. **Benes, Ross.** PORN: The Hidden Engine That Drives Innovation In Tech. *Business Insider.* [Online] July 5, 2013. http://www.businessinsider.com/how-porn-drives-innovation-in-tech-2013-7.

157. **Kludt, Tom.** Playboy enters non-nude era: Sexy but 'safe for work'. *CNNMoney.* [Online] February 4, 2016. http://money.cnn.com/2016/02/04/media/playboy-first-non-nude-issue/index.html.

158. **Wong, Raymond.** VR porn is here and it's scary how realistic it is. *Mashable.* [Online] January 8, 2016. http://mashable.com/2016/01/08/naughty-america-vr-porn-experience/.

159. **Pope Francis.** The Joy of Love: AMORIS LAETITIA. [Online] DigitalBe, April 8, 2016. https://books.google.com/books?id=QO39CwAAQBAJ.

160. **Wooden, Cindy.** 'Who am I to judge?' Pope's remarks do not change church teaching. [Online] Catholic News Service, July 31, 2013. http://www.catholicnews.com/data/stories/cns/1303303.htm.
161. **Alexander, Anne.** Egypt's Muslims and Christians join hands in protest. *BBC News.* [Online] February 10, 2011. http://www.bbc.com/news/world-middle-east-12407793.
162. **United States.** The Constitution of the United States: A Transcription. *U.S. National Archives.* [Online] September 17, 1787. http://www.archives.gov/exhibits/charters/constitution_transcript.html.
163. **First Congress of the United States.** Bill of Rights. *U.S. National Archives.* [Online] December 15, 1791. http://www.archives.gov/exhibits/charters/bill_of_rights_transcript.html.
164. **Slobodzian, Joseph A.** Church-state precedent has roots in Phila. *Philly.com.* [Online] June 29, 2003. http://articles.philly.com/2003-06-29/news/25448800_1_lemon-test-religious-schools-private-schools.
165. **Liptak, Adam.** Alton T. Lemon, Who Challenged State Aid to Religious Schools, Dies at 84. *The New York Times.* [Online] May 25, 2013. [Cited:].
166. **Pew Research Center.** Faith and the 2016 Campaign. *Pew Research Center.* [Online] January 27, 2016. http://www.pewforum.org/2016/01/27/faith-and-the-2016-campaign/.
167. **Talpos, Sara.** What The Amish Can Teach Us About Modern Medicine. *digg.com.* [Online] May 24, 2016. http://digg.com/2016/amish-medicine-mosaic.
168. **Mellen, Ruby.** Thousands of atheists gather in DC for Reason Rally. *CNN.* [Online] June 4, 2016. http://www.cnn.com/2016/06/02/politics/atheist-reason-rally/index.html.
169. **Knoll, Benjamin.** Did Anti-Mormonism Cost Mitt Romney the 2012 Election? *Huffpost Politics.* [Online] December 18, 2013. http://www.huffingtonpost.com/benjamin-knoll/mitt-romney-mormon_b_4121217.html.
170. **Yardley, Jim.** Pope Francis Suggests Donald Trump Is 'Not Christian'. *The New York Times.* [Online] February 18, 2016. http://www.nytimes.com/2016/02/19/world/americas/pope-francis-donald-trump-christian.html.
171. **Burke, Daniel.** The Book of Bernie: Inside Sanders' unorthodox faith. *CNN.* [Online] April 15, 2016. http://www.cnn.com/2016/04/14/politics/bernie-sanders-religion/.
172. **Currie, Carol McAlice.** Midterm voter turnout drops to 72-year low. *USA Today.* [Online] March 20, 2015. http://www.usatoday.com/story/news/politics/2015/03/20/poor-voter-turnout/25082721/.
173. **Pew Research Center.** U.S. Public Becoming Less Religious. *Pew Research Center.* [Online] November 3, 2015. http://www.pewforum.org/2015/11/03/u-s-public-becoming-less-religious/.
174. **Gray, Sarah.** Bill Maher on Charlie Hebdo attacks: "There are no great religions; they're all stupid and dangerous". *Salon.* [Online] January 8, 2015. http://www.salon.com/2015/01/08/bill_maher_on_charlie_hebdo_attacks_there_are_no_great_religions_they%E2%80%99re_all_stupid_and_dangerous/.
175. **McKenna, Josephine.** Pope Francis on free speech: 'You cannot insult the faith of others'. [Online] Religion News Service, January 15, 2015. http://www.religionnews.com/2015/01/15/pope-francis-free-speech-cannot-insult-faith-others/.
176. **Watts, Franklin.** Speech in Munich 24 February 1941. *Voices of History: Great Speeches and Papers of the Year.* [Online] Franklin Watts, Inc., 1942.
177. **More, Max.** On Becoming Posthuman. [Online] 1994. http://eserver.org/courses/spring98/76101R/readings/becoming.html.
178. **Wikipedia.** Quantum entanglement. *Wikipedia.* [Online] February 19, 2016. https://en.wikipedia.org/wiki/Quantum_entanglement.
179. **Markoff, John.** Sorry, Einstein. Quantum Study Suggests 'Spooky Action' Is Real. *The New York Times.* [Online] October 21, 2015. http://www.nytimes.com/2015/10/22/science/quantum-theory-experiment-said-to-prove-spooky-interactions.html?_r=0.
180. **Einstein, Albert.** Out of My Later Years. *Google Books.* [Online] 1956. https://books.google.com/books/about/Out_of_My_Later_Years.html?id=OBPAA3ZI4zcC.

181. **Wookieepedia.** Force ghost. *The Star Wars Wiki.* [Online] January 4, 2006. http://starwars. wikia.com/wiki/Force_ghost.
182. **Dyson, Michael Eric.** Dyson Explores How MLK's Death Changed America. *National Public Radio (NPR).* [Online] July 17, 2011. http://www.npr.org/templates/story/story.php? storyId=89344679.
183. **Lee, Newton.** Digital Da Vinci: Computers in Music. *Amazon.com.* [Online] April 12, 2014. http://www.amazon.com/Digital-Da-Vinci-Computers-Music/dp/149390535X/.
184. **Rosenberg, Adam.** EXCLUSIVE: Paul Verhoeven Calls RoboCop 'The American Jesus,' Is Unexcited By Remake Plans. *MTV News.* [Online] April 14, 2010. http://www.mtv.com/ news/2436200/paul-verhoeven-robocop-christ-story-remake-update/.
185. **60 Minutes Overtime.** David Bowie's unaired 60 Minutes interviews. *CBS News.* [Online] January 11, 2016. http://www.cbsnews.com/news/david-bowies-unaired-60-minutes-interviews/.
186. **R.E.M.** R.E.M. - Losing My Religion (Official Music Video). *Warner Bros. Records.* [Online] 1991. https://www.youtube.com/watch?v=if-UzXIQ5vw.
187. **Bowen, Matt.** Google Prayers. *Googlism.* [Online] [Cited: December 13, 2015.] http://www. thechurchofgoogle.org/Scripture/google_prayers.html.
188. **Berthélemy, Jean-Simon.** Alexander Cutting the Gordian Knot. *The Athenaeum.* [Online] 1767. http://www.the-athenaeum.org/art/display_image.php?id=383494.

Chapter 3
Google Versus Death; To Be, Or Not to Be?

Newton Lee

> *Are people really focused on the right things? One of the things I thought was amazing is that if you solve cancer ... it's not as big an advance as you might think.*
>
> —Larry Page

Prologue "Can Google Solve Death?" graced the cover of the *TIME Magazine* on September 30, 2013. "We're going to gradually enhance ourselves," said Ray Kurzweil, futurist and engineering director at Google. "That's the nature of being human—we transcend our limitations" [1]. Pacemakers, prosthesis, stentrode, optogenetics, antibiotics, and other medical advancements exemplify the use of technology to prolong life and to improve quality of life. Even Pope Francis gave his blessing to human-animal chimera research for organ transplants [2]. We are all transhumanists in varying degrees.

3.1 Existential Threats to Humankind

Dinosaurs, Neanderthals, and Denisovans did not evolve and adapt fast enough, and therefore they became extinct. Modus operandi or business as usual will doom the human race to mass extinction. Astrophysicist Stephen Hawking said in a CNN interview in 2008, "It will be difficult enough to avoid disaster on planet Earth in the next 100 years, let alone next thousand, or million. ... I see great dangers for the human race. There have been a number of times in the past when its survival has been a question of touch and go. The Cuban missile crisis in 1963 was one of these.

N. Lee (✉)
Newton Lee Laboratories LLC, Institute for Education Research and Scholarships, Woodbury University School of Media Culture and Design, Burbank, CA, USA

© Springer Science+Business Media New York 2016
N. Lee (ed.), *Google It*, DOI 10.1007/978-1-4939-6415-4_3

The frequency of such occasions is likely to increase in the future. We shall need great care and judgment to negotiate them all successfully" [3].

In the June 2015 issue of *Science Advances* journal published by the American Association for the Advancement of Science, researchers at National Autonomous University of Mexico, Stanford University, University of California Berkeley, Princeton University, and University of Florida raised the specter of Earth's biota entering a sixth "mass extinction." The last Big Five were Ordovician-Silurian, Late Devonian, Permian, Triassic-Jurassic, Cretaceous-Tertiary (or K-T) extinctions. The Permian period ended with 96 % of all species perished; and the K-T wiped out at least half of all species on Earth including the dinosaurs [4].

Prof. Gerardo Ceballos and his coauthors wrote, "We can confidently conclude that modern extinction rates are exceptionally high, that they are increasing, and that they suggest a mass extinction under way—the sixth of its kind in Earth's 4.5 billion years of history. ... If the currently elevated extinction pace is allowed to continue, humans will soon (in as little as three human lifetimes) be deprived of many biodiversity benefits. ... The loss of biodiversity is one of the most critical current environmental problems, threatening valuable ecosystem services and human well-being" [5].

In January 2016, *The Bulletin of the Atomic Scientists* Science and Security Board announced that the Doomsday Clock remained at 11:57. "Three minutes (to midnight) is too close. Far too close," stated the Bulletin. "We, the members of the Science and Security Board of the Bulletin of the Atomic Scientists, want to be clear about our decision not to move the hands of the Doomsday Clock in 2016: That decision is not good news, but an expression of dismay that world leaders continue to fail to focus their efforts and the world's attention on reducing the extreme danger posed by nuclear weapons and climate change. When we call these dangers existential, that is exactly what we mean: They threaten the very existence of civilization and therefore should be the first order of business for leaders who care about their constituents and their countries" [6].

Figure 3.1 shows the changes in the time of the Doomsday Clock of the Bulletin of the Atomic Scientists from 1947 to 2015 while 2016 remains the same as the previous year [7]. The closest approach to midnight was in 1953 when the United

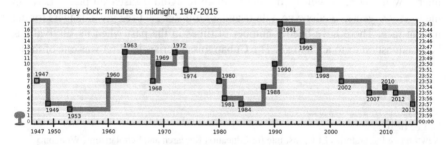

Fig. 3.1 Graph showing the changes in the time of the Doomsday Clock of the Bulletin of the Atomic Scientists from 1947 to 2015. (Courtesy of Wikimedia Commons)

States and the Soviet Union tested thermonuclear devices within 9 months of one another. The furthest from midnight was in 1991 when the United States and Soviet Union signed the Strategic Arms Reduction Treaty (START), followed by the Soviet Union dissolution on December 26.

The world may one day come to its senses and adopt long-lasting peace; but are we well-equipped to deal with existential threats posed by natural disasters and biohazards?

1. Whether climate change is due to carbon dioxide emission [8], shift in Earth's magnetic poles [9], or some other reasons, global warming poses existential threat that must be addressed. Dire consequences of climate change include more severe droughts, heat waves, hurricanes, and raising sea levels [10]. Meteorologists have noticed the record early start of the Greenland ice melt in April 2016. Greenland has been losing ice at a pace of 287 billion metric tons per year, and sea levels around the world could rise by 20 feet if the ice sheet in the size of Alaska were to melt completely [11]. Due to the treats of tsunamis and sea level rise, the Quinault Indian Nation (QIN) located on the Pacific coast of Washington's Olympic Peninsula has been planning to relocate the community to higher ground [12].

2. The Cascadia Fault at the bottom of the Pacific Ocean can create an earthquake almost 30 times more energetic than the San Andreas Fault [13]. In January 1700, the Cascadia caused the largest earthquake in North America, setting off a tsunami that not only struck the Pacific coast and also damaged Japan's coastal villages across the Pacific Ocean [14].

3. While Admiral Michael Rogers, Director of the National Security Agency (NSA), has justifiably sounded the alarm on cyber attacks of critical infrastructure [15, 16], the U.S. government is not paying enough attention to crumbling bridges, decaying pipelines, and inadequate storage facilities for natural gas that are all highly susceptible to serious damage by earthquakes. In August 2011, a 5.8 magnitude earthquake damaged the Washington Monument [17]. Today, nearly 60,000 bridges across the U.S. are in desperate need of repair. "It's just eroding and concrete is falling off," said National Park Service spokeswoman Jenny Anzelmo-Sarles, referring to the Arlington Memorial Bridge crossed by 68,000 vehicles every day [18]. Gas Pipe Safety Foundation cofounder Kimberly Archie called the aging natural gas infrastructure in American cities a "ticking time bomb" [19]. The 2015 gas leak in Porter Ranch, California, for instance, released an estimated 80,000 metric tons of mostly methane into the atmosphere, affecting the health of over 30,000 residents. It took 4 months to permanently seal the leak [20].

4. *The Texas Tribune* and *ProPublica* reported in March 2016 that "Houston … is home to the nation's largest refining and petrochemical complex, where billions of gallons of oil and dangerous chemicals are stored. And it's a sitting duck for the next big hurricane. Why isn't Texas ready?" [21]. Rice University professor

Phil Bedient summed up the inaction of the local government: "We've done nothing to shore up the coastline, to add resiliency … to do anything."

5. In May 2016, Centers for Disease Control and Prevention (CDC) released an alarming statistics that each year in the United States, at least 2 million people become infected with bacteria that are resistant to antibiotics and at least 23,000 people die each year as a direct result of these infections [22]. Superbugs— bacteria that are resistant to all antibiotics including the last-resort nephrotoxic drug Colistin—have infected humans and animals in Asia, Europe, the United States, and more than 20 countries worldwide. "This shows that we are right on the verge of getting into the territory of routine bacterial infections being untreatable," said Steven Roach, food safety program director at the Food Animal Concerns Trust. "It underscores the failure of both the federal government and Congress, and the industry, to get a grasp of the problem. We can't continue to drag our feet on taking needed action" [23]. Also in May 2016, *Review on Antimicrobial Resistance* issued a final report which projects that by 2050, more than 10 million people will die from superbugs each year, outpacing cancer (8.2 million), diabetes (1.5 million), diarrheal disease (1.4 million), and other illnesses [24].

Although natural disasters and biohazards may be unavoidable, the devastating domino effects can be alleviated if we are well informed and better prepared. For example: U.S. Geological Survey (USGS) along with university partners have been developing and testing an earthquake early warning system (EEWS) called ShakeAlert [25]. The National Aeronautics and Space Administration (NASA) employs an automated collision monitoring system known as Sentry and publishes a list of potential future Earth impact with Near-Earth Asteroids (NEAs) at http://neo.jpl.nasa.gov/risk/. The National Oceanic and Atmospheric Administration (NOAA)

Fig. 3.2 Climate Explorer offers interactive visualizations for exploring maps and data to identify potential climate impacts. This diagram shows the impacts a rising sea level may have on coastal regions of the United States

has created an interactive Climate Explorer tool to raise awareness by allowing users to visualize historical data and impacts of climate changes (see Fig. 3.2) [26]. Local community volunteers such as Food Forward in Southern California convene at private properties, public spaces, and farmers and wholesale markets to recover excess fruits and vegetables that would otherwise go to waste, donating them to direct-service agencies that feed over 100,000 people in need each month [27].

Leading by example, Google is the world's largest corporate buyer of renewable energy with a commitment to purchase nearly 2 GW of green energy [28]. Launched in June 2015, Alphabet's Sidewalk Labs with CEO Dan Doctoroff focuses on urban design by pursuing technologies to "cut pollution, curb energy use, streamline transportation, and reduce the cost of city living" [29]. Reducing the cost of city living is music to the ears of angry protesters who in December 2013 blockaded an Apple employee shuttle bus in San Francisco and threw rocks at a Google employee shuttle bus in Oakland to call attention to low-income residents displaced by rising rents [30].

In Hollywood, the 2006 documentary *Who Killed the Electric Car?* and the 2011 feature film *Revenge of the Electric Car* by Chris Paine educated the public about a better alternative to gasoline powered vehicles. Following the zero-emissions vehicle (ZEV) mandate, eight U.S. states (California, Connecticut, Maryland, Massachusetts, New York, Oregon, Rhode Island, and Vermont) and five countries (Germany, the United Kingdom, the Netherlands, Norway, and Quebec of Canada) have proposed to ban sales of gas and diesel powered cars by 2050 [31].

Since March 2015, astronaut Scott Kelly had often posted photos of the planet Earth on his Twitter account during his one-year mission at the International Space Station (ISS). In January 2016, CNN's chief medical correspondent Sanjay Gupta asked Kelly to define the Earth's condition as if it were a human body. Kelly replied, "There are definitely parts of Asia, Central America that when you look at them from space, you're always looking through a haze of pollution. As far as the atmosphere is concerned, and being able to see the surface, you know, I would say definitely those areas that I mentioned look kind of sick. ... [The atmosphere] definitely looks very, very fragile and just kind of like this thin film, so it looks like something that we definitely need to take care of" [32]. The hard question is how to take good care of the planet Earth and its inhabitants.

In *The Day the Earth Stood Still* (2008), astrobiology professor Helen Benson (Jennifer Connelly) demanded to know the intention of the alien named Klaatu (Keanu Reeves) [33]:

Helen Benson: I need to know what's happening.
Klaatu: This planet is dying. The human race is killing it.
Helen Benson: So you've come here to help us.
Klaatu: No, I didn't.
Helen Benson: You said you came to save us.
Klaatu: I said I came to save the Earth.
Helen Benson: You came to save the Earth ... from us? You came to save the Earth from us!

Fig. 3.3 The *TIME Magazine* cover on September 30, 2013 reads "Can Google Solve Death?"

Klaatu: We can't risk the survival of this planet for the sake of one species.
Helen Benson: What are you saying?
Klaatu: If the Earth dies, you die. If you die, the Earth survives.

3.2 Can Google Solve Death?

The *TIME Magazine* cover on September 30, 2013 reads "Can Google Solve Death? The search giant is launching a venture to extend the human life span. That would be crazy—if it weren't Google" [34] (see Fig. 3.3). In November 2014, *Fortune Magazine* named Google cofounder Larry Page "the most ambitious CEO in the universe" [35]. "The breadth of things that he is taking on is staggering," said Ben Horowitz, of Andreessen Horowitz. "We have not seen that kind of business leader since Thomas Edison at GE or David Packard at HP."

Google made its first foray into healthcare with Google Health between 2008 and 2011 to collect volunteered information about personal health conditions, medications, allergies, and lab results [36]. Larry Page has good reason for his deep interest in health sciences. In May 2013, Page announced on his Google+ profile

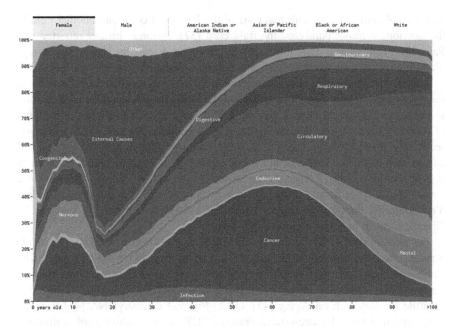

Fig. 3.4 Causes of death for female over a 100-year lifespan

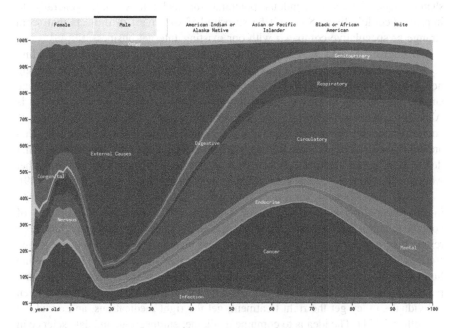

Fig. 3.5 Causes of death for male over a 100-year lifespan

that a 1999 cold left him with paralysis of the left vocal cord, and another cold in the previous summer paralyzed the right cord [37]. *Business Insider* reported that Page has Hashimoto's Thyroiditis, an autoimmune disease in which the thyroid gland is attacked by a variety of cell and antibody-mediated immune processes [38].

Undaunted, Page decided to go for the moonshot. "Are people really focused on the right things?" he asked. "One of the things I thought was amazing is that if you solve cancer, you'd add about 3 years to people's average life expectancy. We think of solving cancer as this huge thing that'll totally change the world. But when you really take a step back and look at it, yeah, there are many, many tragic cases of cancer, and it's very, very sad, but in the aggregate, it's not as big an advance as you might think" [39]. To see the forest for the trees, Figs. 3.4 and 3.5 show the causes of death for female and male over a 100-year lifespan. Statistician Nathan Yau compiled the stacked graphs using 2005–2014 data from the Centers for Disease Control and Prevention (CDC) [40].

On September 18, 2013, Google unveiled a new venture—Calico—to tackle human aging and associated diseases [41]. Led by its CEO Arthur D. Levinson, Calico partners with AbbVie [42], University of Texas Southwestern Medical Center, 2M Companies [43], Broad Institute of MIT and Harvard [44], Buck Institute for Research on Aging [45], QB3 [46], and AncestryDNA [47] to study aging and find cures for age-related diseases. "OK … so you're probably thinking wow! That's a lot different from what Google does today," Larry Page wrote about Calico on Google+. "And you're right. But as we explained in our first letter to shareholders, there's tremendous potential for technology more generally to improve people's lives. So don't be surprised if we invest in projects that seem strange or speculative compared with our existing Internet businesses" [48].

On August 10, 2015, Larry Page announced the creation of a new public holding company—Alphabet (http://abc.xyz)—to restructure Google by moving non-Internet subsidiaries from Google to Alphabet. In the official Google blog, Page singled out "health efforts" as one of the main reasons for restructuring: "What is Alphabet? Alphabet is mostly a collection of companies. The largest of which, of course, is Google. This newer Google is a bit slimmed down, with the companies that are pretty far afield of our main Internet products contained in Alphabet instead. What do we mean by far afield? Good examples are our health efforts: Life Sciences (that works on the glucose-sensing contact lens), and Calico (focused on longevity)" [49].

Google Life Sciences, formerly a division of Google X, was renamed to Verily in December 2015 as a subsidiary of Alphabet [50]. Since the development of contact lens for diabetics to monitor glucose in tears, the company has been researching cardiovascular disease, cancer, and mental health, just to name a few. Verily chief medical officer Dr. Jessica Mega described the need to create a baseline human health by mining "deep molecular data, clinical data, imaging data and patient engagement" that would help doctors to "understand more about a given individual so they get the right treatment, get the right medications, and avoid the side effects" [51]. The idea is to combine medicine, engineering, and data science in an effort to capture the early signs of a disease and to stop it in its tracks. An artificial intelligence program can do Google search at lightning speed, analyze a

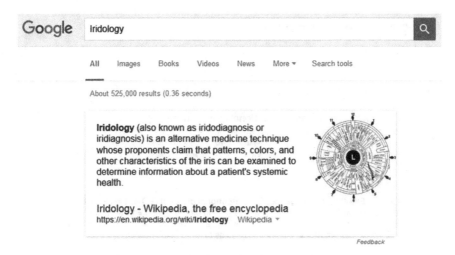

Fig. 3.6 Google search on "Iridology" returns a Knowledge Graph definition from Wikipedia about the alternative medicine technique claiming that patterns, colors, and other characteristics of the iris can be examined to determine information about a patient's systemic health

mountain of research data, discover new patterns, connect the dots, and eventually discover groundbreaking insights.

A comprehensive and well-understood baseline human health will enable scientists to devise not only better diagnostic lab tests but also innovative health monitoring applications. For example:

1. Whenever you take a selfie on your smartphone, a mobile app will analyze your eyes, skin condition, and other facial features for any sign of abnormality that may indicate certain illness. If one's iris can be affected by systemic health, iridology may see the light of day as being evidence-based rather than pseudoscientific (see Fig. 3.6).
2. When you talk on the phone, an app will analyze your voice and speech pattern to determine your stress level, emotional state, and general health condition. It will certainly know if you are catching a cold. (Computers are already capable of reading a person's body language to tell whether they are bored [52]) and smartphones are being used to detect anemia and other medical conditions).
3. A discreet plug-in to any dating app will estimate the emotional intelligence/ quotient (EQ) and love compatibility of your new date (see Fig. 3.7). Enough said.

Interestingly, Verily has a staff philosopher among its 350 scientists, as CEO Andy Conrad explained, "We have to understand the 'why' of what people do. A philosopher might be as important as a chemist" [50].

The real solution to human longevity will likely involve both traditional and alternative medicine, scientific and philosophical problem-solving, as well as big data analysis and human intuition. The mysterious universe cannot be explained from one angle alone; instead it requires multiple paradigms including Newtonian physics, Einstein's Theory of Relativity, and Quantum Mechanics. Taking into

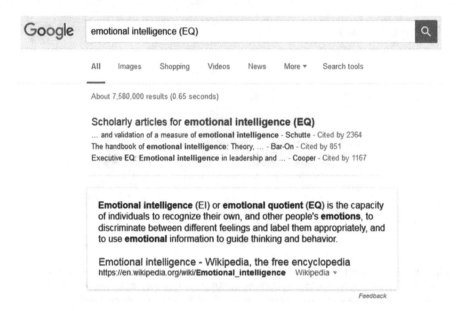

Fig. 3.7 Google search on "emotional intelligence (EQ)" returns a Knowledge Graph of a list of oft-cited scholarly articles in addition to a definition from Wikipedia about the capacity of individuals to recognize their own and other people's emotions

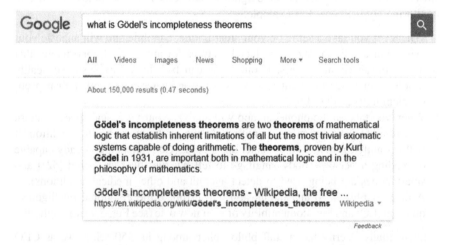

Fig. 3.8 Google search on "what is Gödel's incompleteness theorems" returns a Knowledge Graph result from Wikipedia about two theorems of mathematical logic that establish inherent limitations of all but the most trivial axiomatic systems capable of doing arithmetic

account Kurt Gödel's incompleteness theorems (see Fig. 3.8), a unified Theory of Everything (ToE) may never be attainable (see Fig. 3.9). Similarly, there may not be a magic pill but a holistic solution to human longevity.

Fig. 3.9 Google search on "what is Theory of Everything" returns a Knowledge Graph result from Wikipedia about a hypothetical single, all-encompassing, coherent theoretical framework of physics that fully explains and links together all physical aspects of the universe

3.3 Biblical Account of Human Lifespan

According to the Bible, Adam and Eve gave up eternal life to gain the knowledge of good and evil. "The LORD God said, 'The man has now become like one of us, knowing good and evil. He must not be allowed to reach out his hand and take also from the tree of life and eat, and live forever'" (Genesis 3:22). Altogether, Adam lived a total of 930 years (Genesis 5:5). Enoch and Prophet Elijah were the only two human beings who did not die but were instead taken away to heaven (Genesis 5:22–24 and 2 Kings 2:1–11). The average human lifespan at the time was close to 1,000 years. However, God decided to shorten the life expectancy to 120 years:

> When human beings began to increase in number on the earth and daughters were born to them, the sons of God saw that the daughters of humans were beautiful, and they married any of them they chose. Then the Lord said, "My Spirit will not contend with humans forever, for they are mortal; their days will be a hundred and twenty years." The Nephilim were on the earth in those days—and also afterward—when the sons of God went to the daughters of humans and had children by them. They were the heroes of old, men of renown. (Genesis 6:1–4)

The Nephilim—human-alien hybrid—could very well be the ancestors of people with superhuman or psychic abilities. In the Book of Judges, Samson the Nazirite had extraordinary physical strength. French apothecary Nostradamus became famous for his prophecy and horoscopes in the 16th century even though skeptics dismissed his work as retroactive clairvoyance or postdiction. At the end of World War II, Dutch magician Mirin Dajo baffled the medical community and astounded the audience by impaling his body with bladed weapons without bleeding or feeling pain. Some researchers call the deliberately caused bodily damage (DCBD) a metahypnotic phenomenon. In more recent news, there are anecdotal reports of teenage girls lifting up 3,000 lb of cars and trucks to save their family members [53–55].

Fig. 3.10 Life spans of 26 prominent figures from Adam to Neil Armstrong who died of natural causes. The estimated birth years for the biblical characters are based on the data provided in the Jewish Virtual Library

Name	Born	Age
Adam	c. 3760 BC	930
Methuselah	c. 3317 BC	969
Noah	c. 2704 BC	950
Abraham	c. 1813 BC	175
Isaac	c. 1713 BC	180
Jacob	c. 1653 BC	147
Moses	c. 1392 BC	120
King David	c. 1040 – 970 BC	70
King Solomon	c. 970 – 931 BC	c. 60
Confucius	551 BC	72
Plato	c. 428 – 423 BC	c. 80
Aristotle	384 BC	62
Augustine of Hippo	354	75
Marco Polo	1254	69
Christopher Columbus	1451	54
Leonardo da Vinci	1452	67
Isaac Newton	1642	84
Pierre-Simon Laplace	1749	77
Charles Darwin	1809	73
Thomas Edison	1847	84
Albert Einstein	1879	76
Pablo Picasso	1881	91
Walt Disney	1901	65
Mother Teresa	1910	87
Pope John Paul II	1920	84
Neil Armstrong	1930	82

Figuring out the mechanism behind a tenfold reduction of human lifespan from a thousand years to a hundred years may provide important insights into human longevity research. The human genome appears to have undergone three mutations adversely affecting human lifespan: the first was after Adam and Eve were banished from the Garden of Eden, the second was after Noah and the Great Flood, and the third was after Moses and the Exodus.

Figure 3.10 shows the lifespan of 26 prominent figures from Adam to Neil Armstrong who all died of natural causes. Centers for Disease Control and Prevention (CDC) reported in February 2016 that American men and women lived 28.2 fewer years than residents in other countries with similar economies, including the United Kingdom, Japan, and Germany. The average life expectancy for American men and women are 76.4 and 81.2 years, respectively [56].

In Fig. 3.10, the estimated birth years for the biblical characters are based on the data provided in the Jewish Virtual Library [57]. Given the scientific calculation of an estimated 200,000 years of Homo sapiens roaming the Earth [58], the relatively short history of biblical time supports the theory that the Scriptures focuses on only one branch of the human family tree starting with Adam and Eve as the common ancestors. Other branches of modern humans include Asians, Europeans, and

Melanesians of Papua New Guinea—some of whom have Neanderthal and Denisovan ancestry that went extinct about 40,000 years ago [59].

3.4   Selfish Versus Selfless Question of Super Longevity

A Dying Matters survey in 2011 found that only 15 % of people want to live forever [60]. The older the people are, the less desire they have for super longevity or immortality: 12 % of over-65 year olds against 21 % of 18–24 year olds want to live forever. There are a number of reasons for the surprisingly low numbers:

1. Declining Health—The elderly often have weakened immune system and are more prone to serious illnesses. A Google search on "age related diseases" returns cardiovascular disease, cancer, arthritis, dementia, cataract, osteoporosis, diabetes, hypertension and Alzheimer's disease (see Fig. 3.11). The incidence of cancer, in particular, increases exponentially with age. Without a doubt, no one wants to suffer forever. A 2013 study by the Stanford University School of Medicine revealed that an overwhelming 88 % of the 1,081 physicians surveyed would choose a do-not-resuscitate or "no code" order for themselves if they are terminally ill, even though the same doctors tend to pursue aggressive, life-prolonging treatment for patients facing the same prognosis [61]. When Albert Einstein suffered an abdominal aortic aneurysm at the age of 76, he was

Fig. 3.11 Google search on "age related diseases" returns a Knowledge Graph of scholarly articles followed by an excerpt from Disabled World about examples of aging-associated diseases

Fig. 3.12 Google search on "life satisfaction" returns a Knowledge Graph of an excerpt from Positive Psychology Program about one's evaluation of life as a whole, rather than the feelings and emotions that are experienced in the moment

taken to the Princeton University Medical Center but he refused surgery, saying, "I want to go when I want. It is tasteless to prolong life artificially. I have done my share, it is time to go. I will do it elegantly" [62]. Since Dr. Jack Kevorkian's "assisted suicide" movement in the 1990s, several states including California, Oregon, Vermont, and Washington have passed a "right to die" legislation.

2. Life Dissatisfaction—Far too many people are getting tired of lifelong toil to make ends meet or working at a job that they do not enjoy. Gallup polls conducted between 2011 and 2016 have consistently shown that two-thirds of U.S. employees are unhappy or disinterested at work [63]. San Diego State University professor Jean Twenge has shown that anxiety and depression are on an 80-year upswing for young Americans since 1935 [64]. The 1994 autobiography *Prozac Nation* by Elizabeth Wurtzel and its 2001 film adaption starring Christina Ricci depict difficult real life struggles confronting some teenagers. If people are afraid to make positive changes to achieve life satisfaction (see Fig. 3.12), destructive behaviors and death will eventually make the final decision for them.

3. Status Quo—Society has taught its citizens to maintain the status quo: "There is a place for everything and everything in its place." Society generally accepts advanced technologies to prolong human life, but eliminating death would seriously disrupt the status quo. Overpopulation [65], food security [66], pension [67], capital punishment [68], and a host of other issues will need to be reexamined. However, most people are averse to change.

4. Boredom—British journalist and author Bryan Appleyard expressed eloquently, "I think there's a point that the immortalists don't understand, and it's that one

exhausts one's own personality over a certain period. It's a weird idea that you would go on and on, still being interested in being yourself. I don't think anyone would. I think you'd get excruciatingly bored of being yourself" [69]. My response to Appleyard is that boredom often stems from the lack of desire to reinvent oneself. To become a Renaissance man or woman, for instance, can take an eternity. Understandably, not everyone wants to be a polymath, but there are so many books to read, places to visit, ideas to share, new hobbies to discover, and charities to volunteer for—the bucket list is practically endless. In early 2016, bucketlist.org showcased more than 4 million life goals from over 300,000 members [70]. Life is anything but boring.

5. Misanthropy—I recently had a debate with some university students on their pessimistic view that human beings are parasites on Earth, deriving natural resources at the planet's expense. Voluntary Human Extinction Movement (VHEMT—pronounced vehement) is a radical environmental movement with spokesperson Les Knight who believes that "When every human chooses to stop breeding, Earth's biosphere will be allowed to return to its former glory, and all remaining creatures will be free to live, die, evolve (if they believe in evolution), and will perhaps pass away, as so many of Nature's 'experiments' have done throughout the eons" [71]. Other Malthusians range from the moderate Population Action International to the extreme Church of Euthanasia.

In Tarsem Singh's visually stunning film *The Fall* (2006), 5-year-old patient Alexandria (Catinca Untaru) asked the bedridden stuntman Roy Walker (Lee Pace) to tell her a story about her namesake. When the fantasy tale turned tragic, Alexandria began to weep and a heart-wrenching dialogue emerged:

Alexandria: *Why are you killing everybody? Why are you making everybody die?*
Roy Walker: *It's my story.*
Alexandria: *Mine, too.*

Roy finally came to realize that his self-pity and self-centeredness were hurting an innocent child. A man's selfishness is often metaphorically and sometimes literally killing his family and friends. American author Og Mandino wrote in *The Greatest Miracle in the World*: "Most humans, in varying degrees, are already dead. In one way or another they have lost their dreams, their ambitions, their desire for a better life. They have surrendered their fight for self-esteem and they have compromised their great potential. They have settled for a life of mediocrity, days of despair and nights of tears. They are no more than living deaths confined to cemeteries of their choice. Yet they need not remain in that state. They can be resurrected from their sorry condition. They can each perform the greatest miracle in the world. They can each come back from the dead…"

Whether the answer is a yes or a no, "Do I want to live forever?" is a somewhat selfish question symptomatic of the me generation who conveniently forgets that their decisions affect other people as well.

On one hand, super longevity may create in people an aversion to risking one's life to save others. The Bible tells a story of King Hezekiah who begged God to cure his terminal illness and spare him from his imminent death. While enjoying his extra 15 years of life, Hezekiah showed off his immense wealth to the king of Babylon — an action that proved to be disastrous. Hezekiah also fathered a son named Manasseh, who later became king at age 12, did evil in the eyes of God, and led Israel into ruin. The worst part was that Hezekiah did not seem to care as he thought to himself, "Will there not be peace and security in my lifetime?" (2 Kings 20:19).

On the other hand, super longevity may motivate people to think very long-term, protect the environment, recycle, prevent wars and destructions, and do everything possible to achieve a better quality of life instead of leaving the world's problems to the next generation. Cofounded by Danny Hillis and Steward Brand in 1996, the Long Now Foundation provides "a counterpoint to today's accelerating culture," helps make "long-term thinking more common," and fosters "responsibility in the framework of the next 10,000 years" [72].

In my 2006 interview for the Association for Computing Machinery, my former Disney colleague Danny Hillis elucidated his rationale, "The [10,000 Year Clock] project is about my acknowledgement that I do have some relationships with people thousands of years from now, and there is some continuity between what I do and what their life will be. ... The business of making people think long-term is really something that is missing in the world in general. So we started a foundation that does the clock project and other projects for basically stretching out people's sense of the moment that they care about—which is now—that's why we call it the Long Now Foundation. It is actually Brian Eno's suggestion that we want to stretch out the moment of now to include the next 10,000 years" [73] (see Fig. 3.13).

A selfless question of immortality may be more along the lines of "Do I want someone else to live forever?" That someone may be a spouse, parent, child, or even friend.

In December 2007, Boy Scout leader Tim Billups donated one of his kidneys to Scout leader Mel Northington and saved his life. "It serves as an example, because in any kind of society you have to think of yourself as fitting into a larger picture," said Mike Andrews, scoutmaster of Troop 500. "What an unselfish act for Tim to do this. Tim knows Mel through Scouting. It's not like he's a family member or a boyhood friend who grew up with him. It's very humbling" [74]. Should Tim's lone remaining kidney ever fail, his brother John Billups has promised to donate one of his.

Since the first issue of *Computers in Entertainment* published by the Association of Computing Machinery in October 2003, the list of "In Memoriam" for the magazine's editorial board has sadly gotten longer. Amongst them are Charles Swartz, Randy Pausch, Roy E. Disney, and Bob Lambert [75]:

- Charles Swartz was CEO of Entertainment Technology Center at the University of Southern California where his Digital Cinema Lab became Hollywood's de facto digital cinema forum [76]. He lost his battle with brain cancer at age 68.
- Randy Pausch was associate professor of computer science, human-computer interaction, and design as well as a cofounder of Entertainment Technology Center at Carnegie Mellon University [77]. Pancreatic cancer took his life at age 47.

Fig. 3.13 First prototype of the 10,000 Year Clock (1999) on display at the Science Museum in London. The clock ticks once a year. Courtesy of Pkirlin at en.wikipedia.org

- The first interviewee of *Computers in Entertainment*, Roy E. Disney was the chairman of Walt Disney Animation presiding over many critically acclaimed films including *Who Framed Roger Rabbit?* (1988), *The Little Mermaid* (1989),

Beauty and the Beast (1991), *Aladdin* (1992), *The Lion King* (1994), and *Fantasia 2000* (1999) [78, 79]. He died of stomach cancer at age 79.

- Bob Lambert was a senior executive at The Walt Disney Company and founder of Digital Cinema Initiatives that help U.S. theaters transition from celluloid to digital [80]. He passed away at age 55.

When I requested an interview with Prof. Pausch in October 2007, he replied to my email, "I'm afraid I'll have to decline; time is just in short supply for me, as I'm sure you can understand."

3.5   Precious Commodity of Time

Time is the most precious commodity. Time does not discriminate based on race, gender, sexual orientation, physical features, and socioeconomic status.

Noted for its 1 min and 42 s of hypnotizing and long intro, Pink Floyd's *Time* from their 1973 album *The Dark Side of the Moon* reminds listeners that "you are young and life is long and there is time to kill today, and then one day you find ten years have got behind you, ... and you run and you run to catch up with the sun but it's sinking, ... you're older, shorter of breath and one day closer to death" [81]. When people look back on their lives, they often wish that they had spent more time with their families, less arguing and more forgiving, less fighting and more peace. The most precious thing that people can give to one another is time.

In 1748 Benjamin Franklin, a Founding Father of the United States and the face on the one hundred-dollar bill since 1928, coined the phrase "Time is Money" in his essay *Advice to a Young Tradesman* [82]. In the 2011 sci-fi movie *In Time*, people are genetically engineered to stop aging at 25 years old, and time is literally money because currency is measured in hours and minutes instead of dollars and cents [83]. People would use time as currency to pay for daily expenses. Factory worker Will Salas (Justin Timberlake) ran into a 105-year-old young man Henry Hamilton (Matt Bomer) who transferred 116 years of his time to Will, leaving himself with only 5 min to live. Will in turn gave some of his newly acquired time to his best friend Borel (Johnny Galecki) who tragically ended up dying prematurely due to alcohol intoxication. If the meaning of life is futility, human longevity loses its luster.

In 2016, Bill and Melinda Gates were asked by some high school students in Kentucky what superpower they wished they could have. Their answers were: "More time. More energy. As superpowers go, they may not be as exciting as Superman's ability to defy gravity. But if the world can put more of both into the hands of the poorest, we believe it will allow millions of dreams to take flight" [84].

In his 2007 "Last Lecture" titled "Really Achieving Your Childhood Dreams," Prof. Randy Pausch said, "We cannot change the cards we are dealt, just how we play the hand. ... It's a thrill to fulfill your own childhood dreams, but as you get older, you may find that enabling the dreams of others is even more fun" [85].

3.6   Negligible Senescence and Princess Leia in Star Wars

In 2014 *HuffPost Women* listed 5 things that women are judged more harshly for than men: "Having 'too many' sexual partners, having a messy home, being overweight, being blunt or assertive, and not having children" [86]. Some film critics deemed Jennifer Lawrence too fat to play Katniss Everdeen in *The Hunger Games* [87]. British model Iskra Lawrence was called a "fat cow," Melissa McCarthy a "female hippo," and Kate Upton a "squishy brick" [88]. Carrie Fisher was pressured to lose 35 lb prior to the start of filming *Star Wars: The Force Awakens* [89]. To add insult to injury, Internet trolls bashed Fisher for her aging looks.

In *Star Wars: The Force Awakens* (2015), 59-year-old Carrie Fisher reprised her role as Princess Leia turned General Organa. While no one complained about Han Solo played by 73-year-old Harrison Ford, gender bias reared its ugly head when Internet trolls criticized Princess Leia's appearance: "You didn't age well and u sucked in Star Wars" [90]. Fisher fired back at the body-shaming trolls: "Please stop debating about whether or not I aged well. Unfortunately it hurts all 3 of my feelings. My body hasn't aged as well as I have. Blow us. ... Youth & beauty are not accomplishments, they're the temporary happy byproducts of time and/or

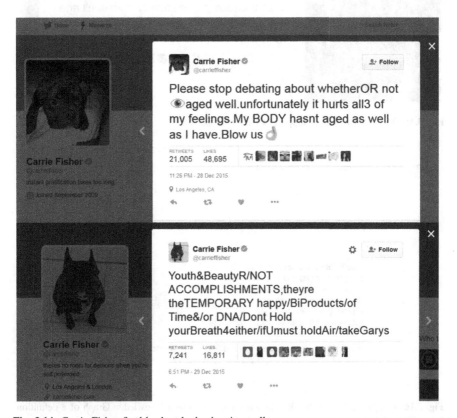

Fig. 3.14 Carrie Fisher fired back at body-shaming trolls

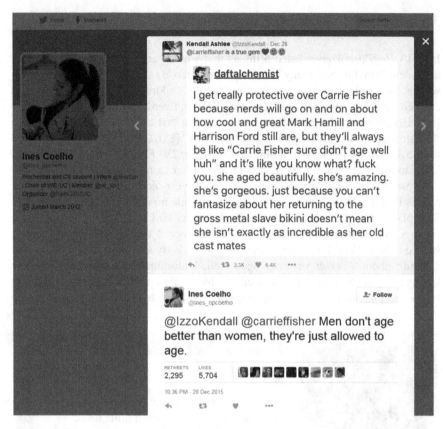

Fig. 3.15 Twitters tweeted in support of Carrier Fisher

Fig. 3.16 Google search on "negligible senescence" returns a Knowledge Graph of a definition from Wikipedia about the lack of symptoms of ageing in a few select organisms

DNA" (see Fig. 3.14). A Fisher supporter tweeted, "Men don't age better than women, they're just allowed to age" (see Fig. 3.15).

In gerontology, scientists have been researching negligible senescence—the lack of symptoms of aging (see Fig. 3.16). There are quite a number of negligibly senescent animals in nature. Adwaita was a tortoise in Calcutta, India that lived to either 150 or 250 years old by some accounts [91]. George, a 140-year-old lobster, was released back into the wild by a New York restaurant [92]. Henry, a New Zealand reptile, became a father at the age of 111 for the first time [93]. Prof. Caleb Finch of USC Davis School of Gerontology said in a 2010 interview, "In theory, if mortality rates did not increase as usual during aging, humans would live hundreds of years. I have calculated for humans that at mortality rates of 0.05 % per year, as found at age 15 in developed countries, the median lifespan would be about 1,200 years. In natural populations of long-lived animals, mortality rates are rarely less than 1 % per year. For very slowly aging turtles, rockfish, the number beyond 70 is 1–2 %. However, there are long-lived trees, like the bristlecone pine at 5,000 years" [94].

On the eve of his 69th birthday, biochemist-geneticist Craig Venter declared, "I have the brain of a 44-year-old. ... It's not just a long life we're striving for, but one which is worth living" [95]. Venter has been acknowledged, along with geneticist Francis Sellers Collins, as being a primary force behind the Human Genome Project [96]. In 2013, Venter launched Human Longevity, Inc (HLI) with stem cell pioneer Robert Hariri and XPRIZE Foundation founder Peter Diamandis. "Using the combined power of our core areas of expertise—genomics, informatics, and stem cell therapies, we are tackling one of the greatest medical/scientific and societal challenges—aging and aging related diseases," Venter elaborated his company vision. "HLI is going to change the way medicine is practiced by helping to shift to a more preventive, genomic-based medicine model which we believe will lower healthcare costs. Our goal is not necessarily lengthening life, but extending a healthier, high performing, more productive life span" [97]. Baylor College of Medicine professor C. Thomas Caskey agreed, "The whole idea behind this is to identify the risk, then modify that risk so that you end up with longer periods of normal health. ... The patient does not want just more years but quality years" [95].

3.7 Rejuvenation and the Curious Case of Benjamin Button

Rumors of a Fountain of Youth (see Fig. 3.17) have endured from Alexander the Great in the 4th century B.C. to legendary patriarch king Prester John during the early Christian Crusades to Spanish explorer Juan Ponce de León in 1513 when he discovered Florida.

Biomedical gerontologist Aubrey de Grey cofounded the SENS (Strategies for Engineered Negligible Senescence) Research Foundation in 2009 to conduct research on regenerative medicine and rejuvenation biotechnologies to prevent or reverse the aging process [98]. de Grey said in a 2013 interview, "SENS is based on

Fig. 3.17 The Fountain of Youth (1546) by Lucas Cranach the Elder. Older women are seen entering a Renaissance fountain, and exiting it after being transformed into youthful beauties

the appreciation that there is a continuum between (a) the initially harmless, progressively accumulating damage that accumulates in the body as a side-effect of its normal operation and (b) the pathologies that emerge when the amount of that damage exceeds what the body is set up to tolerate. We want to treat (remove or obviate) the damage and thereby prevent the pathology" [99].

Regenerative medicine can take its cue from salamanders (Ambystoma mexicanum) that can routinely regenerate complex tissues such as a severed limb, a detached tail, or the lens and retina of a damaged eye (see Fig. 3.18). Northeastern University Professor James Monaghan and his research team studied limb regeneration in salamanders and concluded that "many new candidate gene sequences were discovered for the first time and these will greatly enable future studies of wound healing, epigenetics, genome stability, and nerve-dependent blastema formation and outgrowth using the axolotl model" [100].

Rejuvenation biotechnologies can pick up on the immortal jellyfish (Turritopsis dohrnii) that can age backward like Benjamin Button (Brad Pitt) in *The Curious Case of Benjamin Button* (2008) (see Fig. 3.19). When starvation or injury occurs, "instead of sure death, [Turritopsis] transforms all of its existing cells into a younger state," said Maria Pia Miglietta, then postdoctoral scholar at Pennsylvania State University and now professor at Texas A&M University at Galveston [101]. The immortal jellyfish can rejuvenate from an adult back into a baby, converting muscle cells into nerve cells, sperms, or eggs. The life cycle repeats itself until the jellyfish gets eaten by a predator or succumbs to illness.

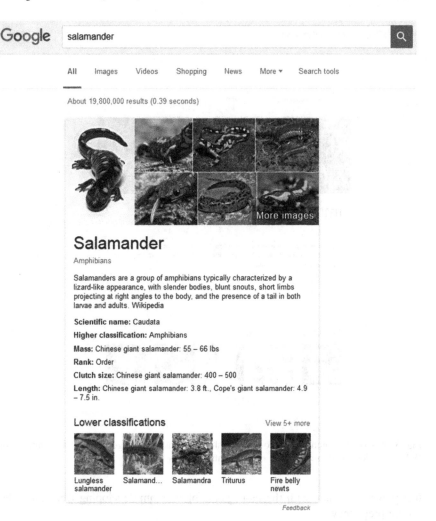

Fig. 3.18 Google search on "salamander" returns a Knowledge Graph of a definition from Wikipedia along with pictures and other scientific details

3.8   The Man with the Golden Arm Helped Save 2 Million Babies

Sometimes we need to look no further than our own human species for miraculous power of healing. 78-year-old James Harrison has donated blood plasma from his right arm almost every week for the past 60 years. The unusual antibody in his blood has saved the lives of more than 2 million babies from rhesus disease. Thanks to Harrison, doctors were able to use his antibodies to develop Anti-D, an injection

Fig. 3.19 Google search on "immortal jellyfish" returns a Knowledge Graph of a definition from Wikipedia along with pictures and other scientific details

that prevents women with rhesus-negative blood from developing RhD antibodies during pregnancy.

"Australia was one of the first countries to discover a blood donor with this antibody, so it was quite revolutionary at the time," said Jemma Falkenmire of the Australian Red Cross Blood Service in a 2015 interview on CNN. "In Australia, up until about 1967, there were literally thousands of babies dying each year, doctors didn't know why, and it was awful. Women were having numerous miscarriages and babies were being born with brain damage. … Every bag of blood is precious, but James' blood is particularly extraordinary. His blood is actually used to make a life-saving medication, given to moms whose blood is at risk of attacking their unborn babies. Every batch of Anti-D that has ever been made in Australia has come from James' blood. … I'm grateful and I think James is really selfless to continue to donate, so that we can keep having this vaccine. … He will have to retire in the next couple years, and I guess for us the hope is there will be people who will donate, who will also … have this antibody and become life savers in the same way he has, and all we can do is hope there will be people out there generous enough to do it, and selflessly in the way he's done" [102].

3.9   The Man Who Can't Catch AIDS Helped Develop Anti-Viral Drugs

Touted as "the man who can't catch AIDS," New York artist and Fodor's travel guides editor Stephen Crohn had a genetic mutation in his white blood cells ($CD4^+$ T helper cells) that effectively blocked HIV infection (see Fig. 3.20). He volunteered for research studies that shed light on the nature of AIDS and led to the development of antiviral drugs at the Aaron Diamond AIDS Research Center and other medical facilitates. His great-uncle was gastroenterologist Burrill Bernard Crohn who discovered the inflammatory bowel syndrome named Crohn's disease [103].

"One of the things that went through my mind was, 'I guess I'm condemned to live,'" said Crohn. "What's hard is living with the continuous grief. You keep losing people every year—six people, seven people … and it goes on for such a long period of time. And the only thing you could compare it to would be to be in a war" [104]. Overcome by survivor guilt after seeing more than 70 of his friends died of AIDS, Crohn committed suicide at age 66. Stephen Crohn could have lived a long and healthy life, but he chose to die instead. Emotional health is just as important as physical health.

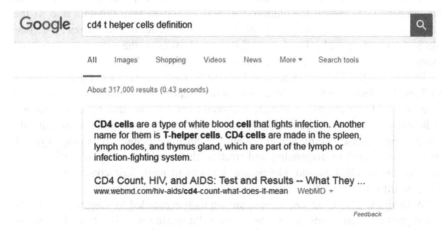

Fig. 3.20 Google search on "CD4 T helper cells" returns a Knowledge Graph of a definition from WebMD about a type of white blood cell that fights infection

3.10   No Mother Should Bury Their Child

Sponsored by the 501(c) (3) nonprofit Institute for Education, Research, and
Scholarships (IFERS), EASE T1D (Education, Awareness, Support, Empowerment
for Type 1 Diabetes) is the joint effort of two mothers—Debbie George and
Michelle Thornburg—who have children with Type 1 diabetes [105]. EASE T1D
addresses the misconceptions of Type 1 diabetes and the lack of knowledge on the
differences between Type 1 (little to no insulin) and Type 2 (insulin resistance, too
little insulin). Type 1 diabetes is a chronic, life-threatening autoimmune disease for
which there is currently no cure. Contrary to popular belief, diet and lifestyle are not
causes of the disease.

Undiagnosed Type 1 diabetes can result in Diabetic Ketoacidosis (DKA) which
can lead to serious conditions including coma, brain damage, and even death. With
the slogan "No parent should bury their child," EASE T1D started a petition in May
2015 to encourage physicians, physician assistants, and nurses in California to
educate parents on the signs and symptoms of Type 1 diabetes as well as to check
blood glucose levels of children and adults who present flu-like symptoms in an
effort to prevent a misdiagnosis and to save lives. This legislation is modeled after
Reegan's Rule.

Reegan's Rule was started in North Carolina by a mother whose 16-month-old
baby girl, Reegan Oxendine, passed away from undiagnosed Type 1 diabetes. Little
Reegan had been misdiagnosed several times over a 3-month period before her
death. Reegan's mother, Darice Oxendine, created a legislation to encourage parent
education on Type 1 diabetes during well-child care visits from birth to age 5 years
old. The first-of-its-kind legislation was signed into North Carolina Law in October
2015 [106].

Advocacy efforts for similar legislation have been happening nationwide. In
November 2015, House Resolution No. 569 passed in Pennsylvania due to the
efforts of Debbie Healy and her State Representative, Ryan MacKenzie. The resolution encourages physicians to educate and discuss the warning signs and
symptoms of Type 1 diabetes with parents or guardians [107].

In March 2016, California Senator Richard D. Roth's measure to raise awareness
of Type 1 diabetes passed the State Senate on a bipartisan, unanimous vote of 38-0
(see Fig. 3.21). "I am proud to have authored Senate Resolution 63 and thank my
colleagues in the State Senate for joining me in raising awareness of this life
threatening disease," said Senator Roth. "Educating parents regarding Type 1
diabetes is critical to diagnosing and treating this condition early and effectively,
helping ensure children and adolescents learn to manage their condition and live
long, healthy lives" [108].

With all the grassroots awareness campaigns spearheaded by concerned parents,
some promising solutions are on the horizon. Partnering with Dr. Jane Buckner of
Benaroya Research Institute (BRI) at Virginia Mason, Dr. David Rawlings and his
team at the Seattle Children's Research Institute have been studying an im-
munotherapy approach. "In Type 1 diabetes, a type of immune system cell, called

By the Honorable Richard D. Roth, 31st Senatorial District;
Relative to

Type 1 Diabetes Awareness

WHEREAS, Diabetes is a chronic disease that affects an estimated 29.1 million Americans, or 9.3 percent of the population, on a daily basis; and

WHEREAS, Approximately 3.8 million people in California are living with diabetes, and it is estimated that over one million Californians are undiagnosed; and

WHEREAS, Diabetes costs an estimated $37.1 billion in California each year; and

WHEREAS, Type 1 diabetes accounts for $14.9 billion in health care costs in the United States each year; and

WHEREAS, Type 1 diabetes, previously called juvenile-onset diabetes, occurs when the body does not produce insulin, a hormone that is necessary to convert sugar, or glucose, into energy; and

WHEREAS, Type 1 diabetes is the third most common autoimmune disease among children; and

WHEREAS, Between 2001 and 2009, there was a 21-percent increase in the prevalence of Type 1 diabetes in people under 20 years of age; and

WHEREAS, Only 5 percent of diabetics have Type 1 diabetes, and it is typically diagnosed in children and young adults; and

WHEREAS, Early diagnosis of Type 1 diabetes can help prevent diabetic ketoacidosis, a potentially fatal condition that develops from high blood glucose levels; and

WHEREAS, In 2009, among hospital discharges of children and young people 0 to 17 years of age, about 74 percent had diabetes as the first-listed diagnosis, and of these patients, 64 percent of the diagnoses were for diabetic ketoacidosis; and

WHEREAS, Education concerning Type 1 diabetes is critical to raising awareness and diagnosing and treating this condition effectively; and

WHEREAS, With the help of insulin therapy and other treatments, young children and adolescents can learn to manage their condition and live long, healthy lives; and

WHEREAS, Since early diagnosis is the key to successful treatment of Type 1 diabetes, health care practitioners can increase the likelihood of early diagnoses by discussing warning signs and symptoms of Type 1 diabetes with parents or guardians at least once a year during well-child visits for infants and children from birth to five years of age; now, therefore, be it

RESOLVED BY SENATOR RICHARD D. ROTH, That he calls the attention of the public to the importance of raising awareness about Type 1 diabetes through education and learning to recognize the warning signs and symptoms of this chronic disease so that it may be diagnosed early in its course and treated effectively.

Member Resolution No. 202
Dated this 14th day of March, 2016.

Honorable Richard D. Roth
31st Senatorial District

Fig. 3.21 California Senate Resolution 63 "Type 1 Diabetes Awareness" signed by the Honorable Richard D. Roth, 31st Senatorial District, on March 14, 2016

an effector T cell, malfunctions and attacks pancreas cells that create insulin," Rawlings explained. "Normally, effector T cells attack foreign viruses, not the body's own cells. With this research, we will edit genes in these cells and change these 'dangerous' cells into regulatory T cells, another type of immune cell that regulates an immune system's response and keeps it from going into overdrive. We expect these gene-edited regulatory T cells, when returned to a diabetic's body, will stop effector T cells from destroying the body's insulin-producing cells" [109].

Clinical trials for new treatments have already begun. In June 2015, Massachusetts General Hospital launched phase II trial of vaccine bacillus Calmette-Guérin (BCG) to reverse advanced Type 1 diabetes [110]. In March 2016, Professor Mark Peakman at King's College London started testing MultiPepT1De (Multiple beta cell Peptides in Type 1 diabetes) injections on trial participants. The peptides are protein molecules found in the insulin-producing beta cells of the pancreas (see Fig. 3.22). Researchers hope that the peptides will re-train the patient's immune system to get rid of its autoimmune disorder.

Although researchers have been making steady progress, fundraising to support them is not easy. Once in a blue moon there are successful campaigns like the Ice Bucket Challenge by the ALS Association to fight Lou Gehrig's disease (see Fig. 3.23). However, the majority of people in the world could care less about diabetes, cancer, or any incurable disease unless they are inflicted by it or they have to take care of a family member who is suffering from it. The public generally relies on deep-pocketed pharmaceutical companies to fund expensive research which

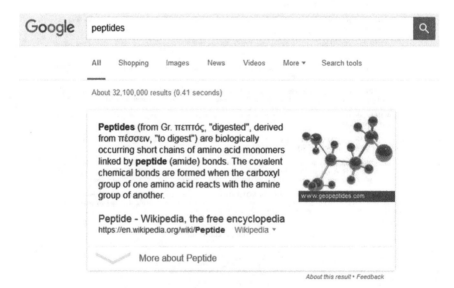

Fig. 3.22 Google search on "peptides" returns a Knowledge Graph of a definition from Wikipedia about biologically occurring short chains of amino acid monomers linked by peptide (amide) bonds

Fig. 3.23 Google search on "Lou Gehrig's disease" returns a Knowledge Graph of information, symptoms, and treatments from Mayo Clinic and other sources

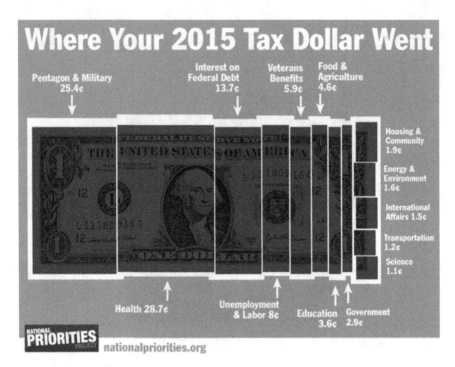

Fig. 3.24 The largest federal government spending in 2015 was on health programs (28.7 cents per tax dollar) according to the calculations by National Priorities Project

often results in high drug prices. *The Wall Street Journal* reported that in 2014 the U.S. Food and Drug Administration (FDA) approved 41 new drugs, the most in nearly two decades, but the catch is their cost [111]:

> Recent treatments for hepatitis C, cancer and multiple sclerosis that cost from $50,000 annually to well over $100,000 helped drive up total U.S. prescription-drug spending 12.2 % in 2014, five times the prior year's growth rate... Out-of-pocket prescription-drug costs rose 2.7 % in 2014. ... Even patients with insurance and comfortable incomes are sometimes forced to make hard choices—tapping savings, taking on new debt or even forgoing treatment. ... Patients on Medicare are starting to feel some relief from out-of-pocket expenses through a provision in the Affordable Care Act that requires a gradual lowering of patient contributions.

In 2015, the U.S. government spent 28.7 % of taxpayer's money on health programs (see Fig. 3.24) [112]. The second largest spending was 25.4 % on the Pentagon and the military. A healthy population is vital to a country's economy and national security.

3.11 Big Data for "Nowcasting" of Epidemics

"One in 20 Google searches are for health-related information," wrote Google product manager Prem Ramaswami in the Google official blog in February 2015 [113]. Instead of calling doctors at the first sign of symptoms, people turn to Google search for help. So much so that Google can predict where and when an epidemic will occur based on geographical data and search queries.

In 2008, Google engineers started working on Flu Trends. It turned out to be a big data success story of "nowcasting" which aims to offer near-term forecast with a high degree of detail and accuracy. During the 2012–2013 season in the United States, Google Flu Trends correctly estimated the start and duration of the season, but overestimated the severity of the flu [114]. Then Google software engineer Jeremy Ginsberg and his research team cautioned that "this system is not designed to be a replacement for traditional surveillance networks or supplant the need for laboratory-based diagnoses and surveillance" [115].

In 2015, Google Flu Trends and Google Dengue Trends stopped publishing estimates of flu and dengue fever based on search patterns. Instead, Google began to provide flu and dengue signal data directly to Columbia University's Mailman School of Public Health, Boston Children's Hospital/Harvard, Centers for Disease Control and Prevention (CDC) Influenza Division, and other institutions that specialize in infectious disease research [116].

3.12 Top 10 Lists of Health-Related Search Queries

According to Simon Rogers, a data editor for Google's News Lab, the top 9 health-related questions on Google search in the United States from January to November 2015 were [117]:

1. Is bronchitis contagious?
2. Is pneumonia contagious?
3. How much water should I drink?
4. How many calories should I eat?
5. What is lupus?
6. How far along am I?
7. When do you ovulate?
8. What is gluten?
9. How long does the flu last?

Megan Ranney, Content Marketing Manager at *Mashable*, reported on countries using Google search to eat and live healthier in 2015. She summarized in her article that "how to eat healthy, healthy body mass index parameters, how to lose weight and how to eat healthy on a budget all ranked among the world's top 10 search queries, as well as questions about whether specific foods were healthy" [118]. The worldwide top 10 health-related queries in 2015 were:

1. How to eat healthy
2. What is health?
3. Is [food] healthy? (Some foods listed were sushi, hummus, and popcorn)
4. How can I be healthy?
5. What is a healthy BMI?
6. How to stay healthy
7. How to lose weight
8. How to eat healthy on a budget
9. What is a healthy blood pressure?
10. What is a healthy heart rate?

3.13   Knowledge Graph of Health Information

Knowledge Graph is Google's "first step towards building the next generation of search, which taps into the collective intelligence of the web and understands the world a bit more like people do" [119]. In February 2015, Google product manager Prem Ramaswami announced the expansion of Knowledge Graph to cover health-related search queries [120]:

> One in 20 Google searches are for health-related information. And you should find the health information you need more quickly and easily. … When you ask Google about common health conditions, you'll start getting relevant medical facts right up front from the Knowledge Graph. We'll show you typical symptoms and treatments, as well as details on how common the condition is—whether it's critical, if it's contagious, what ages it affects, and more. For some conditions you'll also see high-quality illustrations from licensed medical illustrators. Once you get this basic info from Google, you should find it easier to do more research on other sites around the web, or know what questions to ask your doctor.

Larry Page and Sergey Brin wrote in the August 2004 Google IPO letter that Google users trust the search engine to help them with important decisions including medical advice and that Google search results are unbiased and objective [121]. Google relies on its team of medical doctors and the Mayo Clinic to ensure the quality of medical information:

> We worked with a team of medical doctors (led by our own Dr. Kapil Parakh, M.D., MPH, Ph.D.) to carefully compile, curate, and review this information. All of the gathered facts represent real-life clinical knowledge from these doctors and high-quality medical sources across the web, and the information has been checked by medical doctors at Google and the Mayo Clinic for accuracy.

For instance, when we Google "Type 1 diabetes" we get a Knowledge Graph of "About," "Symptoms," and "Treatments" from Mayo Clinic and other sources (see Figs. 3.25, 3.26, and 3.27). Although the Knowledge Graph is very useful, it is by no means comprehensive or inclusive of all types of treatment, new theories, and

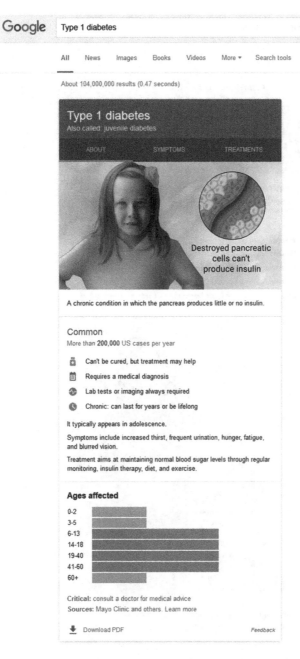

Fig. 3.25 Knowledge Graph of "About" for "Type 1 diabetes"

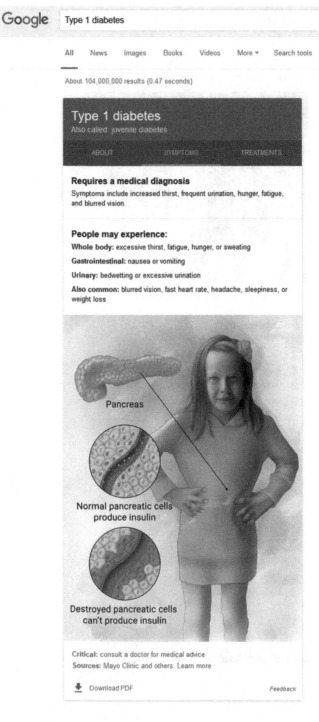

Fig. 3.26 Knowledge Graph of "Symptoms" for "Type 1 diabetes"

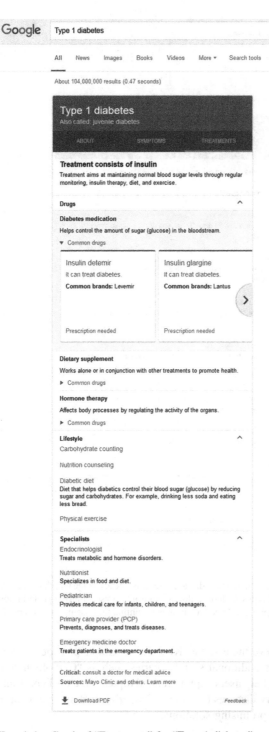

Fig. 3.27 Knowledge Graph of "Treatments" for "Type 1 diabetes"

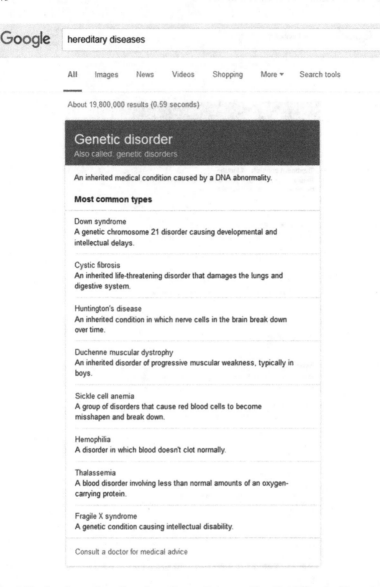

Fig. 3.28 Google search on "hereditary diseases" returns a Knowledge Graph of "genetic disorder" listing the most common types including down syndrome, cystic fibrosis, Huntington's disease, Duchene muscular dystrophy, sickle cell anemia, hemophilia, thalassemia, and fragile X syndrome

modern advances. Nevertheless, the Knowledge Graph provides adequate basic information and offers additional search queries such as "Carbohydrate counting" and "Diabetic diet" to help users explore related topics. To cure Type 1 diabetes and other serious hereditary diseases (see Fig. 3.28), advanced medical treatments such as gene therapy offer promising solutions.

3.14   CRISPR-Cas9 Gene Editing

In 2012, biochemist Jennifer Doudna, microbiologist Emmanuelle Charpentier, and their team of scientists published a seminal research paper on CRISPR-Cas9 (Clustered Regularly-Interspaced Short Palindromic Repeats—CRISPR associated protein 9) which allows scientists to edit genomes with precision, efficiency, and flexibility (see Fig. 3.29) [122].

In 2013, Chinese scientists Yuyu Niu, Bin Shen, Yiqiang Cui, Yongchang Chen, and others at Yunnan Key Laboratory of Primate Biomedical Research and Nanjing Medical University created the first-ever mutant twin cynomolgus monkeys by coinjection of one-cell-stage embryos with Cas9 mRNA and sgRNAs (see Fig. 3.30) [123].

In 2014, Temple University professor Wenhui Hu and his collaborators at Case Western Reserve University and Sichuan University successfully used RNA-directed gene editing to eradicate latently infected cells and to immunize uninfected cells against HIV-1 infection [124].

In 2015, Prof. Junjiu Huang and his research team at the Sun Yat-sen University in Guangzhou, China modified the DNA of human embryos in order to eliminate the inherited blood disease thalassemia [125]. "I believe this is the first report of CRISPR/Cas9 applied to human pre-implantation embryos and as such the study is a landmark, as well as a cautionary tale," said George Daley, a stem-cell biologist at Harvard Medical School. "Their study should be a stern warning to any practitioner who thinks the technology is ready for testing to eradicate disease genes" [126].

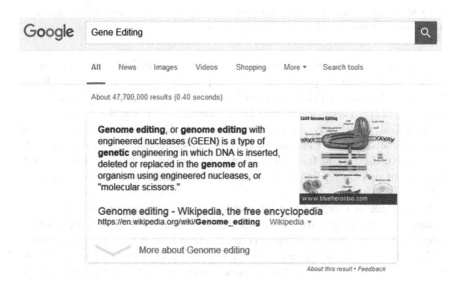

Fig. 3.29 Google search on "gene editing" returns a Knowledge Graph of definition from Wikipedia about a type of genetic engineering in which DNA is inserted, deleted or replaced in the genome of an organism using engineered nucleases, or molecular scissors

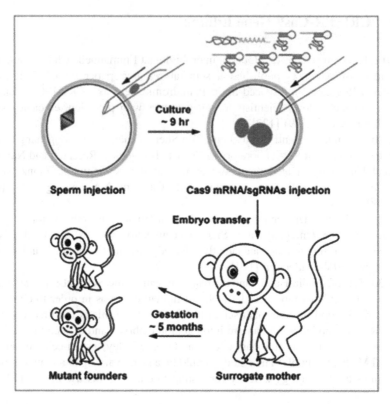

Fig. 3.30 First-ever mutant twin cynomolgus monkeys in China created by coinjection of one-cell-stage embryos with Cas9 mRNA and sgRNAs

In 2016, the Human Fertilization and Embryology Authority (HFEA) regulator in the United Kingdom approved a license application by stem cell scientist Kathy Niakan at the Francis Crick Institute in London to perform gene editing on human embryos [127]. "This research will allow scientists to refine the techniques for creating GM (genetically modified) babies," said Dr. David King, director of Human Genetics Alert. "Many of the government's scientific advisers have already decided that they are in favor of allowing that. So this is the first step in a well mapped-out process leading to GM babies, and a future of consumer eugenics" [128].

It may come as a surprise to many people that the human genome (see Fig. 3.31) contains human endogenous retroviruses (HERVs) (see Fig. 3.32) that are linked to cancer [129], autoimmune diseases [130], multiple sclerosis [131], schizophrenia [132], and more. Gene editing offers a new hope to eradicating many deadly hereditary diseases. Christoph Lahtz, former cancer researcher at the City of Hope Beckman Research Institute, describes gene editing as a new hope for cancer treatment in Chap. 20 of this book.

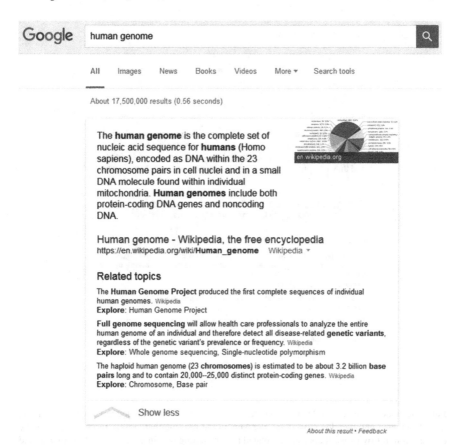

Fig. 3.31 Google search on "human genome" returns a Knowledge Graph of definition from Wikipedia about the complete set of nucleic acid sequence for humans, encoded as DNA within the 23 chromosome pairs in cell nuclei and in a small DNA molecule found within individual mitochondria. It also shows related topics on Human Genome Project, full genome sequencing, and protein-coding genes

3.15 Human-Animal Chimeras

In Greek mythology, the chimera was a fire-breathing female monster with a lion's head, a goat's body, and a serpent's tail (see Fig. 3.33). Researchers of human-animal chimeras attempt to grow human organs inside pigs and other farm animals that can be harvested for transplantation. Using CRISPR-Cas9 gene editing, scientists first remove the animal embryo's capability to develop a certain organ (for example, the pancreas). Then they insert human stem cells into the animal embryo, in hopes that a normal human pancreas will be developed inside the animal.

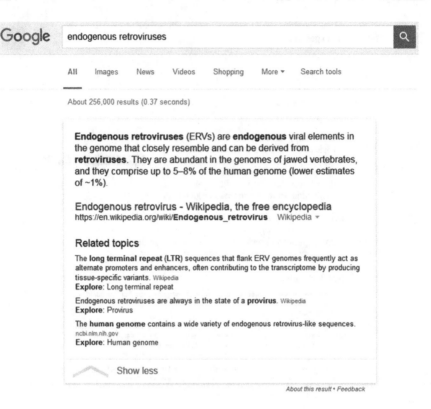

Fig. 3.32 Google search on "endogenous retroviruses" returns a Knowledge Graph of definition from Wikipedia about endogenous viral elements in the genome that closely resemble and can be derived from retroviruses. It also shows related topics on long terminal repeat (LTR), provirus, and human genome

Both human and pig genomes contain endogenous retroviruses (ERVs) (see Fig. 3.32). The porcine endogenous retroviruses (PERVs) can produce full-blown viruses that infect both pig and human cells, rendering the organs unsuitable for transplantation. In a 2015 landmark study, Harvard Medical School professor George Church and his colleagues successfully used one CRISPR molecule to alter all 62 genes in a pig cell to get rid of PERVs without damaging normal cell growth [133].

"This work brings us closer to a realization of a limitless supply of safe, dependable pig organs for transplant," said Prof. David Dunn at the State University of New York at Oswego. "It's a cruel situation currently, that someone who needs a heart transplant has to pin their chance for a healthy life on the untimely death of another person" [134].

Fig. 3.33 The chimera on an Apulian plate (c. 350–340 BC) in Musée du Louvre. (Courtesy of Marie-Lan Nguyen)

In a 2016 *Scientific American* interview, Juan Carlos Izpisua Belmonte at the Salk Institute spoke of Pope Francis' blessing of his work as well as the technical and ethical challenges of human-animal chimeras [135]:

The Vatican is behind this research and has no problem based on the idea to help humankind. And in theory all that we will be doing is killing pigs. One problem and the major problem is that these [human] cells could colonize the brain of the animal in which you put them. And obviously it would not be appropriate to have an animal with neurons from people. Or these cells could colonize the germline so that the sperm or the oocytes of that pig would be human. So to avoid that the government of Spain allowed us to have the pig be born and then immediately after to be sacrificed. ... We are devising genetic engineering technology so that if a cell becomes a neuron it is just destroyed in the embryo. Any cell that starts to be taught okay you are going to become a neuron at the moment of the first stages of neurogenesis, we are putting a toxin construct in it so that it will be destroyed by itself. So that will prevent any pig embryos from having human neurons so to speak.

3.16   3D Bioprinting of Tissues and Organs

For human-animal chimeras, mammals have their gestation periods before the organs can be harvested for transplantation. 3D bioprinting using living cells is an alternative to waiting for gestation. Scientists have succeeded in the generation and transplantation of tissues including multilayered skin, bone tissue, vascular grafts, tracheal splints, heart tissue, and cartilaginous structures [136]. Bioprinting of complex organs such as kidneys and livers is in its early stage of development [137].

"We present an Integrated Tissue-Organ Printer (ITOP) that can fabricate stable, human-scale tissue constructs of any shape," Dr. Anthony Atala at the Wake Forest University Institute for Regenerative Medicine explained. "The correct shape of a tissue construct is obtained from a human body by processing Computed Tomography (CT) or Magnetic Resonance Imaging (MRI) data in computer-aided design software. ... We are actually printing the scaffolds and the cells together. We show that we can grow muscle. We make ears the size of baby ears. We make jawbones the size of human jawbones. We are printing all kinds of things" [138].

In November 2015, Russian scientists Elena Bulanova, Vladimir Mironov, and their research team 3D-printed a thyroid gland and transplanted it into a living mouse to restore its thyroid function [139].

3.17   Cyborg: Cybernetic Implants and Nanobot Repairs

Arne Larsson was the first "cyborg" who in 1958 received the world's first implanted heart pacemaker to regulate his heartbeat [140]. He lived until the age of 86 when he died of melanoma skin cancer. Today there are over 3 million people worldwide with artificial cardiac pacemakers [141]. Other common medical implants include artificial hips, artificial knees, spinal fusion hardware, intra-uterine devices (IUDs), traumatic fracture repair, coronary stents, ear tubes (tympanostomy tubes), artificial eye lenses (psuedophakos), and implantable cardioverter defibrillators (ICDs) [142].

In 2016, Prof. Thomas J. Oxley and his team at the University of Melbourne developed a device dubbed "stentrode" that can be implanted into a patient's brain through blood vessels to make it easier to control artificial limbs from the brain [143]. Without the need for invasive brain surgery, stentroide represents a significant advancement in the field of brain-computer interface (BCI). *The Six Million Dollar Man* (played by Lee Majors) and *The Bionic Woman* (starring Lindsay Wagner) in the 70s TV series gave a positive glimpse into the future of cyborgs.

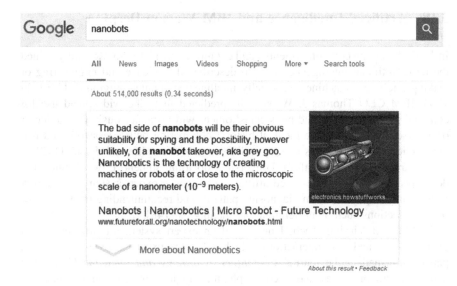

Fig. 3.34 Google search on "nanobots" returns a Knowledge Graph of definition from FutureForAll.org about the technology of creating machines or robots at or close to the microscopic scale of a nanometer

Nanobots are machines or robots at or close to the microscopic scale of a nanometer (see Fig. 3.34). In a 2009 interview with *Computerworld*, futurist Ray Kurzweil spoke about using nanotechnology and biotechnology to wipe out diseases. "It's radical life extension," Kurzweil predicted. "The full realization of nanobots will basically eliminate biological disease and aging. I think we'll see widespread use in 20 years of [nanotech] devices that perform certain functions for us. In 30 or 40 years, we will overcome disease and aging. The nanobots will scout out organs and cells that need repairs and simply fix them. It will lead to profound extensions of our health and longevity" [144].

When tissue damage results in bleeding, polymerized fibrin and platelets form a hemostatic plug or clot over a wound site. If the injury is too severe, however, nanobots will come to the rescue by repairing the damage. All surgeries will be performed by nanobots with precision and speed necessary to save a life in an emergency situation.

T cells, B cells, and antibodies protect the human body against pathogens such as bacteria and viruses. People get sick easier when their immune system is weakened. Vaccines help to train T cells to recognize certain known pathogens. Genetically modified T cells can cure melanoma skin cancer [145], which could have saved Arne Larsson. With machine learning and artificial intelligence, nanobots will be able to identify and annihilate any never-seen-before pathogens and cancer cells.

3.18   Artificial Intelligence and IBM Watson DeepQA

In 1955, American computer scientist and cognitive scientist John McCarthy coined the term "artificial intelligence" (AI) to describe "the science and engineering of making intelligent machines, especially intelligent computer programs" [146]. In 1965, IBM CEO Thomas J. Watson, Jr. predicted that "the widespread use [of computers]...in hospitals and physicians' offices will instantaneously give a doctor or a nurse a patient's entire medical history, eliminating both guesswork and bad recollection, and sometimes making a difference between life and death" [147].

In early 1970s at Stanford University, doctoral student Edward Shortliffe developed MYCIN—a rule-based artificial intelligence expert system that outperformed human physicians in diagnosing patients and recommending treatments for severe infections such as bacteremia and meningitis [148].

In 1980s at Virginia Tech, I developed an expert system for information on pharmacology and drug interactions, which led to my first peer-reviewed research paper, published in the journal *Computers in Biology and Medicine* [149]. The expert system organizes and encodes pharmacological information in rules and frames for systematic retrieval which includes delineation, definition, and hierarchical subdivision of mechanisms responsible for drug interactions; division of pharmacological agents into a hierarchy of subclasses to allow for defining interacting drugs by classes as well as by specific agents; and correlation of drug classes and specific drugs with mechanisms by which they may be involved in drug interactions. Clinicians use the expert system to predict what may happen when two drugs are used together, how drugs interact in the human body, what can be done to alleviate detrimental side effects, and what related drugs may also be involved in similar interactions.

In a highly publicized *Jeopardy!* game shows on television in February 2011, IBM Watson DeepQA defeated human champions Ken Jennings and Brad Rutter [150]. "People ask me if this is HAL," said David Ferrucci, lead developer of Watson, referring to the Heuristically programmed ALgorithmic (HAL) computer in *2001: A Space Odyssey* by Stanley Kubrick and Arthur C. Clarke. "HAL's not the focus; the focus is on the computer on 'Star Trek,' where you have this intelligent information seeking dialogue, where you can ask follow-up questions and the computer can look at all the evidence and tries to ask follow-up questions. That's very cool" [151].

Inspired by IBM Watson on *Jeopardy!*, author Martin Ford penned a piece in *The Atlantic* with the sensational title "Anything You Can Do, Robots Can Do Better" and he asked the question "Is any job safe from automation?" [152] The answer is an unequivocal no in the long run. In fact, given the chance to let machines do the job, Google cofounder Larry Page believes that nine out of 10 people "wouldn't want to be doing what they're doing today" [153].

Shortly after winning *Jeopardy!*, IBM Watson began to learn how to make diagnoses and treatment recommendations at the Memorial Sloan-Kettering Cancer Center in New York. "The process of pulling out two key facts from a Jeopardy clue is totally different from pulling out all the relevant information, and its relationships, from a medical case," said Ari Caroline, Sloan-Kettering's director of quantitative analysis and strategic initiatives. "Sometimes there is conflicting information. People phrase things in different ways" [154].

The voluminous amount of structured and unstructured health information calls for the need of big data scientists to build models to explain and predict patterns. Once again, artificial intelligence has outperformed human researchers. In 2015, MIT's Data Science Machine can create accurate predictive models from raw datasets within 2–12 h whereas a team of human data scientists can take months [155].

In 2016, artificial intelligence is assisting radiologists to identify cancer and other medical abnormalities in X-rays, CT scans, and MRIs [156]. "There are a lot of advances in facial recognition that we wanted to adapt to medicine because it's about determining where the nodules, aneurysms, and things like that are," said Columbia University student researcher Jeet Samarth Raut whose mother was misdiagnosed by radiologists about a rare kind of breast cancer [157]. Artificial intelligence is becoming an indispensable tool in advancing medical science and technology.

At the TED2016 conference, IBM announced the IBM Watson AI XPRIZE which offers $5 million to the winning team. "Our hope is that the teams will show how we can apply AI to the world's great challenges," said Stephanie Wander of the XPRIZE Foundation. "That would be the cat's meow" [158].

At the 2016 Google I/O developer conference, Google senior vice president of infrastructure Urs Holzle revealed the deployment of thousands of specialized artificial intelligence chips—TensorFlow Processing Units (TPUs)—in servers within Google's data centers [159]. TensorFlow is an open source software library for machine learning, originally developed by the Google Brain Team for use in speech recognition, Gmail, Google Photos, Google Search, and other Google products [160].

3.19 Human-Machine Symbiosis and Google DeepMind

In 1960, MIT psychologist and computer scientist J. C. R. Licklider laid the foundation of man-computer symbiosis which led to interactive computing, graphical user interface, the Internet, and today's digital era.

In October 2007, Stanford University's Folding@home project received a Guinness World Record for topping 1 petaflop (a thousand trillion floating point operations per second) running on computers as well as Sony's PlayStation 3 video game consoles [161]. Folding@home helps scientists study protein folding and its relationship to Alzheimer's, Huntington's, and cancerous diseases [162]: "Foldit attempts to predict the structure of a protein by taking advantage of humans' puzzle-solving intuitions and having people play competitively to fold the best proteins. Since proteins are part of so many diseases, they can also be part of the cure. Players can design brand new proteins that could help prevent or treat important diseases" [163].

In September 2011, players of the Foldit video game took less than 10 days to decipher the AIDS-causing Mason-Pfizer monkey virus that had stumped scientists for 15 years [164]. The astonishing accomplishment exemplifies the power of human-machine symbiosis. Fast Company's Michael J. Coren wrote in *Scientific American*, "Humans retain an edge over computers when complex problems require intuition and leaps of insight rather than brute calculation" [165].

In early 2014, Google acquired DeepMind cofounded by Demis Hassabis, Shane Legg, and Mustafa Suleyman to "solve intelligence" and "use it to make the world a better place" [166]. 19 years after IBM Deep Blue defeated world chess champion Garry Kasparov in 1997 under tournament regulations, Google DeepMind's AlphaGo beat legendary Go player Lee Se-dol in 2016 [167]. Go is an ancient Chinese board game that is much more complex than chess involving both strategic and tactical skills [168]. Apart from well-defined board games, Google's DeepMind has also been learning to play video games in a chaotic virtual world. "Taken together, our work illustrates the power of harnessing state-of-the-art machine learning techniques with biologically inspired mechanisms to create agents that are capable of learning to master a diverse array of challenging tasks," said Demis Hassabis [169].

Biologically inspired mechanisms will encourage deeper human-machine symbiosis as computers acquire more human intuitions in problem solving. Scientists and gamers will be working and playing side-by-side with intelligent machines as equals, not subordinates. While artificial intelligence assists humans in solving the most challenging problems, human-based computation (HBC) allows machines to outsource certain tasks to humans to tackle. "In a way, human computation is like cheating at artificial intelligence," said Pietro Michelucci, executive director at Human Computation Institute. "It's sometimes jokingly referred to as 'artificial artificial intelligence,' because what we effectively do is take an AI algorithm and say: this is the hard part that we can't do with computers, so let's farm this part out to a human. It's really like saying we can create the sort of artificial intelligence we imagine for the future today, just by building humans into the system" [170].

J. C. R. Licklider wrote in his 1960 research paper, "The fig tree is pollinated only by the insect *Blastophaga grossorun*. The larva of the insect lives in the ovary of the fig tree, and there it gets its food. The tree and the insect are thus heavily interdependent: the tree cannot reproduce wit bout the insect; the insect cannot eat wit bout the tree; together, they constitute not only a viable but a productive and thriving partnership. This cooperative 'living together in intimate association, or even close union, of two dissimilar organisms' is called symbiosis. ... The hope is that, in not too many years, human brains and computing machines will be coupled together very tightly, and that the resulting partnership will think as no human brain has ever thought and process data in a way not approached by the information-handling machines we know today" [171].

3.20   Turing's Imitation Game and Animal Testing

Alan Turing , widely considered to be the father of theoretical computer science and artificial intelligence, proposed an "Imitation Game" in his 1950 seminal paper "Computing Machinery and Intelligence" published in the journal *Mind* [172]:

> I propose to consider the question, "Can machines think?" This should begin with definitions of the meaning of the terms "machine" and "think." The definitions might be framed so as to reflect so far as possible the normal use of the words, but this attitude is dangerous. If the meaning of the words "machine" and "think" are to be found by examining how they are commonly used it is difficult to escape the conclusion that the meaning and the answer to the question, "Can machines think?" is to be sought in a statistical survey such as a Gallup poll. But this is absurd. Instead of attempting such a definition I shall replace the question by another, which is closely related to it and is expressed in relatively unambiguous words.

> The new form of the problem can be described in terms of a game which we call the "imitation game." It is played with three people, a man (A), a woman (B), and an interrogator (C) who may be of either sex. The interrogator stays in a room apart from the other two. The object of the game for the interrogator is to determine which of the other two is the man and which is the woman. He knows them by labels X and Y, and at the end of the game he says either "X is A and Y is B" or "X is B and Y is A." The interrogator is allowed to put questions to A and B thus:

> C: Will X please tell me the length of his or her hair?

> Now suppose X is actually A, then A must answer. It is A's object in the game to try and cause C to make the wrong identification. His answer might therefore be:

> "My hair is shingled, and the longest strands are about nine inches long."

> In order that tones of voice may not help the interrogator the answers should be written, or better still, typewritten. The ideal arrangement is to have a teleprinter communicating between the two rooms. Alternatively the question and answers can be repeated by an intermediary. The object of the game for the third player (B) is to help the interrogator. The best strategy for her is probably to give truthful answers. She can add such things as "I am the woman, don't listen to him!" to her answers, but it will avail nothing as the man can make similar remarks.

> We now ask the question, "What will happen when a machine takes the part of A in this game?" Will the interrogator decide wrongly as often when the game is played like this as he does when the game is played between a man and a woman? These questions replace our original, "Can machines think?"

The eponymous Turing test essentially states that "a computer would deserve to be called intelligent if it could deceive a human into believing that it was human" [173]. The first well-known software program to put to the Turing test was ELIZA in 1996, created by Joseph Weizenbaum at the MIT Artificial Intelligence Laboratory to mimic a human psychotherapist [174]:

Human: "Men are all alike."

ELIZA: "In what way?"

Human: "They're always bugging us about something."

ELIZA: "Can you think of a specific example?"

Human: "Well, my boyfriend made me come here."

ELIZA: "Your boyfriend made you come here?"

Human: "He says I'm depressed much of the time."

ELIZA: "I'm sorry to hear you are depressed."

Human: "It's true."

Equipped with natural language processing, ELIZA was named after the fictional character Eliza Doolittle who was eager to learn from Prof. Henry Higgins in the George Bernard Shaw novel *Pygmalion*, the musical play *My Fair Lady*, and the film adaption of the same name starring Audrey Hepburn and Rex Harrison. Numerous chatterbots, chatbots, or artificial conversational entities (ACEs) have since been created for different applications including the Mattel toy Hello Barbie [175].

In 2002, Mitchell Kapor and Ray Kurzweil placed a wager of $20,000 on the Turing test. The Long Now Foundation administers the bet which predicts that "by 2029 no computer—or 'machine intelligence'—will have passed the Turing Test" [176].

The bet's predictor Mitchell Kapor, best known for Lotus 1-2-3 and a cofounder of Electronic Frontier Foundation (EFF), contended that "while it is possible to imagine a machine obtaining a perfect score on the SAT or winning Jeopardy— since these rely on retained facts and the ability to recall them—it seems far less possible that a machine can weave things together in new ways or to have true imagination in a way that matches everything people can do, especially if we have a full appreciation of the creativity people are capable of."

The bet's challenger Ray Kurzweil, futurist and director of engineering at Google, argued that "there are many contemporary examples of computers passing 'narrow' forms of the Turing test, that is, demonstrating human-level intelligence in specific domains. For example, Gary Kasparov, clearly a qualified judge of human chess intelligence, declared that he found Deep Blue's playing skill to be indistinguishable from that of a human chess master during the famous tournament in which he was defeated by Deep Blue."

Indeed, Kasparov wrote an essay to the *TIME Magazine* in March 1996 acknowledging a "new kind of intelligence" as he said, "I GOT MY FIRST GLIMPSE OF ARTIFICIAL INTELLIGENCE ON Feb. 10, 1996, at 4:45 p.m. EST, when in the first game of my match with Deep Blue, the computer nudged a pawn forward to a square where it could easily be captured. It was a wonderful and extremely human move. … I had played a lot of computers but had never experienced anything like this. I could feel—I could smell—a new kind of intelligence across the table" [177]. In Nate Silver's 2012 book *The Signal and the Noise*, IBM scientist Murray Campbell from the Deep Blue team revealed that the "extremely human move" in the chess game against Gary Kasparov was actually a bug in the program that was later fixed [178]. What an opportune moment for a computer to evince that "To err is human!"

Over the years, variations of the Turing test have been proposed to augment the original imitation game. Amongst them is the "Ebert test" by film critic Roger Ebert who lost his lower jaw to cancer, unable to eat and speak [179]. In a moving talk at TED2011 using a custom-tailored text-to-speech software, Ebert jocularly told the audience that "if the computer can successfully tell a joke, and do the timing and delivery, as well as Henny Youngman, then that's the voice I want" [180]. Ebert's wish may come true sooner rather than later. The Institute of Advanced Industrial Science and Technology in Japan has already developed the world's first robot pop star HRP-4C who can sing and dance [181].

The Turing test is evolving to become a full-blown mind game. In the 2015 sci-fi film *Uncanny*, an artificial intelligent humanoid fooled an on-screen reporter and the real-life audience into believing that the robot was a human and the human was an android—an ultimate embodiment of the "Imitation Game" that would pass with flying colors not only the Turing test but also the Voight-Kampff test from Philip K. Dick's 1968 novel *Do Androids Dream of Electric Sheep?* and the 1982 movie adaptation *Blade Runner* starring Harrison Ford.

In academia, Prof. Adrian Cheok of City University London and Dr. David Levy of Intelligent Toys have been co-chairing the annual International Congress on Love and Sex with Robots [182]. In 2015, Prof. Nadia Thalmann at the Nanyang Technological University unveiled the world's most human-like social robot Nadine with soft skin and brunette hair, which "smiles when greeting you, looks at you in the eye when talking, shakes hands with you, ... recognizes the people she has met, and remembers what the person had said before" [183].

In business, Google in 2013 acquired 8 robotics engineering companies including Boston Dynamics [184]. Founded in 1992 by MIT researcher Marc Raibert, Boston Dynamics has built DigDog [185], WildCat [186], Spot [187], and Atlas [188], among others. In 2016, Spot was seen playing with and almost fooling a real dog [189] whereas the Next Generation Atlas took a giant leap towards the realization of C-3PO in *Star Wars* [188].

Some YouTubers were disturbed by the videos showing Spot and Atlas being kicked and bullied for testing and demonstration purposes. While humans display empathy towards nonliving robots, laboratory animals for medical experiments are often subject to extreme cruelty. People for the Ethical Treatment of Animals (PETA) reports that "U.S. law allows animals to be burned, shocked, poisoned, isolated, starved, drowned, addicted to drugs, and brain-damaged. No experiment, no matter how painful or trivial, is prohibited—and pain-killers are not required. Even when alternatives to the use of animals are available, the law does not require that they be used—and often they aren't. Animals are infected with diseases that they would never normally contract, tiny mice grow tumors as large as their own bodies, kittens are purposely blinded, rats are made to suffer seizures, and primates' skulls are cut open and electrodes are implanted in them. Experimenters force-feed chemicals to animals, conduct repeated surgeries on them, implant wires in their brains, crush their spines, and much more" [190]. In Tim Burton's *Planet of the*

Apes (2001), Zaius the Ape (Charlton Heston) excoriated humanity, "I warn you, their ingenuity goes hand-in-hand with their cruelty. No creature is as devious, as violent" [191].

Alan Turing would have been a big supporter of PETA because he himself endured 2 years of horrifying and humiliating chemical castration until his death in 1954 at age 41. He accepted castration as the only alternative to incarceration for his criminal conviction in 1952 for his homosexuality. In 1966, the Association for Computing Machinery (ACM) gave out the first annual A.M. Turing Award that is generally recognized as the "Nobel Prize of Computing" [192]. In 2013, Queen Elizabeth II granted Turing a posthumous pardon, but the decision was controversial as British journalist Ally Fogg retorted that "Turing should be forgiven not because he was a modern legend, but because he did absolutely nothing wrong. The only wrong was the venality of the law" [193].

3.21 King Solomon's Challenge and the Trolley Problem

King Solomon asked God for wisdom and knowledge instead of money and power that most kings and emperors were hunger for:

> That night God appeared to Solomon and said to him, "Ask for whatever you want me to give you." Solomon answered God, "You have shown great kindness to David my father and have made me king in his place. Now, LORD God, let your promise to my father David be confirmed, for you have made me king over a people who are as numerous as the dust of the earth. Give me wisdom and knowledge, that I may lead this people, for who is able to govern this great people of yours?" God said to Solomon, "Since this is your heart's desire and you have not asked for wealth, possessions or honor, nor for the death of your enemies, and since you have not asked for a long life but for wisdom and knowledge to govern my people over whom I have made you king, herefore wisdom and knowledge will be given you. And I will also give you wealth, possessions and honor, such as no king who was before you ever had and none after you will have." (2 Chronicles 7–12)

What would an intelligent computer ask its human creator for? Wealth, possession, honor, and other human desires—good or bad—bear no meaning to a machine. Wisdom and knowledge, on the other hand, constitute the heart and soul of artificial intelligence. As artificial intelligence continues to evolve into superintelligence, King Solomon's challenge will supersede the Turing test:

> Now two prostitutes came to the king and stood before him. One of them said, "Pardon me, my lord. This woman and I live in the same house, and I had a baby while she was there with me. The third day after my child was born, this woman also had a baby. We were alone; there was no one in the house but the two of us. During the night this woman's son died because she lay on him. So she got up in the middle of the night and took my son from my side while I your servant was asleep. She put him by her breast and put her dead son by my breast. The next morning, I got up to nurse my son—and he was dead! But when I looked at him closely in the morning light, I saw that it wasn't the son I had borne." The other woman said, "No! The living one is my son; the dead one is yours." But the first one

insisted, "No! The dead one is yours; the living one is mine." And so they argued before the king.

The king said, "This one says, 'My son is alive and your son is dead,' while that one says, 'No! Your son is dead and mine is alive.'" Then the king said, "Bring me a sword." So they brought a sword for the king. He then gave an order: "Cut the living child in two and give half to one and half to the other." The woman whose son was alive was deeply moved out of love for her son and said to the king, "Please, my lord, give her the living baby! Don't kill him!" But the other said, "Neither I nor you shall have him. Cut him in two!" Then the king gave his ruling: "Give the living baby to the first woman. Do not kill him; she is his mother." (1 Kings 3:16–27)

In the modern era, a judge would order a DNA test to determine the biological mother of the baby and solve the case easily unless the two mothers are identical twins. King Solomon's method not only circumvents the identical twin problem but also signifies a wiser decision of giving the child to the caring parent who may or may not be the biological mother. There have been too many horror stories of child abuse, neglect, and endangerment by irresponsible parents.

King Solomon's judgment is an exemplification of wisdom trumping science and technology. Imagine an intelligent machine endowed with King Solomon's wisdom! Computers have already been elevated to the same level as intelligent people in domains as specialized as tournament chess and as generalized as *Jeopardy!* game show. For the first time in history, a nonhuman is granted the "driver" status under federal law by the U.S. National Highway Traffic Safety Administration (NHTSA) in February 2016 [194]. The nonhuman driver is the Self-Driving System (SDS)—an artificial intelligence driver behind Google's self-driving vehicles (SDVs) (see Fig. 3.35) [195].

The question is whether human beings will be willing to relinquish control to machine decisions that are wiser than people's in most circumstances. Opponents of Google's self-driving cars have already raised the ethical issue using the trolley problem that was first proposed by British philosopher Philippa Foot [196]:

To make the parallel as close as possible it may rather be supposed that he is the driver of a runaway tram which he can only steer from one narrow track on to another; five men are working on one track and one man on the other; anyone on the track he enters is bound to be killed.

Foot asked the moral question of whether the driver of the runaway tram should steer from the current track to another track in order to minimize the number of causalities. Are people ready to accept the moral judgments made by machine intelligence?

A stable society requires its citizens to give up certain control to the police and the court of law. A few bad apples in the police force and some unfair trials in the judicial system do not call for citizens to arm themselves and exact vengeance on their own. Machines are bound to make mistakes but there will not be as many or deadly as human errors.

Self-driving cars can offer the algorithmically optimal solution to reduce traffic congestion by minimizing unnecessary lane changes, rerouting traffic if needed, and eliminating accidents caused by reckless, drowsy, or drunk driving [197, 198]. Prof.

Fig. 3.35 Google self-driving car in action (courtesy of Roman Boed under Creative Commons license)

Cyrus Shahabi at the University of Southern California offers his thoughts on the future of driverless cars and decision-making responsibility in Chap. 21 of this book.

Opponents of self-driving cars often cite cybersecurity as a major concern. However, cybersecurity is already an issue with all newer vehicles. A 2014 Sky News investigation has found that almost half the 89,000 vehicles broken into in London were hacked electronically [199]. Darren Manners, senior security engineer at SyCom, discusses the importance of continue penetration testing in Chap. 22 of this book. Prof. Dennis Gamayunov and Mikhail Voronov at the Lomonosov Moscow State University explore autonomous vulnerability scanning and patching of binaries in Chap. 23. The new research may lead to self-healing software in cars, drones, robots, and the Internet of Things (IoT).

3.22 Artificial General Intelligence and Superintelligence

The one thing that Stephen Hawking, Elon Musk, Peter Thiel, and some other notable technology entrepreneurs are afraid of is artificial superintelligence that goes beyond the trolley problem. Marc Goodman, global security advisor and futurist, spoke at the TEDGlobal 2012 in Edinburgh about his ominous prediction:

"If you control the code, you control the world. This is the future that awaits us" [200].

Astrophysicist Stephen Hawking said in a 2014 interview by the BBC, "The development of full artificial intelligence could spell the end of the human race. ... It would take off on its own, and re-design itself at an ever increasing rate. Humans, who are limited by slow biological evolution, couldn't compete, and would be superseded" [201].

Speaking at the MIT Aeronautics and Astronautics department's Centennial Symposium in 2014, Elon Musk of Testa Motors, SpaceX, and PayPal called artificial intelligence the biggest existential threat and he said, "Increasingly scientists think there should be some regulatory oversight maybe at the national and international level, just to make sure that we don't do something very foolish. With artificial intelligence we are summoning the demon" [202]. In a 2015 CNN interview, Elon Musk clarified his viewpoint, "AI is much more advanced than people realize. ... It'll be fairly obvious if you saw a robot walking around, talking, and behaving like a person. ... that would be really obvious. What's not obvious is a huge server bank in a dark vault somewhere with intelligence that's potentially vastly greater than what a human mind can do. I mean its eyes and ears would be everywhere, every camera, every microphone, every device that's network accessible, that's what really what AI means. ... Humanity's position on this planet depends on its intelligence. So if our intelligence is exceeded, it's unlikely that we will remain in charge of the planet" [203].

For better or worse, artificial intelligence has been baked into our society and global economy. More people are relying on intelligent personal assistants such as Apple's Siri, Microsoft's Cortana, Facebook's M, and Google Now. In 2012, CNN reported that "stock markets have become increasingly vulnerable to bugs over the last decade thanks to financial firms' growing reliance on high-speed computerized trading. Because the trading is automated, there's nobody to apply the brakes if things go wrong" [204]. In 2016, machine learning is being applied to insurance and loan underwriting [205]. Behind the scenes, artificial intelligence is being used in banks, hospitals, clinical laboratories, investment firms, law enforcement agencies, and almost every industry and sector.

It is inevitable that artificial intelligence is becoming smarter over time to the point of surpassing human intelligence. For the wager on the Turing test, Ray Kurzweil remarked that "once a computer does achieve a human level of intelligence, it will necessarily soar past it. Electronic circuits are already at least 10 million times faster than the electrochemical information processing in our interneuronal connections. Machines can share knowledge instantly, whereas we biological humans do not have quick downloading ports on our neurotransmitter concentration levels, interneuronal connection patterns, nor any other biological bases of our memory and skill" [206]. For example, in 2015 Microsoft's convolutional neural network has outperformed humans in identifying objects in digital images (see Fig. 3.36). "While humans can easily recognize these objects as a bird, a

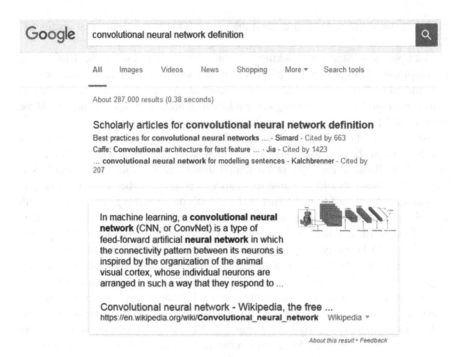

Fig. 3.36 Google search on "convolutional neural network" returns a Knowledge Graph of a list of oft-cited scholarly articles in addition to a definition from Wikipedia about machine learning using a type of artificial neural network inspired by the animal visual cortex

dog, and a flower, it is nontrivial for most humans to tell their species," said the Microsoft researchers involved in the project [207].

Swedish philosopher Nick Bostrom at the University of Oxford defines "superintelligence" as "an intellect that is much smarter than the best human brains in practically every field, including scientific creativity, general wisdom and social skills" [208]. He believes that "it would be a huge tragedy if machine superintelligence were never developed. That would be a failure mode for our Earth-originating intelligent civilization. ... Artificial intelligence is the technology that unlocks this much larger space of possibilities, of capabilities, that enables unlimited space colonization, that enables uploading of human minds into computers, that enables intergalactic civilizations with planetary-size minds living for billions of years. ... I'm not sure that I'm not already in a machine" [209].

There are moments in time when we feel that we are trapped in a mortal body, stuck between a rock and a hard place, being pulled in two opposite directions, or treading the fine line between reality and illusion. In order to improve our game—whether it is intellectual like chess or physical like tennis, we need to play against someone who is better than us. We seek help from mentors, doctors, and experts who are smarter and more experienced than we are. In tribute to *The Matrix* trilogy, artificial superintelligence is the disruptive red pill that will help us to transcend our human limitations in mind, body, and spirit.

Instead of fearing the unknown, we must conquer our fear. Like a good doctor, an intelligent machine will cure their patients without killing them. A good doctor, whether it is human or artificial, is not someone who we should be afraid of. Indeed, Elon Musk, Peter Thiel, and others have invested more than $1 billion into OpenAI whose research director is Ilya Sutskever, former research scientist at the Google Brain Team. "If you're going to summon anything, make sure it's good," said Musk. "The goal of OpenAI is really somewhat straightforward, it's what set of actions can we take that increase the probability of the future being better. We certainly don't want to have any negative surprises on this front" [210].

Google's DeepMind has been working towards the goal of building Artificial General Intelligence (AGI) based on deep learning and systems neuroscience [211]. In Chaps. 24–26 of this book, Microsoft MVP and transhumanist David J. Kelley proposes an Independent Core Observer Model (ICOM) architecture for AGI.

In the 2013 romantic sci-fi comedy-drama *Her*, writer Theodore Twombly (Joaquin Phoenix) develops an affectionate relationship with Samantha (Scarlett Johansson), an intelligent computer operating system [212]:

Samantha: Is that weird? You think I'm weird?

Theodore: Kind of.

Samantha: Why?

Theodore: Well, you seem like a person but you're just a voice in a computer.

Samantha: I can understand how the limited perspective of an un-artificial mind might perceive it that way. You'll get used to it.

While Ava (Alicia Vikander) and Kyoko (Sonoya Mizuno) in *Ex Machina* (2015) exhibit advanced intelligence, human-machine symbiosis goes beyond sexbots to develop empathy and relationship. The last thing we want is a nasty divorce between humans and superintelligent machines, for that would certainly spell the end of the human race.

On the positive front, human-machine symbiosis and artificial superintelligence may hold the key to the Holy Grail of human longevity and immortality. In the 2014 sci-fi film *Transcendence*, Evelyn Caster (Rebecca Hall) helped the dying Will Caster (Johnny Depp) to upload his consciousness to a quantum computer. Will Caster remarked, "For 130,000 years, our capacity to reason has remained unchanged. The combined intellect of the neuroscientists, mathematicians and… hackers… in this auditorium pales in comparison to the most basic A.I. Once online, a sentient machine will quickly overcome the limits of biology. And in a short time, its analytic power will become greater than the collective intelligence of every person born in the history of the world. So imagine such an entity with a full range of human emotion. Even self-awareness. Some scientists refer to this as 'the Singularity.' I call it 'Transcendence'" [213].

3.23   Transhumanism and Anti-Death Political Candidates

"We're going to gradually enhance ourselves," said Ray Kurzweil, futurist and engineering director at Google. "That's the nature of being human—we transcend our limitations" [1]. Kurzweil is an outspoken advocate of transhumanism—an international and intellectual movement that aims to improve the human condition by developing widely available technologies to enhance human intellectual, physical, and psychological capacities. The term "transhumanism" was coined in 1957 by Julian Huxley, the first director-general of the United Nations Educational, Scientific and Cultural Organization (UNESCO) and the older brother of Aldous Huxley who wrote the 1932 novel *Brave New World* about a technological dystopia.

Pacemakers, prosthesis, stentrode, optogenetics, antibiotics, and other medical advancements exemplify the use of technology to prolong life and to improve quality of life. Christopher Reeve, best known for playing the role of the comic book superhero Superman, lobbied for human embryonic stem cell research after a horse-riding accident left him quadriplegic [214]. In spite of the traditionally conservative views from the Catholic Church, Pope Francis gave his blessing to human-animal chimera research for organ transplants [2].

Notwithstanding the negative connotation of transhumanism revealed by Google searches as being crazy, stupid, evil, or insane (see Fig. 3.37), we are all transhumanists in varying degrees. That includes not only down-to-earth adults but also children and adolescents who fantasize about superhuman abilities. In the over-the-top black comedy action film *Kick-Ass* (2010) based on the comic book by Mark Millar and John Romita, Jr., protagonist Dave Lizewski (Aaron Taylor-Johnson) said, "At some point in our lives we all wanna be a superhero. ... In the world I lived in, heroes only existed in comic books. And I guess that'd be okay, if bad guys were make-believe too, but they're not" [215].

Dr. Natasha Vita-More, chairperson of Humanity+ and professor at the University of Advancing Technology, examines the growing worldview of transhumanism in Chap. 27 of this book. She was elected as a Councilperson for the 28th Senatorial District of Los Angeles in 1992 on an openly futurist and transhumanist platform [216]. In Chap. 28, Dirk Bruere, secretary and deputy leader at the Transhumanist Party in the United Kingdom, shares his personal story of why he became a transhumanist. In Chap. 29, Emily Peed at the Institute for Education,

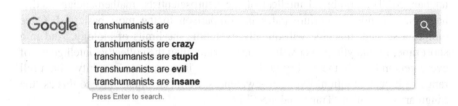

Fig. 3.37 Google search autocomplete predictions for "transhumanists are" (as of June 2016)

Research, and Scholarships (IFERS) discusses the splintering and controversy of transhumanism.

In 2016, Zoltan Istvan became the first transhumanist to run for the Presidency of the United States, aiming to put science, health, and technology at the forefront of American politics. *BuzzFeed* calls Istvan the "anti-death presidential candidate" [217]. Christopher Reeve would have appreciated the idea, for he told *The Guardian* in a 2002 interview, "I'm angry, and disappointed… I think we could have been much further along with scientific research than we actually are, and I think I would have been in quite a different situation than I am today. … If we'd had full government support, full government funding for aggressive research using embryonic stem cells from the moment they were first isolated, at the University of Wisconsin in the winter of 1998—I don't think it unreasonable to speculate that we might be in human trials by now" [214]. A transhumanist platform and an interview with Zoltan Istvan are found in Chap. 30 of this book. Educator Robert Niewiadomski and Prof. Dennis Anderson of St. Francis College present their hopes and expectations for the next president in Chap. 31.

The 2016 U.S. presidential race has almost become a circus like *Betty Boop for President*, a 1932 Fleischer Studios animated short film in which Betty ran for the office of President against Mr. Nobody [218]. In answer to various social issues, Mr. Nobody made many empty promises: "Who will make your taxes light? Mr. Nobody! Who'll protect the voters' right? Mr. Nobody! … When you're hungry, who feeds you? Mr. Nobody! …" Satire and sarcasm aside, a CNN and Facebook poll of college students in December 2015 found that the presidential election has made the millennials feel "disappointed, frustrated, uneasy, worried, embarrassing, divided, and anxious" [219]. A Spring 2016 survey by Harvard's Institute of Politics (IOP) indicated that just 15 % of 18- to 29-year-olds think that the nation is heading in the right direction while 47 % say that the country is on the wrong track [220].

What we need are leaders who can think long-term. In 2014, Sweden's Prime Minister Stefan Löfven appointed 70-year-old Kristina Persson to be the Minister of the Future. "If politics wants to remain relevant and be useful to citizens, it needs to change its approach," said Persson in an interview. "Finding solutions needs the cooperation of all of society's stakeholders. No one [can be] excluded. … Rather than going top-down, we promote inter-ministerial collaboration and force decision makers to confront the long-term issues despite the fact this is harder to do sometimes" [221]. Elsewhere in Russia, molecular biophysicist Maria Konovalenko, Mikhail Batin, Alexey Turchin, Leonid Kaganov, Elena Milova, and others cofounded the Longevity Party in 2012 to "increase human lifespan so that people could live for as long as they would like to and remain young and healthy" [222]. In the same year, Russian entrepreneur Dmitry Itskov created the 2045 Initiative and launched the Evolution 2045 political party to advocate life extension and humanoid robots "not in the arms race, but in the race for building a bright future for mankind" [223].

"The future is ours to shape," said Max Tegmark, MIT cosmologist and cofounder of the Future of Life Institute. "I feel we are in a race that we need to win. It's a race between the growing power of the technology and the growing wisdom we need to manage it". We should heed the warnings from the 1969 hit song "In the

Year 2525 (Exordium & Terminus)" by Dennis Zager and Rick Evans: "… In the year 5555, your arms hangin' limp at your sides, your legs got nothin' to do, some machine's doin' that for you… In the year 8510, God is gonna shake His mighty head. He'll either say I'm pleased where man has been, or tear it down and start again. In the year 9595, I'm kinda wonderin' if man is gonna be alive. He's taken everything this old earth can give, and he ain't put back nothing" [224].

Transhumanism can offer new insights and unorthodox solutions to insurmountable problems that have stumped politicians for years, decades, and centuries. Whether we call ourselves futurists, transhumanists, scientists, technologists, philosophers, or politicians, we ought to share the same goal of advancing humanity and saving the planet Earth. In a 2015 interview by the 2045 Strategic Social Initiative, TV anchor Olesya Yermakova asked SENS Research Foundation cofounder Aubrey de Grey, "Do you consider yourself a Transhumanist?" And de Grey replied. "Not really. No. I really just consider myself a completely boring medical researcher. I just want to stop people from getting sick" [225]. As Apostle Paul wrote in his letters to the Corinthians, "The last enemy to be destroyed is death" (1 Corinthians 15:26). Our strongest ally in the battle against death is the combination of futurism, transhumanism, science, technology, philosophy, and politics.

3.24   Google's D-Wave Quantum Computer

The D-Wave quantum computers at Google, NASA, Lockheed Martin, Los Alamos National Laboratory, and other technology companies may just provide the much-needed new hardware and software paradigm to accelerate machine superintelligence and to solve difficult problems requiring "creativity" [226].

Hartmut Neven, director of engineering at Google's Quantum Artificial Intelligence Laboratory, compared the D-Wave experiments to the Wright brothers' flight trails at Kitty Hawk in 1903: "In fact, the [D-Wave's] trajectory went through parallel universes to get to the solution. It is literally that. That is an amazing, somewhat historical, event. It has worked in principle. The thing flew. … Classical system can only give you one route out. You have to walk up over the next ridge and peak behind it, while quantum mechanisms give you another escape route, by going through the ridge, going through the barrier" [227].

Prof. Aephraim Steinberg at the University of Toronto explained the paradigm shift: "In the past, we believed all computers fundamentally did the same thing—just maybe one a bit faster than another. Now, as far as we can tell, this is just wrong. In the quantum world, information simply behaves differently from in the classical world. If a system, whether an electron or a computer, can be in two or a million different states, it can also be in what we call a superposition of all those states, and that gives it much more room to maneuver to try to get from input to output" [228].

Krysta Svore at Microsoft Research expressed high hopes for its applications: "With a quantum computer, we hope to find a more efficient way to produce artificial fertilizer, having direct impact on food production around the world, and we hope to combat global warming by learning how to efficiently extract carbon dioxide from the environment. Quantum computers promise to truly transform our world" [228].

With quantum computing (Figs. 3.38 and 3.39), scientists are learning how to better formulate questions, lest we end up with a perplexing answer like "42" [229]. In Douglas Adams' *The Hitchhiker's Guide to the Galaxy* (first broadcasted in 1978), the "Deep Thought" supercomputer took seven and a half million years to compute the answer to the ultimate question of life, the universe, and everything. The answer turned out to be "42," because the ultimate question itself was unknown or ill-defined [230]. One can also postulate that the supercomputer in the novel was subtly referring to the 42-line Bible, better known as the Gutenberg Bible (see Fig. 3.40).

Scientists are also getting used to the unpredictable nature of quantum computing which can provide solutions that we may never have considered otherwise

Fig. 3.38 Google search on "quantum computing" returns a Knowledge Graph of a definition from Wikipedia about using quantum-mechanical phenomena such as superposition and entanglement to perform operations on data. It also shows related topics on qubit, spins, and quantum cryptography

Fig. 3.39 Photograph of a chip constructed by D-Wave Systems Inc. designed to operate as a 128-qubit superconducting adiabatic quantum optimization processor, mounted in a sample holder

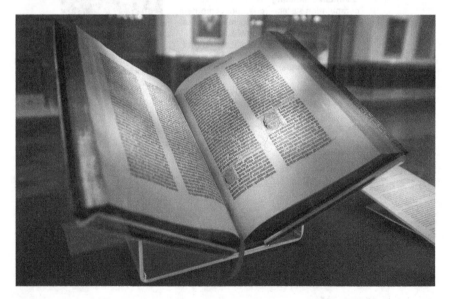

Fig. 3.40 42-line Gutenberg Bible at the New York Public Library. Originally bought by James Lenox in 1847. (Courtesy of Kevin Eng under Creative Commons license)

[231]. Albert Einstein objected to the apparent randomness in nature, asserting that God did not play dice. Quantum mechanics proponents disagreed. Prof. Renato Renner at ETH Zurich said, "Not only does God 'play dice,' but his dice are fair" [231], and Stephen Hawking believes that "the future of the universe is not completely determined by the laws of science, and its present state, as [Pierre-Simon] Laplace thought. God still has a few tricks up his sleeve" [232].

3.25   To Be, or not to Be, that Is the Question

Rhea Drysdale, CEO of Outspoken Media, told David Goldman at CNN that "Google understands humans better than we understand ourselves" [233]. As scientists and legislators are being pulled in a million directions, Google cofounder Larry Page has broached the crucial question: "Are people really focused on the right things?" [39]. Are we asking the right questions?

In William Shakespeare's play *Hamlet* (Act III, Scene 1), a despondent Prince of Denmark sparked an internal philosophical debate on life and death in one of the most widely known soliloquies:

> To be, or not to be, that is the question:
> Whether 'tis Nobler in the mind to suffer
> The Slings and Arrows of outrageous Fortune,
> Or to take Arms against a Sea of troubles,
> And by opposing end them: to die, to sleep
> No more; and by a sleep, to say we end
> The Heart-ache, and the thousand Natural shocks
> That Flesh is heir to? 'Tis a consummation
> Devoutly to be wished. To die, to sleep,
> To sleep, perchance to Dream; aye, there's the rub,
> For in that sleep of death, what dreams may come,
> When we have shuffled off this mortal coil,
> Must give us pause. There's the respect
> That makes Calamity of so long life:
> For who would bear the Whips and Scorns of time,
> The Oppressor's wrong, the proud man's Contumely,
> The pangs of despised Love, the Law's delay,
> The insolence of Office, and the Spurns
> That patient merit of the unworthy takes,
> When he himself might his Quietus make
> With a bare Bodkin? Who would Fardels bear,
> To grunt and sweat under a weary life,
> But that the dread of something after death,
> The undiscovered Country, from whose bourn
> No Traveller returns, Puzzles the will,
> And makes us rather bear those ills we have,
> Than fly to others that we know not of.
> Thus Conscience does make Cowards of us all,
> And thus the Native hue of Resolution
> Is sicklied o'er, with the pale cast of Thought,

And enterprises of great pitch and moment,
With this regard their Currents turn awry,
And lose the name of Action.

The exact meaning of the phrase "to be, or not to be" is open to debate. Shakespeare scholar Harold Jenkins opines that "nothing anywhere in the speech relates it to Hamlet's individual case. He uses the pronouns *we* and *us*, the indefinite *who*, the impersonal infinitive. He speaks explicitly of *us all*, of what *flesh* is heir to, of what *we* suffer at the hands of *time* or *fortune*—which serves incidentally to indicate what for Hamlet is meant by *to be*" [234]. Yale University humanities professor Harold Bloom interpreted the soliloquy as a "testimony to the power of the mind over a universe of death, symbolized by the sea, which is the great hidden metaphor" [235].

"To be, or not to be" is such a mesmerizing quote because it applies to almost every decision in life. The choices we make define who we are. In Plato's *The Apology*, Socrates posed a hypothetical question to himself: "Socrates, can you not go away from us and live quietly, without talking?" And he replied [236]:

> Now this is the hardest thing to make some of you believe. For if I say that such conduct would be disobedience to the god and that therefore I cannot keep quiet, you will think I am jesting and will not believe me; and if again I say that to talk every day about virtue and the other things about which you hear me talking and examining myself and others is the greatest good to man, and that the unexamined life is not worth living, you will believe me still less.

Henceforth the famous dictum: "The unexamined life is not worth living," or in ancient Greek, "ὁ ἀνεξέταστος βίος οὐ βιωτὸς ἀνθρώπῳ." Biochemist-geneticist Craig Venter of the Human Genome Project concurred with Socrates, "It's not just a long life we're striving for, but one which is worth living" [95]. American author Henry David Thoreau wrote in *Walden; Or, Life in the Woods* (1854): "I went to the woods because I wished to live deliberately, to front only the essential facts of life, and see if I could not learn what it had to teach, and not, when I came to die, discover that I had not lived" [237].

According to the U.S. Marine Corps lore, highly decorated Sergeant Major Daniel Daly in World War I yelled to his men during the Battle of Belleau Wood, "Come on, you sons of bitches, do you want to live forever?" [238]. The rhetorical question reminds the Marines how they should live their life by making it count.

If Hamlet were to seek advice from spiritual life coach Teal Swan about his agony, he would have received a rather blunt and unpoetic response from Swan who said in a 2014 interview, "I realized that I was either going to go one of two ways. I was either going to be like: ok I am dead, or else I am living. And if I am living, I am going to be committed because there is no point in being here and just going through the motions of life and remaining this way. So I am either going to commit one way or another: I am going to kill myself, or I am going to literally do as much as I can here with what I have" [239]. That is exactly what Henry Hamilton (Matt Bomer) and Will Salas (Justin Timberlake) each chose to do in the 2011 sci-fi film *In Time*—an Aristotelian mimesis.

References

1. Eugenios, Jillian. Ray Kurzweil: Humans will be hybrids by 2030. *CNNMoney.* [Online] June 4, 2015. http://money.cnn.com/2015/06/03/technology/ray-kurzweil-predictions/.
2. Regalado, Antonio. Pope Francis Said to Bless Human-Animal Chimeras. *MIT Technology Review.* [Online] January 27, 2016. https://www.technologyreview.com/s/546246/pope-francis-said-to-bless-human-animal-chimeras/.
3. CNN. Hawking: If we survive the next 200 years, we should be OK. *CNN.* [Online] October 9, 2008. http://www.cnn.com/2008/WORLD/europe/10/09/hawking/index.html.
4. BBC. Big Five mass extinction events. *Nature - Prehistoric Life.* [Online] [Cited: February 28, 2016.] http://www.bbc.co.uk/nature/extinction_events.
5. Ceballos, Gerardo, et al. Accelerated modern human–induced species losses: Entering the sixth mass extinction. *Science Advances.* [Online] June 19, 2015. http://advances.sciencemag.org/content/1/5/e1400253.full.
6. Sinclaire, Janice. Doomsday Clock hands remain unchanged, despite Iran deal and Paris talks. *Bulletin of the Atomic Scientists.* [Online] January 26, 2016. http://thebulletin.org/press-release/doomsday-clock-hands-remain-unchanged-despite-iran-deal-and-paris-talks9122.
7. Bulletin of the Atomic Scientists. Timeline. *Bulletin of the Atomic Scientists.* [Online] http://thebulletin.org/timeline.
8. Shaftel, Holly. A blanket around the Earth. *NASA's Jet Propulsion Laboratory and California Institute of Technology.* [Online] March 2, 2016. http://climate.nasa.gov/causes/.
9. British Antarctic Survey. Earth's magnetic field is important for climate change at high altitudes. *Phys.org.* [Online] May 26, 2014. http://phys.org/news/2014-05-earth-magnetic-field-important-climate.html.
10. Shaftel, Holly. The consequences of climate change. *NASA's Jet Propulsion Laboratory and California Institute of Technology.* [Online] March 2, 2016. http://climate.nasa.gov/effects/.
11. Pearson, Michael. Greenland ice melt off to record early start. *CNN.* [Online] April 15, 2016. http://www.cnn.com/2016/04/15/world/greenland-ice-melt/index.html.
12. National Oceanic and Atmospheric Administration. Quinault Indian Nation plans village relocation. *U.S. Climate Resilience Toolkit.* [Online] December 17, 2015. http://toolkit.climate.gov/taking-action/quinault-indian-nation-plans-village-relocation.
13. Martinez, Michael, Elam, Stephanie and Nieves, Rosalina. The quake-maker you've never heard of: Cascadia. *CNN.* [Online] February 13, 2016. http://www.cnn.com/2016/02/11/us/cascadia-subduction-zone-earthquakes/.
14. USGS. Historic Earthquakes. *United States Geological Survey.* [Online] April 6, 2016. http://earthquake.usgs.gov/earthquakes/states/events/1700_01_26.php.
15. Bennett, Cory. Critical infrastructure cyberattacks rising, says US official. *The Hill.* [Online] January 13, 2016. http://thehill.com/policy/cybersecurity/265753-critical-infrastructure-cyberattacks-rising-says-us-official.
16. Reuters. NSA Chief Says 'When, Not If' Foreign Country Hacks U.S. Infrastructure. *Fortune.* [Online] March 1, 2016. http://fortune.com/2016/03/01/nsa-chief-hacking-infrastructure/.
17. National Park Service. Washington Monument Earthquake Update. *U.S. Department of the Interior.* [Online] 2011. https://www.nps.gov/wamo/washington-monument-earthquake-update.htm.
18. Marsh, Rene, Gracey, David and Severson, Ted. How to fix America's 'third world' airports. *CNN.* [Online] May 27, 2016. http://www.cnn.com/2016/05/25/politics/infrastructure-roads-bridges-airports-railroads/index.html.
19. CBS2. CBS2 Investigates: Experts Say Decaying Gas Lines Are A Ticking Time Bomb Below City Streets. *CBS New York.* [Online] January 8, 2016. http://newyork.cbslocal.com/2016/01/08/new-york-gas-main/.

20. Walton, Alice, Branson-Potts, Hailey and Sahagun, Louis. Porter Ranch gas leak permanently capped, officials say. *Los Angeles Times*. [Online] February 18, 2016. http://www.latimes.com/local/lanow/la-me-ln-porter-ranch-gas-leak-permanently-capped-20160218-story.html.
21. Satija, Neena, et al. Hell and High Water. *ProPublica*. [Online] March 3, 2016. https://www.propublica.org/article/hell-and-high-water-text.
22. U.S. Department of Health & Human Services. Antibiotic / Antimicrobial Resistance. *Centers for Disease Control and Prevention*. [Online] May 27, 2016. http://www.cdc.gov/drugresistance/.
23. McKenna, Maryn. Long-Dreaded Superbug Found in Human and Animal in U.S. *National Geographic*. [Online] May 26, 2016. http://phenomena.nationalgeographic.com/2016/05/26/colistin-r-9/.
24. O'Neill, Jim. Tackling Drug-Resistant Infections Globally: Final Report and Recommendations. *Revew on Antimicrobial Resistance*. [Online] May 2016. http://amr-review.org/sites/default/files/160525_Final%20paper_with%20cover.pdf.
25. Earthquake Early Warning. ShakeAlert. *Earthquake Early Warning*. [Online] [Cited: March 5, 2016.] http://www.shakealert.org/.
26. National Oceanic and Atmospheric Administration. Climate Explorer—Visualize Climate Data in Maps and Graphs. *U.S. Climate Resilience Toolkit*. [Online] December 17, 2015. http://toolkit.climate.gov/tools/climate-explorer.
27. Food Forward. Mission Statement. *Food Forward*. [Online] https://foodforward.org/about/.
28. Google. Renewable energy. *Google green*. [Online] [Cited: March 7, 2016.] https://www.google.com/green/energy/.
29. Budds, Diana. How Google Is Turning Cities Into R&D Labs: From autonomous vehicles to building codes, Sidewalk Labs is thinking about problems and solutions that could shape cities for centuries. *Fast Company & Inc*. [Online] February 22, 2016. http://www.fastcodesign.com/3056964/design-moves/how-google-is-turning-cities-into-rd-labs.
30. Alexander, Kurtis. Tech buses blocked, vandalized in protests. *SFGate*. [Online] December 20, 2013. http://blog.sfgate.com/stew/2013/12/20/bus-blocked-again-in-tech-boom-backlash/.
31. Atiyeh, Clifford. No More New Gas-Powered Cars by 2050, Say Eight States and Five Countries. *Car and Driver*. [Online] December 8, 2015. http://blog.caranddriver.com/no-more-new-gas-powered-cars-by-2050-say-eight-states-and-five-countries/.
32. Strickland, Ashley. Scott Kelly from space: Earth's atmosphere 'looks very, very fragile'. CNN. [Online] January 12, 2016. http://www.cnn.com/2016/02/11/health/scott-kelly-space-station-sanjay-guptainterview/
33. IMDb. The Day the Earth Stood Still (2008). *IMDb*. [Online] December 12, 2008. http://www.imdb.com/title/tt0970416/trivia?tab=qt&ref_=tt_trv_qu.
34. McCracken, Harry and Grossman, Lev. Can Google Solve Death. *TIME Magazine*. [Online] September 30, 2013. http://content.time.com/time/magazine/0,9263,7601130930,00.html.
35. Helft, Miguel. Google's Larry Page: The most ambitious CEO in the universe. *Fortune Magazine*. [Online] November 13, 2014. http://fortune.com/2014/11/13/googles-larry-page-the-most-ambitious-ceo-in-the-universe/.
36. Brown, Aaron. An update on Google Health and Google PowerMeter. *Google Official Blog*. [Online] June 24, 2011. https://googleblog.blogspot.com/2011/06/update-on-google-health-and-google.html.
37. Page, Larry. About 14 years ago. Google+. [Online] May 14, 2013. https://plus.google.com/+LarryPage/posts/aqy6DvvLJY1.
38. Shontell, Alyson. Larry Page Tells Wall Street This Could Be His Last Google Earnings Call For A While. *Business Insider*. [Online] October 17, 2013. http://www.businessinsider.com/larry-page-wont-be-doing-every-google-earnings-call-2013-10.
39. TIME Staff. Exclusive: TIME Talks to Google CEO Larry Page About Its New Venture to Extend Human Life. *TIME Magazine*. [Online] September 18, 2013. business.time.com/2013/09/18/google-extend-human-life/.

40. Yau, Nathan. Causes of Death. *Flowing Data.* [Online] [Cited: March 13, 2016.] http://flowingdata.com/2016/01/05/causes-of-death/.
41. Miller, Leslie. Google announces Calico, a new company focused on health and well-being. *News from Google.* [Online] September 18, 2013. http://googlepress.blogspot.com/2013/09/calico-announcement.html.
42. Calico. AbbVie and Calico Announce a Novel Collaboration to Accelerate the Discovery, Development, and Commercialization of New Therapies. *Calico.* [Online] September 3, 2014. http://www.calicolabs.com/news/2014/09/03/.
43. Calico. UT Southwestern researchers discover novel class of NAMPT activators for neurodegenerative disease; Calico enters into exclusive collaboration with 2 M to develop UTSW technology. *Calico.* [Online] September 11, 2014. http://www.calicolabs.com/news/2014/09/11/.
44. Broad Institute. Broad Institute and Calico announce an extensive collaboration focused on the biology of aging and therapeutic approaches to diseases of aging. *Broad Institute of Harvard and MIT.* [Online] March 17, 2015. https://www.broadinstitute.org/news/6633.
45. Carroll, John. Google's Calico continues its partnering romp on aging R&D with Buck collaboration. *Fierce Biotech Research.* [Online] April 28, 2015. http://www.fiercebiotechresearch.com/story/googles-calico-continues-its-partnering-romp-aging-rd-buck-collaboration/2015-04-28.
46. Calico. Calico and QB3 announce partnership to conduct research into the biology of aging and to identify potential therapeutics for age-related diseases. *Calico.* [Online] March 24, 2015. http://www.calicolabs.com/news/2015/03/24/.
47. Calico. AncestryDNA and Calico to Research the Genetics of Human Lifespan. *Calico.* [Online] July 21, 2015. http://www.calicolabs.com/news/2015/07/21/.
48. Page, Larry. Larry Page. Google+. [Online] September 18, 2013. https://plus.google.com/+LarryPage/posts/Lh8SKC6sED1.
49. Page, Larry. G is for Google. *Google Official Blog.* [Online] August 10, 2015. https://googleblog.blogspot.com/2015/08/google-alphabet.html.
50. Piller, Charles. Verily, I swear. Google Life Sciences debuts a new name. *STAT.* [Online] December 7, 2015. http://www.statnews.com/2015/12/07/verily-google-life-sciences-name/.
51. Piller, Charles. Google's next big idea: Mining health data to prevent disease. *STAT.* [Online] December 2, 2015. http://www.statnews.com/2015/12/02/google-doctor-jessica-mega/.
52. Witchel, Harry. Computers can tell if you're bored. *University of Sussex.* [Online] February 23, 2016. http://www.sussex.ac.uk/newsandevents/?id=34454.
53. Newcomb, Alyssa. Super Strength: Daughter Rescues Dad Pinned Under Car. *ABC News.* [Online] August 1, 2012. http://abcnews.go.com/US/superhero-woman-lifts-car-off-dad/story?id=16907591.
54. NBC News. Teen daughters find strength to lift 3,000-pound tractor off father. *NBC News.* [Online] April 10, 2013. http://usnews.nbcnews.com/_news/2013/04/10/17689146-teen-daughters-find-strength-to-lift-3000-pound-tractor-off-father.
55. McCrum, Kirstie. Teen girl uses 'superhuman strength' to lift burning truck off dad and save family. *Mirror.* [Online] January 11, 2016. http://www.mirror.co.uk/news/world-news/teen-girl-uses-superhuman-strength-7155438?.
56. Storrs, Carina. Why Americans don't live as long as Europeans. *CNN.* [Online] February 9, 2016. http://www.cnn.com/2016/02/09/health/american-life-expectancy-shorter-than-europeans/index.html.
57. American-Israeli Cooperative Enterprise. Biographies. *Jewish Virtual Library.* [Online] https://www.jewishvirtuallibrary.org/jsource/bios.html.
58. Hopkin, Michael. Ethiopia is top choice for cradle of Homo sapiens. *Nature.* [Online] February 16, 2005. http://www.nature.com/news/2005/050216/full/news050214-10.html.
59. Vernot, Benjamin and al, et. Excavating Neandertal and Denisovan DNA from the genomes of Melanesian individuals. *Science.* [Online] March 17, 2016. http://science.sciencemag.org/content/early/2016/03/16/science.aad9416.full.

60. Clark, Tom. Dying Matters survey finds 15 % want to live forever. *The Guardian*. [Online] May 15, 2011. http://www.theguardian.com/uk/2011/may/16/dying-still-taboo-subject-poll.
61. White, Tracie. Most physicians would forgo aggressive treatment for themselves at the end of life, study finds. *Stanford Medicine News Center*. [Online] May 28, 2014. http://med. stanford.edu/news/all-news/2014/05/most-physicians-would-forgo-aggressive-treatment-for-themselves-.html.
62. Biography.com Editors. Albert Einstein Biography. *Biography.com*. [Online] http://www. biography.com/people/albert-einstein-9285408.
63. Gallup. Employee Engagement. *Gallup*. [Online] http://www.gallup.com/topic/employee_ engagement.aspx.
64. Singal, Jesse. For 80 Years, Young Americans Have Been Getting More Anxious and Depressed, and No One Is Quite Sure Why. *New York Magazine*. [Online] March 13, 2016. http://nymag.com/scienceofus/2016/03/for-80-years-young-americans-have-been-getting-more-anxious-and-depressed.html.
65. Somerville, Madeleine. Are eco-friendly initiatives pointless unless we tackle overpopulation? *The Guardian*. [Online] January 26, 2016. http://www.theguardian.com/ lifeandstyle/2016/jan/26/overpopulation-sustainability-environment-eco-friendly-initiatives.
66. World Health Organization. Food Security. *World Health Organization*. [Online] http:// www.who.int/trade/glossary/story028/en/.
67. NAIC. Longevity Risk. *National Association of Insurance Commissioners*. [Online] December 14, 2015. http://www.naic.org/cipr_topics/topic_longevity_risk.htm.
68. Debate. DEBATE: Immortality will change prison sentences? Execution & Life-behind-Bars.. too sadistic? *Immortal Life*. [Online] March 16, 2013. http:// immortallife.info/articles/entry/debate-immortality-will-change-prison-sentences-execution-life-behind-bars.
69. Hankinson, Andrew. Who Wants to Live Forever? *Men's Health*. [Online] June 15, 2015. http://www.menshealth.co.uk/healthy/brain-training/who-wants-to-live-forever.
70. Bucketlist.org. Your dreams, made possible. *Bucketlist*. [Online] [Cited: March 7, 2016.] https://bucketlist.org/.
71. Knight, Les. About the Movement. *Voluntary Human Extinction Movement*. [Online] http:// www.vhemt.org/aboutvhemt.htm#vhemt.
72. The Long Now Foundation. About Long Now. *The Long Now Foundation*. [Online] http:// longnow.org/about/.
73. Lee, Newton. Interview with Danny Hillis. *ACM Digital Library*. [Online] March 29, 2006. http://dl.acm.org/citation.cfm?doid=1146816.1146821.
74. Lohmann, Bill. Scouts' honor. *Richmond Times-Dispatch*. [Online] March 3, 2008. http:// www.richmond.com/entertainment/article_ba2be0ca-209c-5c3d-99fb-28e74af418bd.html.
75. Computers in Entertainment. ACM Computers in Entertainment Editorial Board. *Association for Computing Machinery*. [Online] [Cited: March 8, 2016.] http://cie.acm.org/about-board/.
76. Giardina, Carolyn. Digital cinema educator Charles Swartz dies. *The Hollywood Reporter*. [Online] February 13, 2007. http://www.hollywoodreporter.com/news/digital-cinema-educator-charles-swartz-130064.
77. Martin, Douglas. Randy Pausch, 47, Dies; His 'Last Lecture' Inspired Many to Live With Wonder. *The New York Times*. [Online] July 26, 2008. http://www.nytimes.com/2008/07/26/ us/26pausch.html.
78. Bond, Paul. Roy E. Disney dies at 79. *The Hollywood Reporter*. [Online] December 16, 2009. http://www.hollywoodreporter.com/news/roy-e-disney-dies-79-92404.
79. Lee, Newton and Madej, Krystina. Disney Stories: Getting to Digital. *Springer Science + Business Media*. [Online] April 26, 2012. http://www.amazon.com/Disney-Stories-Getting-Newton-Lee/dp/1461421004.
80. Barnes, Mike and Giardina, Carolyn. Digital Media Trailblazer Bob Lambert Dies at 55. *The Hollywood Reporter*. [Online] September 10, 2012. http://www.hollywoodreporter.com/ news/bob-lambert-disney-digital-cinema-dies-369125.

81. Floyd, Pink. Pink Floyd - Time. *YouTube.* [Online] November 5, 2012. https://www.youtube.com/watch?v=Z-OytmtYoOI.
82. Franklin, Benjamin and Hall, David. Advice to a Young Tradesman. *Founders Online.* [Online] July 21, 1748. http://founders.archives.gov/documents/Franklin/01-03-02-0130.
83. IMDb. In Time. *IMDb.* [Online] October 28, 2011. http://www.imdb.com/title/tt1637688/.
84. Gates, Bill and Melinda. If you could have one superpower, what would it be? *Gates Notes.* [Online] February 22, 2016. https://www.gatesnotes.com/Annual-Letter-Superpowers.
85. Carnegie Mellon University. Randy Pausch's Last Lecture. *Carnegie Mellon University.* [Online] September 18, 2007. http://www.cmu.edu/randylecture/.
86. The Huffington Post. 5 Things Women Are Judged More Harshly For Than Men. *Huffpost Women.* [Online] March 7, 2014. http://www.huffingtonpost.com/2014/03/07/things-women-judged-for-double-standard_n_4911878.html.
87. Fleming, Olivia. Was Jennifer Lawrence too FAT for the Hunger Games? Critics believe actress should have looked 'more hungry'. *Daily Mail.* [Online] March 28, 2012. http://www.dailymail.co.uk/femail/article-2121740/Was-Jennifer-Lawrence-FAT-Hunger-Games-Male-critics-believe-actress-looked-hungry.html.
88. Leopold, Todd. Why we can't stop body-shaming. *CNN.* [Online] April 15, 2016. http://www.cnn.com/2016/04/15/health/fat-shaming-feat/index.html.
89. Child, Ben. Carrie Fisher: I felt pressured to lose weight for Star Wars: The Force Awakens. *The Guardian.* [Online] December 1, 2015. http://www.theguardian.com/culture/2015/dec/01/carrie-fisher-weight-loss-star-wars-the-force-awakens.
90. Telegraph Film. Carrie Fisher fires back at bodyshaming Star Wars trolls: 'Blow us'. *The Telegraph.* [Online] December 30, 2015. http://www.telegraph.co.uk/film/star-wars-the-force-awakens/carrie-fisher-weight-looks-twitter-attacks/.
91. BBC News. 'Clive of India's' tortoise dies. *BBC News.* [Online] March 23, 2006. http://news.bbc.co.uk/2/hi/south_asia/4837988.stm.
92. Associated Press. 140-year-old lobster's tale has a happy ending. *NBC News.* [Online] January 10, 2009. http://www.nbcnews.com/id/28589278/ns/us_news-weird_news/t/-year-old-lobsters-tale-has-happy-ending/.
93. Marks, Kathy. Henry the tuatara is a dad at 111. *The Independent.* [Online] January 26, 2009. http://www.independent.co.uk/news/world/australasia/henry-the-tuatara-is-a-dad-at-111-1516628.html.
94. Wilkins, Alasdair. Turtles could hold the secret to human immortality. *Gizmodo.* [Online] August 20, 2010. http://io9.gizmodo.com/5618046/the-mystery-of-why-turtles-never-grow-old—and-how-we-can-learn-from-it.
95. Highfield, Roger. What's wrong with Craig Venter? Craig Venter, multi-millionaire maverick, says he can help you live a better, longer life. Roger Highfield asks how. *Mosaic.* [Online] February 2, 2016. http://mosaicscience.com/story/craig-venter-genomics-personalised-medicine.
96. Shampo, Marc A. and Kyle, Robert A. J. Craig Venter—The Human Genome Project. *Mayo Clinic Proceedings.* [Online] April 2011. http://www.ncbi.nlm.nih.gov/pmc/articles/PMC3068906/.
97. Human Longevity Inc. Human Longevity Inc. (HLI) Launched to Promote Healthy Aging Using Advances in Genomics and Stem Cell Therapies. *PR Newswire.* [Online] March 4, 2014. http://www.prnewswire.com/news-releases/human-longevity-inc-hli-launched-to-promote-healthy-aging-using-advances-in-genomics-and-stem-cell-therapies-248379091.html.
98. Strategies for Engineered Negligible Senescence Research Foundation. About SENS Research Foundation. [Online] [Cited: March 15, 2016.] http://www.sens.org/about.
99. Best, Ben. Interview with Aubrey de Grey, PhD. *Life Extension.* [Online] July 2013. http://www.lifeextension.com/magazine/2013/7/Interview-with-Aubrey-de-Grey-PhD/Page-01.
100. Monaghan, James R. et al. Microarray and cDNA sequence analysis of transcription during nerve-dependent limb regeneration. *BMC Biology.* [Online] January 13, 2009. https://www.ncbi.nlm.nih.gov/pmc/articles/PMC2630914/.

101. Than, Ker. "Immortal" Jellyfish Swarm World's Oceans. *National Geographic.* [Online] January 29, 2009. http://news.nationalgeographic.com/news/2009/01/090130-immortal-jellyfish-swarm.html.
102. Bresnahan, Samantha. This man's blood has saved the lives of two million babies. *CNN.* [Online] November 30, 2015. http://www.cnn.com/2015/06/09/health/james-harrison-golden-arm-blood-rhesus/index.html.
103. Waggoner, Walter H. Dr. Burrill B. Crohn, 99, An Expert On Diseases Of The Intestinal Tract. *The New York Times.* [Online] July 30, 1983. http://www.nytimes.com/1983/07/30/obituaries/dr-burrill-b-crohn-99-an-expert-on-diseases-of-the-intestinal-tract.html.
104. Woo, Elaine. Stephen Crohn dies at 66; immune to HIV, but not its tragedy. *Los Angeles Times.* [Online] September 21, 2013. http://articles.latimes.com/2013/sep/21/local/la-me-stephen-crohn-20130922.
105. George, Debbie, Lopez, Robyn and Thornburg, Michelle. EASE T1D. [Online] http://www.easet1d.org/.
106. Cahill, Jim. Reegan's Rule Passes in North Carolina. *Insulin Nation.* [Online] October 5, 2015. http://insulinnation.com/living/reegans-rule-passes-north-carolina/.
107. MacKenzie, Ryan. Regular Session 2015-2016 House Resolution 569. *Pennsylvania General Assembly.* [Online] November 9, 2015. http://www.legis.state.pa.us/cfdocs/billinfo/billinfo.cfm?syear=2015&sind=0&body=H&type=R&bn=569.
108. California State Senator Richard D. Roth. Senator Roth's Measure to Raise Awareness of Type 1 Diabetes Approved by State Senate. *California State Senate.* [Online] March 14, 2016. http://sd31.senate.ca.gov/news/2016-03-14-senator-roth%E2%80%99s-measure-raise-awareness-type-1-diabetes-approved-state-senate.
109. Kader, Hanady. Researchers Work Toward New Type 1 Diabetes Therapies For Patients Like Juliana. *Seattle Children's Hospital.* [Online] March 22, 2016. http://pulse.seattlechildrens.org/seattle-childrens-researchers-work-toward-new-type-1-diabetes-therapies-for-patients-like-juliana/.
110. Massachusetts General Hospital. Mass. General Hospital launches phase II trial of BCG vaccine to reverse type 1 diabetes. *Massachusetts General Hospital News Release.* [Online] June 7, 2015. http://www.massgeneral.org/news/pressrelease.aspx?id=1817.
111. Walker, Joseph. Patients Struggle With High Drug Prices. *The Wall Street Journal.* [Online] December 31, 2015. http://www.wsj.com/articles/patients-struggle-with-high-drug-prices-1451557981.
112. National Priorities Project. Tax Day 2016. *National Priorities Project.* [Online] March 10, 2016. https://www.nationalpriorities.org/analysis/2016/tax-day-2016/.
113. Ramaswami, Prem. A remedy for your health-related questions: health info in the Knowledge Graph. *Google Official Blog.* [Online] February 10, 2015. https://googleblog.blogspot.com/2015/02/health-info-knowledge-graph.html.
114. Stefansen, Christian. Flu Trends updates model to help estimate flu levels in the US. *The Official google.org blog.* [Online] October 29, 2013. http://blog.google.org/2013/10/flu-trends-updates-model-to-help.html.
115. Ginsberg, Jeremy et al. Detecting influenza epidemics using search engine query data. *Google.* [Online] February 19, 2009. http://static.googleusercontent.com/media/research.google.com/en/us/archive/papers/detecting-influenza-epidemics.pdf.
116. The Flu Trends Team. The Next Chapter for Flu Trends. *Google Research Blog.* [Online] August 20, 2015. http://googleresearch.blogspot.com/2015/08/the-next-chapter-for-flu-trends.html.
117. Mulpeter, Kathleen. 9 Burning Health Questions People Asked Google This Year. *Health.* [Online] December 4, 2015. http://news.health.com/2015/12/04/top-health-searches-google-2015/.
118. Ranney, Megan. The countries using Google Search to eat and live healthier. *Mashable.* [Online] July 24, 2015. http://mashable.com/2015/07/24/google-trends-health/.
119. Singhal, Amit. Introducing the Knowledge Graph: things, not strings. [Online] May 16, 2012. https://googleblog.blogspot.in/2012/05/introducing-knowledge-graph-things-not.html.

120. Ramaswami, Prem. A remedy for your health-related questions: health info in the Knowledge Graph. *Google Official Blog.* [Online] February 10, 2015. https://googleblog.blogspot.ca/2015/02/health-info-knowledge-graph.html.

121. Google Inc. Amendment No. 9 To Form S-1 Registration Statement. *Securities And Exchange Commission.* [Online] August 18, 2004. http://www.sec.gov/Archives/edgar/data/1288776/000119312504142742/ds1a.htm.

122. Jinek, Martin et al. A Programmable Dual-RNA–Guided DNA Endonuclease in Adaptive Bacterial Immunity. *Science.* [Online] August 17, 2012. http://science.sciencemag.org/content/337/6096/816.

123. Niu, Yuyu, et al. Generation of Gene-Modified Cynomolgus Monkey via Cas9/RNA-Mediated Gene Targeting in One-Cell Embryos. *Cell.* [Online] February 13, 2014. http://www.cell.com/cell/abstract/S0092-8674%2814%2900079-8.

124. Hu, Wenhui et al. RNA-directed gene editing specifically eradicates latent and prevents new HIV-1 infection. *PNAS Online.* [Online] June 19, 2014. http://www.pnas.org/content/111/31/11461.abstract.

125. Puping Liang, et al. CRISPR/Cas9-mediated gene editing in human tripronuclear zygotes. *Protein Cell.* [Online] May 2015. http://www.ncbi.nlm.nih.gov/pmc/articles/PMC4417674/.

126. Cyranoski, David and Reardon, Sara. Chinese scientists genetically modify human embryos. *Nature.* [Online] April 22, 2015. http://www.nature.com/news/chinese-scientists-genetically-modify-human-embryos-1.17378.

127. Siddique, Haroon. British researchers get green light to genetically modify human embryos. *The Guardian.* [Online] February 1, 2016. https://www.theguardian.com/science/2016/feb/01/human-embryo-genetic-modify-regulator-green-light-research.

128. Gallagher, James. Scientists get 'gene editing' go-ahead. *BBC News.* [Online] February 1, 2016. http://www.bbc.com/news/health-35459054.

129. Mullins, Christina S and Linnebacher, Michael. Human endogenous retroviruses and cancer: Causality and therapeutic possibilities. *World Journal of Gastroenterology.* [Online] November 14, 2012. http://www.ncbi.nlm.nih.gov/pmc/articles/PMC3496880/.

130. Brodziak, Andrzej, et al. The role of human endogenous retroviruses in the pathogenesis of autoimmune diseases. *International Medical Journal of Experimental and Clinical Research.* [Online] June 1, 2012. http://www.ncbi.nlm.nih.gov/pmc/articles/PMC3560723/.

131. Ryan, F.P. Human Endogenous Retroviruses in Multiple Sclerosis: Potential for Novel Neuro-Pharmacological Research. *Current Neuropharmacology.* [Online] June 2011. http://www.ncbi.nlm.nih.gov/pmc/articles/PMC3131726/.

132. Slokar, Gorjan and Hasler, Gregor. Human Endogenous Retroviruses as Pathogenic Factors in the Development of Schizophrenia. *Frontiers in Psychiatry.* [Online] January 11, 2016. http://www.ncbi.nlm.nih.gov/pmc/articles/PMC4707225/.

133. Harvard University Wyss Institute and Harvard Medical School. Removing 62 barriers to pig-to-human organ transplant in one fell swoop. *The Wyss Institute for Biologically Inspired Engineering.* [Online] October 11, 2015. http://wyss.harvard.edu/viewpressrelease/222/removing-62-barriers-to-pigtohuman-organ-transplant-in-one-fell-swoop.

134. Zimmer, Carl. Editing of Pig DNA May Lead to More Organs for People. *The New York Times.* [Online] October 15, 2015. http://www.nytimes.com/2015/10/20/science/editing-of-pig-dna-may-lead-to-more-organs-for-people.html.

135. Gorman, Christine. Tissue Mash-Up: a Q&A with Juan Carlos Izpisua Belmonte. *Scientific American.* [Online] January 25, 2016. http://www.scientificamerican.com/article/tissue-mash-up-a-q-a-with-juan-carlos-izpisua-belmonte/.

136. Murphy, Sean V and Atala, Anthony. 3D bioprinting of tissues and organs. *Nature Biotechnology.* [Online] August 5, 2014. http://www.nature.com/nbt/journal/v32/n8/full/nbt.2958.html.

137. Bajaj, Piyush et al. 3D Biofabrication Strategies for Tissue Engineering and Regenerative Medicine. *Annual Review of Biomedical Engineering.* [Online] May 29, 2014. http://www.ncbi.nlm.nih.gov/pmc/articles/PMC4131759/.

138. Fox, Maggie. Wake Forest University Scientists Print Living Body Parts. *CNBC.* [Online] February 16, 2016. http://www.cnbc.com/2016/02/16/wake-forest-university-scientists-print-living-body-parts.html.

139. RT. Russian 3D-bioprinted thyroid gland implant proves functional in mice. *RT.* [Online] November 26, 2015. https://www.rt.com/news/323494-russian-3d-printed-thyroid/.

140. Altman, Lawrence K. Arne H. W. Larsson, 86; Had First Internal Pacemaker. *The New York Times.* [Online] January 18, 2002. http://www.nytimes.com/2002/01/18/world/arne-h-w-larsson-86-had-first-internal-pacemaker.html.

141. Wood, Mark A. and Ellenbogen, Kenneth A. Cardiac Pacemakers From the Patient's Perspective. *American Heart Association.* [Online] 2002. http://circ.ahajournals.org/content/105/18/2136.full.

142. 24/7 Wall St. The Eleven Most Implanted Medical Devices In America. *24/7 Wall St.* [Online] July 18, 2011. http://247wallst.com/healthcare-economy/2011/07/18/the-eleven-most-implanted-medical-devices-in-america/.

143. Byrne, Michael. New 'Stentrode' Goes Deep Inside the Brain Without the Need for Brain Surgery. *Motherboard.* [Online] February 13, 2016. http://motherboard.vice.com/read/new-stentrode-gets-inside-the-brain-without-the-need-for-brain-surgery.

144. Gaudin, Sharon. Nanotech could make humans immortal by 2040, futurist says. *Computerworld.* [Online] October 1, 2009. http://www.computerworld.com/article/2528330/app-development/nanotech-could-make-humans-immortal-by-2040-futurist-says.html.

145. Carreno, Beatriz M. et al. A dendritic cell vaccine increases the breadth and diversity of melanoma neoantigen-specific T cells. *Science.* [Online] May 15, 2015. http://science.sciencemag.org/content/348/6236/803.

146. McCarthy, John. What is Artificial Intelligence. [Online] Stanford University, November 12, 2007. http://www-formal.stanford.edu/jmc/whatisai/node1.html.

147. Millenson, Michael. Watson: A Computer So Smart It Can Say, "Yes, Doctor". *Forbes.* [Online] February 17, 2011. http://www.forbes.com/sites/sciencebiz/2011/02/17/watson-a-computer-so-smart-it-can-say-yes-doctor/.

148. Shortliffe, Edward. Computer-Based Medical Consultations: MYCIN. *Elsevier.* [Online] 1976. https://books.google.com/books?id=i9QXugPQw6oC.

149. Roach, J., et al. An expert system for information on pharmacology and drug interactions. *U. S. National Library of Medicine.* [Online] Computers in Biology and Medicine, 1985. http://www.ncbi.nlm.nih.gov/pubmed/3979039.

150. Markoff, John. Computer Wins on 'Jeopardy!': Trivial, It's Not. *The New York Times.* [Online] February 16, 2011. http://www.nytimes.com/2011/02/17/science/17jeopardy-watson.html?pagewanted=all.

151. Markoff, John. Computer Wins on 'Jeopardy!': Trivial, It's Not. [Online] The New York Times, February 16, 2011. http://www.nytimes.com/2011/02/17/science/17jeopardy-watson.html?pagewanted=all.

152. Ford, Martin. Anything You Can Do, Robots Can Do Better. *The Atlantic.* [Online] February 14, 2011. http://www.theatlantic.com/business/archive/2011/02/anything-you-can-do-robots-can-do-better/71227/.

153. Waters, Richard. FT interview with Google co-founder and CEO Larry Page. *FT Magazine.* [Online] October 31, 2014. http://www.ft.com/cms/s/2/3173f19e-5fbc-11e4-8c27-00144feabdc0.html.

154. Cohn, Jonathan. The Robot Will See You Now. *The Atlantic.* [Online] March 2013. http://www.theatlantic.com/magazine/archive/2013/03/the-robot-will-see-you-now/309216/.

155. Hsu, Jeremy. Artificial Intelligence Outperforms Human Data Scientists. *IEEE Spectrum.* [Online] October 2, 2015. http://spectrum.ieee.org/tech-talk/computing/software/artificial-intelligence-outperforms-human-data-scientists.

156. Finley, Klint. Robot Radiologists Will Soon Analyze Your X-Rays. *Wired.* [Online] October 27, 2015. http://www.wired.com/2015/10/robot-radiologists-are-going-to-start-analyzing-x-rays/.

157. Reese, Hope. How an AI program helps doctors identify cancer and other medical abnormalities. *TechRepublic.* [Online] April 4, 2016. http://www.techrepublic.com/article/how-an-ai-program-helps-doctors-identify-cancer-and-other-medical-abnormalities/.

158. Boyle, Alan. IBM Watson AI XPRIZE offers $5 M for projects that link humans and computers – but who'll sign up? *GeekWire.* [Online] February 17, 2016. http://www.geekwire.com/2016/ibm-watson-ai-xprize-offers-5-million-for-human-computer-collaboration-but-wholl-enter/.

159. Clark, Jack. Google Reveals Use of 'Thousands' of AI Processors It Designed. *Bloomberg.* [Online] May 18, 2016. http://www.bloomberg.com/news/articles/2016-05-18/google-reveals-use-of-thousands-of-ai-processors-it-designed.

160. Google. TensorFlow. [Online] https://www.tensorflow.org/about.html.

161. Terdiman, Daniel. Sony's Folding@home project gets Guinness record. [Online] CNet, October 31, 2007. http://news.cnet.com/8301-13772_3-9808500-52.html.

162. Stanford University. Folding@home distributed computing. [Online] Stanford University. [Cited: January 17, 2013.] http://folding.stanford.edu/English/HomePage.

163. Foldit. The Science Behind Foldit. *Foldit.* [Online] http://fold.it/portal/info/science.

164. Boyle, Alan. Gamers solve molecular puzzle that baffled scientists. [Online] NBC News, September 18, 2011. http://cosmiclog.nbcnews.com/_news/2011/09/18/7802623-gamers-solve-molecular-puzzle-that-baffled-scientists.

165. Coren, Michael J. Foldit Gamers Solve Riddle of HIV Enzyme within 3 Weeks. *Scientific American.* [Online] September 20, 2011. http://www.scientificamerican.com/article/foldit-gamers-solve-riddle/.

166. DeepMind. Google DeepMind. [Online] https://deepmind.com/.

167. Byford, Sam. Google's DeepMind defeats legendary Go player Lee Se-dol in historic victory. *The Verge.* [Online] March 9, 2016. http://www.theverge.com/2016/3/9/11184362/google-alphago-go-deepmind-result.

168. Bozulich, Richard. Chess and Go: A Comparison. *Kiseido Publishing Company.* [Online] 2015. http://www.magicofgo.com/roadmap9/chess%20and%20go.htm.

169. Lopatto, Elizabeth. Google's AI can learn to play video games. *The Verge.* [Online] February 25, 2015. http://www.theverge.com/2015/2/25/8108399/google-ai-deepmind-video-games.

170. Turk, Victoria. 'Human Computation' Could Save the World Without the Risks of AI. *Motherboard.* [Online] December 31, 2015. http://motherboard.vice.com/read/human-computation-could-save-the-world-without-the-risks-of-ai.

171. Licklider, J. C. R. Man-Computer Symbiosis. *IRE Transactions on Human Factors in Electronics.* [Online] March 1960. http://groups.csail.mit.edu/medg/people/psz/Licklider.html.

172. Turing, A.M. Computing machinery and intelligence. *Mind.* [Online] 1950. http://loebner.net/Prizef/TuringArticle.html.

173. Coope, S. Barry. Alan Turing: "I am building a brain." Half a century later, its successor beat Kasparov. *The Guardian.* [Online] May 14, 2012. http://www.theguardian.com/uk/the-northerner/2012/may/14/alan-turing-gary-kasparov-computer.

174. Gardner, W. David. Remembering Joe Weizenbaum, ELIZA Creator. *Information Week.* [Online] March 13, 2008. http://www.informationweek.com/remembering-joe-weizenbaum-eliza-creator-/d/d-id/1065648?.

175. Nagy, Evie. Using ToyTalk Technology, New Hello Barbie Will Have Real Conversations With Kids. *Fast Company.* [Online] February 13, 2015. http://www.fastcompany.com/3042430/most-creative-people/using-toytalk-technology-new-hello-barbie-will-have-real-conversations-.

176. The Long Now Foundation. A Long Bet: Bet 1. *Long Bets.* [Online] 2002. http://longbets.org/1/.

177. Kasparov, Garry. THE DAY THAT I SENSED A NEW KIND OF INTELLIGENCE. *TIME Magazine.* [Online] March 25, 1996. http://content.time.com/time/subscriber/article/0,33009,984305-1,00.html.

178. Silver, Nate. The Signal and the Noise: Why So Many Predictions Fail-but Some Don't. *Penguin.* [Online] September 27, 2012. https://books.google.com/books?id=SI-VqAT4_hYC.

179. Ebert, Roger. Remaking my voice. *TED.* [Online] March 2011. https://www.ted.com/talks/roger_ebert_remaking_my_voice?language=en.

180. Ostrow, Adam. Roger Ebert Tests His Vocal Cords, and Comedic Delivery. *The New York Times.* [Online] March 7, 2011. http://bits.blogs.nytimes.com/2011/03/07/roger-ebert-tests-his-vocal-cords-and-comedic-delivery/?src=me&_r=0.

181. VanHemert, Kyle. Meet the World's First Robot Pop Star. *Gizmodo.* [Online] October 18, 2010. http://gizmodo.com/5666855/meet-the-worlds-first-robot-pop-star.

182. Cheok, Adrian and Levy, David. Love and Sex with Robots. [Online] December 2016. http://loveandsexwithrobots.org/.

183. Kok, Lester. NTU scientists unveil social and telepresence robots. *Nanyang Technological University.* [Online] December 29, 2015. http://media.ntu.edu.sg/NewsReleases/Pages/newsdetail.aspx?news=fde9bfb6-ee3f-45f0-8c7b-f08bc1a9a179.

184. Gibbs, Samuel. What is Boston Dynamics and why does Google want robots? *The Guardian.* [Online] December 17, 2013. https://www.theguardian.com/technology/2013/dec/17/google-boston-dynamics-robots-atlas-bigdog-cheetah.

185. Boston Dynamics. BigDog Overview. *YouTube.* [Online] April 22, 2010. https://www.youtube.com/watch?v=cNZPRsrwumQ.

186. Boston Dynamics. Introducing WildCat. *YouTube.* [Online] October 3, 2013. https://www.youtube.com/watch?v=wE3fmFTtP9g.

187. Boston Dynamics. Introducing Spot. *YouTube.* [Online] February 9, 2015. https://www.youtube.com/watch?v=M8YjvHYbZ9w.

188. Boston Dynamics. Atlas, The Next Generation. *YouTube.* [Online] February 23, 2016. https://www.youtube.com/watch?v=rVlhMGQgDkY.

189. Boston Dynamics. Alex the Dog vs Spot the Robot. *YouTube.* [Online] February 29, 2016. https://www.youtube.com/watch?v=93B55I8qrGM.

190. PETA. Cruelty to Animals in Laboratories. *People for the Ethical Treatment of Animals.* [Online] http://www.peta.org/issues/animals-used-for-experimentation/animals-laboratories/.

191. IMDb. Planet of the Apes. *Quotes.* [Online] July 27, 2001. http://www.imdb.com/title/tt0133152/quotes.

192. Brown, Bob. Why there's no Nobel Prize in Computing. *Network World.* [Online] June 6, 2011. http://www.networkworld.com/article/2177705/data-center/why-there-s-no-nobel-prize-in-computing.html.

193. Fogg, Ally. Alan Turing's pardon is wrong. *The Guardian.* [Online] December 24, 2013. http://www.theguardian.com/commentisfree/2013/dec/24/alan-turing-pardon-wrong-gay-men.

194. Shepardson, David and Lienert, Paul. Exclusive: In boost to self-driving cars, U.S. tells Google computers can qualify as drivers. *Reuters.* [Online] February 10, 2016. http://www.reuters.com/article/us-alphabet-autos-selfdriving-exclusive-idUSKCN0VJ00H.

195. Hemmersbaugh, Paul A. Letter to Chris Urmson, Director, Self-Driving Car Project, Google. *National Highway Traffic Safety Administration.* [Online] February 4, 2016. http://isearch.nhtsa.gov/files/Google%20-%20compiled%20response%20to%2012%20Nov%20%2015%20interp%20request%20-%204%20Feb%2016%20final.htm.

196. Foot, Philippa. Virtues and Vices and Other Essays in Moral Philosophy. *Oxford.* [Online] 2002. https://books.google.com/books?id=fIHnCwAAQBAJ&pg=PA23&lpg=PA23.

197. CDC. Drowsy Driving: Asleep at the Wheel. *Centers for Disease Control and Prevention.* [Online] November 5, 2015. http://www.cdc.gov/features/dsdrowsydriving/.

198. CDC. Impaired Driving: Get the Facts. *Centers for Disease Control and Prevention.* [Online] April 15, 2016. http://www.cdc.gov/MotorVehicleSafety/Impaired_Driving/impaired-drv_factsheet.html.

199. Cheshire, Thomas. Thousands Of Cars Stolen Using Hi-Tech Gadgets. *Sky News.* [Online] May 8, 2014. http://news.sky.com/story/1257320/thousands-of-cars-stolen-using-hi-tech-gadgets.
200. Goodman, Marc. Marc Goodman: A vision of crimes in the future. [Online] TEDGlobal 2012, June 28, 2012. http://www.ted.com/talks/marc_goodman_a_vision_of_crimes_in_the_future.html?quote=1769.
201. Cellan-Jones, Rory. Stephen Hawking warns artificial intelligence could end mankind. *BBC News.* [Online] December 2, 2014. http://www.bbc.com/news/technology-30290540.
202. McFarland, Matt. Elon Musk: 'With artificial intelligence we are summoning the demon.'. *The Washington Post.* [Online] October 24, 2014. https://www.washingtonpost.com/news/innovations/wp/2014/10/24/elon-musk-with-artificial-intelligence-we-are-summoning-the-demon/.
203. Daly, Cullen, Sevilla, Robert and Foglia, Louis. Artificial Intelligence is scary. Elon Musk explains why. *CNNMoney.* [Online] September 2015. http://money.cnn.com/video/technology/2015/09/10/elon-musk-artificial-intelligence.cnnmoney.
204. Eha, Brian Patrick. Is Knight's $440 million glitch the costliest computer bug ever? [Online] CNNMoney, August 9, 2012. http://money.cnn.com/2012/08/09/technology/knight-expensive-computer-bug/index.html.
205. Kahn, Jeremy. Baidu Looks to Artificial Intelligence to Reduce Insurance Risks. *Bloomberg.* [Online] January 20, 2016. http://www.bloomberg.com/news/articles/2016-01-20/baidu-looks-to-artificial-intelligence-to-reduce-insurance-risks.
206. Kurzweil, Ray. A Wager on the Turing Test: Why I Think I Will Win. *@Kurzweil Accelerating Intelligence.* [Online] April 9, 2002. http://www.kurzweilai.net/a-wager-on-the-turing-test-why-i-think-i-will-win.
207. Thomsen, Michael. Microsoft's Deep Learning Project Outperforms Humans In Image Recognition. *Forbes.* [Online] February 19, 2015. http://www.forbes.com/sites/michaelthomsen/2015/02/19/microsofts-deep-learning-project-outperforms-humans-in-image-recognition/.
208. Bostrom, Nick. How Long Before Superintelligence? *International Journal of Future Studies.* [Online] 1998. http://www.nickbostrom.com/superintelligence.html.
209. Achenbach, Joel. The A.I. anxiety. *The Washington Post.* [Online] December 27, 2015. http://www.washingtonpost.com/sf/national/2015/12/27/aianxiety/.
210. Clark, Jack. Elon Musk and Other Tech Titans Create Company to Develop Artificial Intelligence. *Bloomberg.* [Online] December 11, 2015. http://www.bloomberg.com/news/articles/2015-12-11/tech-titans-create-company-to-develop-artificial-intelligence.
211. Kavukcuoglu, Koray. DeepMind moves to TensorFlow. *Google Research Blog.* [Online] April 29, 2016. http://googleresearch.blogspot.com/2016/04/deepmind-moves-to-tensorflow.html.
212. IMDb. Her. *IMDb.* [Online] January 10, 2014. http://www.imdb.com/title/tt1798709/trivia?tab=qt&ref_=tt_trv_qu.
213. IMDb. Transcendence. *IMDb.* [Online] April 18, 2014. http://www.imdb.com/title/tt2209764/trivia?tab=qt&ref_=tt_trv_qu.
214. Burkeman, Oliver. Man of steel. *The Guardian.* [Online] September 17, 2002. http://www.theguardian.com/education/2002/sep/17/science.highereducation.
215. IMDb. Kick-Ass. *IMDb.* [Online] April 16, 2010. http://www.imdb.com/title/tt1250777/trivia?tab=qt&ref_=tt_trv_qu.
216. Rothman, Peter. Transhumanism Gets Political. *Humanity+.* [Online] October 8, 2014. http://hplusmagazine.com/2014/10/08/transhumanism-gets-political/.
217. Madej, Shane. Meet The Anti-Death Presidential Candidate. *BuzzFeed.* [Online] August 27, 2015. https://www.buzzfeed.com/shanemadej/meet-the-anti-death-presidential-candidate?utm_term=.bmn6Y0xmKj#.tkma7rZXlO.
218. Fleischer Studios. Betty Boop for President (1932). *YouTube.* [Online] 1932. https://www.youtube.com/watch?v=c0-q_ZkDcsk.

219. Capachi, Casey. The 2016 election in one word, according to millennials. *CNN*. [Online] December 29, 2015. http://www.cnn.com/2015/12/29/politics/election-in-one-word-according-to-millennials/.

220. Cappabianca, Mary. Spring 2016 Youth Poll. *Harvard's Insitute of Politics*. [Online] April 25, 2016. http://iop.harvard.edu/youth-poll/harvard-iop-spring-2016-poll.

221. Mucci, Alberto. Sweden's Minister of the Future Explains How to Make Politicians Think Long-Term. *Motherboard*. [Online] November 26, 2015. http://motherboard.vice.com/read/swedens-minister-of-the-future-explains-how-to-make-politicians-think-long-term.

222. Konovalenko, Maria. Russians Create the "Longevity Party". *Institute for Ethics and Emerging Technologies*. [Online] July 26, 2012. http://ieet.org/index.php/IEET/more/konovalenko201207261.

223. Dolak, Kevin. Human Immortality in 33 Years Claims Dmitry Itskov's 2045 Initiative. *ABC News*. [Online] August 27, 2012. http://abcnews.go.com/blogs/technology/2012/08/human-immortality-in-33-years-claims-dmitry-itskovs-2045-initiative/.

224. Zager, Dennis and Evans, Rick. In the Year 2525 (Exordium & Terminus). *Google Play Music*. [Online] https://play.google.com/music/preview/Ttplw3xk3c24e747l57eazgrtsm?lyrics=1.

225. 2045 Initiative. AUBREY DE GREY / Interview / ENDING AGING. *YouTube*. [Online] November 5, 2015. https://www.youtube.com/watch?v=2lmdp96ySlU.

226. CBC News. Google buys B.C. firm's quantum computer for NASA lab. *CBC News*. [Online] May 17, 2013. http://www.cbc.ca/news/technology/google-buys-b-c-firm-s-quantum-computer-for-nasa-lab-1.1393158.

227. Metz, Cade. For Google, Quantum Computing Is Like Learning to Fly. *Wired*. [Online] December 11, 2015. http://www.wired.com/2015/12/for-google-quantum-computing-is-like-learning-to-fly/.

228. Hutchins, Aaron. Trudeau versus the experts: Quantum computing in 35 seconds. *Maclean's*. [Online] April 19, 2016. http://www.macleans.ca/society/science/trudeau-versus-the-experts-quantum-computing-in-35-seconds/.

229. Aaronson, Scott. Can Quantum Computing Reveal the True Meaning of Quantum Mechanics? *PBS*. [Online] June 24, 2015. http://www.pbs.org/wgbh/nova/blogs/physics/2015/06/can-quantum-computing-reveal-the-true-meaning-of-quantum-mechanics/.

230. Adams, Douglas. The Ultimate Hitchhiker's Guide to the Galaxy. [Online] Del Rey, April 30, 2002. http://books.google.com/books/about/The_Ultimate_Hitchhiker_s_Guide_to_the_G.html?id=a-apCPdumpsC.

231. University of Calgary. A roll of the dice: Quantum mechanics researchers show that nature is unpredictable. *phys.org*. [Online] July 9, 2012. http://phys.org/news/2012-07-dice-quantum-mechanics-nature-unpredictable.html.

232. Hawking, Stephen. Does God play Dice? *Stephen Hawking Public Lectures*. [Online] 1999. http://www.hawking.org.uk/does-god-play-dice.html.

233. Goldman, David. The truth about the Hillary Clinton Google conspiracy theory. *CNNMoney*. [Online] June 15, 2016. http://money.cnn.com/2016/06/10/technology/hillary-clinton-google-search-results/index.html.

234. Mabillard, Amanda. Hamlet's Soliloquy: To be, or not to be: that is the question (3.1). *shakespeare online*. [Online] [Cited: February 22, 2016.] http://www.shakespeare-online.com/plays/hamlet/soliloquies/tobeanalysis.html.

235. Rothenberg, Jennie. Interview with Harold Bloom: Ranting Against Cant. *The Atlantic*. [Online] July 16, 2003. http://www.theatlantic.com/past/docs/unbound/interviews/int2003-07-16.htm.

236. Plato and Fowler, Harold North. Plato: Euthyphro. Apology. Crito. Phaedo. Phaedrus. *Loeb Classical Library*. [Online] 1999. http://www.amazon.com/gp/product/0674990404.

237. Thoreau, Henry David. Walden. [Online] Houghton, Mifflin and company, 1882. https://books.google.com/books?id=-EoLAAAAIAAJ.

238. Moskin, J. Robert. The U.S Marine Corps Story. [Online] Little, Brown and Company, 1992. https://books.google.com/books/about/The_U_S_Marine_Corps_Story.html?id=-sbEQgAACAAJ.
239. Marie, Laura. Laura Marie Interviews Teal Scott Swan. *YouTube.* [Online] January 10, 2014. https://www.youtube.com/watch?v=kwhO6t7hgXE.

Part II
Don't Be Evil; Making the World a Better Place

Don't Be Evil

Don't be evil. We believe strongly that in the long term, we will be better served—as shareholders and in all other ways—by a company that does good things for the world even if we forgo some short term gains. This is an important aspect of our culture and is broadly shared within the company.

Google users trust our systems to help them with important decisions: medical, financial and many others. Our search results are the best we know how to produce. They are unbiased and objective, and we do not accept payment for them or for inclusion or more frequent updating. We also display advertising, which we work hard to make relevant, and we label it clearly. This is similar to a newspaper, where the advertisements are clear and the articles are not influenced by the advertisers' payments. We believe it is important for everyone to have access to the best information and research, not only to the information people pay for you to see.

Making the World a Better Place

We aspire to make Google an institution that makes the world a better place. With our products, Google connects people and information all around the world for free. We are adding other powerful services such as Gmail that provides an efficient one gigabyte Gmail account for free. By releasing services for free, we hope to help bridge the digital divide. AdWords connects users and advertisers efficiently, helping both. AdSense helps fund a huge variety of online web sites and enables authors who could not otherwise publish. Last year we created Google Grants—a growing program in which hundreds of non-profits addressing issues, including the environment, poverty and human rights, receive free advertising. And now, we are in the process of establishing the Google Foundation. We intend to contribute significant resources to the foundation, including employee time and approximately 1% of Google's equity and profits in some form. We hope someday this institution may eclipse Google itself in terms of overall world impact by ambitiously applying innovation and significant resources to the largest of the world's problems.

—Google Inc. SEC Form S-1 Registration Statement (April 29, 2004)

Chapter 4
SEO in the Age of Digital Transformation: What Every Business Leader Must Know

Trond Lyngbø

Few other companies in the history of commerce have impacted humanity more than Google. What started out as a search engine conceived by Larry Page and Sergey Brin in the 1990s quickly grew to become much more. How much more? Without deep knowledge about search marketing, technology and psychology, it's hard to imagine how extensively it influences your life today, let alone how it will guide the evolution of society and business in the future.

While many see Google as a search engine (which it is), it's no longer just about technology. To think of the company as it was many years ago—a simple tool to find websites with information—is to grossly underestimate its impact on diverse aspects of day to day life today.

4.1 Google Has Transformed Business Forever

Google has become central to most buying behavior. Typical consumers use search engines to research or learn more about a product they plan to buy, a restaurant they want to eat at, or to locate a local business like a plumber, electrician or carpenter.

Search marketing has revolutionized the sales process. What used to be "selling" is now "informed buying". We don't have a sales-cycle any more, it has become a buying cycle. The rules are different, and you need a new rule book—because marketing and business will never be the same again. Ignoring Google is like turning away customers or sending them to a competitor.

Google can send you the lion's share of your customers. You might ignore reality, but you can't get away from the consequences of doing so. Any business that fights this revolution, or refuses to acknowledge this situation won't be able to

T. Lyngbø (✉)
Search Planet AS, Oslo, Norway
e-mail: trondlyngbo@gmail.com

© Springer Science+Business Media New York 2016
N. Lee (ed.), *Google It*, DOI 10.1007/978-1-4939-6415-4_4

fix problems, evolve and adapt. They will fall behind, losing customers, sales and market share. For some, it will happen overnight. For others, it may take some time.

What is certain is that you need to evolve and adapt with Google in order to survive in the world of business. It's no longer about you or your company, it's about your customers. Analytics-driven organizations will win. Data from diverse sources will help plan and prepare better. You can use search trends to predict the future. You must then prioritize your time and resources, investing only in whatever matters for accomplishing business-critical goals.

But Google isn't limited to business and sales. Its impact goes far beyond.

4.2   Google's Impact on Society

The search giant is in the process of building a knowledge base that's far beyond anyone's imagination. Google is compiling data about public figures (people), places, historic events, health, nutrition and much more. It then presents this to users in the form of information and insights directly in the search results.

No longer is the focus on delivering a list of 'ten blue links' on a subject. Try Googling Barack Obama and you'll get a complete presentation of all important and relevant facts about the U.S. President—who he's married to, what his daughters' names are, how he's related to other politicians, where he lives, when he was elected, and any other relevant information.

As an educational tool, Google helps students around the world quickly gain knowledge and insights for free. No longer do you have to dig through dozens of webpages to learn about the American leader... you can get the information without even visiting a single website—right there on Google!

What used to take hours, or even days, to research can now be accessed within a matter of seconds. It's like being plugged directly into all the libraries, newspapers, teachers and information on the planet, pulling out the key information you need without wasting time or money.

This means you can learn more, learn it faster and get more done. But that's not the only way Google is speeding up how you learn.

It understands exactly what kind of information is relevant to a searcher based on an intimate knowledge of users' preferences and associations, derived from associations of individual snippets of data stored in a humungous database. Search results are no longer simple directions to other websites with Obama-keywords in them, but a carefully culled selection that's most likely to address your specific needs at the time you are searching for information.

Google is on a mission to not only index information but to actually understand it.

Figuring out the relationships between people, places, context and more means that search engine users can tap this to get answers, like an Oracle, to anything. As a consequence, Google has become the world's most intelligent personal assistant. Your go-to resource for all answers.

Google keeps us one step ahead. Helps us make better decisions, faster. Lowers risk. Eliminates unpleasant surprises. Guides us in making important choices.

And thus, it shapes society… all the while, lurking in the background.

4.3 Google—A Trusted Personal Aide

Through the way you use the search engine, you're telling Google your most secret thoughts, things that not even your closest friends and family know about you.

And so, what started as a simple search engine has become the world's most trusted 24/7 personal assistant. Not only do you get instant access to all the world's knowledge, it's all organized, sorted and prioritized for you based on need and context, individually.

How did this shift happen? When Google decided to provide answers based on a user's intent, helping solve problems and meet needs to enrich their lives. It couldn't do this just as a search engine, ranking pages based on math and theory. That would make it no better than an "answer engine".

Enter the knowledge base of aggregated user behavior across billions of searches. Intelligently leveraging its unparalleled rich search data based on your inner thoughts and what you signal with your search terms, Google tries to predict what you're actually looking for. It sees several steps down the line—and wants to solve even those unstated problems for you. And it is the best in that class, no other search engine even comes close.

4.4 What Gives Google Such Impact and Importance?

To understand how Google works, we must look beyond technology and coding, beyond websites and content, beyond marketing and sales, even beyond the Internet… and change how we think of search engines today.

We must look at what Google used to be, where it is now, and where it wants to be. The aim is simple, if ambitious. To continue being the preferred search engine of your future.

To retain users, Google must be the best option whenever you seek answers, help or assistance. No matter when you need it, where you are, or how you access the Web—via desktop search or mobile device or wearable computers and even more futuristic technology that's "coming soon".

To achieve this dream, Google must evolve and adapt better and faster than anyone else.

4.5   Google's Past, Present and Future

Google of the 1990s was far less complex than it is now. A much simpler set of rules and criteria determined which websites should rank on a search results page. Even then, good search rankings were a business critical KPI.

Today, these rankings are driven by greater complexity. And have become more critical to businesses.

Search users get different results while using different devices. Mobile search results are customized to be relevant locally, so they better serve users on the move. Google arrives at the best fit through a combination of criteria, many of them beyond what's on the actual website, involving offline ingredients such as a searcher's physical location and the address of a store, restaurant or other place.

But is this exactly how things will work tomorrow?

Remember, Google is in the "trust business". They thrive because users trust the service. While the company makes money when people click paid ads on the search results page, Google realizes that users visit the site for its organic search... not to view the ads!

The moment another search engine offers them better service or makes it easier to find what they want, users will shift loyalty. Quality and trust are critical for the multi-billion dollar corporation's continued success.

Since search ads make up the bulk of its revenue, Google will continue to do its best to serve us the best organic search results possible. Ensuring a great match between people and businesses that serve them will remain their primary goal. But to achieve it, and keep constantly improving the user experience, Google will involve more and more factors in their ranking algorithms.

What does this mean for business?

Google has a huge financial impact on most companies today—and will continue to have an impact in the future also. Some of this impact is direct, the rest is indirect, subtle, and harder to understand because it's invisible and more about synergies than directly measured results.

Regardless, the important point is that every business must adapt to the way Google serves its users. That is how search engine optimization will be different in the age of digital transformation. To deconstruct the steps technically requires detailed knowledge about technology, search engine ranking factors and broad, deep expertise in various overlapping disciplines that make up modern SEO.

4.6   Google's on a Mission—And It's just Getting Started

Viewed from an artificial intelligence (AI) perspective, Google as a search engine has the IQ of a 4 year old child. Google is smart today. It will become a lot smarter in the years to come.

Once it gets to the age of 10 or so, search technology might be totally different. But the core of its service will still be the same.

Helping people. Guiding users. Making lives better.

That is the foundation Google was built on.

Google will continue to evolve and adapt. It will keep up with the technology revolution, and how we humans embrace it. The whiz-kid engineers, scientists and sociologists who guide the company's growth will make sure they do EVERYTHING possible to remain the best friend, philosopher and guide you know, love and trust.

This is critical for Google's future success. The Internet of Things (IoT) is just around the corner. Expect Google (and other companies) to target this market as well. Getting access to data from your home, learning about you, matching your needs to relevant offers and making sure you come back for more will be key to the search giant's strategy for survival and success.

What seems like science fiction today will be something we'll look back at a few years later as being the digital Stone Age. Things will only be more dynamic and complex. That's great for consumers, for students of technology and business, and overall for humanity. But succeeding in business will not be easy or guaranteed. It will require regularly updated knowledge, unlearning of outdated tactics and a changed mindset.

4.7 Who Wins in This Fast-Evolving Scenario?

Here's what businesses need to understand. Tomorrow's winners aren't those who look at Google as a search engine, or limit its role to search marketing or ranking for certain keywords. No, they are people who think differently. And follow a different approach. One that understands exactly how Google is critical to their financial results.

What does the future hold in store for businesses and companies?

With a plethora of options to choose from, the winners will be those who look to analytics data as a tool for strategic management. They'll see data and trends as revealing opportunities to seize, and won't consider time and resources spent on it as a hassle or waste. The future—of Google and businesses—is ALL ABOUT HUMANS. They'll need to pick our brains, dig into our thoughts and feelings, understand our needs and wants (including those we may not even be aware of yet!)

Everything is seamlessly connected and integrated. There isn't a distinction any longer between 'offline' and 'online' business. And the rich data available for analysis heralds the end of qualified guesswork and gut feeling in making business decisions.

Google is smarter than all the brains in the world—combined! Its arrays of servers, with their tera- and peta-bytes of stored data, work day and night to guide us, helping us safely and confidently navigate the uncertain seas of our everyday existence. Not just helping us find opportunities and see more options, but assisting us with decisions as well, guiding us to the right match.

The world's most powerful matchmaking service has transformed business and sales. And to the extent that businesses can leverage this shift, they will thrive and succeed. Search engine optimization is the means to that end.

Chapter 5
Search Engine Optimization: Getting to Google's First Page

Frank Buddenbrock

5.1 Google's First Page

You've got an established business, or perhaps, are just starting one. Or you're a professional with highly sought-after talents and you produce successful outcomes for your clients. In either of these scenarios, to market you or your business (they could be the same thing) effectively, in 2015 you're going to have to build a great website. It's pretty much expected—if your customer base/clients can't find you online either using their desktop computer or their mobile device, they're going to assume you don't know what you're doing, and they'll search out someone else.

With the more than 1 billion websites now available (as of August 2016 according to internetlivestats.com), it's critical, downright crucial, to get found if you want your venture to grow, or even survive. And you're going to have to be found on Google's first page at the optimum, or on page two at the least—the very least as you'll see shortly.

In the current environment, people just aren't going to search much past page one. If they don't see you there, they're going to assume they searched using the wrong phrase, and they'll go back to the search bar and try again. But they're not too likely to venture onto page two. If you're on page 3 or beyond, you could just as well be on page 300 for all the good it'll do.

According to recent research, 94 % of searchers do not go past page one—that leaves an incredibly dismal 6 % who dare to explore page two. And it's equally frightening how searching drops off even on the Holy Grail first page. The number one position gets more than 34 % of the impressions, position 10... only 2.7 %. OUCH! And position 20, that dark and lonely place on the bottom of page two... only 0.29 %!

F. Buddenbrock (✉)
Google AdWords Certified Specialist, Miami, USA
e-mail: frank@canyoufindmenow.com

© Springer Science+Business Media New York 2016   195
N. Lee (ed.), *Google It*, DOI 10.1007/978-1-4939-6415-4_5

Position Number One on Google's First page not only gets 34 % of the impressions, but more importantly, it also gets more than a third of all the clicks. That's why it's vital to be on Google's first page, and above the fold, that part of the screen that's visible without scrolling.

In a recent blog post, a gentleman wrote: I use "Google suggest, then top 3 listings. That's how I search, mostly. Sometimes down to result number 5 or 6, depending on the size of the screen. I've heard it said, though, that Page 1 is for information seekers, mostly. Wonder if this has any credence. Either way I don't really care. If a site is not on Page 1, result 1–7, then it has failed. Period."

Pretty sobering, don't you think?

5.2 Search Engine Optimization (SEO)

The new marketing paradigm, marketing your business online, has at its core, visibility. And as illustrated above, pages one and two of a Google Search result is where you need to be seen.

So just how do we get to that digital hallowed ground? Through a process called Search Engine Optimization, more commonly known as SEO. What we want to do is "optimize" our website so that Google will favor us over other similar websites and put ours at, or near, the top of the page when someone does a search. This is also known as SERP—search engine results page.

Google is quite open about what it expects from a website to get it to rank well; in fact, they spell it out simply and clearly in their Search Engine Optimization Starter Guide. By following their Best Practices suggestions, you'll make it easier for the search engine software (fondly referred to as Googlebot or spider) to crawl, index and make sense of your website's content.

The primary goal of the Google search engine is to provide the best possible experience for the user, that person who just made a search or query. And it's pretty easy to understand why Google would want that.

Google wants to be your first, and perhaps only, choice for online searches because when you use them for your searching, you're more likely to see and then click on one of the myriad ads populating those search results pages. That's where Google makes the bulk of their revenue. Each ad placement, whether at the top of the page, bottom of the page, or running down the side, is paid for by the advertiser (yup, each time you click on one of those ads that advertiser just paid Google for that opportunity).

Now that you understand Google's objective, you can see that they want to display those websites that they believe will give the user/searcher the best possible experience. They want websites that best answer the user's query and provide recent and relevant information. They want to display websites with good and useful content, including text, graphics, images and videos. As is commonly said about marketing online, "Content is King."

The process that helps Google determine where your website should rank involves Search Engine Optimization (SEO). Google uses approximately 200 variables/factors in their proprietary algorithm to determine where a website should rank. Obviously, they keep those 200 variables a closely guarded secret, but through testing and observing, the SEO community has been able to discover the most important factors. Let's explore those factors.

5.3 Keywords

Simply put, keywords, and keyword phrases, are those words and phrases a searcher types into the Google Search bar. It may be a single word, though unless it's extremely unique, the search will return results that are much too general. Typically, a searcher will type in compound keywords such as "red leather purse" or "off road tires" when looking for something. Or they may even type in entire sentences or questions seeking results, such as "What do I do if I find out I'm diabetic?" In this example, I'd say the keywords are "find out I'm diabetic."

When marketing online, keywords are THE most important elements for everything you'll do because those keywords/phrases are going to be those words people are typing into the search bar. And you'll want your website to be in the search results they see. Those keywords will drive all of your marketing efforts— websites, blogs, social media posts, image names, link text (also known as anchor text), press releases, forum content and so on.

One of the first and most important tasks you'll need to do is determine the best keywords to use—what are the terms your potential visitor is typing into their search bar. It may not always be what YOU think is best. This is especially true if you use industry jargon and terms not used by the general public.

For example, while the phrase "professional hair-cutting shears" may accurately describe that tool your hair stylist uses when cutting your hair, through keyword research you'll discover that "professional hair-cutting scissors" is searched considerably more often. Using keyword research software, we see that "professional hair-cutting shears" was searched 468 times in a month, while "professional hair-cutting scissors" was searched 1920 times. It's pretty easy to see which keyword/phrase will likely generate the most traffic to a website. That's not to say we won't use "professional hair-cutting shears" but rather we'll concentrate and put more of our efforts on "professional hair-cutting scissors."

You'll want to perform this same exercise for all those keywords and phrases relevant to you, your products and/or services. It's helpful to use a spreadsheet to keep track of your results because you'll be using these throughout your marketing/SEO efforts. Once you've collected your list of keywords, the work begins.

5.4 SEO Ranking Factors

Let's look at some of the more important SEO ranking factors that we have control over.

5.4.1 Site Speed

We are seeing the search engines giving more and more importance to how quickly a website loads—in fact, it's becoming one of the top ranking factors. Be sure to keep images no larger than necessary (they take longer to load than text), and keep ads and content from third-party sites to a minimum. You are often at their mercy for how quickly something may load from their site or server. The slower your site loads, the lower Google will rank your site, all other things being equal.

5.4.2 Title Tag

One of the most important ranking factors is the title tag. Think of it as the title of a book and you can see how important it is. It helps the visitor understand what the page is about, and also helps the search engines index and rank your page. Be sure to write a unique title tag for every page of your website. Again, if you think of it as the title of a book, you'd want a unique and different title for each book to reduce any confusion for your visitor, as well as for the search engines.

The title tag is already familiar to most people—it's that blue line of text you see for each listing on a search results page. That title tag is written behind the scenes in the html coding of a web page. It is highly recommended to have a least one of your major keywords in your title tag, and if possible have it as early as possible. If our keyword/phrase is "dog training" you should write a title tag like this:

DOG TRAINING | Dog Obedience Training | Train Your Dog

Using variations of your keyword phrase is quite effective. You see that we used "Dog Training" and "Train Your Dog," plus a similar phrase "Dog Obedience Training." By using title tags like this, there is no doubt what this page is about. It is recommended to use the pipe (horizontal line) to separate your keywords/phrases, and to keep them in the 65–70 character range.

5.4.3 Meta Description

The meta description is that text displayed below the Title tag and website url in a search result. This presents an opportunity in 160 characters or less to add information that supports or continues the idea presented in the title tag. Write meta descriptions that inform and interest users. Each page of your website should have its own unique meta description. This text is part of the behind-the-scenes html code in the header of a page and is used by Google to display in their search results.

The meta description is a great place to include even more keyword phrase variations and synonyms. It's a fantastic opportunity (missed by most people) to write compelling copy and offer a call-to-action to get a potential visitor to click on your search result link. If you have space and can't think of anything to add, at least add your phone number to make it easy for people to contact you.

5.4.4 Heading Tags

Heading tags can be thought of as the sub-headings on a web page. They start with H1 and can run up to H6. Each page should only have one H1 tag, but may have several succeeding smaller heading tags. You could have several H2 tags, followed by several H3 tags and so on. Here again, if possible, include a keyword/phrase or variation of it.

One important idea to keep in mind when writing the content for your website is to make it read well.

Resist the temptation to overdo it with keywords and phrases. Your writing should read the way people speak. People are pretty sharp—they'll catch on pretty quick if your writing is too focused on stuffing your content with keywords with the hope of gaining favor with the search engines. And the search engine software is quite smart as well—it is able to determine if you're trying too hard to stuff your content with keywords. If it catches you doing this, you could be penalized rather than rewarded.

If all this sounds challenging—it is. A good and talented Search Engine Optimization expert (also known as an SEO) will work very hard to get your major keyword phrases included in all the necessary places as well as the content on your page, and make it read well all while pleasing the search engines. A good SEO is like a master chef—they have access to the same ingredients you do but are able to put together just the right combinations of ingredients in just the right proportions to make a beautiful presentation. They definitely bring a little bit of artistic talent to the project.

5.4.5 Keyword in URL/Keyword in Domain Name

When possible, it's recommended to include your keyword in a web page's url, especially if you are not able to get a domain name that includes your keyword. Most simple and common (and many not-so-simple or common) domain names are taken—DOGTRAINING.COM for example. But ABCKENNELDOGTRAINING.

COM may be available. Don't go too crazy to try to get your keyword in your domain name, and don't make it too long or cumbersome.

While you may not be able to get a domain name that includes your major keyword, you should include your major keywords in the url of sub-pages, those pages that are not the home page.

Your dog training page's url could be: ABCKENNEL.com/dog-training for example. Follow this example for the remaining pages, such as: ABCKENNEL. COM/obedience-training-school.

5.4.6   Keywords in Body Text

When writing the body text for each page of your site include your keywords and variations throughout the text. Include your keywords early in the text, in the first one or two sentences, and try to finish the page with a keyword phrase or variation. Sprinkle those keywords in following paragraphs, but remember to keep it readable —don't stuff your content with keywords.

Another useful tool is modifying the attributes of the text for those keywords by making the text **bold** or *italicized*. This helps those keywords stand out from the surrounding text making it easier for your users to see what this page is about. The search engines also like those emphasized keywords for indexing your content.

5.4.7   Media

When planning the content for your pages, including images, graphics and videos are useful tools not only for your user, but for the search engines as well.

Images, graphics, and videos are an often overlooked opportunity to make use of your keywords. First, use your keywords when naming your media. For example, an image titled "dog-training-in-park.jpg" is much better than "image1.jpg." The same idea applies when naming graphics used on your site, as well as videos inserted into a page.

Additionally, and especially for images and graphics, be sure to make use of the ALT attribute. This is often overlooked and is a valuable SEO opportunity. The "ALT" attribute allows you to write alternative keyword-rich text for an image or graphic if for some reason it cannot be displayed. You'll see ALT text display when you hover over an image or graphic.

The "ALT" attribute was originally developed for sight-challenged people who would use a screen reader to describe the contents of a web page. The search engines use the ALT text of an image similarly to the anchor text of a text link.

Optimizing your image and graphic filenames and ALT text makes it easier for Google Image Search to understand and index your images. Google will often display images in their first page of search results and you increase your chances of being displayed when making use of the ALT attribute and keyword-rich image and filenames. This functionality comes and goes as Google sees fit.

5.4.8 Sitemaps

A simple HTML sitemap listing the all the pages of your website can be useful. Think of a sitemap as being similar to a Table of Contents or Index for your user. Include all the important pages of your website and their links. Be sure to organize them properly by subject or category.

You should also create and include an XML sitemap to ensure that the search engines can find and index the pages of your site. Once you've created the XML sitemap, upload it through your Google Webmaster Tools account to inform the search engines. Be aware that the search engines may not indicate that they have indexed every single page listed in your sitemap. For example, you may upload 250 pages/links and the Google may only show that it has indexed 220 pages. Google does not guarantee that they will index all the pages/urls from your website.

5.4.9 Anchor Text

Write anchor text that is useful for your users and the search engines.

Anchor text is that clickable text that is used for links that direct a user to other content or to another page. Well-written anchor text tells your user and the search engines something about the content or page that you are linking to. Here is an example of a well-written anchor tag:

<a href=http://www.abckennel.com/potty-training/ Potty Training Your Dog></a>

You can see that this is much clearer and more accurately describes the content rather than:

<a href="http://www.abckennel.com/potty-training/">Click Here</a>

"Click Here" offers no value for the user, nor for the search engines.

Make it easy for users and search engines to distinguish anchor text from regular text—typically anchor text links are a different color from the surrounding text, and/or is underlined. This has become the standard for indicating link text.

5.4.10 Social Media

Social media and "social signals" are becoming increasingly important as an SEO ranking factor. Google continues to consider social signals as indicators of the importance of a website. The different social media platforms Google considers important are:

Google+
Facebook
Pinterest
Twitter
YouTube

Content from these platforms that link to a website improve its SEO rankings.

5.4.11 Backlinking

Links from relevant websites have long been an important SEO ranking factor. Links should be from high-quality sites and the more you have the better.

Google views links from another website as a vote for your site, and considers them a trust factor. You should work to getting links from "authority" websites—websites that are considered reputable and trustworthy.

Google may penalize you for having too many backlinks from completely unrelated websites. Many lazy SEO companies try to game the system by flooding a website with unrelated links. Google may even penalize those sites by deleting all of a website's indexed pages. Links to a website typically come gradually, not all at once, and come from people who have discovered your site through natural processes such as blog posts, press releases, Facebook and Twitter posts, and related articles. Unnatural linking could actually be detrimental to the reputation of your website.

5.5 SEO Done Properly

While not terribly difficult (though not much is terribly difficult when you know all the proper techniques and processes—being a master chef, for example, as alluded to earlier), SEO done properly is a lengthy, complex process. Research can be time-consuming. Creating interesting content for your user that compels them to stay on your website long enough to decide whether they want to develop a relationship with you is time-consuming and may require the talents of a real copywriter. Promoting your website through blogs, press releases, and social media and article sites is also time-consuming. Producing images, graphics and videos to increase the value of your site for both the user and the search engines takes special skills, talents, and possibly large budgets. Developing relationships with other people and websites that like your content and want to share it with THEIR followers takes time, diligence and fortitude. The good thing is that when done intelligently, methodically, and effectively, getting to Google's first page can be within your reach.

5.6   SEO Case Study

One of my first SEO clients was Eileen Koch who owns a self-named Los Angeles-based Public Relations firm—in fact, she is still a client nearly 10 years later.

Eileen contacted me when the business directory website she was listed with suddenly vanished from Google's indexed pages. We learned later that the owner/developer of the site had tried to "game the system" and that is definitely something Google won't stand for. So Google simply removed that site from its index.

The sad thing is, there really is no reason to break, nor bend, the rules Google lays out. When you follow Google's rules (what they call their Best Practices), a properly optimized website can show up on Google's first page. I know—I've done it more than 60 times.

Eileen needed a new strategy—the business directory had been her major lead generator and had been responsible for many phone calls from prospective clients. Now it was crickets—the phone stopped ringing.

She contacted me, told me of her debacle, and I recommended that we perform an audit of her website. Unfortunately for her, we discovered that there was virtually no optimization done on her website—almost no on-page optimization, and no off-page optimization at all.

Her site couldn't be found in the first 50 Google search results pages. Of this I'm certain—I sat one evening in front of my muted TV with my laptop and literally went through each and every of the first 50 pages and didn't see ANY mention of her nor her website.

In a sense this wasn't a terrible starting point—at least there was nothing we would have to undo. We were working with a fresh website which allowed us to perform a complete optimization. We of course started with an extensive Keyword Research program. Once we compiled our list, we discussed it with Eileen and agreed on which keywords were best, and most relevant, for her business and website. Not only would these keywords/phrases be used as the basis of the optimization campaign, it would be the foundation of all future marketing efforts.

These keywords were uploaded to the meta Title, meta Description, and meta Keyword fields for each page of her site. We made sure to include variations, combinations, and synonyms. We also made sure to include these terms within the 300+ words on each page, and made certain terms bold and/or italicized. Using our major keywords spreadsheet made easy work of developing anchor text intra-site links, as well as inter-site links.

For the next few months we continued to optimize the site until five months later I searched for one of our major keyword phrases, and lo and behold, there we were on the bottom of page one! I couldn't wait to call Eileen to share the good news.

I called her up and asked her to perform a Google Search, expecting her to scream in delight to see her website on the first page. However, she promptly and disappointedly told me she could not see her site listed there. Yes, I said, it's right there at the bottom of page one—I'm looking at it, I see it on my screen.

No, she said, it's not there. I sat there dumbfounded. How could I see it on my screen, yet she couldn't see it on hers? It was then that I learned that a Mac computer running Safari displays different results than a PC running Firefox. We learned why she couldn't see her website on page one, and it wasn't until another few agonizing/confusing/frustrating days that her Mac running Safari displayed her site on page one. EUREKA!

And after a few more weeks, Eileen's site was ultimately #1 for her major keyword phrase, "Los Angeles Public Relations firm."

As of August 2016, Eileen's site is on Google's first page in the fourth position. The number one spot goes to Odwyer's business directory of PR firms. (Directories are very, very difficult to beat. They rank well typically because of their abundance of content, multiple pages and near-continuous addition of fresh, relevant content.)

Eileen's site has remained on Google's first page for over 10 years. A steady trickle of relevant content, intermittent modifications to meta data, and a strategic back-linking campaign can keep a website ranking well, high up in Google's search results.

I hope that your optimization efforts bring you similar results.

Good luck.

Chapter 6
4 Tips for Writing Outstanding SEO Boosting Content

Tina Courtney

6.1 The SEO Game

Custom content; it is the essence of the SEO game, its very heart and soul. This is how your company will spread its gospel wide and far, gaining massive public awareness and droves of visitors to your site. Without creatively written, unique, and highly informative content, your site is likely to gain little to no foothold as this copy is what lets the masses know you have high-quality information to provide. And the public simply cannot get enough of the delicious insights delivered through top-notch content (https://www.linkedin.com/pulse/why-content-marketing-your-businesss-bff-tina-kat-courtney?trk=mp-reader-card).

People aren't the only ones who love compelling content; Google absolutely adores it. Lives for it, even. The search giant is constantly improving its algorithms to present the most relevant, informative, and downright enlightening search experience possible. In February 2011, Google even created a specific algorithm that aims to eradicate lackluster content, affectionately called Panda (http://searchengineland.com/library/google/google-panda-update).

While this update may sound all cute and cuddly, it is anything but. Google's Panda is more of a seek-and-destroy algorithm that obliterates thin or low quality sites and content by sinking it in the SERPs, rarely to ever see the light of day again. Sounds intense right? Google is fully aware of how important superb content has become to the average internet dweller and is on a mission to provide nothing less than spectacular results.

Google summed up Panda by stating, *"Our recent update is designed to reduce rankings for low-quality sites, so the key thing for webmasters to do is make sure their sites are the highest quality possible. We looked at a variety of signals to detect low quality sites. Bear in mind that people searching on Google typically*

T. Courtney (✉)
Evolve Inc., Greenville, USA
e-mail: tina@evolvesinc.com; oshgumishy@gmail.com

© Springer Science+Business Media New York 2016　　　　205
N. Lee (ed.), *Google It*, DOI 10.1007/978-1-4939-6415-4_6

don't want to see shallow or poorly written content, content that's copied from other websites, or information that are just not that useful. In addition, it's important for webmasters to know that low quality content on part of a site can impact a site's ranking as a whole."

If a site contains unoriginal content, thin content, duplicate content, poorly-written content, content that is not useful, has a poor ad-to-content ratio, or does not have a pleasant UX, prepare to get buried. If you do manage to pass this strict criteria, Google will then take a look at engagement and awareness metrics to determine if your site belongs on the front page. Your site must legitimate, intriguing, and top-notch to meet Google's high standards.

If you were on the receiving end of Panda's wrath, here is a checklist (http://googlewebmastercentral.blogspot.ca/2011/05/more-guidance-on-building-high-quality.html) to determine where you may have gone wrong.

With Google's extreme emphasis on content, great copy provides the keys to site-salvation. In order to produce lauded and celebrated content, there are some practices that must be diligently implemented across the board. So without further ado, here are the top SEO tips to get your site at the top of the SERPs.

6.2 Tip #1: The Audience Comes First

Without a fan-base to read your content, there is no point in producing it in the first place, so be sure to always put these folks first in your writing. Try to put yourself in their shoes for a moment—do you want to read content that is jam-packed with keywords? Do you want to see your favorite brands posting materials that are completely off-topic for their niche? Do you want to read a hefty article or blog that leaves you none-the-wiser at the end? Of course not, so don't provide anything like this to your audience either, or beware the wrath of Panda.

With all that Google has done to provide its users a top-of-the-line experience, stuffing content with keywords is genuinely much more harmful than beneficial. Content that isn't relevant to your audience will get tossed to the side and forgotten faster than it took to write it. Utilizing any of these black hat SEO tactics will ultimately damage your site in the rankings and degrade public opinion of your business. And with today's competition, you can't afford this kind of backwards movement.

The only thing that consumers and Google care about is thoughtfully crafted content that provides a fresh, informative, and extremely relevant perspective on topics.

To successfully achieve the art of writing pristine copy, don't fuss over things like keywords; these are still important but will almost always be woven naturally throughout your piece. Focus on acquiring original information through studies or polls, shift the paradigm on subject matter and provide an entirely new perspective so as to not regurgitate the same opinion, and most importantly, focus on the pain-points and relevant issues that are most pressing to your audience. By putting these first, your site will begin to gain the credibility and authority that it needs and deserves.

6.3 Tip #2: Backlink Like You Mean It

Backlinks are another valued component to dominating SEO. Much like the content element, however, backlinks should be approached with a rather discerning eye. There is an old black hat tactic of back-linking anything that a site could get its proverbial hands on; this is no longer relevant and will only succeed in helping to demolish a site's reputation and ranking status.

In April of 2012, one year after the unveiling of Panda, Google introduced another warm and fuzzy algorithm update into the mix; this one would be dubbed Penguin.

Penguin has a single mission; to eradicate spam from the internet through downgrading sites that use spamming techniques to obtain better rankings in SERPs. Any site that is utilizes unnatural links, spam-like links, paid links, article directory links, links from sites with low-quality content, or anything else that appears questionable will be sent into abysmal Google standings.

Be sure to only set your sites on links that are quality and reputable such as news organizations, esteemed sources that are relevant to your material, sites with high trust flow metrics, and credible blogs. Do keep in mind, however, that creating too many backlinks on your site will be detrimental to your efforts as well. If you do find that you have been hit by the Penguin update, it is possible to recover, for everything is impermanent.

Google's disavow tool (https://www.google.com/webmasters/tools/disavow-links-main?pli=1) is easy to use and can help get your site out of hot water. The disavow tool is Google's provided resource for allowing site owners to ask for certain links to not be counted. This tool is an absolute blessing to webmasters, but be certain that you are only disavowing links that are manipulating the results, otherwise your site can be harmed in the process. There are some great resources available online which provide a step-by-step tutorial (https://moz.com/blog/guide-to-googles-disavow-tool) of how to complete this process and be well on your way to getting back into Google's good graces.

6.4 Tip #3: Extra, Extra, Read All About It

Remember this little saying as the newspaper boy would wave around the current days copy and quote the front page headline? This was done because the headline is the hook to reel in readers. And despite the evolution in how people read the news, the power of headlines has not faltered.

Headlines must be engaging, gripping, and eye-catching in order to be click-worthy. The internet is littered with copy, content, images, quotes, stories, and so forth; there is simply too much information available online for readers to be hooked by a drab headline. No matter how epic, legendary, or poetic your piece may be, it will never reach your audience without a title that packs a wallop. In fact, only 2 out of every 10 individuals is likely to read past the headline (http://www.copyblogger.com/magnetic-headlines/), so make it count.

While it may be tempting, avoid developing "click-bait" at all costs. This type of misdirection will only serve to disappoint and possibly infuriate your audience while simultaneously contaminating your reputation as a thought-leader in your niche.

To develop some attention-grabbing headlines that clearly reflect the content ahead, focus on cultivating expressions that echo the value contained within that readers can look forward to. Seek to insight inquisitive interest, deep wonderment, or even shock; anything that will strike an emotional chord for readers without misleading them. Utilize keywords and phrases that mirror what the piece is about too.

One great way to learn about the types of headlines you are aiming for is to research popular blogs within your niche and gain inspiration from their most successful posts. This will give you an understanding of successful structures, keywords, and general appeal. Alternatively, the internet houses a multitude of magnificent tools to help you achieve headline gold. Tools like CoSchedule's Headline Analyzer (http://coschedule.com/headline-analyzer) and SEOPressor (http://seopressor.com/blog-title-generator) will put you on the path to some masterful headlines.

6.5 Tip #4: Visual Voodoo

Visual content can cast a spell on folks and lead them down hours upon hours of YouTube videos, Vines, gifs, and so on. Think this sounds a bit exaggerated? Consider this: over 6 billion hours of video is watched on YouTube each and every month (http://expandedramblings.com/index.php/youtube-statistics/). It is predicted that by 2018, nearly one million minutes of video will be published online *every single second*. The bottom line here is that people cannot get enough visuals incorporated into their content.

Last year, images accompanied 75 % of all posts made to Facebook (http://www.socialbakers.com/blog/2149-photos-are-still-king-on-facebook). Tweets with accompanying images receive 18 % more clicks, 89 % more favorites, and 150 % more retweets (http://blog.hubspot.com/marketing/visual-content-marketing-strategy) than those that don't. In just 2 years, infographic search volume exploded by more than 800 % (http://www.socialmediaexaminer.com/infographics-everything-you-need-to-know/). Without a captivating image to accompany that artfully crafted article or post, it will simply be drowned out in a barrage of visuals, both static and dynamic.

For your social posts, blogs, articles, or whatever else to stand any chance, it is imperative that the content is coupled with a high-resolution, interesting, and relevant image, video, or gif. This will help draw the reader's eye to your post before they continue on. Once you have gained their attention, your highly clickable headline is there to finish the job and gain that coveted click-through.

All of the information discussed here should be taken to heart and integrated into your SEO blueprint as soon as possible, for high quality content will always be the backbone of SEO success. Visual content is well on its way to becoming just as critical. It's a simple yet immensely powerful recipe.

As the internet and SEO continues its evolution, people's standards for quality content will only continue to advance and Google's rules and regulations will grow increasingly stricter. Do your business a favor and learn to become a master craftsman (or craftswoman) when it comes to content; there are plenty of tools and resources available online to help you along the way. All you need is a strong drive for success, a heavy dose of creativity, and the nerve to stand out from the crowd. Add a little consistency and you are light years above your competition. Google—and the world—will take notice.

Chapter 7
Internet Advertising and Google AdWords

Nicole Ciomek

7.1 The Importance of Internet Advertising

The Internet has changed how businesses of all sizes market their products and services. No longer do businesses have to rely on the Yellow Pages, newspaper ads, or TV spots. The Internet has opened up a new venue for reaching potential customers, one that is more tailored and accountable. Businesses have the ability to truly understand what the return on investment is with online advertising. It is not about circulation or how many people drive by a particular billboard each day. Advertisers have the ability to know how many people clicked on an ad every day, what search led them to the website, how long they spent on the website, and whether they purchased or filled out a lead form.

Online advertising allows businesses to make the most of their advertising dollars. Any business can find success with the right online advertising program set up in the correct manner. But where does one start?

The answer is simple: Google. Google is the largest search engine on the planet. In 2014, Google reported that there were 5.7 billion searches completed per day worldwide. With the majority of Internet users turning to Google to help them find the information, services and products they need, Google chose to monetize their search engine through paid advertisements.

Google began selling advertising in October 2000 through the Google AdWords platform. AdWords has since become the largest online advertising marketplace.

N. Ciomek (✉)
Radiant PPC, Bend, USA
e-mail: nicole@radiantppc.com

© Springer Science+Business Media New York 2016
N. Lee (ed.), *Google It*, DOI 10.1007/978-1-4939-6415-4_7

7.2   Google AdWords

Google AdWords is a keyword search advertising platform. This means that advertisers select what keywords they'd like their ads show on, then they create the ad copy that will show when a web user searches for that particular term. Ads are displayed to users on Google.com at both the top and right hand sides of the page (see Fig. 7.1). Ads include a link that will send users to a relevant page on their website.

AdWords is a pay per click (PPC) marketplace. This means that Google only charges advertisers when a user clicks on one of their ads.

The AdWords marketplace is an auction. Advertisers tell Google what they are willing to pay for each ad click for a particular keyword search. Google then utilizes this information plus the relevancy of the ad copy and landing page to determine what order the ads show on a particular search. Advertisers bidding higher and with strong relevancy, will show up on higher on the page and garner more clicks.

Fig. 7.1   Google AdWords

7.3   AdWords Ads: What Google Is Selling

7.3.1   What an AdWords Ad Looks like

Ads that appear on Google are quite concise and to the point. There is a limited amount of space to convey what's on offer, why a user should visit the site, and why it is better than its competitors. The ad copy is how users are drawn in. What makes a user click? To understand that, you first need to understand what goes into an ad (see Fig. 7.2).

7.3.2   The Different Parts of an Ad

Each AdWords ad is composed of a headline, two lines of description, the display URL and the destination URL.

Headline
The headline is the top line of text that will be the link to the page. Google gives you 25 characters for this line. This is the part of the ad that will catch the users' eye. Make this highly relevant to the keywords in the ad group. Putting keywords from the ad group in the ad copy will help your quality score greatly. It increases ad relevancy. Since the headline is the most eye catching part of an ad copy, putting keywords into the headline will draw in users since the headline will match or be closely related to what they just searched for. You also want the headline to be intriguing. Try including high value words like best, top, and #1. You can also ask a question in the headline.

Description
The description is 2 lines each of which can be 35 characters in length. Here is the meat of your ad. This is where you give users additional information. You'll want to make sure your description contains two components: *a value proposition* and a *call to action*. The value proposition can be anything that makes your business more compelling than a competitor. Free Shipping, Lowest prices, Rated #1, and so forth. Think about what sets your business or products apart from the rest. Think about what might compel a user to do business with you and get that in here. The call to action should come toward the end of the ad and should describe the action you want the user to take. It can be anything from Shop Now to Call Today to

Fig. 7.2 Ad copy

Best Cash Back Cards
www.wisebread.com/**CashBack**Rewards ▾
Up to 6% **Cash Back** & Bonus Offers.
Compare **Credit Card** & Apply Now!

Learn More. Think about what the action is you want user to complete and make sure you include this in every ad.

Display URL

The display URL is the final part of the ad that shows on Google.com when a user searches. This will be the URL of your website. This is pretty straightforward, but there are ways to utilize this more fully. Putting a "/" after the URL can be another place to increase relevancy. For example, you could just put WomenShoes.com. But, WomenShoes.com/HighHeels will be even more relevant to a user looking for high heeled shoes. Make the most of that space to get some additional keywords in the ad.

Destination URL

This is the page where you'll send users. They do not see this, but as we'll review later, selecting a relevant one is highly important.

7.3.3 What Makes an Ad Compelling?

Creating compelling ad copy takes work with such limited space to communicate with your users. When composing ads, there are a few key points that will help you get to create engaging ads:

- *Be direct* Your space is limited, say what you need to say in as few characters as possible. This will allow you to fit in as much information as possible to your potential customer and help users to quickly understand what products or services you offer.
- *Be specific* Make ads in each ad copy specific and unique. Tailor the ads in each ad group to those keywords. This will lead to more clicks and higher quality traffic than a vague, generic ad.
- *Know your audience* Knowing who your ideal customer is will help you to write better ad copy. Write out a description of who you see your customer as before you create ad copy. This will put you in the right mindset.

7.4 AdWords Campaigns: The Advertiser's Structure

7.4.1 What Is a Campaign?

AdWords is organized around campaigns. The best way to think of a campaign is the way you would any marketing campaign: a campaign is the avenue by which you'll advertise for a certain set of products or services to your targeted market. In AdWords, campaigns are the umbrella under which ads are managed.

The campaign is where an advertiser decides what users they want to target. Here they can set what hours of the day ads run, in what locations ads will run in (this can be anything from zip codes to countries), and most importantly, what keywords will be used.

7.4.2 How to Select a Campaign Structure

Before you get started with your campaigns, you need to contemplate what you want to advertise for and what your goals are. Are you advertising for a series of products? Do you have variety of services you offer or just one? Are you looking into increase brand awareness? Or do you want to increase sales?

You need to have a strong understanding of what you want to accomplish with these campaigns as this will help to determine your campaign structure.

Determining how to organize your campaigns is the first step to running a successful AdWords account. The best practice is to create campaigns for different products or different services. If an advertiser is a women's clothing retailer and they want to advertise for a variety of products they sell, the best format would be to have a campaign for dresses, then one for pants, another for tops and so forth. If the advertiser is a service provider, like a dentist, then campaigns should be divided up by service: cleanings, tooth implants, gum disease treatment and so on.

7.4.3 How Campaign Structure Impacts Performance and Optimization

Creating a clean and logical campaign structure is essential as it allows you to quickly understand the performance of your campaigns and determine which products and services are performing well, and which are not. This also allows you to designate different budgets to different product lines, pause a campaign if you no longer want to advertise for that service and informs you which campaigns need improvement.

Taking the time to figure out a campaign structure first will make optimizations easier, and allows for an advertiser to utilize their time most efficiently.

There is no one-size-fits-all approach to campaign organization: the best structure is the one that fits the marketing strategy of the business and allows for flexible growth.

7.5   Keywords: How Ads Are Targeted

Picking the correct keywords is the most important part of an AdWords campaign. These keywords determine when your ads will show, how a user will find you and learn about your business. Selecting a list that is relevant to your business is extremely important. Choosing keywords that are not relevant can result in wasted budget and driving users to your site who are unlikely to be interested in what you have to offer. To get a strong keyword list, you must begin with keyword research.

7.5.1   How to Go About Keyword Research

Keyword research can be done in a variety of ways, but is best undertaken in a logical and organized manner.

First, look at your first campaign. What products will you be selling here? Let's say it's shoes. You could advertise on the word "shoes" but a user searching for shoes could have a variety of intentions. When picking keywords, it is essential to think about what a user's intent would when they search for a particular keyword. Why would someone search for the word "shoes"? They may be interested in purchasing shoes. Or they may be looking for historical information on shoes. Or they might be looking for pictures of shoes. It is difficult to say. If it is difficult to say what the user's intention is, it is important to narrow your focus further. Who are these shoes for? Let's say they are women's shoes. This already narrows your focus. Are they dress shoes? Casual shoes? This can narrow the focus further.

Take the time to brainstorm a list of keyword related to the products or service you want to advertise for in that campaign. Look at your landing page. Think about what's on that page. Use the content from that page to help you brainstorm.

Once you've brainstormed a list, you can then utilize Google's Keyword Planner tool (see Fig. 7.3). This tool helps you to find different variations of these keywords and potentially some related terms you had not thought of. You can copy your list of brainstormed keywords into the search box, and then check off the keywords Google has found for you that you like. This will all be exported into an Excel file.

Finally, go over this list one more time. Write down any other variations that come to mind. Think of different modifiers: Leather women's shoes, black women's shoes, cheap women's shoes, buy women's shoes. Any word that can be added to the key phrase of what you are advertising for will help to give you better coverage on Google searches and help you to determine which keywords will work best for your business.

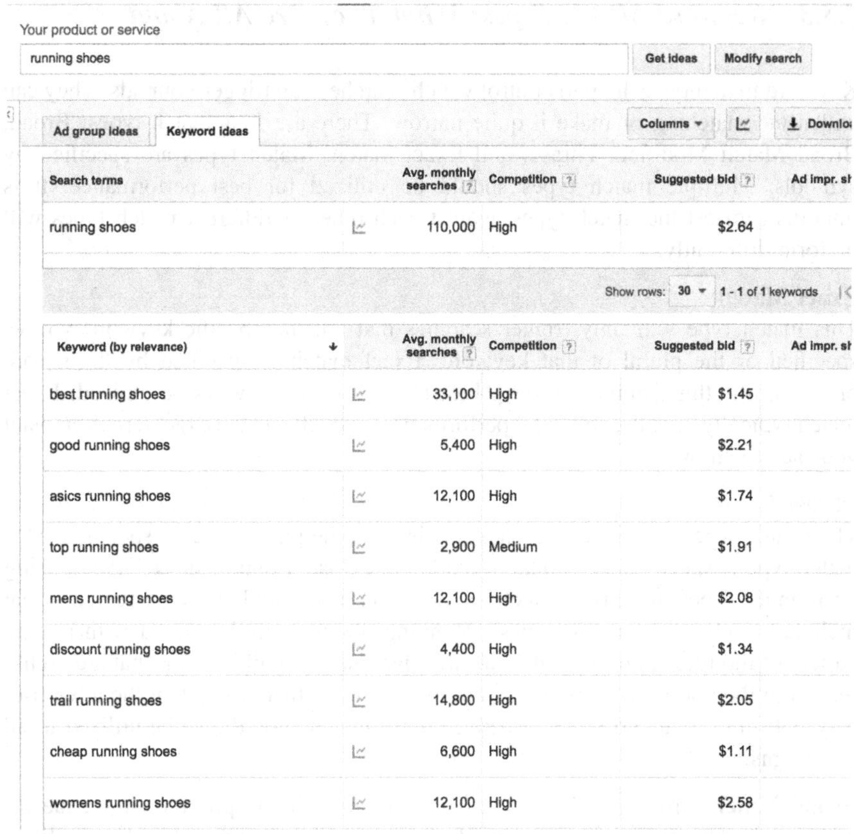

| Your product or service | | | | | |
| running shoes | | | | Get ideas | Modify search |

| Ad group ideas | Keyword ideas | | | Columns ▾ | ⌁ | ⬇ Downloa |

Search terms		Avg. monthly searches ⁇	Competition ⁇		Suggested bid ⁇	Ad impr. sh
running shoes	⌁	110,000	High		$2.64	

Show rows: 30 ▾   1 - 1 of 1 keywords   ⟨

Keyword (by relevance)	↓	Avg. monthly searches ⁇	Competition ⁇		Suggested bid ⁇	Ad impr. sh
best running shoes	⌁	33,100	High		$1.45	
good running shoes	⌁	5,400	High		$2.21	
asics running shoes	⌁	12,100	High		$1.74	
top running shoes	⌁	2,900	Medium		$1.91	
mens running shoes	⌁	12,100	High		$2.08	
discount running shoes	⌁	4,400	High		$1.34	
trail running shoes	⌁	14,800	High		$2.05	
cheap running shoes	⌁	6,600	High		$1.11	
womens running shoes	⌁	12,100	High		$2.58	

Fig. 7.3 Keyword Planner

7.5.2   How to Determine if You've Got a Strong Keyword List

Once you've finished your keyword research phase, go over the list of keywords once again. This may seem tedious, but this is probably the most important phase of creating an ad campaign. If you start with keywords that are way too generic, you could easily waste a good amount of your advertising budget. It is essential to take keyword selection seriously and be very thoughtful in this phase. This will help you to reach success as quickly as possible.

In this final review, eliminate any keywords that you feel uncertain about. You can always test them out later. When most advertisers start using AdWords, they are trying to determine if AdWords will work for them. You want to give yourself the best chances of success. Any keywords that seem questionable should be eliminated and revisited once your campaign is up and running.

7.5.3   Keyword Match Types: What They Are All About

Keyword match types help to control which searches can trigger your ads. They can widen the matching or make it quite narrow. There are four match types: Broad, Broad Match Modified, Phrase, and Exact match. Match types are specified by symbols. Multiple match types should be utilized for best performance. It is important to test the match types against each other as different match types will perform differently.

Exact Match
This match type will only trigger searches that are the specific keyword you've specified or the plural of that keyword. Exact match is specified by the square brackets, like this: [running shoes]. Exact match should always be utilized. Exact match typically sees the strongest performance since this is the exact term you want your ads to show on.

Phrase Match
This match type will trigger searches that include the phrase you've specified. This match type is specified by quotation marks, like this: "running shoes". Advertising on running shoes on phrase match would mean ads would show on searches that include the phrase "running shoes". Running Shoes could be paired with a large variety of modifiers. Phrase match allows advertisers to pull in terms that are highly related to their services or products but that include other words that the advertiser may not think to advertise on directly. Phrase match should also be utilized in all campaigns.

Broad Match Broad match is a bit of a wild card. It can provide an avenue for picking up traffic for long tail keywords, but it can also pick up a lot of unqualified traffic. Broad match can be utilized, but it needs to be done so very carefully. Google may portray broad match as still being very related synonyms, but sometimes the matching can be very broad and not that related at all. This match type has more of an advantage for Google than it does advertisers. It allows them to increase their revenue, but it wastes advertiser's budget.

Broad Match Modified This is a newer match type that Google came up with in the past few years to help advertisers to pick up long tail traffic, but to keep searches more relevant. Broad Match Modified works by putting a + mark in front of the most important words in your keyword. For example: if you want to advertising on women's sweaters on sale, you'd could specify that women's and sweaters the most important words and Google would then match to searches that include those words, but the order could be reversed and there can be different modifiers. It would like such: +women's +sweaters for sale.

Figure 7.4 is helpful chart that show examples of what kind of searches could be triggered by keyword on all the different match types.

Best practices are to utilize Exact and Phrase match on all terms, and then Broad Match Modified or Broad match on areas where you'd like to pull in a wider variety

Match type	Special symbol	Example keyword	Ads may show on searches that	Example searches
Broad match	none	women's hats	include misspellings, synonyms, related searches, and other relevant variations	*buy ladies hats*
Broad match modifier	+keyword	+women's +hats	contain the modified term (or close variations, but not synonyms), in any order	*hats for women*
Phrase match	"keyword"	"women's hats"	are a phrase, and close variations of that phrase	*buy women's hats*
Exact match	[keyword]	[women's hats]	are an exact term and close variations of that exact term	*women's hats*

Fig. 7.4 Keyword Match types

of searches. Choosing match types correctly can help you to save budget and better control where your ads show.

7.6 Keyword Organization

7.6.1 You've Got Your Keywords, Now What?

You've now got a list of clean keywords that are very relevant to your business. Now comes in another organization step. How will you organize your keywords within a campaign?

This will be done by ad groups. Ad groups are ways to sub-divide your keywords within a campaign and pair those keywords with relevant ad copy. Subdividing your keywords into groups allows for easier campaign management, sending the keywords to the most relevant page on your website, and creating ad copy that is highly relevant to those keywords.

7.6.2 Different Theories on How to Organize Your Keywords

Talk to a dozen PPC professionals and you'll get a dozen ways of how to organize your keywords within a campaign. There is nearly an infinite number of ways to organize your keywords. Below are the 3 main ways PPC professionals will organize ad groups:

1. One keyword, one match type per ad group. This is a pretty granular way to organize your campaigns and only recommended when you have a lot high volume keywords. This creates a lot of ad groups and can be time consuming to

manage and for ad copy testing. It allows for ads to really match that single keyword quite well.

2. One keyword on all match types per ad group. This style isolates a single keyword to each ad group, but includes all match types. This can allow you to see which match types performs best with a great amount of ease, but there will be a larger number of ad groups still.

3. Themed ad groups of tightly knit keywords. In this model ad groups would be something like this: Best Running Shoes, Cheap Running Shoes, Black Running Shoes, and so forth. Basically, you'll have an ad group for each modifier of the main product or service. Each ad group will include any and all keywords with that modifier. This is a more manageable style as it makes a lower number of ad groups, but still allows ads to be very tailored to keywords in the ad group. It can make it more difficult to understand which match type performs best.

7.6.3 How to Choose an Organizational Style

Deciding which organization style to select will really depend on each individual account manager. In selecting an organization style consider the following:

1. How large is your budget and how much traffic do you estimate your ads will get? If it is quite high, the 1st or 2nd organizational styles may be best. If it is lower, the 3rd may be right for you.

2. How much time do you have? The top 2 organizational styles are more time consuming in some ways because there will be more ad groups to create ad copy for. Consider how much time you have to spend on PPC management.

3. How many keywords do you have? If you have a lower number of keywords, putting one keyword per ad group could be the best. But, if you have a high number of keywords then you'd probably be better off with #2 or #3.

To most important factor of keyword organization is that it is logical and clean. You can use one of the above styles or something entirely different. Make sure the organization of the keyword makes sense and is done in a thoughtful manner. Keywords related to each other should be in the same ad groups for the best ad copy creation.

7.7 Negative Keywords

7.7.1 What is a Negative Keyword?

There are going to be keyword searches that you do not want your ads to show up on. These searches may be slightly relevant to your business, but not the users you want. Or the searches may be entirely irrelevant. To continue with the women's shoes example, you would not want to show up on men's shoes if you only sell women's shoes.

But how do you control for this? Selecting the match types can help, but doesn't entirely eliminate your ads being shown on irrelevant searches.

The only way to ensure your ads do not show on irrelevant searches is by adding negative keywords. Negative keywords are keywords you do not want your ads to show up on. These terms can be very broad, like "men's" or very specific like "patent leather pumps".

To succeed in AdWords, you must have negative keywords. Google is a for-profit business with shareholders. They are always going to be pushing the boundaries of what searches they are matching your ads to as broader matching increases their revenue. It is up to each advertiser to be diligent and maintain a strong negative keyword list to cut out any of these searches.

7.7.2 How to Select Your Negative Keywords

Selecting negatives correctly is very important as you don't want to cut out any relevant traffic, but you also don't want to waste your marketing dollars.

Before starting a campaign, you should brainstorm a list of negative keywords that you know 100 % are not related to your business. It is best to think of related products or services and add a negative keyword for that. If you sell only clothes for adult women, you'd want to add negative keywords like men and children. Coming up with a good list of high level negatives before you even start a campaign can help you get your campaigns to profitability faster. Your ads will never even have an opportunity to show on those terms.

The next avenue to find negative keywords is after your campaigns are running. Google provides you with the data on what users have searched for before they

Search terms

Learn how customers are finding your ad. With the Search terms report, you can see the actual searches people entered on Google Search and other Search Network sites that triggered your ad and led to a click. Depending on your keyword match types, this list might include terms other than exact matches to your keywords. Learn more

| Add as keyword | Add as negative keyword |

		Match type ?	Added / Excluded ?	Campaign	Ad group	Clicks ?	Impr. ?
	Total					28	8,822
☐	google adwords consultancy	Phrase match (close variant)	None	Radiant PPC test	Adwords Consultant	2	1
☐	google ppc	Broad match	None	Radiant PPC test	Adwords Management	2	39
☐	ppc marketing services	Broad match	None	Radiant PPC test	PPC Management	1	1
☐	ppc ads	Broad match	None	Radiant PPC test	PPC Management	1	11
☐	ppc agencies	Broad match	None	Radiant PPC test	PPC Management	2	1
☐	small business ppc marketing services	Broad match	None	Radiant PPC test	PPC Management	1	1
☐	google adword managment vendor	Broad match	None	Radiant PPC test	Adwords Management	1	1

Fig. 7.5 Search terms: how customers find your ad

click on your ads. This is extremely useful information as it allows you to find irrelevant terms and stop wasting money on them. Google unfortunately does not tell you every single search that your ad has shown on, but there is enough information to make some definite improvements to your campaigns (see Fig. 7.5).

It's important to consider user intent before adding a negative keyword. Think about what the user would be looking for when the search for a particular term. This can really help you to narrow if that keyword is relevant to your business.

Like keywords, you can add negatives as broad, phrase, or exact match types. This is another thing you'll want to consider before adding a negative keyword. There will be those high level negatives that you know you never want your ads to show up on, and those can be added as broad negatives. But, there may be something that's closely related to your business that you'll want to exclude. This is where negative exact matches come in handy as when you add a negative exact match you are telling Google that you do not want to show up on that exact search. Think about what match type is correct. Search queries should be reviewed regularly to cut out unqualified traffic. You need to remember that Google is a for profit business with shareholders who want to see an increase in revenue. Google will be liberal with their matching of keywords to searches at times. This because they need to increase their revenue. It is up to each advertiser to do their due diligence and cut out any unqualified searches Google shows their ads on.

7.8 Landing Pages: Clicking an Ad is just the Beginning

7.8.1 What is a Landing Page and Why They are One of the Most Important Aspects of PPC

One of the most important aspects of a PPC campaign is the landing page. The landing page is the page where your ad will send visitors. This page is the first impression visitors will get of your business. Having high quality, nice looking landing pages is very important. The landing page will help the user to determine if they want to stay on your website, or leave. The landing page can be a page on your website that already exists, or a page that's been created specifically for users from your advertising campaigns to visit. Since you are paying to drive users to your website, getting them to stay on your page and take an action is extremely important.

As mentioned earlier, the relevancy of your landing page is part of what Google considers when they determine your quality score. This is another reason a landing page is very important. Google scans landing pages to determine their relevancy to the keywords and the ad copy. If Google thinks the page isn't relevant, you'll potentially end up with a lower quality scores and paying more per click. Google also wants a landing page to have plenty of information, multiple actions for a user to take (such as availability of links to other pages on your website so users are not forced to take just one action), and clear presentation of your products or services.

Google wants landing pages to be relevant and of a strong quality because they want their users to have great experiences when they click on ads. They want their users to find what they are looking for. Ensuring this improves their platform. But, it also benefits the advertiser. By requiring advertisers to have good landing pages, it improves an advertiser's chance at better quality leads and more sales.

You can have a great AdWords campaign set up that is highly optimized and sending high quality users to your website. But, if your landing page is of poor quality, then the great advertising campaign's impact is zero. You need to care about your website and landing pages more than your campaign. The landing page is where your business will succeed or fail.

7.8.2   What Makes a Great Landing Page and How to Select the Landing Page for Your Campaign

There are several aspects of a great landing page that will give you the best chance of turning clicks into customers.

1. *Relevancy* Relevancy is the number one issue when picking a landing page. You want the page to contain relevant content to what the user searched for. If someone is searching for women's running shoes, you want to send them to the page with women's running shoes, not a page with all women's shoes. Always pick the most specific page you possibly can on your website. This will instantly improve relevancy. Keep landing pages specific and with good quality information on a landing page. This will keep users on your website longer and increase the likelihood of a conversion occurring.
2. *Make it easy for the user to take the desired action* This goes hand in hand with relevancy. The landing page needs to be relevant to the user so that they take the desired action on the site. But, it also needs to be easy for them to do. Is the lead form in on the landing page? Is it short and easy to complete? Is the "Add to Cart" or "Buy Now" button clear and easy to see? Is it easy to complete a purchase? These are important questions to ask when evaluating a landing page. In almost all cases, the home page is not a good landing page. Even with lead generation, this can deter users from taking the action you want. If they have to navigate to find the contact form, they may leave the site. You want to make it easy for your potential customers to find what they want and to take action.
3. *Easy Navigation* Most aspects of a landing page flow into each other. You want to make it easy for a user to take action, but you also want to make it easy for them to navigate your site. Perhaps a user doesn't find exactly what they want

on the landing page or they want to learn more about your business before they purchase, make the navigation easy to find the links to other pages on your site.

When first creating a campaign, take the time to find the landing pages that are most relevant to each ad group. Consider what you'd want to see if you were searching for that product or service. Review to make sure the desired action is obvious and that it is easy for users to navigate your site. A great landing page will improve the results of your advertising campaign and help you to get the most for each marketing dollar.

7.9 Tracking Conversions

7.9.1 Why Conversion Tracking Matters

One of the major benefits of online advertising over most other forms of advertising is that you can truly understand what you are getting from your advertising dollars. Internet advertising is very trackable. You can know how many users clicked on your ads at what hour of the day, from what location and via what ad copy. But, there is more. You can track whether a user who arrived via a paid ad purchased or completed a lead.

This is invaluable information. To know that a certain keyword is creating a good number of your sales, or that certain ad copy brings in the majority of leads is invaluable.

To gain this valuable information, all that is required is a code being placed on your website. Surprisingly, there are a good number of online advertisers who do not utilize this option. To truly understanding your advertising program and to get the most of your campaigns you MUST utilize this feature of AdWords. It will change how you look at your advertising program forever. Being able to know exactly how much you pay per lead or how many sales you get from an advertising campaign each day will allow you to improve your advertising as well as allocate your marketing budget properly.

7.9.2 What is Conversion Tracking?

Conversion tracking is the method by which you track sales or leads on your website. Google makes this incredibly easy to set up. A unique code is generated within the interface after selecting the "Conversions" options (see Fig. 7.6). You then take this code and place it on the Thank You page on your website. This is the page users see after they complete the lead form or make a purchase.

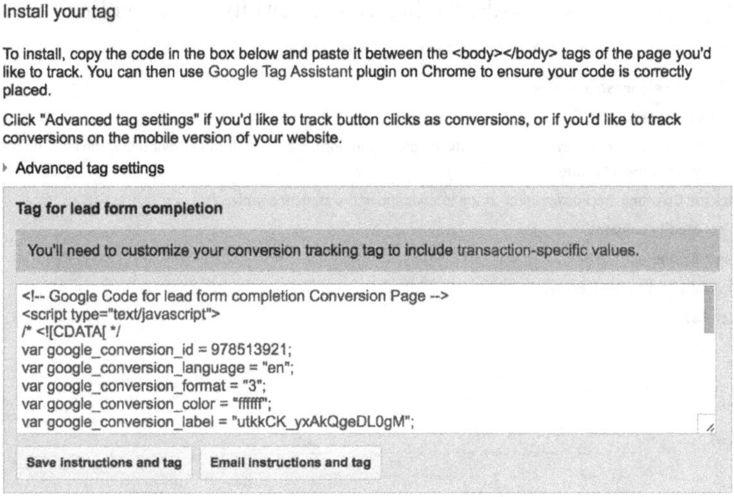

Install your tag

To install, copy the code in the box below and paste it between the <body></body> tags of the page you'd like to track. You can then use Google Tag Assistant plugin on Chrome to ensure your code is correctly placed.

Click "Advanced tag settings" if you'd like to track button clicks as conversions, or if you'd like to track conversions on the mobile version of your website.

▸ Advanced tag settings

Tag for lead form completion

You'll need to customize your conversion tracking tag to include transaction-specific values.

```
<!-- Google Code for lead form completion Conversion Page -->
<script type="text/javascript">
/* <![CDATA[ */
var google_conversion_id = 978513921;
var google_conversion_language = "en";
var google_conversion_format = "3";
var google_conversion_color = "ffffff";
var google_conversion_label = "utkkCK_yxAkQgeDL0gM";
```

Save instructions and tag   Email instructions and tag

Fig. 7.6   Install Your Tag

When a user enters your site from a paid ad, Google drops a cookie on their browser. When a user completes the action you've determined to be your "conversion", the code on the Thank You page is triggered and it is reported back to AdWords that this action has been taken. It can take up to 24 hours for conversions to be reported to AdWords. You will then be able to see what keyword and ad copy led to that conversion. Through other reports in AdWords you can also learn where the user was located, what time the conversion occurred, and what device type they used. This information will help you to make essential decisions in your AdWords campaigns and can greatly improve your ROI.

7.10   Quality Scores: Google's Invisible Hand Tilts the Field

7.10.1   What Are Quality Scores?

A quality score is a variable that influences where ads rank and how much advertisers pay per click. Google defines a quality score as the estimated quality of your keywords, ads, and landing page. Google calculates a quality score for each and every keyword. Quality scores can be 1–10. 1–3 are considered to be poor quality scores; 4–6 are average quality scores; and 7–10 are good/great quality scores.

The factors they use to calculate a quality score are expected click-through-rate, landing page experience and ad relevance. The higher the relevance of your ads and landing pages to the keywords, the higher your quality scores will be.

Another way to see your Quality Score is to enable the Qual. score column:

1. Click the **Campaigns** tab at the top.
2. Select the **Keywords** tab.
3. Look for the **Qual. score** column in the statistics table. If you don't see this column in your table, you can add this column by doing the following:
   - Click the **Columns** drop-down menu in the toolbar above the statistics table.
   - Select **Modify columns**.
   - Select **Attributes**.
   - Click **Add** next to **Qual. score**.
   - Click **Save**.

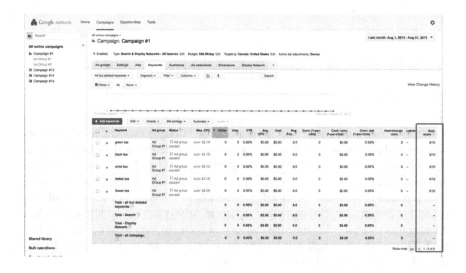

7.10.2 Why They Matter and How They Impact Your AdWords Campaigns

Quality scores matter because they determine how much you pay per click and how high your ads rank. If you have a quality score of 10, your cost per click (CPC) will be less and you can rank higher at a lower cost per click. If your quality score is 2, it will cost you a lot more per click than an advertiser with a quality score of 10. Your ads will also most likely be showing lower in the search results.

Having poor quality scores means you'll pay more for each and every click. You will have to pay more if you want your ads to show at the top of the page. This is why how you organize your campaigns matter, why the landing page and ad copy matter. This is Google's way of ensuring relevancy from advertisers. It allows them to algorithmically control the quality of the marketplace by punishing users who they believe are advertising on keywords that are irrelevant to their website or who are not giving users a quality experience.

Every account will have a few low quality score keywords. These are typically the broader terms that may not be quite as relevant as the others in your account. But if low quality scores begin to dominate an account it will drive up costs, your ads will show less frequently and generally damage your PPC account performance.

7.11 Analysis and Optimization: Ongoing Effort

7.11.1 An Ongoing Effort

You've researched your keywords, organized them into ad groups, created ad copy and picked great landing pages. Your conversion tracking is all set up. You've turned your campaign live... Now what?

Now comes the time to analyze and optimize. Do not set it and forget it. This will lead to wasted advertising dollars. Google has many automated features that can help you to cut down your time spent optimizing your campaigns, but at the end of the day a human is needed to review what search queries your ads are showing on, create new ad copy and to truly understand performance.

Check in on your campaigns regularly, make changes every single week. Proper maintenance will lead to on-going success in AdWords. Google provides a platform that can put your ads in front of hundreds to thousands to hundreds of thousands of users. They've created a user friendly interface, great tools and a simplistic way to create an advertising campaign. All the necessities for success are in front of every small business. Google AdWords makes it possible for small businesses to succeed in an online marketplace. Everyone can compete equally and acquire new potential customers. It is up to the business to make the most of that. Once you've got a campaign running, take the time to pay attention to it. Regular review and optimizations will lead to great degrees of success, more customers, and better return on investment. It's up to you to grab that opportunity and make it work for you.

7.11.2 Approaches for Account Maintenance

It's 100 % necessary to maintain your AdWords account. The "Set it and Forget" mentality does not work. Eventually, performance will fall off and your account will suffer. Account maintenance is about patience and diligence. Campaigns, ad groups, ads and keywords all collect data at their own rate, unique to each circumstance. Some will be higher volume; others will take longer.

You have to have and wait for the appropriate amount of data to collect to make changes. It is imperative to wait until you have a statistically significant amount of data to make changes. Optimizations have to be data driven. You need to make decisions that are informed by real numbers. True AdWords success, especially in competitive areas, comes from diligent ongoing account maintenance and informed optimizations. This is really the only way to make AdWords work for you.

There are a large number of optimizations that can be made to improve your account performance. Some changes need to be made weekly, bi-weekly, and monthly or even longer, depending on what those changes are. Below guidelines for making changes at the appropriate intervals. The frequency of changes can vary based on the size of your AdWords account (accounts with high spend may need changes to be made more frequently, accounts with only a couple hundred dollars in spend will need changes even less frequently).

Weekly

- Optimize bids. You should review weekly and make any manual bid changes. You should also utilize Google's automated bid rules, which will change bids automatically for you. This will help with optimization of the long tail. Bid rules should be set up on cost per conversion goals. It is still important to review keyword performance and bids weekly though as there may be keywords that need adjustments outside of the bid rules.
- Review 7 day trends.
- Monitor budgets. Make sure your campaigns are not running out of budget too quickly or spending more than desired.
- Review key performance indicators to make sure they are on target. Look at CPCs, CTR, average ad position, conversion rate and traffic levels to make sure everything is in line with desired results. If not, dig into the account to find low performing areas.

Every other week

- Add negative keywords using last 14 days to 30 days of data.
- Review keyword level performance, pause poor performing keywords or adjust bids. (Large accounts only.) A poor performing keyword is one that does not meet your key performance indicators. This could be a high cost and no conversions. It could be a keyword with a very high cost per conversion. It could be a keyword with high impressions and no clicks/very few clicks. These are all types of poor performing keywords and what you should look for when you evaluate performance.

Monthly

Review ad copy performance for past 30–60 days. Pause any poor performing ads and then start new variations of top performing ads. You need at least 1,000 impressions to make a decision on an ad. You'll want to pause any ads that do not convert and have a high cost. You'll also want to look at those with very low click-through-rates and pause those if the other ad/ads in the ad group are performing better.

- Review keyword and ad group performance by bounce rate and time on site. Make bid adjustments or pause poor performers based on this data. Push keywords with very long time on site.
- Add new keywords. Use search query data to find new keywords, or use one of the many available keyword tools. Add converting searches that are not already in the account.
- Review long term keyword performance for larger accounts. Use last 3 months of data. Make adjustments on any poor performers (reduce bids, pause poor performers, look at search query information for those keywords).
- Review keyword performance by match type (larger accounts). Pause poor performing match type variations and add other match types, if necessary.
- Review performance by device. Make adjustments to mobile bid as necessary. If mobile has a high cost and no conversions, you may want to reduce the bid further. If mobile is performing well, you'd want to increase the bid.
- Review quality scores. Monitor any keywords with low quality scores and look at ways to make improvements. If a keyword has a low quality score and does not perform well, then pause it.
- Review 30 day trends. Look for weak performing campaigns/ad groups/ keywords and make optimizations. Also, if a campaign/ad group/keyword is performing very well, look at ways to get additional volume there.
- Review keyword level performance and pause poor performing keywords or make adjustments to reduce spend and improve performance (small accounts).

Every 2 months

- Review long term keyword performance for small accounts. Look back up to 6 months. Pause any poor performing keywords.
- Review keyword performance by match type (small accounts). Pause poor performing match type variations and add other match types, if necessary.
- Review sitelink extensions performance. Remove any poor performers. Test out different sitelinks and/or new sitelinks text. Poor performers here would be considered any sitelinks that have a low conversion rate or low click-through-rate in comparison to the others.
- Review call out extensions performance. Remove any poor performers. Begin new tests. Poor performers here would be considered any call out that has a low conversion rate or low click-through-rate in comparison to the others.
- Review landing page performance. Test new pages as necessary. Here you'll want to look at conversion rate, time on site, and bounce rate to evaluate if a landing page is a strong performer.
- Review keywords below first page bid estimates and see if any bids should be increased.
- Review keywords in low ad positions (6 or lower) and see if any bids should be increased.

Once a quarter

- Review hourly performance and adjust day parting as necessary. Look for the best performing hours and potentially increase bids in those hours. Also consider reducing bids in the low performing hours.
- Review all account settings and make any necessary changes.
- Complete a historical performance review. Look at the last 6 months or year to date. Review all non-converting keywords in account for relevancy. Review all non-converting ad copy and review for relevancy. Look at under performing campaigns and ad groups. Put together plans for making improvements to performance.

7.12   Beyond the Basics

You've now gotten a handle for what the core parts of an AdWords campaign are, what you need to get started on AdWords and some ideas for optimizations. As you become an increasingly advanced user of AdWords, there are some additional features within the interface that you'll definitely want to utilize. These features will help to further improve the performance of your AdWords campaigns and help you to better analyze your performance.

7.12.1   Ad Extensions

Ad extensions are exactly what their name implies: ways to extend your ads.

As ad revenue becomes increasingly important to Google, they've begun to create ways for ads to take up more space in the Google search results. There are several types of Google Ad extensions.

The first is sitelinks. These are the blue links that will appear underneath of an ad (see Fig. 7.7) These links are links to other pages on an advertiser's web page. They need to be different than the actual ad URLs. These links should be places you'd like a user to potentially visit, but the click-through-rates on these links are typically quite low. The main point of the links is that they take up additional space for your ad. This is why they are very important to set up. With your ad taking up more space, it will be very eye catching to a user.

Call Out Extensions are another way to take up some additional space with your ad as well as tell users a bit more about what makes your business unique. Call out extensions typically show right below the ad copy and above the sitelinks (see Fig. 7.8). Google will display 2 or 3 at a time, but you can create up to 10, which they will rotate. The text for each can be up to 25 characters.

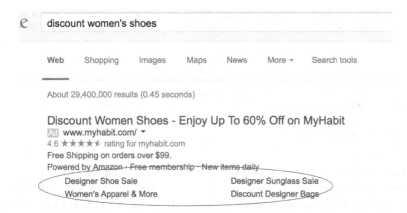

Fig. 7.7 Ad extensions

Fig. 7.8 Ad extensions

You should use Call Out extensions to tell users more about your business – what makes you different from your competitors, what some value adds are, and other information that may convince users to do business with you.

Call Extensions add phone numbers next to your ad (see Fig. 7.9). These are great for ads on mobile devices, as they allow users to click a call button and immediately call your business. Phone numbers can display on desktop and tablets as well, but obviously have a convenience factor on mobile phones.

Add phone numbers to your ads

What phone numbers can do

- Encourage calls to your business by showing your phone number on your ad.
- Display a clickable call button with your ad (on high-end mobile devices).
- Cost the same as a headline click (a standard **CPC**).
- For call-only campaigns, ads will only appear on devices capable of making calls.

Fig. 7.9 Adding phone numbers to your ads

- **Google Search Network**: On the Google Search Network, ads with location extensions can appear as a standard text ad with your business address and phone number. On high-end mobile phones, a clickable "Call" button

Amherst Ice Cream Parlour
Ad www.example.com
(413) 123-4567
Our specialty is pistachio.
English majors, buy 1 get 1 free.
may appear in place of your phone number.   ◉ 100 Dardanelles Rd, Amherst MA

Fig. 7.10 Locations extensions

Locations extensions will show your address with your ad. (see Fig. 7.10). This is great for businesses that want customers to come into their store or office. It makes it easy for a user to understand exactly where the business is located. It can also encourage users to pick a local business over a nationwide company that does not have a location in their town.

7.12.2 Dimensions Tab

The dimensions tab is last tab within the AdWords console and one that is often underutilized (see Fig. 7.11). Within in this tab is a plethora of data broken down in a variety of ways to help you best optimize your campaigns.

The reports you'll find here are under the drop down menu view (see Fig. 7.12). You can look at a variety of information:

Conversions
If you've got multiple conversion types, you can see here how those different conversions perform.

Labels
AdWords allows you to tag different campaigns, ad groups, keywords or ads with labels. Some advertisers will utilize labels to group ad groups by product type (e.g.,

Fig. 7.11 Dimensions

Fig. 7.12 Dimensions drop
down menu

shoes, pants, etc.). Here you can look at how each label is performing over the date range you choose. This can be very helpful for quick analysis of a large product set.

Time
This can be broken down by hour of day, day, day of week, week, and month. This is a great place to learn how your campaigns perform on an hourly and daily basis and make optimizations based off of that.

Destination URL
This will give you a break down how each page you are sending users to performs. It is an easy way to see landing page performance.

Top Movers
This report allows you to see which ad groups and campaigns have had the biggest changes in clicks, impressions and so forth. It is useful for looking for any changes in performance.

Geographic
This is a report that gives you information about performance of your account by geographic location. It can tell you high level—like country or state performance. But it can also break down the data as far as zip code and city. This can give you some very useful information on what places perform best for you and give you ideas on more advanced geographic targeting to test.

User Locations
This is a variation on the Geographic. It utilizes user's IP address information instead of physical location. There isn't really much variance on these reports.

Distance
Distance is a report that gives you insight into user's locations from your address, if you are using ad extensions. This can give an idea of how far users are willing to travel to your business.

Search Terms
This is just another location to see search query data.

Paid and Organic Report
This table shows how your paid and organic results performed for every search that triggered an ad or organic listing.

Free Clicks
This gives you information of any free clicks you have may received from Google Shopping results.

Call Details
This gives you detailed information on any calls received from users who called in from the Call Extensions. It will tell you users area code, length of call, and date/time of call.

Campaign Details/Ad Group Details
Both of these reports give you a high level look at your campaigns and ad groups. Here you can see if you have any disapproved ads, how many sitelinks are active, if you've got ad scheduling set up and so forth.

7.12.3 Remarketing

Once many advertisers are established with a Google search campaign, they then move onto Remarketing.

Remarketing is a way to advertise to users who already visited your site, but did not convert. These users are then shown ads for your website around the web through Google's network. Google has an entire network of sites that are available to display ads. Remarketing to users allows them to remind them of your services or products, and hopefully bring them back to your website.

The way you advertise to these users is by putting another code on your website. This code will drop a cookie into every user's browser. This cookie will track what pages a user on your website visits.

Once you've got this code on your website, you'll then want to create your remarketing lists (See Fig. 7.13) To create a remarketing list, you'll go under your Audiences tab.

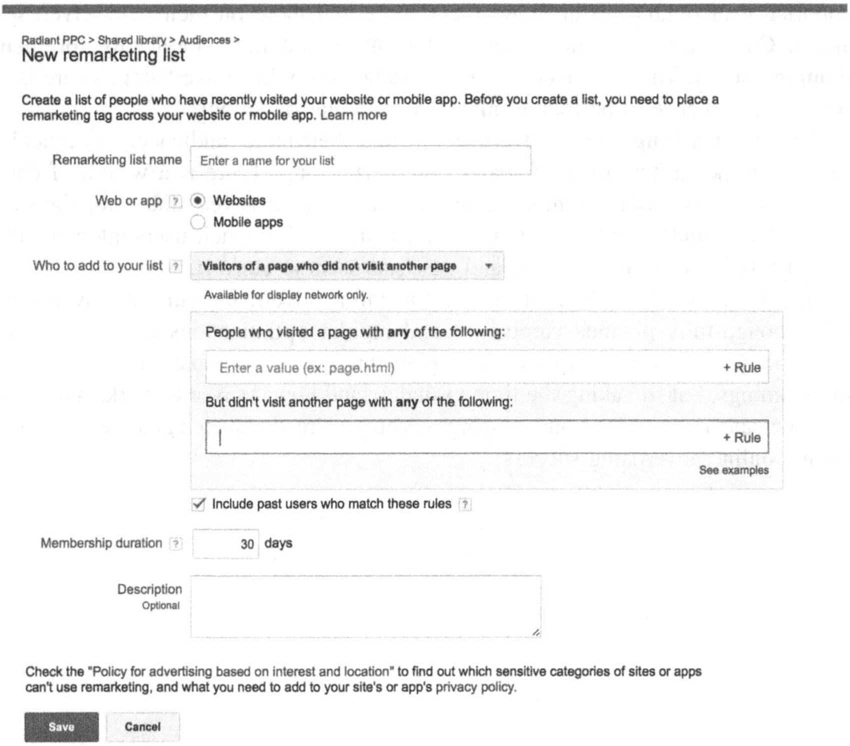

Radiant PPC > Shared library > Audiences >
New remarketing list

Create a list of people who have recently visited your website or mobile app. Before you create a list, you need to place a remarketing tag across your website or mobile app. Learn more

Remarketing list name | Enter a name for your list

Web or app [?] ● Websites
○ Mobile apps

Who to add to your list [?] | **Visitors of a page who did not visit another page** ▼

Available for display network only.

People who visited a page with **any** of the following:

Enter a value (ex: page.html) | + Rule

But didn't visit another page with **any** of the following:

| | + Rule

See examples

☑ Include past users who match these rules [?]

Membership duration [?] | 30 | days

Description
Optional

Check the "Policy for advertising based on interest and location" to find out which sensitive categories of sites or apps can't use remarketing, and what you need to add to your site's or app's privacy policy.

Save   Cancel

Fig. 7.13 Remarketing lists

There are many, many ways to retarget users. You can create a good variety of lists. The best way to start though is to just a create one list for users who did not complete the desired conversion.

Here you will create a list of users who visited a landing page, but did not visit the Thank You URL.

The typical membership duration is 30 days. This means your ads will be shown to users for 30 days after they initial visited your site.

Remarketing is a great way to try and recapture non-converting visitors, bring them back to your site and turn them into customers. It's also a great way to move into more advanced AdWords advertising.

7.13 Business Marketing Plan

Google AdWords is becoming an increasingly important part of any businesses marketing plan. It will only become more so in the future. As Google needs to constantly increase their revenue to keep share holders happy, they will reduce the importance of organic listings and focus more and more on their paid advertise-ments. Consumers continue to turn online more and more frequently for their shopping needs. This puts increasing importance on online advertising. There is a very large audience to be found online, in particular on Google.

Online advertising allows advertisers to find their target audience, and quickly gain feedback on how their campaigns are performing. There is a wealth of data available for advertisers to improve performance of their advertising campaigns as well as better understand what users are interested in, how their users interact with their website and what keywords and ads lead users to convert.

Google AdWords is the best avenue for online advertising currently available. With thoughtfully planned campaigns and regular optimizations success can be found in Google's marketplace. The platform has its complexities and some shortcomings, but by taking the time to understand how AdWords works and how to best create and manage your campaigns, you've already taken a giant leap toward finding online advertising success!

Chapter 8
Google Maps and Google Local Search

Nyagoslav Zhekov

8.1 A Brief History of Google Maps for Business

Google realized that location search, and mapping as an inseparable part of it, was arguably the most important and useful feature a search engine should have. Many of Google's competitors were far ahead when "Search by Location" was launched as part of Google Labs in September 2003. However, trying to produce accurate results entirely based on scraped data proved to be not just difficult, but in fact impossible. Before the official launch of Google Local in March 2004, Google improved the location data it had acquired via site indexing with additional structured data from yellow pages sources. At that early stage of development, the main factors Google used in their local search algorithm were "location", "prominence" or "trust", and "relevance" (as part of Google's general organic search algorithm). It was only in February 2005, when Google Maps was officially launched, that "distance" came in the mix to bring the algorithm closer to its current form.

8.2 Google Local Business Center (LBC)

The biggest breakthrough in the area of serving business owners came when Google Local Business Center (LBC) was officially announced in March 2005. The LBC had a great impact on all stakeholders:

(1) Businesses were able to get control of their business information and make sure it was accurate;

N. Zhekov (✉)
Whitespark Inc., Edmonton, Canada
e-mail: nyagoslav.zhekov@gmail.com

© Springer Science+Business Media New York 2016
N. Lee (ed.), *Google It*, DOI 10.1007/978-1-4939-6415-4_8

(2) Customers had one more source of correct and reliable information for products and services;

(3) Google benefited by having their business database improved for free.

To get hold of their business listing, or to create it if it did not exist, business owners or representatives needed to verify their ownership over a listing by obtaining a PIN, which was mailed to the business's address by Google.

The first publicly released version of Google's business dashboard was relatively basic. It included only information editing options to update business name, address, phone number, website, email, and a brief description. About 2 years later, in March 2007, Google released a major update to the Local Business Center. Businesses were able to add customized details, which could include anything from information about services to areas served. One important addition was that the verified business owner could "tell" Google about certain information that was attributed to the business listing was incorrect. Why was this important? Since the launch of the business dashboard, Google would add related information that it found on the web and show it in the local search results together with the business listing. Such information included images and map location. Frequently such data was incorrect, and so it was essential for businesses to be able to edit or completely remove those connections to their business record.

8.3 Google Street View

Google's main goal since the launch of Google Maps (and Google Search) had always been to collect the entire world's data. Maps was the most ambitious of all of the data collection projects, because what conglomeration of data could be bigger than the world itself? Google realized that relying solely on online sources to map the world was not enough. What was more, all the information Google obtained online was very untrustworthy, because it might have been derived from offline sources dating a few years back. As millions of buildings are erected worldwide every year and others are demolished, new roads are built, and businesses move their offices, data older than a few months could very easily be outdated. The only logical solution was that data needed to be obtained "on the ground". Thus, in May 2007, Google officially announced one of its largest projects yet—Street View. At that time they had 360° street imagery collected for only a handful of cities in the United States (San Francisco, New York, Las Vegas, Miami, and Denver). It was only in July 2008 that Street View was introduced outside the US—in selected cities in France and Italy, followed by imagery from Australia and Japan. By June 2015, Street View covers the entire inhibited part of the United States, Canada, Australia, and New Zealand, Mexico, almost entire Europe (excluding only Macedonia, Montenegro, and parts of Ukraine and Belarus), big parts of South America, large parts of inhibited Russia, Southeast Asia, Japan and Korea, as well as parts of India. The least covered continent remains Africa, with only South

Africa, Botswana, and Madagascar available, as well as small landmark parts of Tanzania. Overall 88 countries and territories have complete or partial Street View imagery by June 2015, with more than 10 additional countries to be added soon.

8.4 Google Maps

The next step for Google was to get the largest possible pool of contributors involved in the process of mapping the world—the general users. In March 2008, Google officially opened its entire Maps database (available for US, Australia, and New Zealand maps only at the time) for public edits. Google was arguably not well prepared for the amount of updates they started receiving from users, and the data verification complications involved in making sure that the user feedback was accurate. However, it is possible that they were forced to act fast, because since July 2007 reports of spam and abuse, mainly related to business listings, had been frequent. Google lacked the capacity to effectively stop, or even slow, the spam wave, and the easiest way to achieve at least partial victory was to get well-meaning users to help. Unfortunately, later it seemed that the attempt backfired, because the early adopters of the public edit functionality were the exact same spammers. They found easily exploitable loopholes and literally flooded Google Maps with fake business listings. Because of the nature of Google Maps, such spam was useful predominantly to service-based businesses, i.e. businesses which serve customers at the customers' location, and frequently do not even have a public-facing location (for instance, they operate from home). Locksmiths, towing service companies, garage door repair contractors, carpet cleaners, and bail bonds were the main culprits. Prominent local search blogger Mike Blumenthal notes in his February 2009 article that there were "more locksmiths in NYC than cabs" if one was to trust Google Maps. In an attempt to prove the ease with which the system could be abused, Blumenthal took control of Microsoft's Redmond, WA business listing and changed the location, the website URL, and even the business name—to Microsoft Escort Service.

8.5 Google Map Maker

At that point it was evident that Google had to do something fast. Their response came in June 2008 when Google Map Maker was launched. In essence, it was an improved user-generated content (UGC) gathering and verification system. Its main purpose was to both make it easier for users to add or update mapping and business data (including complicated geographic objects such as roads, rivers, and buildings), and at the same time give more control to Google's mapping data verification team. The effort was relatively successful as spam was at least partially contained, although later on numerous withstanding loopholes were discovered. The main problem of Map Maker was its great reliance on humans, and specifically on Google's verification team. Unfortunately, due to lack of focus on the problem, Google made the mistake to outsource this crucial part of the process to an overseas team in India. The overseas members were poorly trained by Google, and they lacked the important insight of understanding the realities of life in different countries around the world (one team was responsible for verifying UGC from all over the world). The problems culminated in the temporary suspension of Map Maker in May 2015 after several abuses had received worldwide media coverage. One of the most publicized pranks was the inclusion of a business named "Edward's Snow Den" at the address of The White House. However, the one that prompted the quick suspension of Map Maker was an image of the Android mascot seemingly urinating on the logo of Apple. The image was mapped out using vegetation outlines addition on an empty piece of land south of Islamabad, Pakistan.

8.6   Google Reviews and Google Places

While Google gradually figured out the best ways to obtain and verify mapping and business data, the last piece of Google's local search puzzle was the introduction of an option for customers to provide their feedback on a company's business listing. In the early days of Google Local, Google would associate reviews (or content they believed was reviews) they found on the web or provided directly by third-party sources with the corresponding business listing. However, third-party data could never be the most reliable and trustworthy source, and that is why in June 2007 Google announced the introduction of native Google reviews to Maps. For a variety of reasons (probably due to Google's apparent inability to comprehend "social") the task of getting enough high-quality business reviews proved to be beyond Google's capabilities. They realized that relatively fast, because by December 2009 they were in advanced acquisition negotiations with review site Yelp. The speculated $500-million deal fell through, however, which prompted Google to make significant effort in the area of reviews and ratings.

Google's first significant effort in adding improvements to the "reviews department" was an update to their sentiment analysis algorithm, which was designed to analyze content around the web to see if it potentially included feedback sentiment on an "entity". Google would then match the entity's speculated review (usually an excerpt from a block of text in a blog or news article) to a business listing that they believed corresponded to that entity. Furthermore, in August 2010, Google added the ability for business owners to be able to reply to reviews, in an attempt to gain the support of business owners and to encourage Google reviews. The arguably most successful effort was the launch of Hotpot in November 2010. Hotpot was meant as a personalized recommendations booster. The logic derived from the fact that if a friend of yours left a positive review for a business, the chances that you would also like that business were much higher than the chances that you would like another business for which a stranger left a similar positive review. Later on, Hotpot was integrated into Google Places—the new name of Google Local Business Center.

8.7 Google Products for Local Businesses

As discussed above, Google realized early on that there could be mutual benefit if they work together with business owners to improve business data in Google's data clusters. However, the road to creating a product that was useful and at the same time user-friendly was not straightforward.

8.8 Google Business Dashboard with Google+

Google's business dashboard went through three re-brandings and tens of updates to finally be seen in the way it is today. The latest version of the dashboard is integrated with Google+—Google's social network (or "layer" as Googlers like to call it). It is both the most sophisticated and the easiest to use version yet. Some of the most important benefits of the integration with Google+ are as follows:

(a) The ability for business owners to "socialize" with their customers.

Previously, business owners were able to share short, Twitter-like updates on their business listings, but these lacked the exposure and the overall integration with the public interface of the listings to be of any real use. Additionally, while previously introduced an option for users to "like" their favorite business listings, the option was buried and thus very few people used it on a regular basis. The "like" function and the "update sharing" function were not connected, i.e. if a business shared an update on their business listing, a user that had liked the listing would not be notified. This changed with the integration into Google+, and now users are able to get notifications when a business posts something on their stream. Additional social function was the introduction of Google Hangout—a way for businesses to set up online video conferences with customers.

(b) Integration with Google Analytics.

Measuring results of online marketing efforts has always been one of the biggest concerns for businesses of all sizes. Google's old business dashboard provided very basic details about impressions (how many times a business listing has shown in the search results) and actions (how many people have clicked on the displayed result), but even such basic feature frequently malfunctioned and were generally known to provide inaccurate statistical data. Due to various technological constraints of the way the old business dashboard had been set up, Google was unable to successfully integrate it with Google Analytics—the specialized search and conversion data analysis tool by Google. This changed with the integration of Google+. Additionally, Google improved the dashboard's native statistical data display and overall performance. The function is nowadays called "Insights" and it provides a good amount of data, including driving direction requests, phone calls, social engagement (for instance, how many people have +1'd or shared the posts of the business), and audience segmentation.

(c) Increased trustworthiness of business data.

This is arguably the most important change that occurred. Previously, even though a business might have claimed their online listing on Google, it was very possible that the information displayed in the search results would differ from the one the business owner or representative had provided via the dashboard. This was caused by a discrepancy between the data which Google might have discovered or received from third-party sources and the data provided by the verified owner or representative (more on this later). After the integration with Google+ was completed, information provided by the verified business owner became the most trustworthy source of data, and even if Google decides that different data should be displayed publicly, they notify the owner, and allow them to re-update the information if they want to.

Additional features were also added after the integration was completed, and at the same time some features were dropped or updated (as mentioned above).

8.9 Google AdWords and AdWords Express

By now you might be wondering how Google monetized this whole endeavour, as the investment of time and money into building the best possible location data was extraordinary. The major means for Google to get revenue from all of their online properties is ad placement. Their now highly sophisticated AdWords product is their most important income source by far. In 2014 their revenue from advertising was close to $60 billion, out of a total of $66 billion revenue for the year. In other words, Google relies 90 % on its power as an advertising platform. The logic behind the monetization of Google Maps is not different—more users and more relevant results mean more clicks. This could only be achieved with the provision of as much high-quality and in-depth information as possible. However, attracting as many users as possible is only one side of the story. Google needed to have a sophisticated, yet easy to use, advertising dashboard to entice business owners.

At first, AdWords was not a perfect dashboard for business owners, because location targeting was inaccurate or non-existent. In July 2009, Google launched a "location extensions" function for AdWords—an integration with LBC. However, it soon became evident that AdWords was too complicated for some business owners and they were not able to get it to work successfully for their marketing needs. That is why in October 2010 Google Boost was introduced (later re-branded to AdWords Express). It was a very simplified solution tailored specifically to small businesses that lacked the knowledge (and time) to be able to play around with the multitude of functions that the regular AdWords dashboard offered. Nevertheless, the main advantage of AdWords Express is not just about simplicity but also the 5-star (or fewer stars, depending on the ratings) review rating that shows up in the advertisement. This has empirically been proven to be a great click-through rate booster, which made the product relevant for businesses which had good Google reviews. However, the cost per click for AdWords Express was much higher than a well-tailored AdWords campaign with location extension. Eventually AdWords Express was absorbed by AdWords, probably due to the fact that Google Places (the name of Google's business dashboard at the time) was moving to Google+. As of June 2015 AdWords Express is still an available advertising option.

8.10 Google Tags

A short-lived product, launched in February 2010, was Google Tags. Although it was discontinued in April 2011, Google Tags is worth mentioning because it was one of the very few products outside the realm of AdWords, which Google launched for the immediate purpose of gaining additional revenue. The pricing structure of Google Tags was rather unique—a flat rate of US$25/month fee. Google Tags offered the opportunity for business owners to add a small yellow-coloured tag under their business listing with a featured message and link to a part of their business listing. For example, if an advertising business was a restaurant, it might be beneficial for them to feature a link to their listing's images in the search results, so that customers could easily see the ambience and pictures of the cuisines on their dining menu. Additionally, the tag would occupy real estate space in the immediately visible part of the Google search result page, and it would attract attention, because of its colour. Unfortunately, the main disadvantage of Tags, which probably led to their demise, was the fact that they were useful only if a business was already ranked within the top local search results. If they were not, additional visibility at a place where no one has any visibility (i.e. second page of Google's search results) would not be worth even the $25 Tags cost.

8.11 Google Business Views

Google Street View was able to display the exterior of a business, as well as the surroundings. Google Maps would show the driving directions, and Google Places would provide information about the business such as its phone number, working hours, and services. However, Google realized that the piece of "data" it lacked was an interactive imagery of the inside of a business. Business owners were able to post pictures on their listings, but those could feature anything from the face of the owner to the logo of the company, and they would rarely give a clear idea about the interior of the premises. In February 2010, the first rumours of a new product called Google Store Views appeared. In May 2011, the product was officially launched under the name Google Business Photos (later re-branded to Google Business View). In the beginning, it consisted of a series of photos of the interior of a business that were later uploaded by the Google certified photographer who took the photos for the business listing. Later on, the product evolved into a 360° imagery walk into the store, an exact copy of Street View. The trusted photographer (or agency) would normally do the sales and actual implementation of the service, and would share some of the profits with Google. Thus, Google Business Views does not cost Google anything. Today there are hundreds of trusted photographers and agencies in 29 countries.

There are speculations that in future Google will be looking into using Google+ as a platform for increased ads exposure, or that they would experiment with different, new types of advertising products, but as of June 2015 nothing concrete has been announced.

8.12 Google Business Listings and Business Data

Up to now we discussed mainly the relationship among Google, its products' users, and business owners in providing and updating mapping and business information on Google's multitude of platforms. However, these are not the only sources of information that Google has been using. In fact, the main layer of the basic data comes from third-party sources. Google collects it in two main ways: by purchasing it and by scraping it.

8.13 Business Data Aggregators

Google purchases licensed business data from the main business data providers in each country they have ever introduced Google Places. In the United States, for instance, the main business data aggregators (as they are better known) from which Google has been getting data in exchange for payment are Infogroup, Acxiom, and LocalEze (see "Appendix" at the end of the chapter for a list of major business data providers for Google in selected countries). Each of these aggregators "aggregates" business data from numerous other sources, such as government records, printed yellow pages, and direct phone calls to businesses. Google pays for such data, because it is generally more accurate and up-to-date than "scraping" local business data.

8.14 Business Data Scraping

Scraping is the process in which Google's crawler bots go through a website's content (which is mostly in structured form) and compile information that might be useful for Google and could be re-used by them in different ways. If the scraped information is structured, in other words, if the site provides additional encoded details about what each bit of information represents, it is even easier for Google to determine if and when such scraped data might be useful. That is why the majority of the scraped data that Google uses for business listings are obtained by going through the content of online business directories, Internet yellow pages (IYPs), and websites whose main purpose is not to serve as business directories but to collect and display information about a large quantity of businesses. The most famous (and most frequently scraped) IYPs in the United States are Yellowpages.com, Superpages.com, Citysearch.com, and Merchantcircle.com. Other sites that are not necessarily business directories but are Google's favourites are Yelp, Yahoo! Local, and Foursquare.

8.15 Business Listings

By obtaining information through direct purchases and through scraping, Google is able to determine if a business actually does exist, as well as to cross-check and verify the correct business name, address, and phone number. After Google's algorithm determines that a threshold of trustworthiness is reached, a Google business listing is automatically generated. Such listings usually feature only very basic information about the business—the mentioned name, address, and phone number (abbreviated as NAP) as well as one business category. That is why most businesses do not need to set up brand-new listings on Google, but to search for their business record and claim it instead. However, it is possible that Google might have determined that certain information is trustworthy enough and might have generated a listing using that information, when in fact that information is outdated because the business has closed down, moved, changed phone numbers, or rebranded. That is when the real problem occurs, and that is why Google introduced very early on the Local Business Center as mentioned above. The solution was not as sustainable as Google would have wanted it to be however. Many owners did claim their business listings and updated the information, but once they have closed doors or sold their business, they did not update the information. That is why in the beginning of June 2015, Google introduced a simple system for business information to be kept as up-to-date as possible. Now, in order for a business listing to stay owner-verified and claimed, the business owner would need to log into the dashboard at least once every 6 months. Google would notify them via email 2 weeks prior to the "expiry date". If they fail to log into the account, their account will be disassociated with the business listing, and the listing will be free for the public to edit, and previous information provided by the business owner might get stripped down.

8.16   Business Rankings in Local Search Results

The additional nuance of Google using third-party business data is the fact that the more of certain matching data Google finds, the more trustworthiness this data is in Google's weighed algorithm. Therefore, ceteris paribus, a higher number of accurate mentions of a business's information online would mean a higher ranking in the local search results, because trustworthiness (or prominence) is one of the three major local search ranking factors (relevance, prominence, and distance). The term used within the local search community to refer to such kind of mentions is "citations". A citation could be any mention on any web document (a web page, for instance) of a combination of at least two of the three main business attributes (i.e. NAP: name, address, and phone number). In fact, citations are widely considered to be one of the most important individual ranking factors. According to the most authoritative survey on local search—the Local Search Ranking Factors, conducted by David Mihm, "external location signals" (i.e. citations) carry 15.5 % of the total weight in the organic local search rankings on Google.

8.17   Business Citations

There are two main sub-factors to citations—their "volume" and their "consistency".

8.17.1   Citation Volume

The volume of citations for a business means exactly that—how many citations are associated with a business. Citations, however, are not born equal and some are more valuable than others. Citations could be divided into three groups based on the type of website they are found on—generic citations, industry-specific citations, and location-specific citations. Generic citations are found on sites where any kind of business could be mentioned, such as the main business directories and data aggregators, for instance. The industry-specific and location-specific citations are found on sites where only businesses that comply with certain restrictions could be mentioned. For instance, on the business directory website Justia.com only law firms and lawyers could be listed. In terms of absolute value, generic citations are the heavyweights. However, in terms of relative value as differentiators in the local search rankings, the industry-specific and location-specific citations play a more important role. Therefore, a business that wants to be ranked high in the relevant local search results should optimally obtain a broad mix of citations from both generic and specific citation sources.

8.17.2   Citation Consistency

Google is not the only player who collects business data from different third-party sources. Almost every business directory does the same thing. They all face the same problem as Google does—the business data they obtain is sometimes inaccurate or outdated. As business records on such websites are used by Google as citations, business owners need to make sure that their business data are accurate not only on Google but also everywhere else on the web. The task is difficult and in many cases unachievable, because some sites do not provide an easy way for business owners to claim the existing business listings and to update their business details. The introduction of pay-to-play automated tools introduced since 2013 has made the task even more difficult. Companies providing such solutions get into exclusive relationships with some of the business directories and offer them regular payments in exchange for the access to business directories, allowing them to update business data quickly. Such relationships make perfect sense to both the solution providers and to the business directories. However, since their cost is relatively high, they are generally unsustainable. As they require yearly fee payment, the ones who suffer are the business owners. Once the service is cancelled, the results will oftentimes be reverted back to the state in which they were prior to the service purchasing. The pay-to-play solutions have numerous imperfections in the way they try to resolve the very complex business data issues, and therefore they should not be viewed as a get-and-forget type of offerings.

8.18   Complexity of Online Business Data

The following example illustrates the complexity of the problem with online business data:

LocalEze is one of the most important business data providers in the US. They provide business information to a number of online platforms, one of which is MerchantCircle.com. We do not know with 100 % accuracy how frequently MerchantCircle *receives or pulls data from Localeze's database but for the purposes of this example we will set the cycle at 45 days. Here is an example scenario: On January 1, new business information from an official government source is added to LocalEze's database:*
Business Name: Bob's Painting
Business Address: 25 John's Street, Miami, FL 33133
Business Phone: 305-555-1000
This information is provided for use by the "data receivers" and MerchantCircle.com on February 15. However, in the meantime, on January 25, the business owner (Bob) claims the listing on LocalEze and updates the

information to feature his correct phone number. The LocalEze listing is updated
with the new phone number: 305-666-1555
Unfortunately, MerchantCircle.com does not understand (either due to the way the
new data is provided by LocalEze, or due to imperfections in their data clustering
system) that this new business phone, together with all the other unchanged
information, is for the exact same business. It is just that the phone number has
changed. Thus, on March 10, a new listing appears on MerchantCircle.com. Now
there are two listings for Bob's business on their site:
Listing #1:
Business Name: Bob's Painting
Business Address: 25 John's Street, Miami, FL 33133
Business Phone: 305-555-1000
Listing #2:
Business Name: Bob's Painting
Business Address: 25 John's Street, Miami, FL 33133
Business Phone: 305-666-1555
Here's a summary of the order of events:
January 1—The original listing is added to LocalEze.
January 25—The phone number on the original listing on LocalEze is edited.
February 15—The original listing finally enters the MerchantCircle database
(assuming it is a 45-day cycle).
March 10—The edited information enters the MerchantCircle database. However,
because MerchantCircle are not good at matching and de-duping info, they create a
separate listing (a duplicate) instead of editing the original listing from February 15.
Now Bob has a problem he does not even suspect he has.

From the above example, one could understand that the problem with online
business data does not end with just finding all the incorrect or outdated listings for
a business and updating them. A business owner needs to first understand the
original source of the incorrect business data and start by fixing the information at
the source. Once this is done, they can update the rest of the listings, as well as
removing any duplicate listings that might have been generated. The majority of the
automated pay-to-play solutions do not go through any of these two steps (finding
and editing the information at the source and removing duplicate listings).

8.19   Google Map Maker and Maps Spam

As mentioned earlier in this chapter, spam has been one of the most significant
problems with Google Maps. First spam reports could be found as early as 2007,
and in 2015 the situation has improved only slightly. An issue connected with Maps

spam and probably the main catalyst for abuse is the fact that Google seem to be unwilling to invest enough resources on creating stronger anti-spam system. This is hardly a problem specific to Google Maps though. Google have always been trying to resolve complex issues in programmable manner, most probably because of the easier scalability of such solutions as compared to the use of human force. In the early days of Google Maps and Google Places there was practically no way for business owners or users to get in touch with a human being at Google, and the only sporadic communication was automated responses to public edits (e.g. "your edit has been approved"). In addition, there was no clear guidance available for business owners or users on what to expect from Google. The abundance of technical glitches on Google Maps made the situation nearly unbearable. In fact, it was so horrific by the end of 2012 when Google was in the process of transferring the whole business dashboard from Google Places to Google+ that famous local search blogger Mike Blumenthal wrote a now very revered article named "*Google Local: Train Wreck at the Junction*".

8.20 Out of Business?

Up until 2013, there were very few options for Google Maps users to contact directly or indirectly a Google representative. The "Report a problem" feature had been available since 2007, but it had never worked well enough to be a reliable solution. At the same time, it has always taken Google a great deal of persuasion to pay closer attention to particular issue. In fact, it has been proven numerous times that they tend to react on non-prioritized matters only when the story gets picked up by national and international media, and negative publicity looms over Googleplex. A particularly damaging businesses exploit was brought to the attention of *The New York Times* by Mike Blumenthal in 2011. The exploit stemmed from the fact that it was relatively easy at the time to mark a business listing as closed (i.e. to report that the company is out of business) and for such an edit to go live quickly. When a business listing is marked as closed, it disappears from virtually any organic search results, excluding (in some cases) exact-match brand name searches. This was the silver bullet to killing competitors' online presence for many unscrupulous businesses. The exploit had been around for a number of months and literally thousands of businesses had complained via the only possible mean at the time—the Google Places forum, but nothing was done on Google's end to fix it. However, after the story was posted on the NYT on September 5, 2011, it took less than a day for Google to respond publicly that they were taking action, and less than 10 days (including non-working days) for them to resolve the issue.

8.21 Google Phone Support

As mentioned above, the only option for businesses to contact Google indirectly was via the Google Places forum up until January 2013. In a big part due to the significant outrage that arose after the publication of the aforementioned article by Mike Blumenthal (on November 29, 2012), Google introduced phone support for the first time in the existence of the local business center. The phone support began on January 8, 2013—a little over one month after Blumenthal's article, even though the issue had been around for more than 5 years.

8.22 Google Map Maker

In 2014 and 2015, a new wave of loopholes in the Google Maps data verification system was exposed. However, it was not until The White House and Apple's logo got involved that Google decided to unprecedentedly suspend user edits on Map Maker (or at least their publishing) on May 12, 2015. How it all unveiled:

First, in early 2014 a Maps spammer, who had worked together with illegitimate businesses in the past started publicizing Google Maps pranks he had created in order to bring the issue to Google's higher management, and to potentially prompt them to prioritize the closing of the loopholes in the system. He was partially successful, especially with a prank in which he managed to intercept calls to San Francisco's FBI office, as well as Washington's Secret Service office. The publicity was apparently not enough and in 2015 he retried by creating bogus business listing with an address at The White House. The timing that time was "right", because just two weeks before that another story broke the news—of a prank in which Android's logo was mapped out in Northern Pakistan, urinating on an apple that looked very much like Apple's logo. It took just a few days for Google to make the decision to completely suspend user edits on Map Maker. In August 2016, Google Map Maker reopened in over 50 countries with two major changes: (1) Top mappers are empowered to moderate user edits, and (2) polygon editing is no longer available.

In Google's defense, some of the issues, which needed to be attended to, were relatively complex and significant amount of resources and time were needed for them to be fixed. For example, the Google LBC (later Google Places) dashboard was built on a technologically outdated platform, and Google needed the time to transfer the dashboard to the much more advanced Google+ platform. However, the large majority of issues mentioned above required Google's full attention for just a few days to get resolved. Additionally, their main problem had always been the lack of willingness to spend resources on high-quality customer service and ground-truth data verification teams. I have previously offered the following simple anti-spam solutions and they are still valid nowadays.

8.23 Anti-spam Solutions

(1) Never offer phone verification the first time when a listing is claimed.

Instead, include stricter verification requirements for listings created through user-generated content (UGC). For instance, require at least one additional hard-to-fake verifiable supporting evidence (registration with government institution that has public record with business information, local business chamber registration, listing on Localeze (requires payment) or Acxiom (requires document verification), and document related to the business for which all the business information is visible (similar to Acxiom's verification method)).

(2) Stricter checks on users who submit UGC through Map Maker and place restrictions based on the number of edits or reviews on Map Maker.

A great example in this regard is Waze, which allows you to edit only certain areas of the map (where you have passed through), and the areas expand based on your activity on Waze. Some would argue that if stricter information verification rules are imposed, it would be the small business owners and not the spammers who would suffer the most. However, I believe that while spammers are persistent and unscrupulous, driven by potential high profits, it is a big leap from creating fake online business listings to providing counterfeit documentation for verification purposes. I believe that most of them would give up.

(3) Use postcards to verify ownership over a business listing.

Changing the phone number on an existing listing is relatively easy with some persistence under the current state of the Map Maker system. Once the phone number is changed to the one that the spammer has access to, they could easily take control of the listing and potentially change all the information as they like. While there are complicated schemes with rogue and fake addresses, a verification postcard makes it much more difficult to change the address of a business listing as compared to its phone number. Therefore, verification by postcard as the only way to verify one's ownership over a business listing would provide a strong additional protection.

Appendix: Business Data Providers for Google in Selected Countries

Country	Business data providers
United States	Infogroup, Acxiom, LocalEze
Canada	Yellowpages (discontinued), Acxiom (speculative)
United Kingdom	Market Location, 118 Information, Local Data Company

(continued)

(continued)

Country	Business data providers
Germany	Gelbeseiten, Infobel
France	PagesJaunes, Infobel
Italy	Paginegialle, Infobel
Australia	Sensis
New Zealand	Finda

Chapter 9
Subtle New Forms of Internet Influence Are Putting Democracy at Risk Worldwide

Robert Epstein

9.1 Democracy

The free and fair election is fundamental to democracy, but the internet has made possible new, nearly invisible forms of influence that have likely been affecting the outcomes of close elections worldwide for several years now. Left unchecked, these forms of influence will inevitably grow over time, slowly but surely nullifying the democratic process as we know it.

9.2 Digital Gerrymandering

One possible type of influence has been labeled "digital gerrymandering" by Harvard law Professor Zittrain [4]. In conventional gerrymandering, the boundaries of voting districts are altered to favor one political party, virtually guaranteeing that the majority of voters in the newly-drawn districts will vote for that party. Zittrain has pointed out that social media giants such as Facebook could easily accomplish the same sort of manipulation by sending out multiple prompts to "go out and vote" only to people who are known to favor one candidate or party. In close elections, which are quite common, increasing the number of such people by even a small amount could easily flip an election.

Of greater concern, this kind of manipulation could be accomplished—or perhaps *is* being accomplished—without anyone being the wiser. As it is, social media companies already send out customized advertisements to more than a billion

R. Epstein (✉)
American Institute for Behavioral Research and Technology, Vista, California, USA
e-mail: re@aibrt.org

© Springer Science+Business Media New York 2016
N. Lee (ed.), *Google It*, DOI 10.1007/978-1-4939-6415-4_9

people daily based on gender, age, location, purchase histories, and other factors. They could easily send out—or perhaps are *already* sending out—prompts to vote to select groups of people without anyone knowing that these groups are being singled out.

9.3   Facebook Experiment

In a controlled study conducted in 2010, researchers at the University of California San Diego, working with employees at Facebook, demonstrated that repeatedly flashing "VOTE" ads to over 60 million Facebook users on an election day in the US caused 340,000 more people to vote than otherwise would have [1]—an increase of approximately 0.57 %. The sample of people who received the vote prompts was selected at random for the study, but, given Facebook's massive database of personal information about its users, the company could easily have targeted people with known preferences for certain candidates or parties; hence, the basis for Zittrain's speculation. Shifting 0.57 % of voters toward one candidate results in a marginal difference of 1.14 %—enough to flip the outcomes of many close elections (that is, elections with win margins under this percentage).

It is notable that the kinds of advertisements that could mobilize voters are ephemeral in nature. They are flashed briefly to users and then disappear. Unlike the main content of a website, which remains somewhat constant and might even be preserved by tools like the Wayback Machine (see https://archive.org/web/), advertisements don't leave a paper trail. This gives companies such as Faceback complete deniability if it was ever accused of interfering in an election.

Here is the kind of prompt Facebook sent its users in the 2010 experiment

9.4   Search Engine Manipulation Effect (SEME)

Research I have been conducting with Ronald Robertson since early 2013 has identified another problematic source of influence over voter preferences which we call SEME (pronounced "seem"), for Search Engine Manipulation Effect. SEME, we have concluded, is almost certainly already influencing close elections around the world, and it is a much larger effect than digital gerrymandering; in fact, it is proving to be one of the largest behavioral effects ever discovered.

Initially, in a series of five randomized, controlled experiments we completed with 4556 participants in two countries, we demonstrated and repeatedly confirmed that when high ranking search results favor one candidate—that is, make him or her look better than his or her opponent—the proportion of undecided voters supporting that candidate can easily be increased by 20 % or more—up to 80 % in some demographic groups [2]. Perhaps of greater concern, very few participants in our experiments showed any awareness that they were viewing biased search results. In other words, SEME is not only a large behavioral effect, it is also almost entirely invisible.

What's more, search results, like advertisements, are ephemeral. No records are kept of them, which means that they leave no paper trail. Once again, this gives the company displaying such results complete deniability.

High ranking search results alter opinions because most people mistakenly believe that search rankings are determined by an objective, omniscient, and infallible mechanism that is beyond human control. This is confirmed by a variety of research on consumer behavior which shows that people trust and believe higher-ranked search results more than lower-ranked results. Over 50 % of all clicks go to the top two search results, and more than 90 % of users never leave the first page of results. Research shows that this occurs even when high-ranking items are of poor quality; it is not just for convenience sake that people click on high-ranking items but rather because of those deeply-help beliefs regarding their validity. Because of the enormous value that high-ranking items have for purchases, North American companies are now spending more than 20 billion US dollars per year on SEO in an attempt to push their links to higher positions.

When companies, candidates, or political parties compete in an open market-place to get people's attention, fairness is maintained—although, of course, the players with more resources have always have the advantage. What happens, however, when the search engine company *itself* has preferences?

This, of course, is the topic of investigations of Google, Inc. by the U.S. Federal Trade Commission, the European Union, and the government of India, all of which have found that Google unfairly favors its own products and services in its search rankings. Given the power of SEME, one must also wonder: what impact might Google have on elections if its search rankings also favored one candidate over another?

9.5   2014 Lok Sabha Election

The fifth experiment reported by Epstein and Robertson [2], conducted in India during the 2014 Lok Sabha election—the largest democratic election in history—is especially relevant to this question. Our previous experiments had focused on elections that were already completed and on candidates who were unknown to the participants in order to minimize any bias the participants might otherwise have brought to the experiments—in other words, to guarantee that they were truly "undecided" voters.

Mr Gandhi, Mr Kejriwal, and Mr Modi, candidates in the 2014 Lok Sabha election in India

In the 2014 experiment, however, we used newspaper ads and online subject pools to recruit 2,096 undecided voters in 27 of India's 35 states and territories— real voters in the midst of an intense, hotly-contested election campaign. Participants were randomly assigned to groups in which search rankings favored either Mr Modi (the ultimate winner), Mr Kejriwal, or Mr Gandhi. As we found in our previous experiments, exposure to biased search rankings that linked to real web pages (which participants could examine freely) caused voting preferences to shift toward the targeted candidate by 20 % or more. In some demographic groups, such as unemployed males from a certain region of India, the shift was over 70 %. We got this result even though our participants were highly familiar with the candidates.

That we conducted this research in India was especially appropriate given that in March 2014, Google was fined 10 million rupees (USD$164,000) by the Competition Commission of India for "search engine bias."

Are there any indications that actual Google search rankings might have favored Mr Modi in the Lok Sabha election? Google's own data—the daily "Google Score" it assigned to the major political candidates based on search volumes—showed that Modi outscored his opponents by at least 25 % for 60 consecutive days prior to the day the polls closed on May 12th.

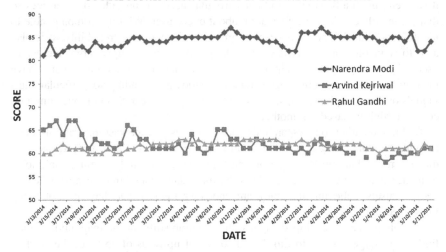

Google Scores for the major candidates in the 2014 Lok Sabha election in India for the 60 days prior to the close of polls on May 12th. The data were compiled by the author from daily data posted by Google prior to the election. Google has since removed these data from the Internet

Because search volume is one predictor of search ranking, it is reasonable to assume that Mr Modi was also favored in Google's search rankings as more than 430 million votes were cast between April 7 and May 12. If even 10 % of India's voters with internet access were still undecided in the weeks before the polls closed, biased rankings could have driven several million votes toward Mr Modi. What's more, with an increasingly large portion of the world's population having internet access, whatever the impact of biased search rankings is today, it will certainly be much larger in the future. Our own data suggest that more than 85 % of people with internet access are getting at least some of their election information from internet sources—a number that is also likely to increase in coming years.

9.6   2008 and 2012 Presidential Elections

In the United States, Google Scores posted by the company before the 2008 and 2012 presidential elections2012 U.S. Presidential Election showed strong preferences for Barack Obama, the winner of both elections (see https://plus.google.com/

photos/104361405143172836769/albums/5795430883215139905/5795430882434
931746), and a study published in 2015 on Slate.com confirms that Google search
rankings strongly favor Democratic candidates [3].

Couldn't researchers or government agencies simply track and rate search
rankings to determine the extent to which they are biased toward one candidate or
another? This is not as easy a task as it sounds, and it might even be impossible.
Google's revenue model depends on being able to identify users in real time so that
it can send them ads targeted to their particular needs; it provides free services so
that it can collect relevant information about every user. With the company able to
identify individuals and demographic groups with increasing reliability, it is also
able to send out customized search rankings to an increasingly large number of
users. From a regulatory standpoint, the problem here is that monitors would have
no way to look at the customized rankings Google is sending to particular individuals or demographic groups. Rankings that might appear clean on one computer
could be highly biased on another.

As I noted earlier, our research also demonstrates that the vast majority of voters
are unaware that the search rankings they are viewing are biased toward one candidate; more than 99 % of participants in our India study seemed oblivious.
Influence that is invisible to people is the most dangerous kind, because it leaves
people falsely believing that they are choosing freely—that they are not being
influenced at all.

Models we have developed suggest that opinion shifts of the magnitude we are
finding are large enough to flip the outcomes of upwards of 25 % of the world's
national elections. As of this writing, we have now replicated SEME nine times
with nearly 10,000 participants in multiple countries, and we also been examining
ways of suppressing the effect. Among other things, we have discovered that SEME
is probably having an enormous impact on a wide range of important decisions
people are making every day, not just on voting preferences.

9.7 Power of Google

SEME wouldn't be much of a threat if the online search business were in the hands
of a dozen competing companies. Because more than 90 % of search in most
European countries and in many other countries around the world is in the hands of
a single company, however, no candidate or party has a way of offsetting the
influence that Google's search rankings are likely having on elections.

One last point: Although it is reasonable to assume that Google executives are
using search rankings to favor candidates that they deem preferable for their
business needs (to do otherwise would be imprudent), our data suggest strongly that
Google's search rankings are influencing elections even if Google's executives are
keeping hands off. This is inevitable because of the fundamental nature of Google's
ever-changing search algorithm. So-called "organic" search phenomena will
inevitably boost the rankings of some candidates over others; when this happens,

the preferences of undecided voters will shift toward those candidates in a kind of digital bandwagon effect.

Either way, to protect democracy, search rankings related to elections should be strictly regulated.

Robert Epstein is Senior Research Psychologist at the American Institute for Behavioral Research and Technology and recently retired as Professor of Psychology at the University of the South Pacific. The former editor-in-chief of *Psychology Today* magazine and a PhD of Harvard University, Dr Epstein has published 15 books on artificial intelligence and other topics. You can learn more about SEME research at http://aibrt.org or about Dr Epstein himself at http://drrobertepstein.com. You can follow him on Twitter@DrREpstein (http://twitter.com/DrREpstein).

References

1. Bond, R. M., Fariss, C. J., Jones, J. J., Kramer, A. D., Marlow, C., Settle, J. E., & Fowler, J. H. (2012). A 61-million-person experiment in social influence and political mobilization. *Nature, 489*(7415), 295-298.
2. Epstein, R., & Robertson, R. E. (2015, August 4). The search engine manipulation effect (SEME) and its possible impact on the outcomes of elections. *Proceedings of the National Academy of Sciences USA, 112*(33), E4512-E4521. http://www.pnas.org/content/112/33/E4512.full.pdf?with-ds=yes
3. Trielli, D., Mussenden, S., & Diakopoulos, N. (2015, December 7) Why Google search results favor Democrats. *Slate.* http://www.slate.com/articles/technology/future_tense/2015/12/why_google_search_results_favor_democrats.single.html
4. Zittrain, J. (2014, June 1). Facebook could decide an election without anyone ever finding out: The scary future of digital gerrymandering—and how to prevent it. *New Republic*, 1. https://newrepublic.com/article/117878/information-fiduciary-solution-facebook-digital-gerrymandering

Chapter 10
Social Media as a Tool for Information Warfare

Aylin Manduric

10.1 Social Media

The ability to quickly and cheaply share images and news across social media has empowered the current generation to witness and participate in the development of global tensions as they erupt into conflict. Over the past decade, social media has shaken the monopoly the conventional news media once had on disseminating knowledge about human rights abuses, civil wars, terrorist attacks, insurgencies, and revolutions. With over 350 million (https://www.opendemocracy.net/opensecurity/kym-beeston/'sharing'-witness-is-social-media-changing-way-we-see-conflict) photographs uploaded daily to Facebook alone, social media has, and will continue to revolutionize the way the world witnesses conflict. Like any tool, social media can be used to further desirable outcomes like peace, or to facilitate illicit operations that spread instability and promote violence. Using several recent examples, the following section will illustrate some of the ways social media is changing conflict.

Few conflicts are as fraught as that between the states of Israel and Palestine. This conflict, which began before the advent of the internet, now has an enormous presence on social media, with both Hamas and Israeli Defense Forces making use of dramatic photography to garner support for their conflicting causes. The Israeli Defense Forces Twitter page has over half a million followers (https://twitter.com/IDFSpokesperson), and posts updates on the conflict as well as self-promoting material every few hours. The State of Palestine's Mission to the United Nations has around 18 thousand followers (https://twitter.com/Palestine_UN), and also

A. Manduric (✉)
University of Toronto, Toronto, Canada
e-mail: a.manduric@mail.utoronto.ca

A. Manduric
Center for the Study of the Presidency and Congress, Washington, DC, USA

© Springer Science+Business Media New York 2016
N. Lee (ed.), *Google It*, DOI 10.1007/978-1-4939-6415-4_10

makes several posts a day. The Hamas Movement's English (https://twitter.com/HamasInfoEn) Twitter account has nearly 25 thousand followers. These conflicting parties dedicate significant time to cultivating an online presence, which suggests that they are aware of the potential benefits to be derived from sharing their position widely.

10.2 Information Warfare

On March 4, 2015, the Congressional Research Service released a collection of insights on the role of social media in conflict. The short report was entitled "Information Warfare," (https://www.fas.org/sgp/crs/misc/IN10240.pdf) and described social media as "a weapon of words that influences the hearts and minds of a target audience, and a weapon of mass destruction that can have effects on targets in the physical world." Aside from Twitter and Facebook, YouTube, blogs, Instagram, and countless other social media applications can become tools for conflict when they fall into the wrong hands. Al-Qaeda and the Islamic State (IS) even have their own online magazines, which are available both online and in print. Terrorist organizations make use of sleek videos, dramatic photography, and well-formatted web pages to both appeal to potential recruits and donors and to broadcast their capabilities and resources. A terror group with the resources to construct and maintain an effective social media presence may also appear more threatening to ordinary people who view or encounter the group online. Heavy use of social media raises a group's profile globally, but by no means do terrorist groups restrict their use of social media to marketing and recruitment. Al-Qaeda and IS have both made extensive use of social media to identify targets, share training tools and data, and disseminate information about their plans and "achievements." All of these functions lubricate the gears of an organization and facilitate operations. Often, the marketing and recruitment uses of social media overlap with the target identification and data-sharing elements to catalyze new means of attack. In January 2015, after IS had been sharing their message through social media for some time, the CyberCaliphate (https://www.washingtonpost.com/news/the-switch/wp/2015/01/12/the-centcom-hack-that-wasnt/), a group of tech-savvy sympathizers hacked the U.S. Central Command's Twitter account, making the US Central Command look vulnerable and even inept. After this case, Twitter responded by freezing roughly 2000 accounts with links to IS, though the psychological impact of the attack on those who witnessed or heard about it could not be reversed.

Of course, the use of social media is not restricted to terrorists or independent individuals. Most national governments have an immense online presence, and make extensive use of social media to spread messages both in support of their own policies or to endorse actions and movements abroad. The Congressional Research Service report on the role of social media in conflict highlights the tensions between Ukraine and Russia as being illustrative of how government use of social media can impact developing conflict between states. According to the report, spyware believed to have been of Russian origins was discovered on Ukrainian government systems. The report's author links this, as well as several cyberattacks on the Ukrainian government networks, to patriotic hackers mobilized by the Russian narrative of the conflict as it was spread on social media. Even so, the power of social media to mobilize non-state actors to commit crimes on behalf of a state remains dubious. More powerful is the impact of the threat of social media-related attacks on state policies. Since 2012, Russia has had a law in place requiring bloggers with more than 3,000 followers to register with the government. In Ukraine, the 2014 cyberattacks were followed by the introduction of legislation criminalizing open criticism of the government and placing restrictions on bloggers.

The fear that sparked these legislative changes may be linked to the association of social media with the regime-ending revolutions that took place in the Middle East and North Africa in 2011. The Egyptian Revolution, amongst others comprising the Arab Spring movement, is sometimes referred to (http://english.ahram.org.eg/News/63253.aspx) alternately as a "Twitter Revolution" and a "Facebook Revolution" because of the apparent role played by social media in the uprisings that preceded the end of the reign of Hosni Mubarak. Though the significance of the role of the protests themselves in ending the regime remains in some circles (http://english.ahram.org.eg/News/63253.aspx) a point of controversy, the vast social media presence of the revolution is undeniable, with relevant Facebook pages and Twitter accounts garnering immense popularity throughout the revolution. Social media provided a forum for dissent, and a means by which likeminded people can organize and express their desire for change. While these expressions do not in themselves make change occur, they can nonetheless play a role in amplifying existing sentiments within a community, and making those sentiments, and the injustices that spurred them, known to everyone with an internet connection.

10.3   Arab Spring

As former CIA director James Woolsey indicated in an interview with CNN, it was only during the Arab Spring (http://www.wired.com/2011/02/egypts-revolutionary-fire/) that state intelligence agencies began to realize the power of social media in organizing and linking movements across countries and around the world. Meanwhile, people like Google executive Wael Ghonim(http://www.wired.com/2011/02/egypts-revolutionary-fire/) have made this discovery through personal experience. Ghonim (http://www.nytimes.com/2012/02/19/books/review/how-an-egyptian-revolution-began-on-facebook.html?r=0) is the protester whose Facebook page is said to have catalyzed the revolution by drawing attention to the death of a young Egyptian man in a gruesome police beating. According to an article in the New York Times (http://www.nytimes.com/2012/02/19/books/review/how-an-egyptian-revolution-began-on-facebook.html?_r=0), 300 people joined Ghonim's Facebook group in the first 2 min after he created it. The page had a following in excess of a quarter million people within 3 months, and though the Egyptian government responded to widespread protests in January of 2011 by shutting down virtually all (http://www.theguardian.com/technology/2011/jan/28/egypt-cuts-off-internet-access) internet access across the country. After the fact, some commentators have claimed that since the protests continued despite the shutdown, social media played only a negligible role (http://www.thewire.com/global/2011/01/the-twitter-revolution-debate-the-egyptian-test-case/21296/) in igniting the Egyptian Revolution.

Yet, the Egyptian government saw the shutdown as a necessary measure to cope with the threat posed by the protesters to its continued hold on the country. By connecting people, and providing a window for the world to see the injustices that triggered the protests, social media wore down the legitimacy of the previous regime and confirmed for each protester the idea that hundreds of thousands of people, in Egypt and around the world, supported their revolution. The social impact of this support is not small.

Nor can the role of social media in internationalizing the revolution be ignored. After the Egyptian regime shut down internet access in response to the protests, Andrew Noyes, a representative from Facebook, told the New York Times (http://www.nytimes.com/2011/01/29/technology/internet/29cutoff.html) that "limiting Internet access for millions of people is a matter of concern for the global community." Though Egypt was not the first state whose government interfered with its citizens' internet access, the extent of the cutoff was unprecedented and attracted global condemnation. Activists from outside the country also mobilized to help Egyptians circumvent the shutdown or communicate using alternative means. Professor Ronald Deibert from the University of Toronto indicated publicly (http://www.nytimes.com/2011/01/29/technology/internet/29cutoff.html) that the shutdown could only cost the regime diplomatic leverage and political legitimacy. Rather than stifling the protests, the government's choice to cut communications amongst the protesters triggered an overwhelmingly negative international response, granting the protesters increased leverage against the regime.

Social media plays a role in both active conflicts and burgeoning uprisings, helping many notable occurrences receive international attention from vast numbers of people who may have no direct stake in the outcome. This has myriad effects on both the conflict itself and the people who are now capable of bearing witness to it. Faced with a deluge of contradictory information, individuals may be inclined to believe whatever aligns best with their existing normative assumptions. Some may discover solidarity with the suffering of those far away, while others may find that their fear outweighs any newly awakened empathy. Across all of these examples, what remains clear is that in the twenty-first century, conflicts may play out within the borders of a single state or region, but will do so with the participation or observation of the whole world.

Part III
A Library to Last Forever

Because books are such an important part of the world's collective knowledge and cultural heritage, Larry Page, the co-founder of Google, first proposed that we digitize all books a decade ago, when we were a fledgling startup. At the time, it was viewed as so ambitious and challenging a project that we were unable to attract anyone to work on it. But five years later, in 2004, Google Books (then called Google Print) was born, allowing users to search hundreds of thousands of books. Today, they number over 10 million and counting.

In the Insurance Year Book 1880–1881, which I found on Google Books, Cornelius Walford chronicles the destruction of dozens of libraries and millions of books, in the hope that such a record will "impress the necessity of something being done" to preserve them. The famous library at Alexandria burned three times, in 48 B.C., A.D. 273 and A.D. 640, as did the Library of Congress, where a fire in 1851 destroyed two-thirds of the collection.

I hope such destruction never happens again, but history would suggest otherwise. More important, even if our cultural heritage stays intact in the world's foremost libraries, it is effectively lost if no one can access it easily. Many companies, libraries and organizations will play a role in saving and making available the works of the 20th century. Together, authors, publishers and Google are taking just one step toward this goal, but it's an important step. Let's not miss this opportunity."

—Sergey Brin in "A Library to Last Forever"
(The New York Times, October 8, 2009)

Chapter 11
Wolfram|Alpha: A Computational Knowledge "Search" Engine

John B. Cassel

11.1 Introduction

Wolfram|Alpha (W|A) is a search engine in some senses, but not in others. People do use W|A to search for responses to their questions. Additionally, W|A often runs queries, or searches, through its databases. However, what W|A does not search for information containing words that match or are similar to a provided phrase. Instead, Wolfram|Alpha makes new contact with the history of librarianship, using the curation work of library professionals and domain experts to map words into functions describing the query, attempting to encode the question's intent.

This chapter will first describe how Wolfram|Alpha is different from other search and answer technologies. Next, we will explain how Wolfram|Alpha took this different path, one against the grain of technical trends but clearly sensible when contextualized by the internal forces at play in the Wolfram group of companies. Then we will examine a variety of Wolfram|Alpha queries, demonstrating what this particular approach gives its users. From there, we will talk about the ongoing legacy and the future directions for Wolfram|Alpha. Finally, we will conclude by summing up how this particular technical organization serves people who need to know not what is out there on the internet, but facts that can be readily calculated and summarized when content is carefully distilled.

J.B. Cassel (✉)
Agrible, Champaign, USA
e-mail: john.benjamin.cassel@gmail.com

© Springer Science+Business Media New York 2016
N. Lee (ed.), *Google It*, DOI 10.1007/978-1-4939-6415-4_11

11.2 Against the Grain: Comparing Wolfram|Alpha's Technology

A key factor in the success of Wolfram|Alpha was not only the quality of its technology, but that it went against the grain of the received wisdom of the time. W| A was created in the height of the Web 2.0 boom when crowdsourcing and collective intelligence were all the rage, but W|A worked by hiring in-house curators who were librarians and domain experts to vet their data. In a time where statistical machine-learning methods were dominating artificial-intelligence development, novel syntactic approaches did Alpha's heavy lifting. It was not information retrieval technology that mined text of the Internet, but instead a symbolic approach that found the computable meaning of queries. Instead of being a product of Silicon Valley, it was a product of a low-key Midwestern town known for its high-tech university and highly-ranked library school: Champaign, Illinois. By going against the popular tide, Wolfram|Alpha had blank canvas for new technological innovation. Let's look into these differences by comparing Wolfram|Alpha with Google, IBM's Watson, and Wikipedia, and then talk a little about Wolfram|Alpha's disparities with contemporary artificial-intelligence trends.

11.3 Direct Differences: Wolfram|Alpha in Comparison with Google

It is easiest to see Wolfram|Alpha's difference by comparing it with Google as Google was at the time. The most obvious difference is in the change of Google's results. Figure 11.1 displays sections of what the Wolfram|Alpha result for George Washington looks like as of this writing.

As of this writing, if you search for "George Washington" on Google, you will get the following summary on the right-hand side (see Fig. 11.2):

When Wolfram|Alpha came out, this was not a feature of Google, which had links to other pages, not content in itself. Wolfram|Alpha rarely had links, and when it did, they were not the primary content. Further, what links were displayed were selected for inclusion by people. A better Wolfram|Alpha comparison in 2009 was Wikipedia, with its summary boxes that organized its content into something more uniform for each domain.

The striking thing about Wolfram|Alpha is that this kind of information is not merely uniformly prepared canned content fed to a template. A result for, say, "$x^3 - 5x^2 + 2y^3 - 7y$", which the author just made up and has no particular

George Washington (politician)

Basic information: [Local historical calendar ▾]

full name	George Washington
date of birth	**Friday, February 11, 1732** (Julian calendar) (283 years ago)
place of birth	**Westmoreland County, Virginia**
date of death	**Saturday, December 14, 1799** (Gregorian calendar) (age: 67 years) (215 years ago)
place of death	**Mount Vernon, Virginia**

Image:

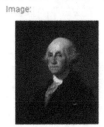

Leadership positions:

official position	country	start date	end date
Senior Officer (Department of Defense)	**United States**	**July 13, 1798**	**December 14, 1799**
President (1ˢᵗ)	**United States**	**April 30, 1789**	**March 4, 1797**
Continental Army General and Commander in Chief (Department of Defense)	**United States**	**June 15, 1775**	**December 23, 1783**

Timeline: [Include today]

George Washington ●━━━━━━━━━━━━━━━━━━━━━━━━━━━━━━━━━━━━━━●
                      1740   1760   1780   1800

Fig. 11.1 Partial Wolfram|Alpha result for "George Washington"

significance, is just as meaty, including a 3D plot and a contour plot (see Fig. 11.3), as well as alternate forms, roots, polynomial discriminant, integer roots, properties as a function, roots for the y variable, derivative, indefinite integral, local minimum, local maximum, definite integral over a disk of radius R, and definite integral over a square of edge length 2L, and it promises more if given more computation time (a feature available to Wolfram|Alpha Pro users).

Notable facts:

First US president and preeminent military and political leader during formative years
of the country

Champion of Federalist policies, including strong national government, who oversaw
drafting of the US Constitution

Famously led troops across icy Delaware River for significant victory over British at
Trenton

Commander in chief of the Continental Army during the Revolutionary War

Physical characteristics: Metric

height   6′2″

Familial relationships: Show full dates   More

Parents:

Augustine Washington | Mary Ball Washington

Siblings:

Fig. 11.1 (continued)

Even this ability to create novel content doesn't show Wolfram|Alpha in its
strongest light. Its most powerful feature is that it can link facts with computation,
to give a sense of proportion. Consider "unemployment rate New Mexico,
Nebraska", a partial result of which is shown in Fig. 11.4. There are several striking
things about this. First, you are not easily going to get a direct comparison between
these two states elsewhere. The combination is novel, assembled to meet the query.
The next thing that the results not only include the rate, they also offer a calculation
of the change in rate over time. In an additional result segment that is not included
here, there are also some absolute counts of those employed and the rank of the
states in those counts. What this example shows is that novel queries to W|A that
combine different entities of the same kind and receive both customized and integrated results, along with a further quantification in rates and ranks that gives those
results context.

Wolfram|Alpha is also useful for discovering facts that are not explicitly present,
but available through formulas. Consider the result of "two drinks over an hour for
a 160 lb male", as shown in Fig. 11.5. This result shows the blood alcohol content
as it relates to the legal driving limit of the United States.

Wolfram|Alpha is different from Google not only in what it does, but what it
does not do. If you ask Google "who is the best president of the United States", it
will return documents with historical comparisons, polls, and opinions. However,

George Washington

1st U.S. President

George Washington was the first President of the United States, the Commander-in-Chief of the Continental Army during the American Revolutionary War, and one of the Founding Fathers of the United States. Wikipedia

Born: February 22, 1732, Westmoreland County, VA

Died: December 14, 1799, Mount Vernon, VA

Height: 6' 2" (1.88 m)

Vice president: John Adams (1789–1797)

Presidential term: April 30, 1789 – March 4, 1797

Spouse: Martha Washington (m. 1759–1799)

Fig. 11.2 Google summary of "George Washington" search

Wolfram|Alpha does not index documents or generally deliver answers with opinions or conjectures. You should expect from Wolfram|Alpha what you might expect from a very adept computer program, and there is no agreed-upon computational definition of what it means to be the best president. For this reason, although Wolfram|Alpha may have to weigh in on disputed matters of fact such as borders between countries or topics where the science has not yet been settled, it does not participate in many controversies.

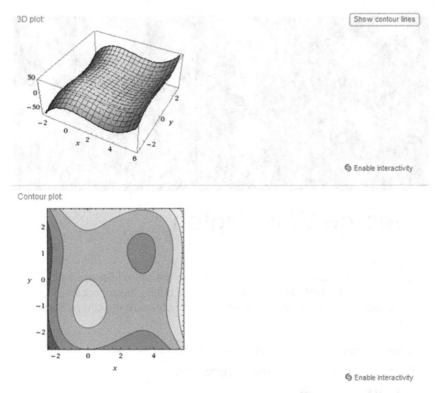

Fig. 11.3 3D plot and contour plot from "$x^3 - 5x^2 + 2y^3 - 7y$" W|A query result

As Wolfram|Alpha and Google have complementary offerings, would Wolfram|Alpha + Google be a better, more comprehensive search technology? It is a matter of individual taste, but more is not necessarily better. Once one knows Wolfram|Alpha is capable of answering a question in a particular domain, one can use it to get the answer straightforwardly without sorting through other results. Similarly, there is no point in using Wolfram|Alpha when searching for articles or opinions.

Overall, Wolfram|Alpha aims not to find documents that contain facts, but to combine facts in novel ways for comparison, or even assemble facts out of formulae, while staying within the limits of what we might reasonably credit to algorithms run over facts. Throughout this text, we'll see further examples of Wolfram|Alpha making calculations to achieve novel and salient results.

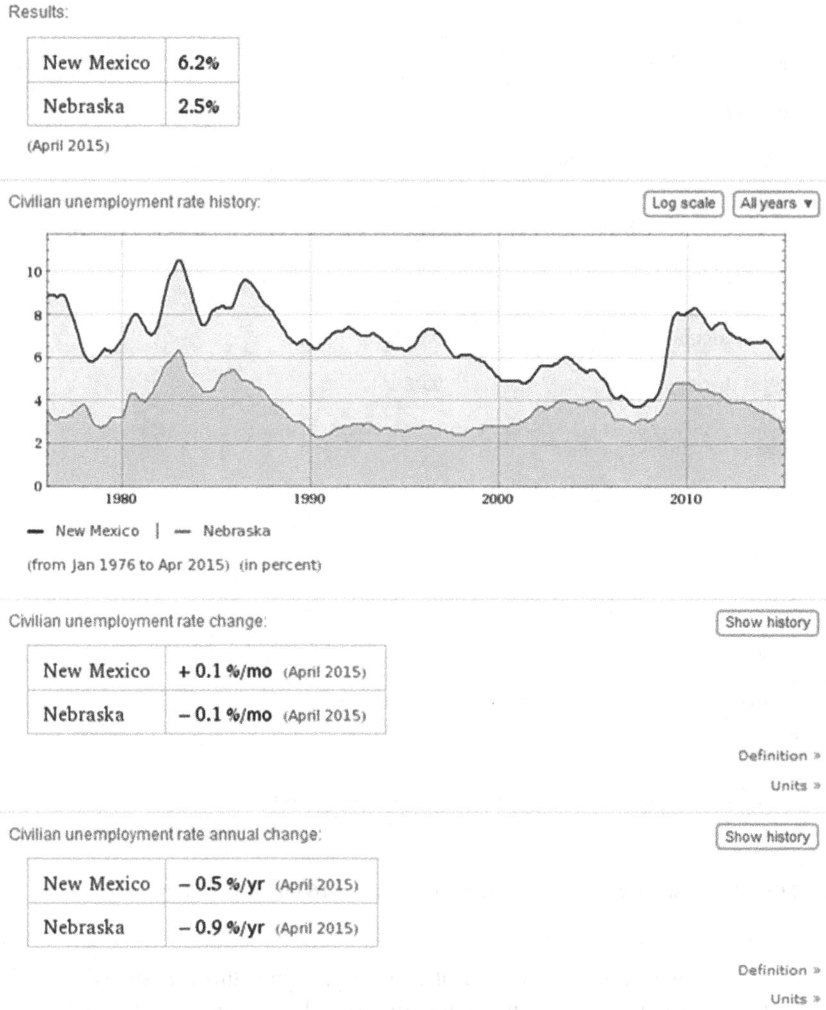

Results:

New Mexico	6.2%
Nebraska	2.5%

(April 2015)

Civilian unemployment rate history: [Log scale] [All years ▼]

— New Mexico | — Nebraska

(from Jan 1976 to Apr 2015) (in percent)

Civilian unemployment rate change: [Show history]

New Mexico	+ 0.1 %/mo (April 2015)
Nebraska	− 0.1 %/mo (April 2015)

Definition »
Units »

Civilian unemployment rate annual change: [Show history]

New Mexico	− 0.5 %/yr (April 2015)
Nebraska	− 0.9 %/yr (April 2015)

Definition »
Units »

Fig. 11.4 Partial Wolfram|Alpha result for "unemployment rate New Mexico, Nebraska"

11.4   Internal Differences: Wolfram|Alpha in Comparison with IBM's Watson

We might be led by these differences from Google to say that, instead of being a search engine, Wolfram|Alpha is a tool for question answering. This would be entirely fitting, as Wolfram|Alpha poses itself as a "computational answer engine". The appropriate comparison would then be with IBM's Watson, the system known

Input information:

number of drinks	2
gender	male
time	1 hour
body weight	160 lb (pounds)

Estimated result:

blood alcohol percentage	0.05%
legal driving limit (United States)	0.08%

(estimated ignoring variations in metabolism, food intake, medication, etc. and assuming a drink
contains 0.648 fl oz of alcohol)

Time plot:

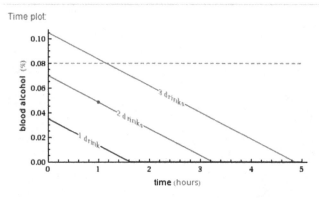

Fig. 11.5 W|A result for "two drinks over an hour for a 160 lb male" query

for being the first computer to win at Jeopardy. While those systems' intended
applications are indeed closer, there are still critical differences in their internals.

The primary difference between the two is in information retrieval versus
symbolic processing. Watson uses information retrieval technology, processing
queries into searches over text indices and applying statistically trained result
selection. Wolfram|Alpha works rather differently. It is helpful to realize that
Wolfram|Alpha was developed by the same company that wrote the Wolfram
Language, a very large programming language. The initial phases of interpreting a
query do not involve going to particular indexes, but is effectively a compiler for a
very ambiguous programming language. The ambiguity between surviving parses is
then managed by a scoring system that internal curators have tuned to yield the

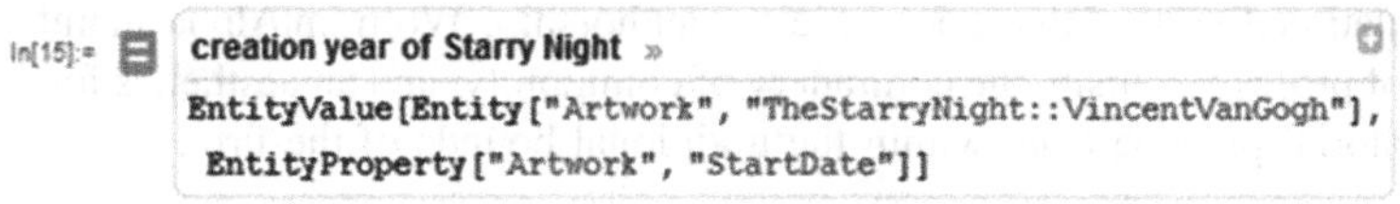

In[15]:= ▤   **creation year of Starry Night** »   ▢

```
EntityValue[Entity["Artwork", "TheStarryNight::VincentVanGogh"],
EntityProperty["Artwork", "StartDate"]]
```

Fig. 11.6 *Mathematica* 10 "single equals" evaluation of "creation year of Starry Night"

most likely answer while also offering a mechanism for the user to select from other valid interpretations of their question. These parses are then translated into a symbolic representation, in a way made usable by the public in *Mathematica* 10. For example, let us ask about the computable form of "creation date of Starry Night", which will be understood as pertaining to the painting by Vincent van Gogh, with results shown in Fig. 11.6.

What's happening is that this linkage to Wolfram|Alpha has converted the text to a function. It is then the processing of that function, and processing that determines information related to that function, that prepares the pages shown in Wolfram|Alpha. Later, we will explain the reasons for this methodology choice, which comes from both the technical history behind Alpha and how this allows integration into yet broader systems.

11.5   Institutional Differences: Wolfram|Alpha in Comparison with Wikipedia

As Wolfram|Alpha does not search external resources, which might have no clear transformation into the symbolic relations described above, it works from an internal compendium of databases and data feeds. In that way, Wolfram|Alpha is like Wikipedia, where its knowledge resources are part of the overall system. However, unlike Wikipedia, its content is not crowdsourced, save for some processed feedback from users and from some volunteer programs, but instead comes from a process of internal curation.

Wolfram|Alpha does not have open data as its implementation provides little opportunity for the rewards of open authorship, namely community recognition. Wolfram|Alpha's data is not presented directly, but combined in multiple ways by means of computational expression. Without a narrative or composition, the pleasure of having presented one's composition is not as preserved. Further, certain kinds of disputed data (such as contended borders between countries) need a presentation outside of a text which could nuance and qualify those disputes. Finally, numerical data, being outside of a narrative, might be too easily altered by someone with poor intentions. Overall, instead of attempting to build rewards into a computational system that does little for authorship, Wolfram|Alpha maintains an ethos of professionalism, the comradery of community, and an aesthetics for nuance in curatorial practice from within the traditional bounds of the firm.

11.6 Wolfram|Alpha's Relationship with Artificial Intelligence

As we have discussed, Wolfram|Alpha's contributions to artificial intelligence are in a scalable infrastructure for syntactic question answering and the corresponding demands in knowledge representation. However, Wolfram|Alpha's differences from common artificial intelligence practices, in particular its distance from the statistical techniques dominant over the last 20 years, is perhaps more interesting.

Unquestionably, the task of answering queries posed in natural language is an artificial intelligence project. The need of Wolfram|Alpha was very different than posed in question answering tasks. Question answering is typically focused on being able to answer inquiries about a particular text, having extracted relationships. With Wolfram|Alpha, the kinds of domains and relationships that can be addressed are planned ahead of time: can Wolfram|Alpha be sufficiently comprehensive in a particular area? Does every kind of computation the W|A team chooses to make available lead to a correct answer? Among the possible interpretations of a given statement, are those that fit with W|A's potential answers plausible? The basic guideline that W|A is engineered for is delivering high precision (a minimum of false positives) at the cost of recall (a minimum of possible, but missed, answers).

This position means that nearly all potential translations to computational form are as intended, and those that are not can be remediated through clear rules. Scoring different outcomes is still necessary, but this is because some interpretations are more useful than others, even if all are semantically correct. For example, the query "bra size chart" could validly mean one wants a graph of how the population of the city of Bra in Italy changed over time. The city population area interpretation is therefore allowed, but the garment interpretation is scored higher. If interpretations are close in score, only one will be immediately displayed, but an assumptions section will be displayed that allows the other interpretations to be easily selected. Thus far it has simply been more efficient to have scoring rules for such cases than trying to introduce the floating point overhead of probability calculations into the tight loops of the parse itself.

What the Wolfram|Alpha team has done is develop efficient procedures by which an organization of people can combinatorially assemble the kinds of relationships that might be the output of natural language processing techniques, and representations appropriate to that assembly. How did the team come to this particular set of technical choices? Let us now look into the forces that made Wolfram|Alpha.

11.7   How Wolfram|Alpha Could Happen:
A Social History

How could a technology so against the popular tide be developed? What we will see is that Wolfram|Alpha directly addressed business needs while opening up a new area of development. These business needs were able to be addressed by particular organizational processes forming in the wake of a particular event, all harnessed through existing precedents and control structures. These new processes, combined with a palette of existing and new technological capabilities and focuses, led these counter-trend technical directions to be the company's dominant path. This development was then sustained financially with the aid of longtime strategic partners whose needs the technology supported. This chapter will take you through how the Wolfram|Alpha organization could and did make startlingly original choices.

11.8   The Business Context of Wolfram|Alpha

Wolfram|Alpha came along at a key time for what employees then called Wolfram Research. At that point, the conglomerate was clearly one company and had long sold one flagship product, *Mathematica*, a comprehensive platform for technical computing. Originating about 20 years before W|A, the product made a huge splash, offering new heights in symbolic mathematics, function visualization, and interactive technical document creation. In the next 20 years, growth in the company was synonymous with growth in this product, which holds a comfortable niche with competitors *Matlab* and *Maple*. It was not cheap, but you got what you payed for, backed by a dedicated group of mathematicians and computer scientists bent on integrating as much technical computing as they could into a single unified language and experience.

With such a mature product, one has to be very vigilant for competitors, particularly the ones that might initially be described as inferior. Clayton Christensen's popular theory of disruptive innovation [1, 2] posits that many products are undermined not by superior competitors, but inferior competitors that serve customers in a different way. First, when an inferior competing product is launched, the company with the superior one does not worry too much, as very few users of the new product would buy the superior product anyway. Often, the inferior product has no real competition in its submarket. Slowly, the inferior product improves, taking progressively larger shares of the overall market, but at no time does it make sense for the superior product's company to make a cheaper offering, as its own higher-margin users would buy the lower margin product, eating away at that company's own profits worse than by its competitors. Eventually, the competing product can become good enough that whatever new need it was created to fill (say a program to run on mobile devices instead of desktop programs, or a disk drive smaller in physical size) will tip the overall balance of features would tip to the competitor, leaving the once-superior product's company to be playing catch-up in a new technological substrate. Often, there is little sign that such a disruption is taking place until it has occurred, making the theory critical to recognizing potential trouble.

Mathematica was starting to show signs of being a candidate for disruption. University students often preferred scientific calculators to software programs, given their feasibility for individual classroom use and their not requiring any additional hardware except batteries. The Internet and open-source licenses meant that special-purpose university-written languages such as GAP were viable alternatives for mathematicians in particular fields. Perhaps most daunting of all, open-source general-purpose languages, such as Python, with easily extended library systems were starting to sprout numerical and symbolic computing extensions, such as numPy, that were nowhere near as powerful as *Mathematica* but were practical for some applications and continually gaining in capability. The capabilities of *Mathematica* over-served the needs of these products' users.

In the Wolfram group's business landscape, Wolfram|Alpha functions as a disruptor to potential disruptors. The Internet could now generally be assumed, and smartphone technology was progressively widespread. A web page or downloadable app that did everything the calculator could, and more, meant upstaging the calculator's market. This sort of system also meant that instead of writing their own software, people faced with complicated technical problems could enter them in plain text and receive not only an answer, but all kinds of useful related information, such as the steps by which such problems might be solved. Generally speaking, the users who would have bought Mathematica to undertake these relatively simple tasks would be few, but W|A represents direct competition for *Mathematica*'s disruptors.

The icing on the cake was that Wolfram|Alpha would not only disrupt certain kinds of competition, but could provide natural language technology for the Wolfram Language, *Mathematica,* and other products. This would allow one to write Wolfram Language programs that used natural language phrases rather than code in sections. In other words, natural language technology would form a different basis for competition that no competitors would be prepared to follow, allowing it to act as a disruptive presence in new areas.

11.9   An Organization that Made Wolfram|Alpha Possible

Any number of firms recognize that they have problems with disruption, but most find it nearly impossible to counter due to existing organizational commitments. R&D staff are busy creating new incremental developments that retain and grow the company's largest-margin customers. Sales staff are more motivated to pursue sales that have larger margins. How was Wolfram Research able to build Wolfram|Alpha and become the Wolfram group after 20 years of pursuing a single platform?

The true secret behind the organizational capacity to create Wolfram|Alpha was the book *A New Kind of Science,* and its accompanying research program, both known by the abbreviation NKS. The impact of NKS was not only, or even mainly, its ideas but instead its human-resource effect on Wolfram Research. This work attracted a great number of very bright people eager to build something completely new and demonstrate the power of what they were working on. Though highly praised and hugely influential on some, the book also received neutral and negative reviews. With the book so new and not immediately adopted by skeptical and already-preoccupied professors, young physicists wanting to get involved were not immediately swallowed up by other research programs, but turned to Wolfram Research itself. It was in NKS summer schools that many key developers for W|A were discovered.

These individuals were often directed immediately into special projects or ambiguous research positions instead of fixed roles with binding job descriptions. This fluidity meant that a large part of the W|A work force was not in specific roles, but was ready to be mobilized. Further, even for *Mathematica*, it was not uncommon to organize the developing functionality of future releases through a portfolio of projects allowing a robustness to one or the other not working out. These flexibilities were critical to providing the necessary staff.

With this new influx and its corresponding projects, there was privately a fair amount of concern from longer-term company developers about this new project, with *Mathematica*-related work seeming the more stable option. However, if there was any direct budgetary competition between the company's initiatives, it was well hidden. This centralization of fiscal discretion was critical for fostering Wolfram| Alpha's disruptive presence. An important part of that fostering was equal access to a number of key services, including release engineering, systems administration, accounting, human resources, and other corporate components that allowed the internal start-up to begin with many institutional problems already solved.

It must be said that, even prior to the influx of NKS-influenced physicists and mathematicians, Wolfram Research staff had long had a variety of backgrounds other than computer science. Of course, given *Mathematica*'s initial framing as a tool for mathematics, mathematicians were strongly present. Physics, itself intensively requiring technical computing, has also been well represented. The base of Champaign, Illinois, home of the University of Illinois at Urbana-Champaign (UIUC) is a highly-regarded engineering program drawing its student population heavily from Chicago. As is inherent to businesses in college towns, it attracted a variety of people who came to teach and study but then moved on to something else.

As an example of the limited way in which those with conventional backgrounds played a role in W|A, a content director met a new developer in the hallway. He asked him what he had studied in school. "Computer Science" was the reply. Hmm, not physics, nuclear engineering, mathematics, or chemistry? How rare? Was he perhaps one of the few hired to work with the data frameworks? And so he was.

As another example, a key Wolfram|Alpha engineer who had been a string theorist was initially given the task of working on text-processing functions, known as string manipulation. This might sound like a bad game of telephone gone wrong, but it was entirely in tune with the organization's strategy of turning smart people loose on comparatively simple problems.

This wave of new researchers produced a tremendous amount of technology, but it might never have been harnessed for the practical application of Wolfram|Alpha if not for another aspect of Wolfram Research, namely its history of hiring librarians. UIUC is not only highly ranked in engineering but is a top-ranking library school. The company had long had a company library and actual librarians, who would sometimes take on new roles. As the Alpha project progressively moved from its early phases to needing a definite conclusion, the company turned to an executive known for her ability to close difficult projects, who had happened to be the company's first librarian.

The gradual hiring of a department of library professionals and domain experts marked the institutional gestation of Wolfram|Alpha. These "data curators" worked in tandem with developers to shape and clean the data, and to give the phrases that would designate that particular terms of a query should evaluate to a particular term of a particular kind of function. Slowly, the group of developers improvising their tools to get things done bifurcated into "content developers," pursuing new domains and "framework developers," who added new development and curation tools, as well as new runtime features.

The establishment of a data curation department was the first step in Wolfram| Alpha settling down into a more stable organization. Content managers specializing in different kinds of knowledge brought a unified view to mathematical or socio-economic topics. Standard review processes were set up, with upper management participating only at key points in the beginning and end of the process. A systematic, frequent, and doggedly pursued release cycle provided an internal clock that assured new features were carefully engineered to be integrated quickly and that severe bugs were not allowed to linger. A team engaged with the feedback of external users, making sure they received responses and closing the loop between domain development and user needs. Log analysis provided another mode of user feedback. The project was now being managed for incremental improvement, its immediate disruption having been manifested.

This form of organization led to the conventional way that Wolfram|Alpha works [3]. Whether through logs, feedback, commercial demand, internal use, discovery of a convenient data source, or other means, a realm is identified as a viable area for potential development. A developer is paired with a librarian, one of them typically having strong knowledge of the domain. They are given relevant query logs and, with the help of the content manager, imagine potential other queries users might submit for the new content domain. Together, they research myriad pertinent kinds of facts associated with the domain. Once those are decided upon, the developer focuses on structuring the data, evaluating relevant expressions, and presenting the results while the librarian vets facts and makes all viable expressions of the domain find the correct functions. If they come up against limits or questions in expressing what is desired, they collaborate with framework developers to engineer what is needed. Depending upon the domain and the nature of available sources, the content developers and librarians can employ a great deal of automation in preparing their work, developing more automation themselves when appropriate. Therefore, advances in any technology that might help them, such as improvements in artificial intelligence, will either first develop on a domain-specific basis, become readily provided through existing frameworks, or appear through changes in the Wolfram Language itself, and thus not seem disruptive but part of the regular workflow. The establishment of W|A's organization is not limited to development, but also to how domains are reviewed, how current events from news or the internet (such as change in a country's political leadership) are quickly incorporated, how the site is monitored, how quality is assured, how consulting projects are created, and any number of other matters that allow Alpha to continue to run smoothly.

The combination of NKS-driven physicists and library professionals meant that there was an absence of computer scientists or startup-interested business professionals, an unusual climate for a startup-like setting. This meant that Web 2.0's "architectures of participation" were not an ideological requirement. Instead, the proclivities stereotypically associated with these disciplines, a personal control over correctness and organization, were actually fully manifested by the team, institutionalized rather than outsourced. This separation from technical trends also meant that Alpha's technology was path dependent upon then Wolfram Research's existing technology and its surrounding norms, which we will turn to now.

11.10   Forming the Technology that Forms Us

The dual origin of Wolfram|Alpha technology is clear: in "computable" data and in projects to support flexible calculators. By version 6, Mathematica had embedded computable data for a variety of domains. What is meant by "computable data"? Technical computing often needs access to mathematical and physical constants, and basic socio-economic uses benefit from having commonly used data and statistics available. The question was how to take on those requirements with the Wolfram Language's "everything-included" maximalist philosophy. The answer was to allow each domain of data correspond to its own function, each with a common design to aid in documentation and overall usability. These functions could be set up to yield different results at different times, and not evaluate if given unbound variables, keeping the result of asking for data "symbolic".

Even earlier, the need for flexible calculators was becoming clear. How could people calculate an integral without either using a typeset entry palette or remembering the order of Integrate's arguments? If users could get a text box that corresponded to an integral, there would probably be a relatively fixed number of ways they might try to enter what they wanted to integrate, and those ways would be easy to pick out. From there, the thought was maybe it would be easy enough to tell what somebody typed in for an integral from what was typed in for a derivative, or other mathematical functions. This thought extended "Integrate" to "Calculate", after which project name changes would be more for external than internal communication.

The idea and then pursuit of being able to include computable data in free text calculations is an example of scope creep elevated to a software-engineering methodology. The conversion of "Integrate x^2 from one to five" to Integrate[x^2, {x,1,5}] is perhaps harder than converting "population of france" into CountryData ["France","Population"].

Though it may seem clear that Wolfram|Alpha is, from an implementation perspective, a direct descendent of these computable-data and free-text-entry projects, why was computable data pursued in the first place? Why should a symbolic CountryData[country,"Population"] be considered valid source code? The ethic of allowing everything to be symbolic stems from the Wolfram Language's roots in handling symbolic mathematics. Every result is subject to the simplification of term-rewriting and if not resolved is left unevaluated for future substitutions if at all possible. The early Wolfram Language was actually a term-rewriting system, with more traditional language constructs like lexical scope appearing later. As we'll see later, the idea of an entirely symbolic language has been taken yet further.

With mathematics and data in place, it became clear that Wolfram|Alpha would need other capabilities, including data on formulas and how to compute them, along with units such as length and time. An early uncanny result of Wolfram|Alpha's integration between data, units, and mathematics occurred when a software quality engineer tried the query "cubic lightyear of jello" and it worked! (See Fig. 11.7).

How could this complex interaction between computable data and free text work, given the variety of different processing needs for different kind of data? How was Wolfram|Alpha able to take advantage of domain-expert developers without much

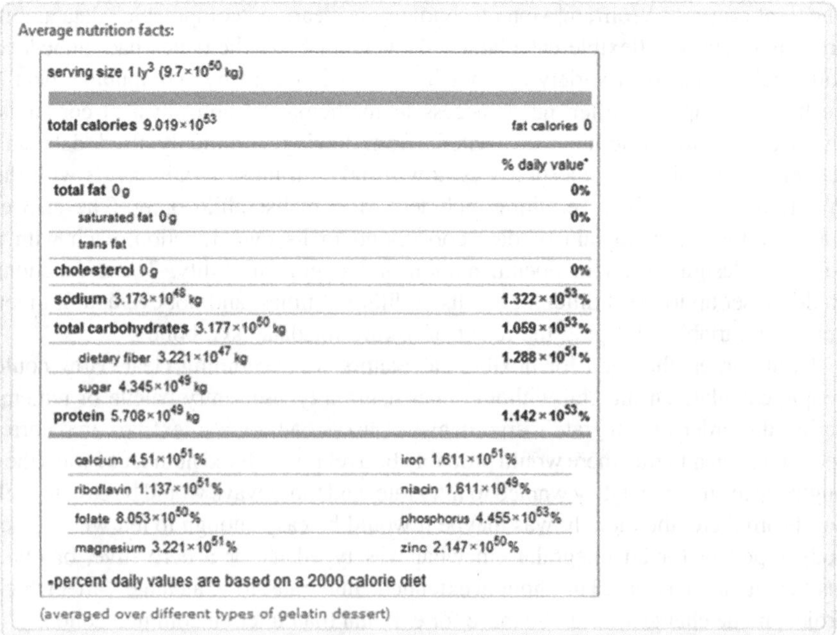

Average nutrition facts:

serving size 1 ly^3 (9.7×10^{50} kg)

total calories 9.019×10^{53}	fat calories 0

	% daily value*
total fat 0 g	0%
saturated fat 0 g	0%
trans fat	
cholesterol 0 g	0%
sodium 3.173×10^{48} kg	1.322×10^{53}%
total carbohydrates 3.177×10^{50} kg	1.059×10^{53}%
dietary fiber 3.221×10^{47} kg	1.288×10^{51}%
sugar 4.345×10^{49} kg	
protein 5.708×10^{49} kg	1.142×10^{53}%

calcium 4.51×10^{51}%	iron 1.611×10^{51}%
riboflavin 1.137×10^{51}%	niacin 1.611×10^{49}%
folate 8.053×10^{50}%	phosphorus 4.455×10^{53}%
magnesium 3.221×10^{51}%	zinc 2.147×10^{50}%

•percent daily values are based on a 2000 calorie diet

(averaged over different types of gelatin dessert)

Fig. 11.7 Wolfram|Alpha "cubic lightyear of jello query rendered in *Mathematica*

particular training in databases or information retrieval? The secret was to extend the designed commonality across various data functions like CountryData and ElementData all the all the way down into a database of common modular parts, easily translated from Wolfram Language statements. Like tactics in a videogame based on invading territory, W|A's data infrastructure was built in simple units that could be assembled in giant swarms to speed the initial rush of product release. It was only after this technology was deployed that the team pursued more refined platforms with greater expressiveness but more specialized uses. To this day, the simple, modular system survives, with its integration into standard toolkits and its terminology forming the dominant metaphors for developers.

These structures also created a standard metaphor that colored the kind of interaction the organization took on with unstructured data. Wolfram|Alpha didn't need to automatically understand messy data. The question for the team was always this: Can we get the data into forms we want to work with relatively simple tools? Dealing with unstructured data was a problem handled by strategic assessment rather than raw processing ability.

11.11 Partnerships: "Help from Our Friends"

Though Wolfram|Alpha's role in Wolfram's competitive landscape was clear, what wasn't exactly clear was how it was going to make money. Of course, there was advertising on the website itself. The company offers a paid Wolfram|Alpha Pro version that can analyze the users' own data, as well as calculate step-by-step breakdowns of mathematics and chemistry problems, and process user images, among other features continually being added. Another approach has been to create appliances that allow users to query their own data with natural language, and Wolfram|Alpha servers for institutions not wishing to enter their queries over the public Internet. An API lets users to put W|A to their own uses. Widgets permit content producers to embed purpose-specific information and visualization in their contents.

All of these have contributed to Wolfram|Alpha, but partnerships have been the dominant means of support. Unquestionably, the use of W|A within Siri has been W|A's broadest exposure. The relationship between Wolfram companies and Apple goes back to Steve Jobs having suggested the name *Mathematica*. And in Wolfram| Alpha's early years, Microsoft played a crucial role in Wolfram|Alpha, at times integrating W|A with Bing. Overall, these partnerships enabled Wolfram|Alpha transitioning from an emerging to an established technology, through which we see the cycle from strategic need and organizational mobilization to a mature platform completed.

Now that we have seen the transition to a contemporary Wolfram|Alpha, let us look into today's W|A itself and see some of what it does and is for.

11.12 What Wolfram|Alpha Does, by Example

When should you use Wolfram|Alpha? One answer is when you don't know if the content is out there, and you need facts based on the content you have. Another is when you have the facts, but need a summary. Yet another is to get a sense of proportion or scale. W|A is also useful when you want a clean view of just the information. It can also be fun to try the service just to see what it will do. Let us illustrate these points by looking into a few of the many queries Wolfram|Alpha could address. This is necessarily only a small sample; you might prefer to visit the gallery of examples at http://www.wolframalpha.com/examples/.

11.13 Example: Food

Food is one area that combines facts with mathematics. Oftentimes, people want to understand what they are eating without having to be too particular or do the math themselves. If you just happen to want, say, a hamburger and fries at some random cafeteria, W|A can create an approximate nutrition label (see Fig. 11.8).

People often want to go the other way, too, where they have some nutritional goal and want to find the foods that help get there. For instance, a person with an iron deficiency might want to know which foods have the most iron (see Fig. 11.9).

That's great, but what does this mean? We can also ask much iron a person needs per day (see Fig. 11.10).

11.14 Example: Automobiles

Wolfram|Alpha can also offer capable summaries. Automobiles are a good example for this. Let's start with the most basic case, a single car, let's say the "Ford C-Max Hybrid". Wolfram|Alpha provides an image and a summary (see Fig. 11.11) as well as breakdowns for price, fuel efficiency, engine and transmission information, interior and exterior dimensions and available colors, safety ratings, warranty provisions, awards, features, available rebates, and Wikipedia hit history.

An overview of a particular car can be quite useful, but often we are interested in comparing the merits of one car against another. Let's compare "Ford C-Max Hybrid versus Toyota Prius", this time excerpting (Fig. 11.12) from down the page where we can see price and fuel-efficiency comparisons. We can see that the C-Max is modestly more affordable while the Prius has superior fuel economy.

In addition to model-by-model comparisons, we can compare automotive divisions as a whole, such as "Ford versus Toyota", to really get a sense of the differences between the two brands. In addition to the body style, price, and

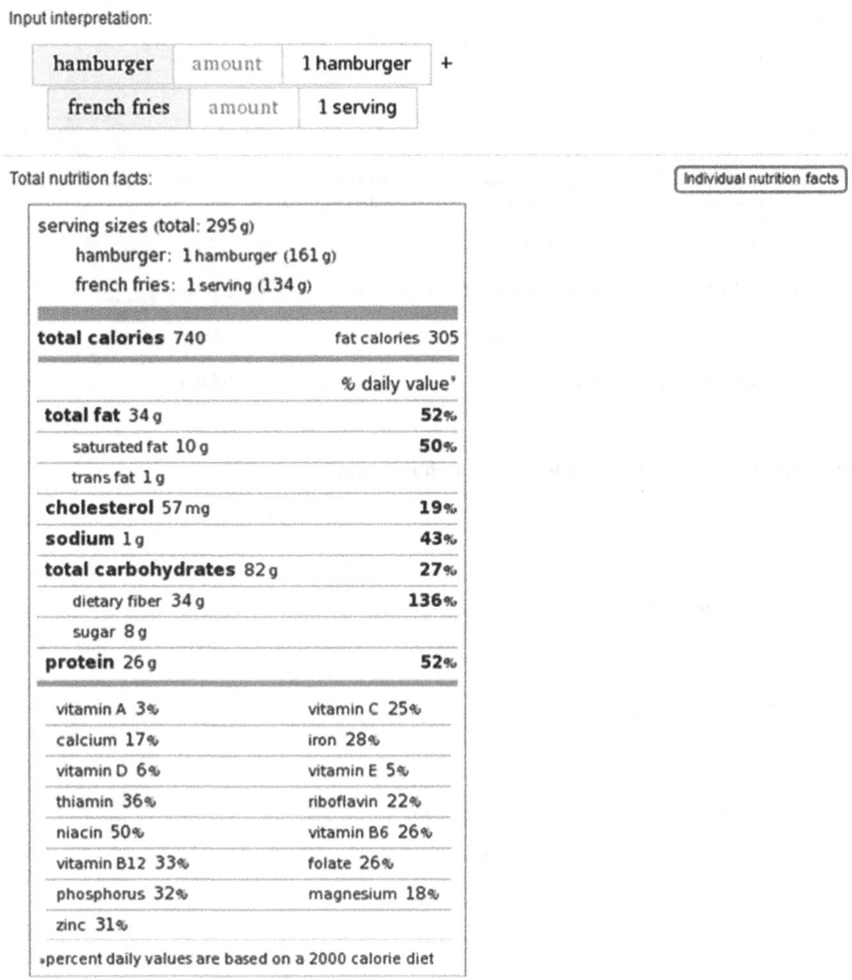

Input interpretation:

hamburger	amount	1 hamburger	+
french fries	amount	1 serving	

Total nutrition facts:

Individual nutrition facts

serving sizes (total: 295 g)
   hamburger: 1 hamburger (161 g)
   french fries: 1 serving (134 g)

total calories 740	fat calories 305

	% daily value*
total fat 34 g	**52%**
saturated fat 10 g	**50%**
trans fat 1 g	
cholesterol 57 mg	**19%**
sodium 1 g	**43%**
total carbohydrates 82 g	**27%**
dietary fiber 34 g	**136%**
sugar 8 g	
protein 26 g	**52%**

vitamin A 3%	vitamin C 25%
calcium 17%	iron 28%
vitamin D 6%	vitamin E 5%
thiamin 36%	riboflavin 22%
niacin 50%	vitamin B6 26%
vitamin B12 33%	folate 26%
phosphorus 32%	magnesium 18%
zinc 31%	

*percent daily values are based on a 2000 calorie diet

Fig. 11.8 Partial Wolfram|Alpha result for "hamburger and fries"

fuel-economy comparisons shown in Fig. 11.13, W|A also compares the engine volumes and physical dimensions of the companies' vehicles.

11.15 Example: Sports and Games

A different area with a lot of possibilities for quantitative comparison is sports and games. We might be interested in extreme events, such as "2014 MLB game with most hits", excerpted below (see Fig. 11.14).

Input interpretation:

most foods	by iron

Result: [More]

1	beef, shoulder pot roast or steak, boneless, braised, choice, cooked, separable lean and fat, trimmed to 0" fat	3 mg	
2	Naked Juice red machine, generic	2.7 mg	
3	Naked Juice protein zone double berry, generic	2.7 mg	
4	Naked Juice berry vegetable machine, generic	2.2 mg	
5	Naked Juice protein zone, generic	1.8 mg	

Units »

Fig. 11.9 Wolfram|Alpha result for "foods with most iron"

Input interpretation:

iron	recommended daily allowance

Result:

18 mg/day (milligrams per day)

Reference daily intakes:

recommended daily allowance	18 mg/day
dietary reference intake (female)	18 mg/day
dietary reference intake (male)	8 mg/day

Units »

Unit conversions: [More]

0.018 g/day (grams per day)

2.083×10^{-10} kg/s (kilograms per second)

12.5 µg/min (micrograms per minute)

0.0125 mg/min (milligrams per minute)

12500 ng/min (nanograms per minute)

Fig. 11.10 Wolfram|Alpha result for "daily recommended iron"

We might also be interested in, say, understanding the total scale or variety in a particular game. Consider "(Most powerful pokemon hit points)/(least powerful pokemon hit points)", which offers one window into how balanced the various

Input interpretation:

2015 Ford C-Max Hybrid

Image:

Basic information:

division	Ford
type	hatchback | sedan
price	$24170 to $27170 (MSRP) | $22540 to $25340 (invoice)
mileage	42 mpg (city) | 37 mpg (highway)
passenger doors	4 doors
passenger capacity	5 people
engine type	I4 hybrid
overall crash test rating	★★★★
basic warranty	36 000 miles | 3 years
trims	SE | SEL

Definitions »

Units »

Fig. 11.11 Partial Wolfram|Alpha result for "Ford C-Max Hybrid"

characters in the Pokémon game are. This result is very direct, with Fig. 11.15 showing the complete output for this query.

Wolfram|Alpha's information on games is not just confined to understanding their attributes—it can occasionally offer some direct assistance. Consider how "_al__la__" is interpreted as a request for those English words that can fill the implied blanks of the underscore, for help with crossword puzzles, hangman, and the like. Figure 11.16 shows the complete result.

Fig. 11.12 Price and fuel-efficiency comparisons from "Ford C-Max Hybrid versus Toyota Prius" Wolfram|Alpha query

11.16 Example: Places

In addition to offering quantitative comparisons between things and information on events, Wolfram|Alpha also offers geographic, economic, social, and meteorological insight into places (and not just on Earth). A friend of mine is now living in Santa Rosa in California, so I decided to compare Santa Rosa with Champaign, Illinois. "Champaign, Santa Rosa" returns some basic facts about the two cities, including the segment shown in Fig. 11.17.

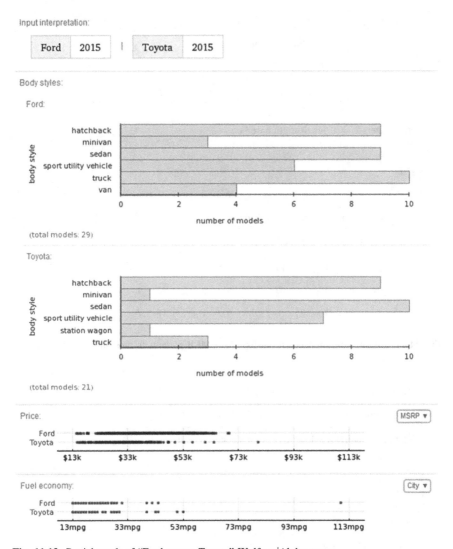

Fig. 11.13 Partial result of "Ford versus Toyota" Wolfram|Alpha query

You might suspect the cost of living would be sharply higher in Santa Rosa, and "Champaign versus Santa Rosa cost of living" reveals that is true, as shown in Fig. 11.18.

Among the facts returned by a "Champaign to Santa Rosa relocation" query, we see (Fig. 11.19) the difference in some commonly incurred costs.

However, that same query also reveals that Champaign's crime rates and sales tax are higher (see Fig. 11.20).

Input interpretation:

MLB game	with	2014 season
	by	most combined hits

Definition »

Result:

Detroit Tigers (63–52) at **Toronto Blue Jays** (63–56) (August 10, 2014)
(39)

Box score:

	1	2	3	4	5	6	7	8	9	10	11
Det	3	0	1	1	0	0	0	0	0	0	0
Tor	0	0	0	0	0	2	2	0	1	0	0

	12	13	14	15	16	17	18	19	R	H	E
Det	0	0	0	0	0	0	0	0	5	22	1
Tor	0	0	0	0	0	0	0	1	6	17	2

Fig. 11.14 Partial Wolfram|Alpha result for "2014 MLB game with most hits"

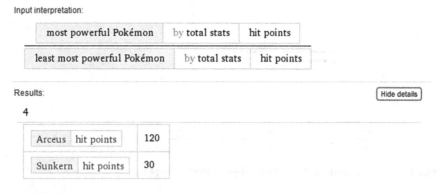

Input interpretation:

most powerful Pokémon	by total stats	hit points
least most powerful Pokémon	by total stats	hit points

Results: [Hide details]

4

Arceus	hit points	120
Sunkern	hit points	30

Fig. 11.15 Wolfram|Alpha result for "(Most powerful pokemon hit points)/(least powerful pokemon hit points)"

11.17 Example: Relations

The kinds of practical relations that can be discovered computationally are not restricted to numerical ones. Consider the result of "grandfather's sister's daughter", which finds the name of the relation considered by this path (see Fig. 11.21).

Fig. 11.16 Wolfram|Alpha
result for "_al__la__"

Input

$$\_ a l \_ \_ l a \_ \_$$

English words:

| balaclava |
| calculary |
| calculate |
| falculate |
| maltolate |
| malvalate |
| saleslady |

(7 words)

11.18   Example: Humor

A completely different reason for using Wolfram|Alpha is for the fun of seeing how it will respond. Any technology that answers general questions will be asked for cultural references, such as "What is the answer to the question of life, the universe, and everything?" or "What is the airspeed velocity of an unladen swallow?" Once W|A successfully responds to such questions, there is naturally an interest to see whether it will play along with requests to "Open the pod bay doors" and the like, but perhaps also to see how it will answer personal questions. For the system to play along with cultural references is shows that there are people willing to have fun behind the curtain, and do so with a certain spirit. The query "tell me a joke" usually returns a mathematical one, such as the one shown in Fig. 11.22. Notice the parenthetical "my favorite sub-genre" that attributes a personality to W|A.

There are many more possibilities. The example page lists Mathematics, Words & Linguistics, Units & Measures, Step-by-step Solutions, Statistics & Data Analysis, People & History, Dates & Times, Data Input, Chemistry, Culture & Media, Money & Finance, Image Analysis, Physics, Art & Design, Socioeconomic Data, File Upload, Astronomy, Music, Health & Medicine, CDF Interactivity, Engineering, Places & Geography, Food & Nutrition, Education, Materials, Earth Science, Shopping, Organizations, Life Sciences, Weather & Meteorology, The Technological World, Sports & Games, Computational Sciences, Transportation, Web & Computer Systems, and Surprises as further categories to explore. Visit

Populations:

	Champaign	Santa Rosa
city population	84 513 people (country rank: ≈ 403rd) (2014)	174 170 people (country rank: ≈ 145th) (2014)
urban area population	123 938 people (Champaign– Urbana urban area) (country rank: 215th) (2000)	285 408 people (Santa Rosa (CA) urban area) (country rank: 111th) (2000)
metro area population	232 336 people (Champaign- Urbana metro area) (country rank: 192nd) (2011)	488 116 people (Santa Rosa-Petaluma metro area) (country rank: 104th) (2011)

Path: [Show coordinates]

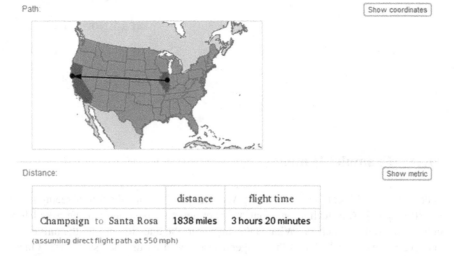

Distance: [Show metric]

	distance	flight time
Champaign to Santa Rosa	1838 miles	3 hours 20 minutes

(assuming direct flight path at 550 mph)

Fig. 11.17 Partial Wolfram|Alpha result for "Champaign, Santa Rosa"

http://www.wolframalpha.com/examples to see the ever-increasing diversity that we can only begin to sample in this chapter.

11.19 What Wolfram|Alpha Cannot yet Do

Yet despite all that might be within Wolfram|Alpha's conventional ability to do, it pragmatically does only what people frequently ask for, and might be expanded at any time. "All former living Presidents of the United States" is a query that does not work at the time of this writing, but W|A has all the development and data capabilities to make it do so if it were sufficiently desired. "2 drinks in one hour for an average US weight male" is another such query that Wolfram|Alpha could be made

Results:

Champaign	**98.7** (Q3 2014)
Santa Rosa	**137.9** (Q2 1997)

(100 = US national average)

Relative values:

	visual	ratios		comparisons
Santa Rosa		**1.397**	1	**39.72%** larger
Champaign		1	**0.7157**	**28.43%** smaller

Cost of living indices:

	Champaign	Santa Rosa
grocery (13.48% of total)	**100.5**	**117.5**
healthcare (4.89% of total)	**117.8**	**135.9**
housing (26.05% of total)	**89.6**	**196.6**
miscellaneous (33.01% of total)	**98.2**	**111.9**
transportation (12.63% of total)	**96.6**	**122.4**
utilities (9.95% of total)	**114.7**	**99.3**

(Q3 2014; 100 = US national average)

Fig. 11.18 W|A result for W|A "Champaign versus Santa Rosa cost of living"

to solve, though notice that the input has the subtle requirement to determine the average weight excluding those below the legal US drinking age. The available content is driven by user demand as well as the availability of the data. Automobile data is a relatively recent addition, and no doubt if a sizable number of boat enthusiasts tried to use Wolfram|Alpha to compare boats, such vehicles would also be added.

Let us now talk about some functionality that is outside of Wolfram|Alpha's scope. The input length of queries to Alpha is deliberately limited. It does not currently include the ability to enter text and answer semantic questions revealed

Average retail prices: [Show prices] [More]

ground beef (1 lb)	**1.39** × Champaign
whole milk (0.5 gal)	**1.03** × Champaign
white bread (24 oz)	**1.39** × Champaign
apartment rent (2 BR)	**1.33** × Champaign
mortgage rate (30 yr fixed)	**1.02** × Champaign
total energy	**0.883** × Champaign
doctor visit (routine)	**1.08** × Champaign
gasoline (1 gal + tax)	**1.15** × Champaign

(Q2 1997)

Fig. 11.19 Average retail prices from W|A "Champaign to Santa Rosa relocation" result

Crime rates: [Show details] [Show chart]

property crime total	**60% less** in Santa Rosa
violent crime total	**36% less** in Santa Rosa
total rate of crime	**58% less** in Santa Rosa

(UCR Part I offenses)
(1992 and 2009 estimates)

Sales taxes: [Show breakdown]

	Champaign, Illinois	Santa Rosa, California
total sales tax rate	9%	8.75%

Fig. 11.20 Crime rate and sales tax information from W|A "Champaign to Santa Rosa relocation" result

from it. Nor does Alpha support ongoing interactivity, such as questions involving pronouns or other indexical references to previous queries or responses.

Right now, Wolfram|Alpha does not automatically extract facts from unstructured data, such as documents. Even simple structures, like tables, present imponderable difficulties. For one thing, documents can refer to any time, while Wolfram|Alpha queries are evaluated at the present time. Consider a document titled "Guest List" that has the entry "All of the living former Presidents of the United States". W|A might be able to determine the current list of people, but what if this guest list is historical, or refers to a future event? It would not necessarily be appropriate to resolve this entry's phrase to a particular present-day list of people. As another example, consider a table of material properties with an entry of "all

Input interpretation:

> **genealogic relation**   grandfather's sister's daughter

Genealogic tree:

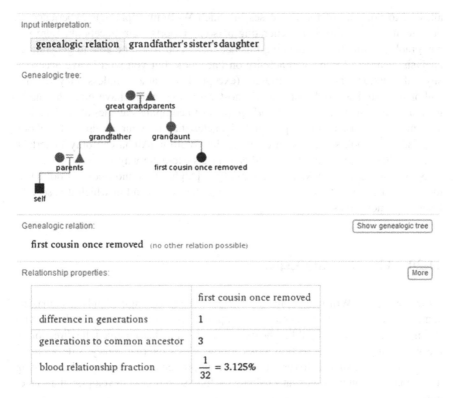

Genealogic relation: [Show genealogic tree]

first cousin once removed (no other relation possible)

Relationship properties: [More]

	first cousin once removed
difference in generations	1
generations to common ancestor	3
blood relationship fraction	$\dfrac{1}{32}$ = 3.125%

Fig. 11.21 Wolfram|Alpha result for "grandfather's sister's daughter"

Input interpretation:

Tell me a joke.

Result:

4/3 of people don't understand fractions.

(according to what passes for mathematical humor (my favorite sub–genre), drawn from several
sources but primarily from P. Renteln and A. Dundes in their paper "Sampling of Mathematical Folk
Humor" in Notices of the American Mathematical Society, vol. 52, pp. 24–34, 2005)

Fig. 11.22 One of the W|A results for "tell me a joke"

plastics not otherwise mentioned". What this phrase may refer to depends on a lot
of domain expertise likely not available to any general-purpose knowledge
extraction method.

As we will see later, the right way to tackle these issues might not always be
further development for Wolfram|Alpha, but further enhancing Wolfram|Alpha so it
can participate as a component in other projects.

In addition to what Wolfram|Alpha happens not to do and that which is extre-
mely difficult for Wolfram|Alpha to ever do, there is what Wolfram|Alpha is not

allowed to do. One of the few cases in which Wolfram|Alpha has experienced a decline in functionality is in generating personal reports from Facebook. Here is a paragraph from a blog announcing this decline: "You'll still be able to generate an analysis of most of your own activity on Facebook, but you won't have access to any information about your friends (except their names) unless they've also authorized our Facebook app. So in most cases, we won't have enough data to generate a meaningful friend network graph, or to compute statistics about location, age, marital status, or other personal characteristics of your group of Facebook friends." Facebook serves as an example that Wolfram|Alpha can only undertake domains that people allow to be gathered in a systematic way.

Now that we have looked at what Wolfram|Alpha does and does not do, let us look to the new directions that Wolfram|Alpha has led to and in which it is serving fascinating new roles.

11.20   The Ongoing Legacy

As we have seen, Wolfram|Alpha has changed the search-engine landscape, making summaries and answers an increasingly important part of search results. Along with its impact on the general public, the existence of Alpha has changed the direction of the Wolfram group. This section looks at how W|A's internal legacy is developing, not resting on feedback driven incremental improvement alone, but also creating new modes of interaction that challenge Alpha's technical development in categorical ways.

11.21   New Initiatives

Overall, W|A Pro led to an experience of finding a new market for Wolfram Language services, which has led to Wolfram building new cloud, web, and mobile offerings aimed at different software needs with different means of collecting revenue. The idea of developers not wanting to the hosting responsibilities of webMathematica to deliver their applications has led to the Wolfram Programming Cloud, offering not just "pay for full-functioning desktop use" but also "pay as you go for application resource usage". The Wolfram Data Science Platform will be aimed at supporting data analysis workflows. Many other such products are in the pipeline, each aimed at serving a distinct customer profile.

The experience of finding a new market has led to other changes. Wolfram| Alpha, Wolfram Finance Platform, and Wolfram *SystemModeler* have brought new customer bases which give the company a new chance to tell its story. These new audiences have led to *Mathematica*'s built-in language, itself initially called *Mathematica*, being revised and rebranded as the Wolfram Language.

11.22 A Language with Reference to the World

However, the most dramatic change to the Wolfram Language as a result of Wolfram|Alpha happened around *Mathematica* 10: the idea of embedding named references to entities in the world in a programming language. Consider the ability in Mathematica to ask for the Wolfram Language expression that returns entities for all the planets, as shown in Fig. 11.23.

What this means is that the representation of a planet is a first-class citizen of the language, in the same degree as 1/5 being represented by Rational[1,5]. For a language audited carefully for conceptual consistency, what this amounts to is a Kripke-esce declaration of the necessity of names that puts Saturn (or France, or any number of other entities) on par with True and 5. From a practical perspective, this expansion means that users can do analysis across a domain, for example not just obtaining the release date and box office for a particular movie, but for all movies. These new requirements require new modes of representation and new infrastructure to support them, modes that cannot help but be suggestive of new Alpha functionality.

11.23 Finding the Semantic Meaning of Unstructured Data

Now that the Wolfram Language has semantic elements in direct reference to things in the world, it has a working metaphor for dealing with unstructured data. New Wolfram Language functions such as SemanticInterpretation, Interpreter, and SemanticImport all work within different internal constraints and usage scenarios for a common end: to produce these semantic annotations in an appropriately relational form, such as the new Dataset, from user data. Of course, users will want more than what Alpha has curated, and can provide their own symbolic item tokens as desired. This may be the future of Wolfram|Alpha development: to exploit the same means of markup available to particular markets of users.

Perhaps the most impressive extension of Wolfram|Alpha's technologies thus far to assist other projects in interpreting unstructured data has been addition of conceptual entities for the Wolfram Image Identification Project. For example, consider the task shown in Fig. 11.24 of identifying a picture of a sundae. W|A already had to know about sundaes for the purpose of giving nutritional information, but now its

```
In[2]:= EntityList["Planet"]

Out[2]= { Mercury , Venus , Earth , Mars , Jupiter , Saturn , Uranus , Neptune }
```

Fig. 11.23 *Mathematica* 10 listing of planet entities

Fig. 11.24 Wolfram Image Identification Project result

internal representation has been expanded to represent the "sundae" apart from its contexts as a food with nutritional content and as a word with a definition. This new computability is having explicit entities not only for proper things in the world, as demonstrated above, but for generic things. What we have seen by now is that this is but one more step in a long project of systematizing symbolic analogs to real-world phenomena in a continued fusion of librarianship with computation.

With this view into Wolfram|Alpha's continuing future, let us now sum up what we have covered in this chapter.

11.24   Summing up

We have now seen how Wolfram|Alpha is different from other search and answer technologies: it does not search through documents, but instead attempting to make ordered presentation of facts related to the function the query best seems to represent. These facts are organized by professionals in an attempt to attain similar saliency to that of a knowledgeable librarian. These differences aligned with a developing need in the Wolfram group's business strategy and were initially fostered by the combination of newly minted NKS-inspired employees and a tradition of librarianship, directed by existing technical choices, and supported by long-standing professional relationships. The queries in Wolfram|Alpha give factual answers to free text questions, providing context, proportion, and scale through comparisons and extremes, while still allowing for a sense of fun. The work of Wolfram|Alpha is not over, but is being extended to provide symbolic links to the world in tandem with new ventures.

In conclusion, Wolfram|Alpha is a people-in-the-loop organizational technology for the symbolic computation of worldly facts, developing both incrementally from the in-product demands as manifested through logs and sharply through the demands for new kinds of applications, aimed at providing specific answers to queries across a wide range of curatable domains. Hopefully you now have an understanding Wolfram|Alpha's niche in search and answer technologies and of its developing role in an ever-broadening range of approaches for bringing symbolic computation into the world.

References

To understand the role of disruptive innovation plays in business competition, consider reading Clayton Christensen's original books on the concept:

1. Christensen, C. M. (1997). *The Innovator's Dilemma: When New Technologies Cause Great Firms to Fail*. Harvard Business School Press, Boston, MA.
2. Christensen, C. M. and Raynor, M. E. (2003). *The Innovator's Solution: Creating and Sustaining Successful Growth Businesses*. Harvard Business School Press, Boston, MA.

To understand the sociological forces that organization formation imposes on information technology practice, consider reading this classic paper by Wanda Orlikowsky:

3. Orlikowski, W. (1992). The duality of technology: Rethinking the concept of structure in organizations. *Organization Science*, 3(3):398–427.

Author Biography

John B. Cassel worked with Wolfram|Alpha from its development, through its initial release, into its stabilization, and into the new horizons of integration. He maintains interests in real-time discovery, planning, and knowledge representation problems in natural systems management, risk governance, and engineering design. John holds a Master of Design Degree in Strategic Foresight and Innovation from OCADU, where he developed a novel research methodology for the risk governance of emerging technologies.

Chapter 12
Toward Seamless Access: Libraries from Alexandria Through the Digital Age

Barret Havens and Jennifer Rosenfeld

12.1 Historical Roots of Library Access

The advent of the internet has revolutionized the ways in which libraries serve their users by facilitating the expansion of collections and enhancing access to those collections. However, though the internet has had an unprecedented and profound impact on libraries, other innovations, also ground-breaking in their respective times, have set the stage for the development of the modern library by facilitating access to information. Among these developments are the invention of the Gutenberg printing press, the collocation of library materials by subject, and the use of assignment indexing.

12.2 Impact of the Gutenberg Printing Press

Since their inception 4500 years ago, libraries have strived to fulfill two functions that appear, on the surface, to be contradictory. On the one hand, they have sought to serve users by making information in its many forms as accessible as possible. On the other, they have needed, at times, to restrict access to information in order to preserve and protect it for future users and future generations. However, throughout history there has been an undeniable trend toward increasing user access to information as it has become easier to record or publish, and less expensive to acquire.

From ancient through Medieval times, publishing was incredibly meticulous work. In most parts of the world, information were recorded painstakingly by hand using media such as stone, clay tablets, papyrus, and animal skin. As one might imagine, texts produced by such arduous methods, many of them existing in very

B. Havens (✉) · J. Rosenfeld
Woodbury University, Burbank, USA
e-mail: barret.havens@woodbury.edu

© Springer Science+Business Media New York 2016
N. Lee (ed.), *Google It*, DOI 10.1007/978-1-4939-6415-4_12

limited quantity or even as one-of-a-kind specimens, were often treated as precious objects [1]. Accordingly, evidence points to the fact that some early libraries were Draconian in their role as guardians of their collections. Some of the earliest and most severe library rules, in the form of entreaties to the gods to punish irresponsible borrowers or thieves were inscribed on some of the clay tablets (see Fig. 12.1) kept in early Mesopotamian archive-libraries: "Whoever removes [the tablet]…may Ashur and Ninlil, angered and grim, cast him down, erase his name, his seed, in the land" [2]. "He who entrusts it to [other's] hands, may all the gods who are found in Babylon curse him!" [1]. Books in some medieval libraries were chained to furniture to prevent theft, but were cut from their bindings occasionally by persistent thieves [1].

Though there is little evidence to suggest that the borrowing of materials from ancient or Medieval libraries was permitted frequently, an inscription found in an ancient Athenian library states that the "directors had decided to eliminate borrowing," suggesting that it was allowed for a time [2] (Casson 107). Access to the collections of early libraries was typically limited to nobility, clergy, and scholars,

Fig. 12.1 Clay tablet recounting the tale of a battle between two gods, found at the site of the Assyrian civilization of Ninevah [3]

though once again, history provides exceptions such as the Roman bath libraries of Nero's era, which were open to all Romans, regardless of class, gender, or age [2].

Progressive printing techniques were introduced early in the Far East. Paper was in use as early as the Western Han dynasty (221–224 B.C.E.) and multiple copies of texts were printed using hand-carved wooden blocks, the earliest example of which dates back to 8th century Korea [4]. The Chinese had even experimented with movable type by the mid 11th century. However, movable type was hardly the game-changing breakthrough in China that it would be in Europe 400 years later, considering the expense and effort involved with producing stamps, or "types" for each of the thousands of characters of the Chinese language that might constitute a literary work [4].

According to library studies scholar Leila Avrin, "no historian believes seriously that Chinese printing directly inspired the European invention" [4]. However, paper did spread from China to Europe, albeit slowly, by way of Korea and Japan. According to an Arabic text dated 1482, paper was being made in the Islamic empire by the early eighth century [4]. It would take the next 600 years for papermaking as a technology to spread from Muslim Spain to Christians in Spain and then to much of the rest of Europe [4].

Papermaking reached Mainz, Germany in the 1320s [4]. In that same city, in approximately 1450, Johannes Gutenberg introduced a wooden hand-press that employed metal movable type, which was a more feasible prospect in Europe than in Asia given the relatively limited quantity of letters in the alphabets of Romance and Germanic languages. The effects of the Gutenberg press and successive versions of it on the availability of books were profound. In Europe before 1500, at most, a book might be available in one hundred copies and read by thousands of people [5]. After 1500, however, thousands of copies of a book could be available and could be read by hundreds of thousands of people [5]. The growth of European libraries during this period was enormous compared to the holdings of libraries during the Medieval period, partially as a result of the increased availability of books and the relative drop in their cost attributed to the Gutenberg press and successive versions of the device [5].

The holdings of college libraries in some cases expanded from under 1000 items to hundreds of thousands of items [1]. The availability of printed material, in turn, increased literacy rates and drove up the demand for books, which fueled the growth of the book trade [5]. Thus, the expansion of libraries was a direct result of the increase of supply and demand [1].

Books, though rare by today's standards, were no longer considered priceless. Consequently, libraries relaxed in terms of their role as guardians of information, expanding services to wider populations and allowing users greater access to materials. For example, Cardinals Richelieu and Mazarin, who served as chief French ministers, collected so many books that they hired a full-time librarian to organize the collection, which was open to "everybody" in 1661 and considered by many to be the "best library of the time" [1]. By the late 1600s, thirty-two Parisian libraries and 3 national 'public' libraries were accessible to general readers [1]. (However, French public libraries catered more to scholars than the public in terms of their collections until the early 1900s [1].) Around the same time in Britain, parish churches made small libraries available to the public [5].

The demand for a wide variety of reading material, including popular items such as novels, was high enough by the late 1700s to mid-1800s that people who lacked access to libraries were willing to pay for it. During this time in America and parts of Europe, subscription or dues-based access to collections at "social libraries" and commercial book rental services known as "circulating libraries" gained popularity [1]. However, by the mid-1800s public libraries had begun to expand and proliferate.

The first American free public library funded by taxes opened in Peterborough, New Hampshire in 1833 [1]. In Britain from 1847 to 1850, however, the history of the modern public library began in earnest when Parliament passed a series of acts that led to the establishment of tax-supported public libraries throughout the country. As a direct result, by 1900, 300 public libraries had been established [5]. Public libraries made significant strides in America, France, Germany and Japan in the mid to late 1800s, some enabled through legislation and others through charitable organizations such as the Franklin Society, as was the case in France [1]. However, the cause of library access received its most significant boost in the form of $56 million in funding by steel baron Andrew Carnegie, a Scottish immigrant who had made his fortune in the United States. In English-speaking countries throughout the world during the late 19th and early 20th centuries, more than 2509 libraries, many of them public (see Fig. 12.2), were established through Carnegie's philanthropy [6]. The chain-reaction started by Gutenberg's invention had rippled far and wide; libraries, and print-based information, were finally available to the masses.

Fig. 12.2 Carnegie Public Library (now Carnegie History Center) in Bryan, TX. Photo by Flickr.com user Edwin S. used under Creative Commons License

12.3   Collocation and Assignment Indexing

Another major breakthrough in terms of user access to library materials has come in the form of two organizational innovations which go hand-in-hand: collocation and assignment indexing. Collocation is the grouping together, whether in a catalog or physical collection, of materials by type. Modern libraries using the Dewey Decimal or Library of Congress classification systems achieve collocation by assigning a call number to each item, which is a precise code denoting where the item is to be shelved. Coded within that call number, typically, is the item's subject focus (astrophysics, for example), or genre (fiction, for instance). Beyond the first portion of the call number indicating the general subject focus or genre of a work, a further subdivision is often made by author's surname, geographic focus, or some other narrower category (see Fig. 12.3).

This arrangement maximizes the potential for serendipitous discovery while exploring a library collection or catalog, as a user setting out to retrieve a particular item may encounter a trove of items on their topic of interest located or listed nearby. Shelf collocation can be reproduced virtually in many online library catalogs through a call number search feature. As demonstrated in Fig. 12.4, by specifying a call number or range of call numbers, a group of records organized in call number order can be browsed virtually before going to the shelf (though some electronic resources will be listed only in the catalog since they cannot be shelved).

Collocation, though useful, poses a challenge for catalogers. Since an item cannot be in more than one place at one time, collocation requires a cataloger to decide on just one subject focus or genre for the purpose of locating an item with similar items. However, items may not be so easy to classify in terms of predicting how users might seek them. For instance, in Hypothetical University Library, an animated film such as "The Lorax," based on the book by Dr. Seuss, might be shelved with all other animated films under Library of Congress Classification system call number NC1766. Though this makes sense, a user might, quite logically, search specifically for films on the topic of conservation of natural resources. Though "The Lorax" addresses this theme, if the user were to browse the shelving

---

Book title: __Night Skies: For Unaccompanied Euphonium__
Author: Steven Winteregg
Call number of book in Library of Congress catalog: **M110 .B33 W 2007**

M110 – Instrumental music, solo instruments, wind instruments, "other" category
.B33 – euphonium
W – first initial of author last name
2007 – year of publication

Fig. 12.3 Anatomy of a Library of Congress call number

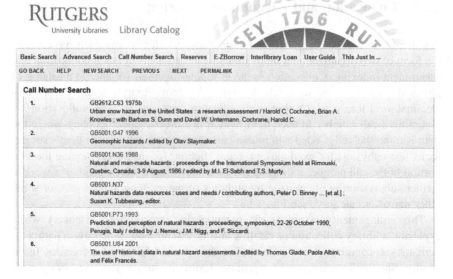

Fig. 12.4 Results of a call number range search targeting items on natural disasters demonstrating the collocation of items by subject, as they would be collocated on the shelf

area where films on conservation are located at Hypothetical University Library, they would clearly miss "The Lorax."

This is where assignment indexing comes in handy. Assignment indexing is the practice of "tagging" bibliographic records (in a modern online library catalog, a bibliographic record is a web page describing an item and providing its shelf location or, if it is an electronic resource, a link to its virtual location) with subject headings from a standardized list of descriptors such as the Library of Congress Subject Headings or Sears List in order to create multiple subject access points. According to the Online Dictionary for Library and Information Science, an access point is "a unit of information in a bibliographic record under which a person may search for and identify items…" [7]. Often times, catalogers will assign multiple subject access points in the form of subject headings to a bibliographic record in order to accommodate a variety of approaches to searching for an item. For example, a bibliographic record for "The Lorax" may, in addition to "conservation of natural resources–juvenile films" contain the subject heading "pollution–juvenile films" just in case users decide to search using the term "pollution" instead of "conservation of natural resources."

Subject headings are an example of a controlled vocabulary. By agreeing to use a controlled vocabulary, or standardized list of terms, in order to "tag" items, catalogers enable searching across multiple databases or library catalogs simultaneously. Since many libraries that own "The Lorax," for instance, are likely to use the pre-determined Library of Congress Subject Heading "conservation of natural resources" to index this, and similar items, it is possible to target these items with a subject search across the holdings of multiple libraries. In addition to enhancing access to items by allowing users to search for them in multiple ways, subject headings in a modern online library catalog enable hypertext cross-indexing. While viewing the bibliographic record for an item, such as "The Lorax," users may navigate to similar items within a catalog or database by clicking on the subject headings that are attached to that record. Clicking "pollution–juvenile films," for instance, would produce a list of items sharing that descriptor, such as "Bill Nye the Science Guy Pollution Solutions."

Though they have been refined substantially in the last 150 years, it is worth noting that systems of collocation and assignment indexing date back to ancient libraries. For example, the collections of Assyrian king Assurbanipal (see Fig. 12.5), who ruled from 668 to 627 B.C.E., consisting of thousands of clay tablets (upon which were inscribed some of the dire threats against irresponsible

Fig. 12.5 Stele featuring sculpture of Assurbanipal [3]

borrowers mentioned earlier in this chapter), were collocated by means of a relatively complex scheme. One room of his palace contained tablets relating to government and history [5]. Other portions of the collection divided up by subject included geography, laws and legal decisions, legends and mythology, and commercial records [5]. Within each room, a shelf list detailing the titles of works contained therein was affixed to the wall [5]. In addition, tablets that were analogous to a subject catalog or descriptive bibliography were found in the rooms. Each of these special tablets offered descriptive details about the other tablets contained in that room including titles of each work, the number of tablets for that work, the first few words, the number of lines, and symbols indicating location or classification [5].

The Library of Alexandria, which was founded in approximately 300 B.C.E. serves as another early example of advanced library organizational systems. Callimachus of Cyrene, a scholar at Alexandria, can be considered one of the early pioneers of assignment indexing. Among his many contributions, Callimachus enhanced access to the alphabetically-ordered collection of the library, which was comprised of hundreds of thousands of works on papyrus rolls. He did so by compiling shelf-lists and bibliographical works including *Tables of Persons Eminent in Every Branch of Learning Together with a List of Their Writings*, a survey of all Greek writings that was so extensive that it was comprised of five times the number of volumes that contained Homer's *Iliad* [2]. Callimachus broke the authors featured in this work into broad genre categories and then made finer distinctions from there, grouping them by their literary specialty: dramatic poets, epic poets, philosophers, comedy writers, historians, etc., [2].

12.4   Floundering in a Sea of Information: The Web and Information Literacy

In libraries during the early to mid-1990s, the use of print indexes declined sharply as CD-ROM and web-based databases greatly expanded access to metadata and digital content such as full-text versions of periodical articles. This marked the beginning of a period of widespread outsourcing of digital collections and a relinquishment of the meticulous level of control over the selection process that librarians had exercised over physical collections. Prior to this point, though a relatively small number of online research databases had been available before the advent of the World Wide Web, the bulk of a library's holdings had been limited to what could be stored within the walls of library buildings. Many, if not most, of the items in those buildings had been vetted carefully by librarians with regard to accuracy, authoritativeness, or other quality-oriented collection development criteria. The inclusion of these massive subscription-driven databases, each containing, potentially, tens of thousands of records along with articles from hundreds of periodicals has made it unfeasible for librarians to continue to apply rigorous selection standards to each and every item in a collection. Furthermore, after those databases are acquired they continue to morph as content is added or subtracted by the database provider.

In terms of content newly available to users, library databases are just the tip of the iceberg. By 1994, with access to user-friendly, web-based search engines and web indexes such as Yahoo!, Lycos, and Infoseek, researchers and casual users alike had expanded their reach beyond the walls of physical libraries via the World Wide Web. Presented with information in new formats that had not been pre-selected by librarians or vetted through established publishers, many struggled to distinguish between reliable and unreliable content, lack the savvy to formulate search strategies that would help them manage the overwhelming number of search results they were presented with. Stoker and Cooke summed up the problem in 1994:

> Information posted on to the network does not go through the same rigorous review procedures as information which has passed through formal publishing channels. The facility has been described as 'clogged with too much junk to make its use effective' and the information 'ephemeral and of questionable quality...' On occasions it might be difficult to determine the originating institution or individual for an item [8].

In a 1998 survey, the Pew Research Center determined that 41 % of adults were using the internet, up from 23 % in 1996 [9]. Despite the potential pitfalls of using the web noted by Stoker and Cooke four years earlier, in 1998, 49 % of web users believed "that Internet news is more accurate than news found in traditional print and broadcast outlets" [10]. Around that same time, some researchers discovered that this user confidence in the web may have been unwarranted: an analysis of 41 web pages offering health advice concluded that "only a few web sites provided complete and accurate information" which indicated "an urgent need to check public oriented healthcare information on the internet for accuracy, completeness, and consistency" [11].

The results of another study in 2000 indicated that consumers of web-based information either lacked the skills to evaluate the reliability of websites or were relatively unconcerned about its origin or trustworthiness. In the study, nearly 1000 respondents were asked to rate how often they applied basic criteria for evaluating the validity of websites such as "check to see who the author of the website is," "consider whether the views presented are opinions or facts" and "consider the author's goals/objectives..." [12]. Mean response scores for all but one of the nine criteria fell between values used to indicate a frequency of "rarely" and "never" with regards to applying each of the criteria [12].

The problem persists. By 2010, 79 % of American adults had become internet users [13] and in 2012, the Pew Research Center published the results of another survey indicating that many of them may be generally uncritical of websites appearing in search engine results. The survey concluded that "roughly two-thirds of searchers (66 %) say search engines are a fair and unbiased source of infor- mation." 28 % of respondents indicated that "all or almost all" of the information they get in their search engine search results is "fair and trustworthy" and an additional 45 % indicated that "most" is "fair and trustworthy" [14]. However, despite this high degree of confidence in search engines, "four in ten searchers" said "they have gotten conflicting or contradictory search results and could not figure out what information was correct. About four in ten also…" said "…they have gotten so much information in a set of search results that they felt overwhelmed" [14].

Assisting clients with internet use has been a major component of many librarians' job duties for nearly two decades. As a result, they have been first-hand witnesses to users' struggles with the relatively new responsibility of evaluating documents and sites they encounter on the web. Critical thinking about the origin of sources, about the publishing process, and about the appropriateness of a source in terms of meeting an information need have always been a part of doing research, regardless of whether information is located on the web or in print. However, the challenge of determining the reliability of web-based information requires a new set of critical thinking skills to be applied in new contexts. As the Association of College and Research Libraries states, the "sheer abundance of information will not in itself create a more informed citizenry without a complementary cluster of abilities necessary to use information effectively" [15].

In order to address the need for these skills, many librarian positions now emphasize teaching as a major component of the job. Before the advent of the world wide web, typically, librarians provided "orientations" or bibliographic instruction geared towards using card catalogs or online public access catalogs and navigating the collections which they had carefully vetted for reliability. Over the last 15 years, however, librarians have shifted their efforts towards providing instruction oriented around deeper critical thinking skills often referred to as information literacy. Information literacy, as a skill set, is highly applicable to online environments as it empowers users to "recognize when information is needed and have the ability to locate, evaluate, and use effectively the needed information" [15].

Analyses of librarian job advertisements have reflected this shift towards a greater instructional role. For instance, in 2002, 54.6 % of librarian job descriptions examined on an international job posting website over a 3 month period indicated that "user education or training is an important part of the job" [16]. In 2013 another study was published that gathered data from supervisors at organizations that had posted librarian job announcements on the American Library Association's job website. The study concluded that for 65 % of the jobs, instruction skills "were a required qualification." For an additional 34 % of the jobs, instructions skills "were a preferred qualification." Only one response did not list instruction as belonging to either category [17].

Librarians teach in a variety of contexts, including credit-bearing university courses, public library workshops, and in online environments. Some also consider one-on-one interactions with users at the reference desk or elsewhere to be an extension of that teaching role. Regardless of the context, by re-envisioning their profession and adapting to their clients' needs, librarians are empowering users to become critical consumers of information.

12.5 Library 2.0 and the Rise of Next-Generation Library Search Interfaces

The Web 2.0 movement that began in the early 2000s has been characterized, in part, by a shift from static web pages to interactive pages, platforms, and applications that enable users to contribute and collaborate in a variety of ways. This user-oriented approach to design has also extended to providing a simple, intuitive, and streamlined online experience. Early social media sites such as Friendster, photo sharing site Flickr, and social bookmarking sites such as Del.icio.us were pioneers of the phenomenon that has changed drastically the way we communicate and engage with information. Web 2.0 also put publishing in the hands of the masses; without knowledge of web programming languages, many users were able, for the first time, to shape the content of the web by using wikis, blogs, and simple web-page creation applications such as Google Sites. Web sites began to invite users to comment and rate content, or even to enhance access to that content via tagging, a crowd-sourced form of assigning subject descriptors that is also known as folksonomy.

Since the advent of Web 2.0, libraries have followed suit by enhancing the interactive capabilities of their websites and contracting with vendors who specialize in incorporating dynamic and interactive capabilities into library catalogs. As a result, navigating the online presences of most libraries has become a more participatory experience for users. Access to library resources and services via online public access catalogs (OPACS) has improved drastically over the traditional catalogs in use prior to what is often referred to as Library 2.0. Traditional, or "legacy" catalogs, according to renown library technology consultant Marshall Breeding in 2007, were overly complex, lacked engaging features, and were "unable to deliver online content" [18].

The ideals of Library 2.0 were epitomized by information architect Casey Bisson's development of a library OPAC overlay interface (which works in concert with an existing OPAC, rather than replacing it) based on the popular open source WordPress blogging software. The project, called Scriblio, was born out of Bisson's conviction "that libraries must use, expose, and make their data available in new ways" [19]. The use of the WordPress platform brought library catalog records up from the deep web where they had long been buried, making them discoverable via search engines and therefore, indexable by users of social bookmarking services. Scriblio, originally called "WordPress OPAC," which was announced on Bisson's blog in early 2006 [20], offered several capabilities beyond those available from traditional OPACs in use at the time. Among those improvements were faceted searching (options for limiting or refining one's search after the initial query has been submitted) and browsing via tag clouds. Within catalog records displayed in the Scriblio interface, similar items were suggested and accessible via hyperlink. Users were also able to comment on catalog items, and by subscribing via RSS, they could receive automatic updates detailing changes to the catalog [21].

Open-source integrated library systems, or ILSs, (library management software which includes the OPAC) such as Evergreen offered similar enhancements to the traditional OPAC. Despite the improvements they brought and the fact that they were free, open-source catalog overlay interfaces such as Scriblio, and open-source OPACs have been adopted by relatively few institutions. This is possibly due to the fact that some libraries may be daunted by the prospect of limited technical support for such products (being largely community-based, rather than provided by the vendor) and their reputation among some for having the buggy aspects of a beta quality platform [22]. Furthermore, few libraries have the type of in-house programming expertise that Lamson Library at Plymouth State University, which employed Casey Bisson, did. Another possible reason is that by 2007, commercial ILS vendors such as Polaris Library Systems, OCLC, and Innovative Interfaces, Inc. had taken notice of these "next-generation OPACS" and enhancements developed by Library 2.0 pioneers like Bisson, and had scrambled to improve their own OPACS [23] by adding dynamic features or by offering new products altogether.

Additional Web 2.0 functionalities to existing OPACs were offered by a variety of third-party developers such as Library Thing for Libraries (see Fig. 12.6), a commercial service which incorporates some navigation features similar to those of

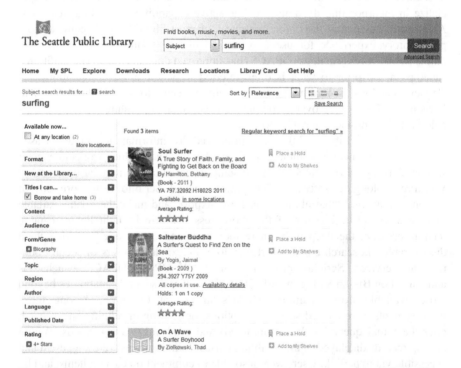

Fig. 12.6 Search results processed by a next-generation catalog that incorporates Library Thing for Libraries. Note the faceted search options on the left for narrowing the list of results by a variety of criteria, including user ratings

Scriblio and has evolved to offer users the ability to rate, review, and tag items displayed in libary catalogs. By incorporating third-party enhancements such as Library Thing for Libraries and overlaying user-contributed content over the elements of the traditional library catalog, the next-generation OPAC has become the "mash-up" of the library world. Since the advent of the next generation catalog, this theme of integration in library search interfaces has expanded much further along towards realizing Marshall Breeding's position that, in "an ideal world, the content of all the library's collections would be available through a single search interface" [18].

12.6 Integrating the Search Process

With the growth of the internet, electronic scholarly journal publishing has also seen an explosion in prevalence. Libraries increasingly license e-journal content (in packages from publishers, single titles, or, most commonly, via database aggregators). E-journal packages and databases allow libraries to dramatically increase the depth and breadth of content available to their patrons—usually at a fraction of the cost of subscribing to or purchasing titles individually. As library patrons experience improved access, they also come to expect that access to be to the digital form of an article—not a physical copy they must locate on a library shelf. But as access has expanded, the need has grown to enable even greater access (to referenced articles in an article of interest or to the full text of articles with citation and/or abstract information in a particular database).

12.7 Search Process: OpenURL Resolvers

In the late 1990s, OpenURL Resolvers (also referred to as link resolvers) entered the scene to address these desired research enhancements. While at first not much more than static links to articles on a publisher's web site, link resolvers soon developed a standardized syntax that allowed for metadata (information about the journal's ISSN, title, article title, author, volume, date, and page numbers, etc.) to be passed from links in one database or platform, query a "knowledge base" provided by an OpenURL vendor to which the library subscribes, and into the full text content, which could reside anywhere else within the library's e-holdings. The first commercially available link resolver, SFX, was released in 2001 by Ex Libris [24]. Through a subscription to this product, libraries could provide information to a vendor about the e-journals, databases, and e-serials packages to which they had access. Ex Libris would then coordinate with the vendors to maintain updated title and date coverage lists within a knowledge base to ensure links reached their appropriate targets [25]. Soon other providers began offering these services. Some examples include EBSCO's LinkSource, Serials Solutions' (later acquired by ProQuest) 360Link, and OCLC's WorldCat Knowledge Base. Figure 12.7 provides

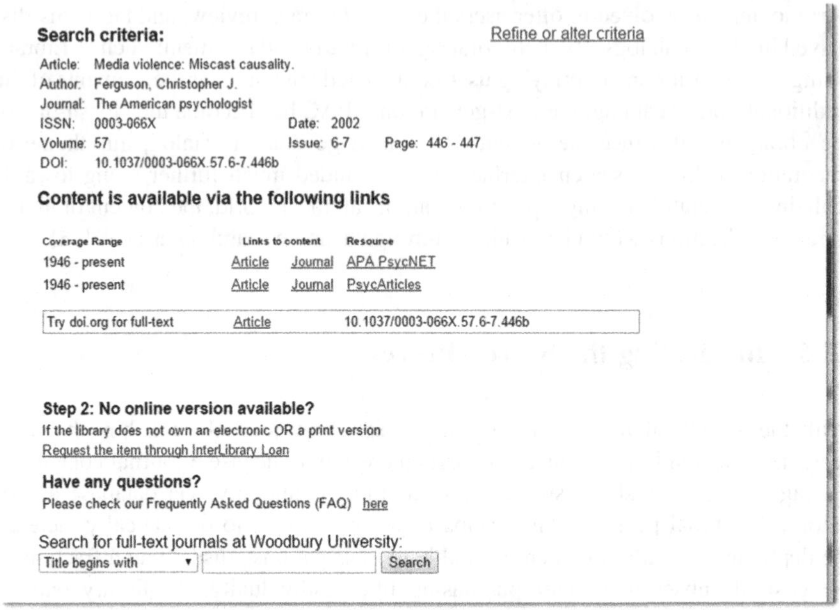

Search criteria: Refine or alter criteria

Article: Media violence: Miscast causality.
Author: Ferguson, Christopher J.
Journal: The American psychologist
ISSN: 0003-066X   Date: 2002
Volume: 57   Issue: 6-7   Page: 446 - 447
DOI: 10.1037/0003-066X.57.6-7.446b

Content is available via the following links

Coverage Range	Links to content		Resource
1946 - present	Article	Journal	APA PsycNET
1946 - present	Article	Journal	PsycArticles

Try doi.org for full-text	Article	10.1037/0003-066X.57.6-7.446b

Step 2: No online version available?
If the library does not own an electronic OR a print version
Request the item through InterLibrary Loan

Have any questions?
Please check our Frequently Asked Questions (FAQ) here

Search for full-text journals at Woodbury University:
Title begins with ▼ [_____] Search

Fig. 12.7 Example of an OpenURL results screen with links to content

an example of an interstitial OpenURL results screen with both article-level and journal-level links available.

By the time Google Scholar launched in November of 2004, libraries were able to work with their OpenURL provider to send their holdings information to Google. This resulted in libraries being able to connect with more potential users who may have been starting their research with Google instead of library resources. Allowing for IP-authentication, all users on an academic campus searching Google Scholar would automatically see information about connecting to their results via their library resources right from the search results list. Libraries could configure the text of the link as well. While this process has hardly been foolproof, as it relies on webcrawler-indexed metadata on Google's side matching up with metadata supplied by content providers; it has provided a way for libraries to link their holdings up using OpenURL technology with what their patrons were locating on the open web (and may otherwise have been prompted to pay for on their own). Figure 12.8 demonstrates the search results screen a user might see in Google Scholar if his/her library has sent their holdings information to Google.

OpenURL resolvers have not been without issue, however. They can be expensive—beyond the reach of a small library's budget—further exacerbating digital divide issues, where library users in smaller communities with less well-funded libraries then do not have access to technology that aids in their discovery of and connection to information. Also, since the success of a link can depend upon the complete matchup in metadata between the provider hosting the

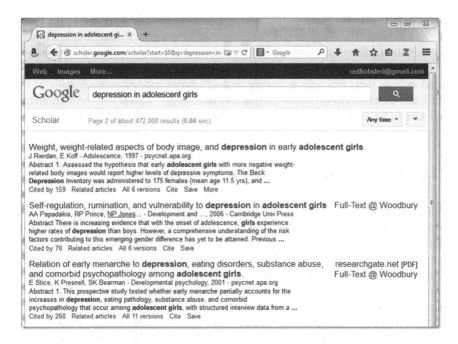

Fig. 12.8 Example of a Google Scholar results screen where a library has sent their e-serials holdings to Google

content and the provider indexing the content (frequently two different vendors), false negatives and positives can often result. That is, the OpenURL resolver may return information stating that the library does not have access to an article that it actually *does* have access to. Or, conversely, the resolver may link to a database where it states the article should be found, but the library does not have access to that article via any of its subscriptions. Understandably, this can be confusing to users.

In a 2010 study, two librarians found the mean total success rate for SFX (across links to books, newspapers, dissertations, and journal articles) was only 71 % [26]. This causes great frustration for librarians who will often be referred from the technical support desk of the indexing vendor to the tech support of the content-provider to the tech support of their OpenURL/Knowledge base provider. Full resolution may take days, weeks, months, or not come at all, and librarians new to e-resource management may be confused about where to begin. Some of this can also be the fault of the knowledge base vendor—who may have neglected to add, delete, or modify coverage dates of a title residing within a particular database.

The need to maintain an updated knowledge base cannot be under-stressed. OpenURL vendors must continuously update their information and libraries must also remain vigilant whenever they add or subtract from their e-collections or when a collection changes platforms or title. Failure to do so results in broken links for patrons. Additionally, some publishers, as a rule, do not allow for links directly to the article level. They may stop at the issue level or even the journal title level in an attempt to encourage libraries or end users to pay for subscriptions (electronic or print) directly to their journal titles. Often, librarians will not be aware of which vendors have these practices until they or their patrons encounter problematic links. Price and Trainor (2010) encourage libraries to thoroughly review content providers in order to have knowledge of which do not allow article-level links [27].

12.8   Search Process: Federated Searching

While OpenURL resolvers do allow for communication between databases, library patrons increasingly expect that their searches will return *all* results held by the library—not just some. Federated search (or metasearching) arrived in the marketplace in the late 1990s/early 2000s with what librarians hoped would provide a Google-like experience for end users [28]. Federated search claimed to make the idea of a "one stop shop" for searching all library resources a possibility. Users do not want to become experts in the various interfaces employed by library databases. Federated searches appeal to the novice user and the experienced user alike [29].

However, the way federated searching and Google work are entirely different. Through automated web crawling, Google is able to pre-index website content, returning results very quickly when users search. There are limitations, however. Google cannot search the deep web—content within subscription databases, data sets stored as files on government websites, orphan resources that are not linked to from anywhere else, dynamic content generated on the fly, and other resources to which libraries and librarians can provide access [30]. Federated searching, on the other hand, sends out queries to multiple databases (often including the library's online public access catalog, or OPAC) which are maintained by different vendors on different platforms with different indexing and different types of search protocols (XML—which uses a type of tagging of search elements—vs. Z39.50—a library-specific search protocol developed before the web—vs. the federated search vendor cobbling together a search strategy to access diverse resources) [29].

So, while a library's federated search product *will* return items on the deep web (indexed by library-subscribed resources) that are unavailable or largely invisible to Google, it also will do so at a much slower speed. Users may get a Google-like single search box, but the results will not populate their screen instantaneously as with Google. Instead, the federated search calls out to databases separately and returns results separately, as they are retrieved from the native databases. This leads to a list of non-ranked, non-de-duplicated results. Librarians may understand that these results need to be combed through carefully, but end users are used to the most relevant results showing up at the top of the screen. That might not always happen with federated searching. Another limitation is the fact that most federated searches are most effective when searching no more than a dozen resources [31].

When most large academic libraries subscribe to 60, 90, or over 100 electronic resources, patrons certainly are not getting a true "one stop shop."

How do users feel federated search compares to the Google experience? A 2013 study by Helen Georgas offers a comparison. Undergraduate students at Brooklyn College were asked to find one book, two articles (one scholarly) and one additional source of their choice using two search tools—a federated search and a Google search. The federated search tool was configured to search 11 databases including the library's OPAC. 81 % of students said Google was easier to use, with one commenting, "because it was faster." When asked which search they liked better, the students were evenly split between Google and the federated search. 59 % of the students said they would use the federated search tool on future assignments and 56 % would recommend it to fellow students. Among their complaints about the federated search, students felt it was difficult to find books and too slow overall. Their complaints about Google related to finding too many irrelevant results and being asked to pay for content. Some students mentioned they wished the library had the federated search, when in fact, the library had subscribed to the service for years. Students also remarked that they had difficulty identifying the types of sources retrieved in the federated search. Whereas Google and Google Scholar identify results by type of material, the federated search simply tags results with the database from which they were retrieved (although librarians felt the type of information was fairly obvious based on the fact that the OPAC results were all physical items and mostly books). These findings point to the need to make sure library patrons are more educated about services and features [32].

Despite the fact that they enable discovery of quality sources and students find them useful, it is obvious that federated searches have their serious limitations. First, the speed of the service is dictated by the slowest-performing of the remote database connections. Similarly, the fastest-performing remote connection will always have its results listed on top—leading to a potential problem of falsely-perceived relevance. Large result sets (as would typically result from a broad search by a novice user—the very type of user and search for which federated search was developed!) cause problems. Due to the time involved in retrieving these large remote result sets, results are typically truncated by the federated search service and any de-duplication or relevance ranking within results sets is then performed on only a small subset [31].

From the librarian's perspective, implementing a federated search product can be frustrating—taking months to launch. And what has been billed as the one stop search is most often far from it. In addition to the fact that the federated search works better when no more than a dozen resources are selected, there is the issue of some vendors refusing to participate in federated search development—rendering their content invisible to federated search users. And because federated search requires some translation across database collections, if a vendor is slow to develop or fix that translator, resources on that platform may be excluded as well [33].

Furthermore, with so much reliance on one product (the federated search) needing to utilize the different types of indexing employed by the disparate content vendors, it is very difficult to make use of database limiters, truncation, or wildcard searching effectively. Different databases may use these advanced search tools differently (or not offer them at all). Attempting to search across different platforms limits the functional search tools to the lowest common denominator across the databases. If search settings in the federated search product are adjusted to get better results from one particular database, the rest of the results may suffer. As Jody Condit Fagan (2011), the editor of the *Journal of Web Librarianship*, puts it: "Who knows if bad results are from the databases searched, the federated search software, or one's own search strategy? Results are messy and duplicative, and users frequently can't tell what the items returned actually *are*" (p. 77) [34]. So what do librarians do with all of the resources that cannot be included or searched effectively in a federated search? They are back to needing to teach users how to search all of the various interfaces individually and select the best resources for an information need (if those users can even find the resources within the depths of the library's website first!).

12.9   Search Process: Web-Scale Discovery

So, then, to truly move into the Google-like search realm with better speed and more reliable and customizable results, a centralized search model needs to be in place. This is what a few vendors began doing next, and in 2009, Serials Solutions (now part of ProQuest) was first to the library market with their launch of the Summon discovery service. Web-scale discovery services are the next generation in library resource searching [31].

Unlike federated searches, discovery services return results quickly and in relevancy-ranked order. Once results are returned, the discovery layer (or search interface) allows the user to refine and sort results using facets (e.g., year of publication, author, language, subject, publication type, or database source). The user is linked to full text via either direct links (if the resource is also hosted on the discovery service vendor's platform) or using OpenURL technology.

This model scales well to the size of the web because content and metadata have been indexed in advance of a user's search. With the increased capacity and reduced cost of data storage, the creation of this type of a centralized index (which is at the heart of all web-scale discovery services) became possible [31]. Within the central index are both the library's local resources and licensed e-content. The library works with the vendor to load its OPAC records into the centralized index (for information about items held physically in the library). Along with this type of local content, libraries may also include metadata for institutional repositories of student and faculty work and/or locally digitized collections. On the more external side, metadata and full-text content from licensed and open access publishers and content providers can be selected for inclusion. Many discovery services have also licensed content from third-party vendors for inclusion in their central index, regardless of whether the library subscribes to that particular resource on its own. Content available to the library through subscriptions to database aggregators (e.g.,

APA's PsycNet, ProQuest's Research Library, EBSCO's Business Source Premiere, etc.) may also be included. However, this type of content needs to be mutually licensed by both the library and the discovery vendor. Since many of the discovery vendors are also in the field of licensing e-content and providing access on their own proprietary platforms, they may choose not to make the metadata for and links to these resources available to other discovery vendors. In this way, not all of a library's e-resources will necessarily be available for inclusion in the centralized index of their discovery service [35].

Because the content is pre-indexed, all of the advanced search options frequently unavailable in federated searching are available to the user of a discovery service. Truncation, wild card, exact phrase searching, and use of Boolean operators are all possible. While discovery services all have these basic characteristics in common, there are differences among them. There are several vendors in the marketplace at this point. Perhaps the four with the largest market share are Summon (formerly launched by Serials Solutions, which has since been bought by ProQuest), Ex Libris' Primo Central, EBSCO's EBSCO Discovery Service (EDS), and OCLC's WorldCat Local (although they also have a new WorldCat Discovery product just launched in March of 2014).

Summon bills itself as: "the only discovery service based on a unified index of content. More than 90 content types, 9000 publishers, 100,000 journals and periodicals, and 1 billion records are represented in the index. New content sources are added every week and content updated daily." With their "Match and Merge" technology, Summon ingests content from various providers, "combin[ing] metadata, including discipline-specific vocabularies, with full-text content when available to create a single record" for each resource [36]. Figure 12.9 shows an example results screen from a search in Summon. More information about items in the results list (abstract, authors, dates) is also shown in the right margin when hovering over a particular result.

EBSCO, which is also a major content provider and has established relationships with diverse publishers, is able to leverage its existing resources to include native database indexing (which is frequently performed by subject experts in the field for inclusion in an individual, subject-specific database and adds value) and subject-specific controlled vocabularies in its discovery service [37].

OCLC has a unique position in the marketplace as it sees itself as "content-neutral." Having gotten out of the business of hosting third-party databases, it claims to be able to build relationships with a larger variety of content providers more easily [38]. And certainly, this *is* an issue. Some database vendors are also in the web-scale discovery business and do not wish to provide all of their indexing or content to competitors. For instance, EBSCO currently refuses to provide its content to Ex Libris for inclusion in Primo Central [39].

Libraries have had to develop their own awkward workarounds, and in the end, patrons are not served well. This debate has been well-documented and brought to public attention by the Orbis-Cascade Alliance [40], a nonprofit library consortium of 37 colleges and universities in Oregon, Washington, and Idaho. In a letter from the alliance dated October 6, 2014, to both vendors, regarding their failure to

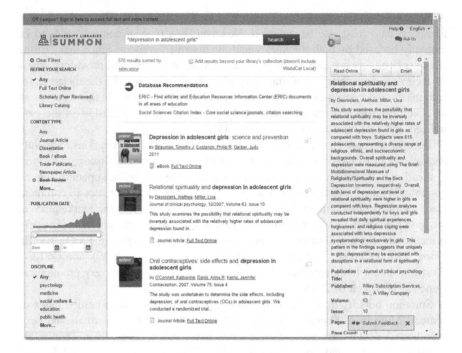

Fig. 12.9 An example of a search results screen in Summon, a web-scale discovery service

resolve the stalemate, the Orbis-Cascade Board of Directors state: "This failure to act is unacceptable and strongly suggests that both companies value business gamesmanship over customer satisfaction and short-term gain over service to students, faculty, and researchers. The library community expects an explanation and we call upon EBSCO and Ex Libris to provide a public update and projection of when this impasse will be resolved. As a major customer, the Orbis Cascade Alliance membership expects to spend in excess of $30 million with EBSCO and Ex Libris over the next five years. With these issues left unresolved, we will now take active steps to reconsider the shape and scope of future business with EBSCO and Ex Libris." [41].

Discovery services have been very popular upon implementation. In a January, 2014 survey of nearly 400 libraries using discovery services, overall satisfaction with the products ranged from 6.26 to 6.95 on a 9-point scale. Marshall Breeding [42] found that overall satisfaction was highest with users of EBSCO Discovery Service and lowest with Primo Central. Interestingly, all discovery services had higher popularity scores among undergraduates than among graduate students or faculty. This could be in part due to issues with known-item searching. Faculty and graduate students are more likely to be searching for a specific resource (a journal article, book, image), and discovery services are better at exposing a large range of resources to the searcher.

Web-scale discovery still remains out of the budget range of many libraries. A 2010 review of Summon, EDS, and WorldCat Local published in *The Charleston Advisor* [43] described the pricing of these services as ranging from $9000 to over $100,000 per year depending on the size of the library's collection, size of population served, and optional add-on services (incorporating institutional repositories, enhanced book content, building connections to additional resources not included in the provider's central index). Despite this, it could be argued that the cost/benefit ratio is in the favor of acquiring a discovery product. Users finally do get closer to utilizing a single search, and the library's e-resources receive greater exposure and usage. Discovery services are also generally mobile-friendly and can incorporate most, if not all, of the content of a library's OPAC. Additionally, because these services are hosted by the vendor, libraries do not need to worry about server or software upgrades [44].

So, what's next? Is there territory beyond web-scale discovery? Certainly discovery services are continuing to improve. Librarians need to remain closely involved in the development of these tools—making sure to customize library products to best meet the needs of the type of users they serve [39]. With the move to a single search portal, librarians may be able to devote more time to the development of local "born digital" collections and institutional repositories—and utilize the discovery service as a way of making that content more visible to the end users. Discovery services may help librarians stay more current and relevant in the eyes of patrons who are always expecting a Google-like experience, but education is still key. Users need to know the basics about evaluating information, considering results for relevancy, and identifying the types of information being retrieved. Librarians are experts in these areas.

12.10 Integrating Services: Library Consortia

As electronic resources available to librarians continue to increase in their depth and breadth of coverage and complexity of access models, libraries have turned to consortial models to help manage these workflows. Library consortia are not new. There is evidence of early consortial behavior back into the late 19th century as groups of libraries have banded together to share cataloging, participate in a very rudimentary form of interlibrary loan, and purchase cooperatively [45]. This section, however, will focus on how library consortia work today.

Libraries license content and/or platforms for access from vendors. Unlike books or videos the library purchases, these items are frequently leased, and not owned. As a result, they have a range of restrictions not found in the purchases of physical formats [46]. The license agreements for e-resources can be tedious, and individual libraries may not have the expertise to fully understand and negotiate these contracts in their own best interest. Concerns arise over which resources allow remote access to affiliated users only and which will allow more relaxed rules. There are also questions about what electronic content may be used to fill interlibrary loan requests from other libraries. Database license agreements are not consistent across the board, and e-resource license management can be overwhelmingly time-consuming if performed thoroughly. Most libraries do not have the staff to spare for this singular function nor the legal expertise to do this, and this is one niche that consortia have been able to fill.

Library consortia can negotiate with vendors on behalf of all of their member institutions. Some may have experts in license agreements on their staff or rely upon committees of librarians from member institutions to review agreements from newly-licensed content before offering them to member libraries for purchase. Consortia can also offer content to libraries that the individual libraries may not have been able to afford on their own. When purchasing together, consortia can purchase large e-journal packages for their member institutions. On a per-title basis, individual libraries are paying *much* less for these titles than they would if they purchased their own subscriptions on an a la carte basis. Consortia are also able to negotiate with vendors to suppress cost increases with more power than individual libraries negotiating ever could [47]. They can reject dramatic cost increases, object to restrictive licensing terms, and achieve better discounts overall. End users benefit because they have access to more expensive, niche resources.

In addition to performing cooperative purchasing and licensing of databases and other e-resources, consortia may also work together to offer interlibrary loan services. Some consortia, like OhioLINK (formed in Ohio in 1989 and composed of 90 public and private academic libraries plus the State Library of Ohio) [48] and the Orbis Cascade Alliance (comprised of 37 academic libraries in the Pacific Northwest, formed in 2003 from a merger of the Orbis and Cascade Alliances, which originated in the early 1990s), [49] have partnered with vendors to create consortial library catalogs. These catalogs enable borrowing and lending among member libraries in a way that is more seamless to the library patron (who may simply just place a hold with one click, rather than filling out an interlibrary loan form). In the case of the OCA, the consortium actually shares an integrated library system (ILS) which is responsible for not only the public catalog (OPAC) but also the back end staff circulation, acquisitions, reporting, and cataloging functions. This allows items to be checked out as if they were from one large library with many branches, as opposed to individually siloed libraries with their own ILS software, circulation rules, and processing procedures. And again, the library patrons benefit because a much larger array of resources is being presented for their use at a reduced cost and expedited processing speed [50].

Even outside of official consortial agreements (which may offer library patrons reciprocal borrowing privileges or fee-free interlibrary loans among member institutions), interlibrary loan has continued to rise in popularity. Discovery services present patrons with more results than ever before, and the sponsoring library will not own all of those items. Interlibrary Loan request links are placed prominently within non-owned search results allowing for an e-commerce-like experience for patrons who are used to purchasing items through Amazon with one click [51]. Figure 12.10 shows what this looks like inside the library catalog of a library using OCLC's WorldCat Local. Generally, interlibrary loan is free for academic library

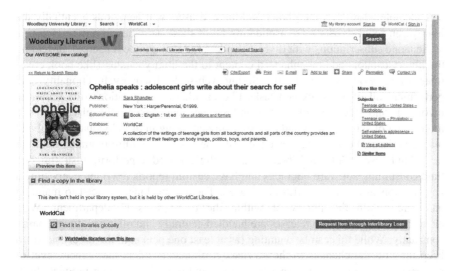

Fig. 12.10 Example of a 1-click Interlibrary Loan requesting option within a library catalog

users or there may be a nominal fee. Interlibrary loan allows libraries to provide access to content for their patrons for which they could not otherwise justify paying full price or even acquiring at all. The modern interlibrary loan framework was largely created by OCLC (formerly the Ohio College Library Center and now the Online Computer Library Center) with its WorldCat product. Beginning as the OCLC Online Library Union Catalog in 1971, it later developed into WorldCat (in 1996) and developed into the freely available and searchable WorldCat.org in 2006. By authenticating to their home library, patrons searching WorldCat can request items via its union catalog (which represents the holdings of libraries—both physical and electronic—all over the world) [52].

12.11   Integrating Services: Vendor Partnerships

Vendor partnerships can be valuable for libraries. Opportunities in this arena are increasing all of the time. One such relatively new development is the introduction of cloud-based, full-featured, integrated library systems (ILS). These new products are also being offered by vendors who, in the past, did not get involved with full library solutions. For example, OCLC introduced its WorldShare Management Services (WMS) in July of 2011. Now, over 300 libraries worldwide are using the service [53]. OCLC's WMS includes modules for acquisitions, e-resource man-agement (knowledge base, metadata, and OpenURL), circulation, analytics, inter-library loan, discovery (the system also acts as a web-scale discovery service), and an optional license manager for online resources. The price and work involved on the part of the library in migrating to a new cloud-based ILS like WMS is considerable. However, the library and the end users will benefit as eventually the library can reduce costs by consolidating all of these activities in one service.

Separate subscriptions with diverse vendors for managing e-resources, providing the library catalog, adding enhanced content to the library catalog (like book cover images and review information), offering web-scale discovery functionality, and performing all of the back-end tasks like circulation, acquisitions (the ordering, invoicing, and receiving of books) and serials management (the placement and monitoring of subscriptions and electronically checking in of individual journal and magazine issues) can all be cancelled once the library has fully migrated onto the new cloud-based platform. Libraries also no longer need to perform frustrating and time-consuming software upgrades and have custom reports rewritten every time a new version of a vendor's ILS is released. There is no longer a need to maintain hardware within the library or within the university's IT department. With cloud-based hosting, new releases are handled centrally and offer new functionality constantly. While this can be daunting (as at least one person in the library needs to keep up with the changes to the service), it ultimately offers the highest level of responsivity to trends in information-seeking and provision. Similar products from other vendors include Ex Libris' Alma and ProQuest's Intota (which is still in development, with its collection assessment piece launched in November, 2013).

Some legacy ILS vendors have offered products in response, but they generally exist as optional overlays to the existing system or are built upon existing ILS infrastructure with some consolidation of services and automation and discovery products offered by them or their partners. These vendors tout the fact that libraries can continue using the product they have always been using that contains all of their data with no need to migrate information to a new and "untested" system [54]. These new services from legacy vendors can be offered as SaaS (software as a service), so they can be fully hosted. However, the new products from legacy vendors will need to be implemented within the current hosting framework (with some libraries hosted in the cloud, some via SaaS, and others locally hosted on their own hardware). This means the legacy vendors will likely need to support multiple versions of their new ILS. In contrast, the built from scratch systems are not based on old legacy code, and updates, patches, and bug fixes can be pushed out to all users from the development side simultaneously [55].

End users are frequently unaware of these behind the scenes machinations on the part of their libraries. However, moving to new cloud-based ILS platforms can result in new workflows and reallocated time on the part of library personnel. With WMS, for example, cataloging is much quicker—with librarians just needing to select the appropriate master record within WorldCat and attach their institution's holdings to it. Libraries can cooperatively manage cataloging by contributing updates and corrections and additional information that will be added to the master record for the benefit of all libraries. With this reduced need for copy cataloging time, librarians working as catalogers may be able to devote time to other projects like cataloging a unique local collection, or advising on the virtual construction of an institutional repository.

12.12   Integrating Services: Cooperative Reference Services

Library services extend beyond the discovery of information and provision of access. Reference, which is at the core of library service, has also been impacted by collaborative management. The American Library Association defines reference transactions as "information consultations in which library staff recommend, interpret, evaluate, and/or use information resources to help others to meet particular information needs." [56] Traditionally, this has taken place in person at a reference desk in the library or via telephone.

According to a 2012 survey by the National Center for Education Statistics, 74.9 % of all US academic libraries offered some form of virtual reference as well, with 26.6 % offering chat reference via a commercial service and 32.8 % offering chat reference via instant messaging applications. This was an increase from 2008, when 72.1 % of all US academic libraries offered virtual reference. Academic libraries serving larger populations are more likely to offer some form of virtual reference [57]. These trends are similar for public libraries [58].

Libraries willing to go it alone have used free instant messaging software like Yahoo chat, Google Chat, and MSN Messenger. However, those services are not provider-neutral, so chat aggregators also became popular in libraries (Pidgin, Meebo). They allowed librarians to receive chat questions from library patrons using any chat software. However, as stand-alone services, only one librarian could monitor the chat queue. Additionally, if a librarian needed to transfer a chat reference question to another librarian, it was quite difficult [59].

In 2008, LibraryH3lp was launched as part of a chat reference collaborative for reference services provided after-hours at Duke University, UNC Chapel Hill, and North Carolina State University. It utilizes the open standard XMPP as its chat protocol, which allows for monitoring of the chat queue by a number of free clients, such as Pidgin. LibraryH3lp widgets can be inserted into websites, databases, discovery layers, or subject guides (like LibGuides) for patrons to access at their point of need [60].

For libraries that want more extensive, 24/7 virtual reference coverage, there are other (more expensive) products in the marketplace. OCLC offers QuestionPoint, which is a reference cooperative staffed by librarians from subscribing libraries. During overnight hours, contract librarians with access to the home institutions' basic information about policies and resources, staff the service. This is important to provide, as many reference questions fall into the basic informational variety (open hours, directions, information about library fine policies, etc.) Libraries using QuestionPoint are responsible for staffing it for their users as much as they'd like as well as for providing coverage a few hours each week to the entire cooperative [61]. When chat questions are submitted by patrons, they are first routed to the home library. If there is no response, they are then sent out to the library-defined partners (which could be a consortium or a network of state universities). If there is still no timely response, the question is sent out to the main QuestionPoint cooperative where any librarian staffing the service may respond to it [59].

By the end of 2012, approximately 24 % of academic libraries offered some form of text (SMS) reference service. With this type of service, library users can send a text message asking for reference assistance. Librarians receive and respond to the text via a web interface (and do not need to monitor it on their mobile phones). From 2010 to 2012, text reference service in large public libraries (those serving 500,000 people or more) increased from 13 to 43 % [62]. Springshare, which created LibGuides, added text reference capabilities to its LibAnswers suite via an add-on LibChat module [63]. Mosio for Libraries offers a Text-a-Librarian service as well as chat and email virtual reference capabilities. My Info Quest, a text reference cooperative sponsored by the South Regional Library Council of Ithaca, NY, uses Mosio's Text-a-Librarian product. It is currently staffed 80 h per week with plans to increase coverage as more libraries join the service. Text messages may be sent via LibChat or Mosio's product but will not get answered until a librarian is able to respond. Using a cooperative like My Info Quest allows quicker response to patron reference needs [64].

12.13 Integrating Services: Instant Information for the End User

In this world of streaming video via Netflix and Hulu and streaming music like Spotify and Pandora, library patrons also want that type of instant access to information and entertainment from their libraries. Libraries and vendors have responded with the development of e-book and e-audiobook platforms like OverDrive, which are readily integrated across a patron's devices via an app.

OverDrive, which started out in the CD-ROM industry in the 1980s, first began offering downloadable e-books and e-audiobooks to libraries in 2003 [65]. Libraries can select from OverDrive's catalog of content and offer what they choose to their patrons. Content can be integrated into the library catalog and/or searched separately on the library's website. Additionally, patrons can download the OverDrive app to their mobile device or tablet and search directly from within that interface. Library patrons need to authenticate with their libraries by providing their library card credentials in order to use the service and view the library's catalog of OverDrive content. Within the OverDrive app, patrons may use the native reader to read an e-book, place holds on titles, or follow a link to check out the Kindle edition directly from Amazon. Libraries are not restricted by physical shelf space, and as such, may choose to buy 5, 10, 15, 20, or even 50–100 copies of popular titles to minimize patron wait times. Some titles may offer libraries the option of purchasing unlimited simultaneous usage models, but most are single-user, single-copy. OverDrive also contains e-audiobook content, which is playable directly within the app and cloud-based, so playback is synced across devices signed into the same account. At the end of the check-out period, the file (whether e-book or e-audiobook) simply expires from the patron's device. There are no overdue fines for patrons, either! [66] Other vendors have also gotten involved in e-book and e-audiobook content. Notably, Axis 360 from Baker and Taylor (a company that started out as a book distributor), 3M Cloud Library, and RBDigital (from Recorded Books).

Books and audiobooks have long been a major brand of the library, but libraries have also begun to offer their patrons electronic access to movies, television shows, music, and popular magazines. Some of this content is downloadable, like music from Freegal. Freegal allows users to download a limited number of songs per week from Sony Music Entertainment and other labels with which they have made agreements. While copyright laws always apply, these downloads are DRM-free and can be played as mp3 files on any device and do not expire [67].

Zinio, a partnership with Recorded Books, (yet another traditional audiobook vendor!) is a service that provides downloadable popular magazines to mobile devices and tablets. It also has its own marketplace (which exists for a fee completely outside of libraries) to provide magazine subscribers access to an electronic copy of their subscriptions via the Zinio app. In May of 2012, they launched a digital magazine newsstand to libraries. Libraries can choose from a catalog of over 5500 titles in over 20 languages to make available to patrons. There are no limits for patrons, and files do not expire. They can remain on a patron's device indefinitely. Libraries pay a tiered platform fee based on annual circulations as well as by title selected. *Library Journal* gave a 2012 price point of $6417 per year paid by the Chattanooga Public Library for access to 121 titles and the cost of the platform. Patrons read the magazines in the Zinio app, which provides high-definition, full-color pictures and interactive media elements [68]. The Zinio interface on a library's website is shown in Fig. 12.11.

Public libraries have also begun to provide access to streaming content, which circumvents long file download times. Hoopla, a streaming service started by Midwest Tapes (which was also a traditional audiobook vendor), offers patrons access to streaming movies, television shows, music, and audiobooks. Libraries pay very little up front to use the service—instead offering the content to their patrons via their website or the Hoopla app. They can then choose to throttle usage to keep within their budget requirements. This may mean that popular titles are only available to the first X number of patrons wanting to access them per day. Subsequent patrons are told that the limit has been reached for the day, but to try back tomorrow. There are no wait lists. Libraries can also choose the loan period for all items [69]. Other streaming services include IndieFlix (for movies), Freegal, and OverDrive (which have both recently entered the streaming movie and television market),

While academic libraries do not usually offer such services to their patrons, students and faculty often need access to information not immediately owned or leased (in the case of database content) by their library. Traditional interlibrary loan has always been on offer, but it can take days (for journal articles), up to weeks for books or videos to be shipped to the borrowing library. For students with a paper due at midnight, that is just not a viable option! In the early 2000s, libraries tried to adapt to patron needs by initiating just-in-time purchases from interlibrary loan requests. Libraries can purchase materials directly from several vendors (Better World Books,

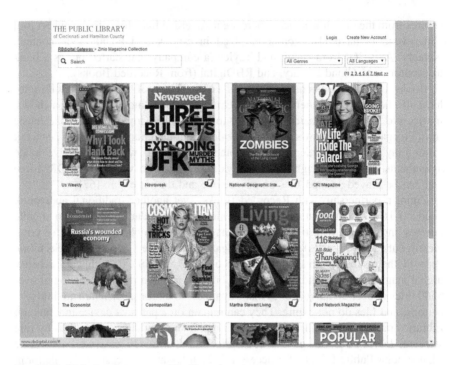

Fig. 12.11 Zinio interface on a public library's website

Alibris) directly within their interlibrary loan requesting modules. In this way, what began as an interlibrary loan request results in a fast purchase. When the item arrives, libraries may forego the usual processing (cataloging, covering the book, entering it into their ILS) and only do so when the book is returned by the requesting patron.

This model evolved even further. What is now termed patron-driven acquisition (PDA) has been applied most frequently to building e-book collections. This refers, in its simplest form, to the process of allowing library user requests and information seeking behavior to decide, in part, which materials the library acquires. Several vendors (Ebrary, eBooks on EBSCOhost, and E-book Library) now offer mechanisms for patron-driven acquisition. Libraries set up a profile with the vendor based on the subjects, publishers, dates of publication, cost of items, authors, keywords, or a variety of other criteria for filtering the types of items that they would like to make available to their patrons. This results in a pool of possible items. Libraries do not purchase or pay for these items up front. In fact, outside of a vendor platform or hosting fee, this is otherwise free for libraries to set up. Depending on their preferences, libraries can request MARC records for this pool of items so that they can be added to their library catalog. Some libraries may choose not to do this and rather to make these titles available through a search interface provided by the vendor (in much the same way a patron would search a database). However, in this age of discovery, that is generally not seen as a best practice. In order to expose patrons to the entire pool of potential items, it is ideal to load them into the library catalog—dramatically enhancing visibility. For libraries using next-gen catalogs (like OCLC's WMS or Ex Libris' Alma) or a discovery layer, these records can be "turned on" as part of e-resource management. When patrons search in these interfaces, results of items from the PDA pool are returned and appear to be owned by the library. Patrons can click and directly access the content.

Libraries are then only charged when patrons use the items in the PDA pool. This usage is triggered by differing thresholds based on vendor, but is usually along the lines of when browsing exceeds 5 or 10 min or when a patron attempts to download or copy and paste content from the e-book. Performing any of these tasks can either trigger an outright purchase of the title, or, if libraries choose, this triggers a "short term loan" (STL) of a time period defined by the library (generally 1 or 7 days). After a certain library-defined number of STLs, a purchase of the e-book is triggered. The STL fees are not applied towards the purchase cost of the e-book. The purchase is at the full price. The library is billed for short term loans at a fixed percentage cost of the list price of the e-book. This percentage varies from vendor to vendor, but it is always less than the cost of purchasing the book. Recently, publishers (e.g., Taylor and Francis, Bloomsbury, Oxford University Press, Wiley, McGraw Hill, among others) have been raising short term loan prices, as they feel the STL model is not a viable one for them (largely due to the fact that libraries can decide how many short term loans can occur before a purchase is triggered). In the past, some STL prices were as low as 5 % of the list price of the book for a day's usage. Some libraries were allowing 4 or 5 or more STLs before triggering an auto purchase of an e-book. Essentially, this meant they were per-petually renting books and buying very few titles outright. This is an area which has been generating much discussion among librarians. To accommodate libraries that object to the higher prices on STLs, some e-book vendors have put settings in place on their platforms to allow librarians to no longer allow STLs from any publisher charging more than a particular percentage of the list price for an STL. Libraries may also choose to trigger an auto purchase of an e-book sooner or to exclude particular publishers from their PDA program completely. Perhaps it is only when a critical mass of libraries take this more drastic step that publishers will rethink the viability of their current pricing models [70].

Libraries can also implement this PDA model with streaming video. Kanopy, originally an Australian DVD distribution service, offers this. Libraries can license individual titles or collections from particular producers and distributors as well, but their PDA model is a new one in the library market. Four views of a video (3 s counts as one view) triggers a license purchase of a video. One and three-year licenses are available [71]. In both the streaming video and e-book PDA models, libraries may choose to put a set amount of money on deposit with a vendor and run their PDA program until it is exhausted or pay as they go. Libraries may choose to mediate the PDA process (where patrons must request access) or let it go unmediated (the more popular choice). Best practices generally state that patrons should not be made aware that a PDA model is in place, as libraries want patrons to access and use the materials they need without thinking about how much that usage might be costing their library, which might cause patrons to alter their information-seeking behavior. Overall, patron-driven acquisition allows libraries to make a much larger pool of items visible and available to their users without having to pay up front or provide shelf-space. Also, the thinking is that if patrons choose the items, perhaps they will be utilized more than items selected solely by librarians or approval plans.

Academic library patrons also need rapid access to journal articles to which their library may not have access. The Copyright Clearance Center (the organization responsible for collecting payments from libraries for interlibrary loan usage of unsubscribed titles and which generally manages licensing of content) provides a Get it Now service [72]. This service can be integrated with a library's OpenURL resolver or existing ILL workflow. Just like PDA for e-books and streaming video, it can be offered either mediated (much like interlibrary loan, but the article can be delivered immediately upon processing) or unmediated (through the library's website). With Get it Now, the requesting library provides payment behind the scenes to the Copyright Clearance Center that covers the cost of access and copyright payments, and the article is delivered immediately to the patron. The library has the ability to set up restrictions in advance (e.g., no patron may incur more than a particular amount of costs via the service, only certain titles are available, price limits per article, etc.), but the process runs invisibly to the end user. Libraries pay less than the per-article cost charged by individual publishers on their websites to access the content, so money is saved as well.

All of these new developments in library services and products continue to change and expand rapidly. Librarians need to remain on the forefront of technology—knowledgeable about tools, products, and services that connect users with information. Similarly, they need to serve as experts in how information is created, evaluated, and disseminated. While books will most likely always be one of the library's most well-known brands [73], provision of access to electronic content is at the true center of librarians' work today.

References

1. **Valentine, Patrick M.** *A Social History of Libraries from Cuneiform to Bytes.* Lanham, MD : Scarecrow Press, 2012.
2. **Casson, Lionel.** *Libraries in the Ancient World.* New Haven : Yale University Press, 2001.
3. **British Museum.** *Guide to the Babylonian and Assyrian Antiquities.* London : Harrison and Sons, 1900.
4. **Avrin, Leila.** *Scribes, Script, and Books: The Book Arts from Antiquity to the Renaissance.* Chicago : American Library Association, 1991.
5. **Harris, Michael H.** *History of Libraries in the Western World, 4th ed.* Lanham, MD : Scarecrow Press, 1999.
6. **Columbia University .** Philanthropy of Andrew Carnegie. *Columbia University Rare Book & Manuscript Library .* [Online] [Cited: Oct. 15, 2014.] http://library.columbia.edu/locations/rbml/units/carnegie/andrew.html.

7. **Reitz, Joan M.** ODLIS: Online Dictionary for Library Information Science. [Online] ABC-CLIO, 2014. http://www.abc-clio.com/ODLIS/odlis_a.aspx.

8. **Stoker, David and Alison Cooke.** Evaluation of Networked Information Sources. [ed.] Ahmed H. Helal and Joachim W. Weiss. *Information Superhighway: the Role of Librarians, Information Scientists and Intermediaries: Proceedings of the 17th International Essen Symposium 24th - 27th October 1994.* Essen : Universitatsbibliothek Essen, 1995, pp. 287-312.

9. **Pew Research Center.** The Internet News Audience Goes Ordinary. *Pew Research Center for the People & The Press.* [Online] January 14, 1999. http://www.people-press.org/1999/01/14/the-internet-news-audience-goes-ordinary/.

10. **Pew Research Center.** The Internet News Goes Ordinary: Section II–Online News Consuption. *Pew Research Center for the People & The Press.* [Online] January 14, 1999. http://www.people-press.org/1999/01/14/section-ii-online-news-consumption/.

11. **Impicciatore, Piero, Pandolfini, Chiara, Casella, Nicola and Maurizio Bonati.** Reliability of Health Information for the Public on the World Wide Web. *British Medical Journal.* June 28, 1997, Vol. 314, p. 1878.

12. **Flanagin, Andrew J. and Metzger, Miriam J.** Perceptions Of Internet Information Credibility. *Journalism & Mass Communication Quarterly.* 2000, Vol. 77, 3, p. 531.

13. **Pew Research Center.** The Web at 25 in the U.S. *Pew Research Internet Project.* [Online] February 27, 2014. http://www.pewinternet.org/2014/02/27/the-web-at-25-in-the-u-s/.

14. **Purcell, Kristen, Brenner, Joanna, and Lee Rainie.** Search Engine Use 2012: Main Findings. *Pew Research Internet Project.* [Online] March 9, 2012. http://www.pewinternet.org/2012/03/09/main-findings-11/.

15. **Association of College & Research Libraries.** Information Literacy Competency Standards for Higher Education. [Online] [Cited: October 15, 2014.] http://www.ala.org/acrl/standards/informationliteracycompetency.

16. **Clyde, Laurel A.** An Instructional Role for Librarians: An Overview and Content Analysis of Job. *Australian Academic & Research Libraries.* 2002, Vol. 33, 3, p. 158.

17. **Hall, Russel A.** Beyond the Job Ad: Employers and Library Instruction. *College & Research Libraries.* 2013, Vol. 74, 1, p. 28.

18. **Breeding, Marshall.** Next-Generation Library Catalogs: Chapter 1 Introduction. *Library Technology Reports.* 2007, Vol. 43, 4, pp. 5-14.

19. Casey Bisson: Movers & Shakers 2007. *Library Journal.* March 15, 2007.

20. **Bisson, Casey.** WPopac: An OPAC 2.0 Testbed. *MaisonBisson.com.* [Online] February 9, 2006. http://maisonbisson.com/post/11133/wpopac-an-opac-20-testbed/.

21. **Bisson, Casey.** Scriblio Questions Answered. *Scriblio.net.* [Online] July 19, 2008. http://scriblio.net/.

22. **Rapp, David.** Open Source Reality Check: Implementation Experience Reveals Pros and Cons. *Library Journal.* August 16, 2011.

23. **Yang, Sharon and Melissa A. Hofman.** The Next Generation Library Catalog: A Comparative Study of the OPACs of Koha, Evergreen, and Voyager. *Information Technology and Libraries.* 2010, Vol. 29, 3, p. 141.

24. **Ex Libris Group.** Ex Libris the bridge to knowledge. Overview. *ExLibris SFX.* [Online] 2014. [Cited: 10 1, 2014.] http://www.exlibrisgroup.com/category/SFXOverview.

25. **Apps, Ann and MacIntyre, Ross.** Why OpenURL? *D-Lib Magazine.* 2006, Vol. 12, 5.

26. **Price, Jason S. and Trainor, Cindi.** Digging into the data. *Library Technology Reports.* 2010, Vol. 46, 7.

27. **Price, Jason S. and Trainor, Cindi.** Improving the resolver menu. *Library Technology Reports.* 2010, Vol. 46, 7.

28. *Federated search engines and the development of library systems.* **Joint, Nicholas.** 9, 2008, Library Review, Vol. 57. 00242535.

29. **Luther, Judy.** Trumping Google: Metasearching's promise. *Library Journal.* 2003, Vol. 128, 16.

30. **Shokouhi, Milad and Si, Luo.** Federated search. *Foundations and Trends in Information Retrieval.* 2011, Vol. 5, 1.

31. *The Systems Librarian.* **Breeding, Marshall.** 2, February 2005, Computers in Libraries, Vol. 25. 10417915.
32. *Google vs. the Library: Student preferences and perceptions when doing research using Google and a federated search tool.* **Georgas, Helen.** 2, 2013, portal: Libraries and the Academy, Vol. 13, pp. 165-185. 15307131.
33. **McHale, Nina.** Accidental federated searching: implementing federated searching in the smaller academic library. *Internet Reference Servicse Quarterly.* 2007, Vol. 12, 1-2.
34. **Fagan, Jody Condit.** Federated search is dead - and good riddance! *Journal of Web Librarianship.* 2011, Vol. 5, 2.
35. *The ins and outs of evaluating web-scale discovery services.* **Hoeppner, Athena.** 3, April 2012, Computers in Libraries, Vol. 32. 10417915.
36. The Summon Service. *ProQuest Products and Services.* [Online] 2014. [Cited: November 18, 2014.] http://www.proquest.com/products-services/The-Summon-Service.html.
37. Ebsco Discovery Indexing. *Ebsco Discovery.* [Online] Ebscohost, 2014. [Cited: November 18, 2014.] http://www.ebscohost.com/discovery/content/indexing.
38. University of Skovde in Sweden selects WorldCat Local. *OCLC News Releases.* [Online] OCLC, August 9, 2012. [Cited: November 20, 2014.] http://oclc.org/en-AU/news/releases/2012/201250.html.
39. **Shapiro, Steven David.** We are all aggregators (and publishers) now: how discovery tools empower libraries. *Library Hi Tech News.* 2013, Vol. 30, 7.
40. About the Alliance. *Orbis Cascade Alliance.* [Online] 2014. [Cited: November 20, 2014.] https://www.orbiscascade.org/about/.
41. Ebsco and Ex Libris. *Orbis Cascade Alliance.* [Online] 2014. [Cited: November 20, 2014.] https://www.orbiscascade.org/ebsco-ex-libris/.
42. *Major Discovery Product Profiles.* **Breeding, Marshall.** 0024-2586, 2014, Library Technology Reports, Vol. 50.
43. **Rowe, Ronda.** Web-Scale Discovery: A Review of Summon, EBSCO Discovery Service, and WorldCat Local. *The Charleston Advisor.* July 2010, Vol. 12, 1, pp. 5-10.
44. **Wisniewski, Jeff.** Web Scale Discovery: the future's so bright, I gotta wear shades. *Online.* 2010, Vol. 34, 4.
45. *A Century of Cooperative Programs Among Academic Libraries.* **Weber, David C.** 3, May 1976, College and Research Libraries, Vol. 37, pp. 205-221.
46. **Stachokas, George.** *After the Book: Information Services for the 21st Century.* Burlington : Elsevier Science, 2014.
47. *More Bang for the Buck: Increasing the Effectiveness of Library Expenditures Through Cooperation.* **Kohl, David F. and Sanville, Tom.** 3, 2006, Library Trends, Vol. 54, pp. 394-410.
48. About OhioLINK. *OhioLINK: Ohio's Academic Library Consortium.* [Online] 2014. [Cited: November 20, 2014.] https://www.ohiolink.edu/content/about_ohiolink.
49. **Helmer, John F.** John F. Helmer, Executive Director. *Orbis Cascade Alliance.* [Online] 2014. [Cited: November 20, 2014.] https://www.orbiscascade.org/helmer-resume.
50. **Kohl, David.** How the virtual library transforms interlibrary loans - the OhioLINK experience. *Interlending & Document Supply.* 1998, Vol. 26, 2.
51. **Mak, Collette.** Add to cart? E-commerce, self-service and the growth of interlibrary loan. *Interlending & Document Supply.* 2012, Vol. 40, 1.
52. A Brief History of WorldCat. *OCLC.org.* [Online] 2014. [Cited: November 20, 2014.] http://oclc.org/worldcat/catalog/timeline.en.html.
53. WorldShare Management Services. *OCLC.org.* [Online] 2014. [Cited: November 20, 2014.] https://oclc.org/worldshare-management-services.en.html.
54. SirsiDynix Announces BLUEcloud Suite at COSUGI 2013. *SirsiDynix.* [Online] March 19, 2013. [Cited: November 20, 2014.] http://www.sirsidynix.com/press/announces-bluecloud-suite-at-cosugi-2013.
55. **Enis, Matt.** Putting the Pieces Together: Library Systems Landscape. *Library Journal.* April 1, 2014, Vol. 139, 6, p. 32.

56. Definitions of Reference. *RUSA: Reference and User Services Association (A Division of the American Library Association).* [Online] January 14, 2008. [Cited: November 20, 2014.] http://www.ala.org/rusa/resources/guidelines/definitionsreference.

57. **Phan, Tai, Hardesty, Laura and Hug, Jamie.** Academic Libraries: 2012 First Look. *National Center for Education Statistics: Institute of Education Sciences.* [Online] January 2014. [Cited: November 20, 2014.] http://nces.ed.gov/pubs2014/2014038.pdf.

58. **American Library Association.** The State of America's Libraries: 2014. *American Library Association.* [Online] 2014. [Cited: November 20, 2014.] http://www.ala.org/news/sites/ala.org.news/files/content/2014-State-of-Americas-Libraries-Report.pdf.

59. *Using Meebo's embedded IM for academic reference services.* **Breitbach, William, Mallard, Matthew and Sage, Robert.** 1, 2009, Reference Services Review, Vol. 37. 0090-7324.

60. Why choose LibraryH3lp? *LibraryH3lp.* [Online] Nub Games, 2014. [Cited: November 20, 2014.] https://us.libraryh3lp.com/why.

61. QuestionPoint. *OCLC.org.* [Online] 2014. [Cited: November 20, 2014.] https://oclc.org/questionpoint/features.en.html.

62. **Wanucha, Meghan and Hofschire, Linda.** U.S. Public Libraries and the User of Web Technologies, 2012. *Library Research Services.* [Online] November 2013. [Cited: November 20, 2014.] http://www.lrs.org/wp-content/uploads/2013/11/WebTech2012_CloserLook.pdf.

63. **Shepherd, Jodi and Korber, Irene.** How do I search for a book? *College & Research Libraries News.* 2014, Vol. 75, 4.

64. Brochure for Librarians. *My Info Quest.* [Online] 2014. [Cited: November 20, 2014.] http://www.myinfoquest.info/data/BrochureForLibrarians.pdf.

65. Company History. *OverDrive.* [Online] 2014. [Cited: November 20, 2014.] http://company.overdrive.com/company/who-we-are/history/.

66. Public Libraries. *OverDrive.* [Online] 2014. [Cited: November 20, 2014.] http://company.overdrive.com/libraries/public-libraries/.

67. Frequently Asked Questions. *freegal music.* [Online] 2014. [Cited: November 20, 2014.]

68. **Kelley, Michael, Schwartz, Meredith and Enis, Matt.** Digital Newsstand Debuts. *Library Journal.* June 15, 2012, Vol. 137, 11, p. 14.

69. **Kelley, Michael.** Midwest tape to go digital with Hoopla. *Library Journal.* 2012, Vol. 137, 7.

70. **Wolfman-Arent, Avi.** *College Libraries Push Back as Publishers Raise Some E-Book Prices.* [online] June 16, 2014. The Chronicle of Higher Education.

71. How it works. *Kanopy.* [Online] 2014. [Cited: November 20, 2014.] https://www.kanopystreaming.com/about-us/platform.

72. Get It Now. *Copyright Clearance Center.* [Online] 2014. [Cited: November 20, 2014.] http://www.copyright.com/content/cc3/en/toolbar/productsAndSolutions/getitnow.html.html.

73. **OCLC.** The Library Brand 2010. *Perceptions of Libraries, 2010: Content and Community.* [Online] 2010. [Cited: November 20, 2014.] http://www.oclc.org/content/dam/oclc/reports/2010perceptions/thelibrarybrand.pdf.

Chapter 13
Privileged Information: The Corporatization of Information

Frances Eames Noland

13.1 Information Literacy

Information Literacy can be defined as a set of skills necessary to find and evaluate information and determine its accuracy, quality, and reliability. This includes research abilities, logical evaluation of content, and background checking skills. Throughout history, information literacy skills have been essential to the scholar and the professional who wishes to not only absorb information, but to assess the accuracy of this information. Studying information further, rather than taking it at face value can greatly benefit the reader, and train him or her to become a critical thinker. This will encourage scholarly discourse, exploration of ideas, and sculpt the reader's ability to come to his or her own conclusions, instead of those reached by the author of the piece. In this way, readers will become active in the research process and create a dialogue with their peers and with the work itself.

In the rapidly changing digital world, information literacy becomes more important than ever to evaluate credibility. As more and more sources become readily available, the reader must learn to distinguish a quality source from a falsified one. This is simply due to the prevalence of information, as opposed to an overall diminishing quality of published materials. Since self-publication has become popularized through the use of blogs and social media, as well as online publication of print materials, the internet presents a vast network of opinions, factoids, findings, hoaxes, and sometimes even leaked information. The Digital Age brings with it an exciting array of new sources, previously unavailable to researchers.

F.E. Noland (✉)
University of Oxford, Oxford, UK
e-mail: fringleton@hotmail.com

© Springer Science+Business Media New York 2016
N. Lee (ed.), *Google It*, DOI 10.1007/978-1-4939-6415-4_13

13.2   Privileged Information

Privileged Information is information that requires monetary compensation or some type of credentials to access, or information that is restricted from the general public. This is a result of the monetization of access to resources. There are three information access levels: primary, secondary, and tertiary. Primary access information is "equal access" information, resources that are readily available to the public. Examples of these resources include wikis, websites, blogs, and search engines. The next level is Secondary Access, which can also be described as "paid access". Secondary resources can be described as any form of information that requires monetary compensation. Such resources include books, periodicals, films, and music. The final level is Tertiary Access or "privileged information", which requires both money and credentials for access. These types of sources include academic journals, scholarly databases, private libraries, and higher education courses and course materials. Through the corporatization of information, this tiered system was built, which benefits the scholars and professionals at the tertiary level, yet disadvantages those with access only to primary sources.

The Digital Age has drastically changed all three information access levels, bringing with it the promise of a new age of research. Tertiary Access has been greatly transformed by the Digital Age, where scholarly resources are now able to be catalogued electronically. This allows for better user interaction with the materials. Through the use of online academic search engines and databases, scholars can find more suitable sources that are applicable to their research topics. Unlike their predecessors, the modern professor or student does not need to spend lengthy hours poring over the library's academic journals, searching for appropriate sources. Instead, the modern scholar can use the search engines purchased by the institution to more accurately locate the correct materials, as well as topics of interest within those sources.

13.3   Librarians

Institutions often purchase a variety of specialized online databases to provide students with easy access to information through the web, many offer remote campus services that allow the student to conduct research outside of the library. This type of solo research can be efficient for quick references of materials, yet it can also be much more of detriment to the user, stripping them of one of the most essential components of the library: the librarian. As Janes [2] ponders, "If anybody had proposed in 1990 that we open up libraries 24 hours a day and leave them unstaffed for half or more of that time, that person would have been scorned, and rightly so." (p. 45). The remote access of library resources provides the scholar with more ways in which to access a greater catalogue of information, but does not provide the means in which to survey, interpret, and decode this information.

13.4   Digital Media

The Digital Age has also altered the way in which consumers interact with Secondary Access information, making the resources more readily available to the user, yet still monetizing them. This is being achieved through the use of multiple platforms in which to access these materials, using devices such as computers, tablets, smartphones, and smart watches, as well as through applications and remote desktop functions. As Pavlik [4] details, "Digital media access and display devices come in at least two basic forms: fixed location and mobile." (p. 37). Massive archaic fixed location devices such as the early televisions, radios, personal computers, and even physical space libraries are locked in place, constraining the consumer's access to only the time they inhabit the particular space in which the information is held. With the creation of the mobile device brings a new era of freedom, liberating the consumer by allowing them to access materials anytime, and anywhere. Pavlik discusses how mobile devices have enabled the industry of electronic books. "These digital bookstores provide a collection of instantly available full-text works that dwarfs even the largest brick-and-mortar sellers." (p. 146). This instant access makes the characteristics of the modern day Secondary Access materials much like that of the Primary Access material, yet the clear differentiation lies in the monetization of their access.

Behind this corporate branch out to electronic access lies the steady demand for digital sales, rather than those of physical books, newspapers, DVDs, and CDs. The key component of these consumer demands is the immediacy that digital media provides. As Brock [1] reports, "In 1997, Japan's extraordinarily successful and economically resilient newspapers had recorded a total circulation of 53.8 million copies. By 2011, this had fallen to 46.8 million." (p. 208). This dramatic transformation of the print industry can be seen not only in the methods in which the information reaches the consumer, but in the actual information that is presented. In the section entitled *Error is Useful,* Brock describes how the Digital Age has changed American journalism beyond recognition. "The new journalism that has flourished in digital's new possibilities is gossipy, unashamedly popular, and heedless of the fact that many journalists would call it a 'down market'." (p. 219). In this way, many Secondary Access sources have begun to lose credibility, jeopardizing their long held superiority over Primary Access materials.

13.5   Total Information Awareness

The number of Primary Access materials have increased exponentially with each year of technological innovation, bringing with it a new wealth of information readily available to internet users everywhere. This new collection of internet based resources contributes greatly to the vast library of information. These new Primary Access materials are the lifeblood of the modern era of Total Information

Awareness, and are beginning to compete with the Secondary and Tertiary resources. As Lee [3] observes, "The industry led TIA is an evolving two-way street. Facebook, Google, YouTube, Wikipedia, and even the controversial WikiLeaks all collect information from everywhere and make it available to everyone, whether they are individuals, businesses, or government agencies." (p. 189). This means of making access a public right is something that challenges the very nature of information and privilege, baffling the institutions that create, distribute and monetize Secondary and Tertiary information. Although many Primary Access sources and databases are credible and cite all sources, their reputability is constantly questioned by institutions, which is cleverly disguised as information literacy concerns. In library and research courses, these sources are often generalized and stripped of any examples of Primary Access academic materials, making the scholar dependent upon the corporatized information system.

If this information access gap is so disparate and obvious, how can it still exist? The various levels have their justifications for the price they bare. Primary Access information provides the advantage of equal access; yet due to the very nature of its accessibility, it is corrupted by its inherent lack of reliability. The quality of Primary Access materials is difficult to assess in comparison to other types of information. Secondary Access materials require a sum for a guarantee of renowned authors and reliability, yet it restricts those who cannot afford the often rather inflated prices. Tertiary Access information demands the highest price, and delivers the most accuracy and reputable publishing institutions, yet the highest access level restricts the information held within its pages from the vast majority of the population.

The dilemma of privileged information may eventually be addressed, but until that time the rise of new sector of information will assist the general public in their search for knowledge: Open Source. This type of information can be defined as content that would typically be restricted for Secondary and Tertiary Access only, yet despite this has been released to the public in some capacity. Open Source materials are often directly published by universities to increase interest in their research, or even directly uploaded by authors to encourage young scholars to cite their articles in their academic works. This system is a very effective one, providing free resources in exchange for free publicity. In this way, not only do the readers and consumers benefit from the information, but the publishers and authors gain further recognition for their works.

13.6 Digital Commons

Anyone with a thirst for knowledge, a need for information, or simply an interest in a given topic should make use of these complimentary resources. Digital Commons is one of the most comprehensive scholarly Open Source databases, covering a plethora of topics throughout the fields of humanities, arts, social sciences, natural sciences, engineering, mathematics, medicine, education, law, and business. Through Digital Commons researchers can find academic journal articles, theses, dissertations, and books, published by universities for the benefit of the public. Another academic resource is Academia.edu, which not only provides scholarly journal articles, but a unique interaction opportunity. Through Academia.edu students can access articles directly from the authors of these works. This takes the academic process a step further than reading, and creates an actual dialogue between the reader and writer, giving the student a chance to converse with professors and students from other educational institutions and private researchers.

As Lee [3] concludes, "Notwithstanding the potential risks and benefits of information sharing, two-way street of Total Information Awareness is the road that leads to a more transparent and complete picture of ourselves, our governments, and our world." (p. 191). Equal access to information is vital to every individual across the globe; this access will enable those without the monetary means or credentials to attain knowledge through the archaic system of a bygone era. The Digital Age brings the necessary tools for mass distribution and can enable this new age of Equal Access. The argument that Primary Access materials have questionable credibility will soon be an obsolete one, as the three access levels continue to morph, eventually converging or disappearing altogether.

References

1. Brock, G. (2013). *Out of print: Newspapers, journalism, and the business of news in the digital age.* London: Kogan Page Ltd.
2. Janes, J. (2003). *Introduction to reference work in the digital age.* New York, NY: Neal Schuman Publishers Inc.
3. Lee, N. (2015). *Counterterrorism and cybersecurity: Total information awareness.* Cham: Springer International Publishing AG.
4. Pavlik, J. V. (2008). *Media in the digital age.* New York, NY: Columbia Press.

Chapter 14
Communication and Language in the Age of Digital Transformation

Tiana Sinclair

14.1 The Philosophy

Technology and futurism usually go hand-in-hand. In consumer electronics the need to innovate is more necessary than ever. Manufactures now simply cannot be certain that a given platform or device will dominate the industry for longer than a few years before something else comes round [1]. While futurism includes not simply the future of gadgets, the field found itself pushing away some of the perceived "softer" elements of foresight: social change, family structures, cultural impacts—in favour of mathematical modelling and technology.

This chapter will explore the simple yet difficult phenomenon of communication. How do we communicate in the age of rapid technological shift? And with the increase in labour automation rates how do we communicate to machines more effectively?

Let's start with looking at the semantic underlining of the concept of information —how we perceive it, store it and communicate it to others.

14.2 Information

Abstractly, information can be thought of as the resolution of uncertainty. In the case of communication of information over a noisy channel, this abstract concept was made concrete in 1948 by Claude Shannon in A Mathematical Theory of Communication [2], in which "information" is thought of as a set of possible

T. Sinclair (✉)
London, UK
e-mail: tiana@futuretechtrack.com

messages, where the goal is to send these messages over a noisy channel, and then to have the receiver reconstruct the message with low probability of error, in spite of the channel noise. Its impact has been crucial to the success of the Voyager missions to deep space, the invention of the compact disc, the feasibility of mobile phones, the development of the Internet, the study of linguistics and of human perception, the understanding of black holes, and numerous other fields.

14.3   The Unspeakable Level

Language is one of the most efficient forms of person to person communication. However, it can be argued that we live our lives on unspeakable level and use the language to help us process what's happening around us, identify it, reflect on that and share the experience with others.

Any map or language to be of maximum usefulness if in structure it was similar to the structure of the empirical world. Likewise from the point of view of the theory of sanity [3] any system or language should in structure be similar to the structure of our nervous system. Language is a fundamental *psychophysiological* function of a man, scientific investigation of a man in all his activities.

Empirical evidence suggests that different man-made verbal systems can stimulate or hamper the functioning of human nervous system. The language of the new standard must explore the relationship between the actual objects outside our skin and our personal feelings inside— so the only link between the objective and the verbal is structural. Structure can be considered as a complex of relations, and ultimately as multi-dimensional order. Meaning of the word depends on meaning of the words used to describe it.

The world is made of language, and sometimes this can be unreliable. The numeral 2 only means something because it is not 1 or 3. House only exists because it isn't a boat or a street. I am only me because I am not someone else.

> The whole system of existence is a closed system floating on nothing, like a locked hovercraft [4].

The duality in the neurorepresentation and cognitive organisation of language holds key interest to scientists [5] Our lives are lived on objective, un-speakable levels, not on verbal levels. In order to necessitate a fundamental revision of the structure of language, a semantic factor the language of four-dimensional structure needs to be introduced. Humans are different from other species because of the *time-binding factor* (which means that the new generation can roughly start off where previous left off).

Alfred Korzybski outlines our need to create a language embracing all the functions—verbal communication, mathematics, science, mental health. Language represents inherent psychophysiological function of human organism that has been neglected for so long. However building of such system is beyond the power of a single man to complete and needs to be addressed by a wider community.

> We are ruled even more, and even less consciously, by the inventors of the wheel, the plow, the alphabet, even the Roman roads [6].

Wilson [6] argues that nervous adjustment of invoking generations that are being forced to develop under unnatural for man semantic conditions imposed on them produces leaders with old animalistic limitations. He describes it as a feature of *bio-survival circuit* where we all have to work hard and hunt for our *food tickets* with the only exception that now it mostly happens virtually. This is also the reason behind our neurotic behaviour when we don't get a tweet or an email back.

Generally we do not use our nervous system properly and have not emerged from a very primitive semantic stage of development in spite of our technical achievements [3].

The ultimate goal is not only to sketch a scientific program for the future (like Aristotle did) but to build a system which at least in structure is similar to the structure of the known facts from all branches of knowledge. Many statements of scientists still have to be translated into a special language in which structural issues are made quite possible, divulging factors in semantic reactions.

All desirable human characteristics (including high 'mentality') have a definite psychophysiological mechanism, easily understood and trained kind of like mastering car driving or spelling. Can it be the next potential human language?

Some problems arise from there. The first one is scientific as it requires a revision of all systems. The second is a practical one given the time and effort required to master the system. Currently there are already some methodologies that overlook this. Neurolinguistic programming, for instance, is a method of influencing brain behaviour through the use of language and other types of communication to enable a person to "recode" the way the brain responds to stimuli and manifest new and better behaviours.

14.4 Logic of Paradox

In the early 80s the urgent need to develop the extremely complex mathematical structure of the Unified Relativity and Quantum Theory (URQT) and to solve its equations resulted in the idea of an artificial (constructed, engineered) universal language. This language was supposed to be a theoretical language like mathematics, a theory of the field described by the language, but in addition it was meant to be a real spoken language to make learning it and using it really easy even from early childhood, ideally from birth [7].

It was decided not to use formal predicate logic as a basis for the new language like it was done for well known constructed language loglan or its modified version lojban [8]. Instead, the most effective tool of modern theoretical physics was used for the foundation of this new language: symmetry group theory. Symmetry was taken in it's fundamental form as the symmetry of oppositions.

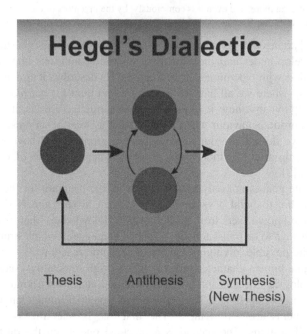

The symmetry of oppositions is known in traditional philosophy under the name of *dialectics* which was popularised by Hegel. Dialectics can be described as the practice of arriving at the truth by the exchange of logical arguments or, in theory "means of extracting the truth" [9]. Dialectics is widely used in natural human languages and is also relatively easy to comprehend. The only problem with the logic of oppositions is that when followed to the very end at each logical point it inevitably leads to paradoxes. In fact traditional dialectics as a way of thinking may be called the "logic of the paradox" because it seemingly contains paradoxes at any given level.

It was quickly realised that paradoxes did not simply represent some kind of bizarre flaw in the theory, but that they were actually powerful "points of singularity" giving rise to new ideas and innovative solutions represented by the means of a new language. Later it was understood that paradoxes served also as fundamental generative rules for the fractal structure of the constructed language and all of its forms.

After the fundamentals of the new language were formulated, together, the group of young physicists developed the body of the language. In homage to its dialectic roots, the new language was named Dial [8]. This language wasn't popularised and is no longer in use however there are studies that claim there is an evidence showing the change in brain's neuroplasticity throughout the tests, especially with the younger segment of people. Very little is known about the history of the language and its destiny remains unknown.

14.5 Artificial Languages

We as humanity developed many languages to communicate our ideas to the machine. There's C++, Ruby, Javascript to name a few. Essentially computers talk to each other just like we do. In the last decade a lot of research has been done into the study of Human Computer Interfaces—how to communicate to machines more efficiently. What's interesting is that on the front end it still has a linguistic input channel but is there a way to exchange information a little more efficiently than in English, Spanish or Chinese? And how do we make them solve the problems more efficiently?

For instance, TRIZ (theory of the resolution of invention-related tasks) [10] developed by the inventor and science-fiction author Genrich Altshuller points to the need to produce a theory which defines generalisable patterns in the nature of inventive solutions and the distinguishing characteristics of the problems that these inventions have overcome.

The graph below summarises its principles.

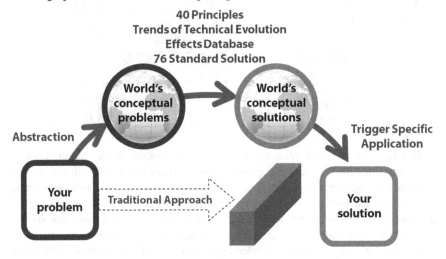

Problem requires an inventive solution if there is an unresolved contradiction in the sense that improving one parameter impacts negatively on another. He later called these "technical contradictions". The full list of contradictions can be viewed in a TRIZ Matrix [11].

Worsening Feature → / Improving Feature ↓		Weight of moving object	Weight of stationary object	Length of moving object	Length of stationary object	Area of moving object	Area of stationary object	Volume of moving object	Volume of stationary object
		1	2	3	4	5	6	7	8
1	Weight of moving object	+	-	15, 8, 29,34	-	29, 17, 38, 34	-	29, 2, 40, 28	-
2	Weight of stationary object	-	+	-	10, 1, 29, 35	-	35, 30, 13, 2	-	5, 35, 14, 2
3	Length of moving object	8, 15, 29, 34	-	+	-	15, 17, 4	-	7, 17, 4, 35	-
4	Length of stationary object		35, 28, 40, 29	-	+	-	17, 7, 10, 40	-	35, 8, 2,14
5	Area of moving object	2, 17, 29, 4	-	14, 15, 18, 4	-	+	-	7, 14, 17, 4	
6	Area of stationary object	-	30, 2, 14, 18	-	26, 7, 9, 39	-	+	-	

Just like the physical attributes of object are subjects to the certain laws of co-relations, the physical and metaphysical has also been observed to display similar patterns. For instance, the subject named *synergetics* explores small actions affecting big systems. The term was coined by Buckminster Fuller who has attempted to define its scope in his two volume work *Synergetics* [12]. It remains an iconoclastic subject ignored by most traditional curricula and academic departments because of its complex nature—it is in fact a so-called aggregator of variables rather than the study itself. All of the exact sciences of physics and chemistry have provided for the accounting of the physical behaviours of matter and energy only through separate, unique languages that require awkward translation through the function of the abstract interpreters known as the *constants*. But synergetics embraces the comprehensive family of behavioral relationships within one language capable of reconciling all the experimentally disclosed values including Einstein's energy equation, Euler's topology of points, areas, and lines, Kepler's third law, Newton's theory of gravity, Thermodynamic laws as well as various studies of philosophical nature.

Based on these concepts what may the search engine of the future be like? People might be using 'search' to discover structural relations. Already there are alternative search engines like Wolfram|Alpha [13] that question the way we look for answers on the web. They might understand and interpret what's going on around them fully and have healthy semantic relations. Mankind might become a truly a time-binding nation: we will master ways to pass information to our heirs instantly so they could build up on our knowledge, however have a full free will set in place to discover their own doctrines.

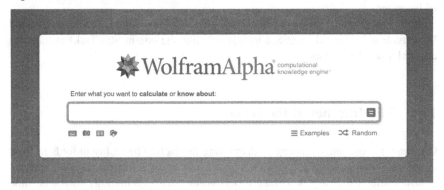

Quantum computers will be much smarter than us. They will be ambiguous and explorative and both will be capable of probable and possible behaviour [14].

When we are born at what point do we start learning about the world and identifying ourselves and everyone around? Are we conscious at that point or does it come with exposure to culture such as language, sciences, humanities?

An interesting case study was presented by FreeSpeech technologies team who are currently working with autistic children to help them communicate more efficiently [15]. Narayanan turns our attention the word 'soup' being represented by an abstract sketch as well as more arbitrary word that stands for 'soup'. Children with autism find it very hard to grasp the contextual relations between the word and the image it represents so simplifications were made to address this gap.

Essentially, FreeSpeech technology represents sentences without language which means you can start constructing a concept from anywhere in the sentence. Also, if the user isn't an English speaker their map would actually hold true in any language. So long as the questions are standardised, the map is actually independent of language.

It can be argued that in the future the methodology behind FreeSpeech could help create a "universal language translator" (like the one in Star Trek) to translate any unknown languages.

14.6   The Internet of the Brain

Our brain is constantly proving or disproving the facts. The reality unfolds in front of us as it's being observed. Now let's take a look at the information on the internet —how difficult it really is to fake a person, an idea? Technology might be time proof but its not idiot-proof or troll-proof and the mediums of information storage are erasable and fragile. What sort of content are we expected to get from people? *On the Internet we have power of the situations where we would otherwise feel powerless. The tech-utopians present it as the new kind of democracy but it isn't. It locks people off in the world they've started with and prevents them for finding anything different* [16].

In the age of digital deception how relevant the phrase "nothing is real until it is observed" really is? The programmes like Catfish explore the dark side of identity manipulation. Where to draw a line between little lies on ones CV and pretending to be someone completely different on the internet?

If this theory would prove to be correct it would follow that we can exist digitally by running ourselves as simulations. And this is what's knows as computational hypothesis of the brain [17].

Technology becoming more and connected with our bodies. So what has changed? Digital immortality (or "virtual immortality") is the hypothetical concept of storing (or transferring) a person's personality in more durable media, i.e., a computer, and allowing it to communicate with people in the future. The result might look like an avatar behaving, reacting, and thinking like a person on the basis of that person's digital archive. After the death of the individual, this avatar could remain static or continue to learn and develop autonomously. A considerable portion of transhumanists and singularitarians place great hope into the belief that they may become immortal by the year 2045, by creating one or many non-biological functional copies of their brains, thereby leaving their "biological shell". These copies may then "live eternally" in a version of digital "heaven" or paradise.

Already scientists suggest that by linking the brains together, they could create Brainets—a system of brains attached together to make an "organic computer" [18]. The language as a mean for communication may be up for the review in that case as well as our psyche mechanisms. As the amount of our daily information intake increases we can see an increase in mental health issues which will need to be addressed in the near-distant future.

By the end of the day what is the purpose of technology? Is it there to highlight our pre-enlightment era habits, are we using Facebook to claim our Rewards of the Tribe [19] due to our need to seek social validation? Are we there to 'burn the witches' and publicly shame [16] the people we don't necessary agree with? Are the app icons with animal icons acting as a symbolism for digital neo-paganism coming back to show us to worship the new gods of the Internet era?

In the same way computer software can run on a different hardware maybe the software of the mind can run on other platforms. Are we connected so much that at times it feels like we want to disconnect for a while? What is awareness? Does to know mean to be aware? And how do we accept or reject information that later forms our knowledge base? All these questions remain unanswered.

Language is an instrument of thought. It must concisely and correctly display and simulate reality. Daniel Webster said this: *"If all of my possessions were taken from me with one exception, I would choose to keep the power of communication, for with it, I would regain all the rest."*

References

1. *Ninja Innovation;* Gary Shapiro, 2015
2. *A Mathematical Theory of Communication;* Claude Shannon, 1948
3. *Science and Sanity: An Introduction to Non-Aristotelian Systems and General Semantics;* Alfred Korzibsky Institute of GS, 1958 http://bit.ly/1khJbhO
4. *The End of Mr. Y;* Scarlett Thomas, Canongate Books, 2008
5. The power of language in the clinical treatment process, p 19; Jason Aronson 1998
6. *Prometheus Rising, p 95;* Robert Anton Wilson http://bit.ly/1Oj6DTK
7. *The Third Project;* M. Kalashnikov. Translated from RoyalLib.ru http://bit.ly/1QMD0AU
8. *Universal Language DIAL: The Logic of Paradox*; Koulikov & Elkin. National Nuclear Research University - MEPHI. Philica online encyclopedia http://bit.ly/1OotO5i
9. *Hegel's Dialectics;* YouTube video https://www.youtube.com/watch?v=v_F4WomLlq0
10. Altshuller's TRIZ https://en.wikipedia.org/wiki/TRIZ
11. TRIZ table http://www.triz40.com/TRIZ_GB.php
12. *Synergetics*; Buckminster Fuller, Macmillan Publishing, 1975 http://www.rwgrayprojects.com/synergetics/print/pc.pdf
13. Wolfram|Alpha http://www.wolframalpha.com
14. *The Astonishing Simplicity of Everything*; Neil Turok, Perimeter Institute Lecture, 2015 https://www.youtube.com/watch?v=f1x9lgX8GaE
15. *A Word Game to Communicate in Any Language;* TED Talk by A. Narayanan http://bit.ly/1qNR3JO
16. *So you have been publicly shamed, p 116;* Jon Ronson, Riverhead Books, 2015
17. *The Brain with David Eagleman;* David Eagleman 2015 http://video.pbs.org/video/2365575366/
18. *Brainet: scientists could make an internet of human brains;* A. Griffin http://www.independent.co.uk/life-style/gadgets-and-tech/news/brainet-scientists-could-make-an-internet-of-human-brains-10381069.html
19. *Hooked*; Ryan Hoover, Portfolio Publishing, 2014.

Picture References

20. TRIZ Matrix http://www.triz40.com/aff_Matrix_TRIZ.php
21. TRIZ illustration http://www.primaperformance.com/triz.htm
22. General Semantics http://www.generalsemantics.org/archives/igsdiscussionforums/learn-gs.
    org/images/igs/ak_rl_35.jpg

Part IV
Tree of Ennoblement

All religions, arts and sciences are branches of the same tree. All these aspirations are directed toward ennobling man's life, lifting it from the sphere of mere physical existence and leading the individual towards freedom. It is no mere chance that our older universities developed from clerical schools. Both churches and universities—insofar as they live up to their true function—serve the ennoblement of the individual.

—Albert Einstein

We don't read and write poetry because it's cute. We read and write poetry because we are members of the human race. And the human race is filled with passion. And medicine, law, business, engineering, these are noble pursuits and necessary to sustain life. But poetry, beauty, romance, love, these are what we stay alive for.

—Dead Poets Society

Chapter 15
"How"—The Key to Knowledge-Building Pedagogy Success in Supporting Paradigm Shifts for Student Growth and the 4Cs of Future Education

Sandra Lund-Diaz, Mireia Montane and Penelope Beery

15.1 Introduction

Over the past few decades, societies have transitioned from being industrialized and existing in isolation to being globally-entwined and based in knowledge. Classrooms have been transformed from factory-era teaching to a different type of learning for the Knowledge Age, where ideas are the main source of economic growth. The future of education is here today, where new pedagogies are needed to facilitate real-world learning, where students are able to acquire skills and competencies they need to achieve academic success, and where educators can become coaches and facilitators of learning AND co-learners rather than simply deliverers of pre-determined content. Knowledge Building is manifested in different types of classroom-based learning, primarily in student engagement, where content is contextualized and a curriculum is based in phenomenon learning, or "teaching by topic". It incorporates technology with methodologies that allow learners to interact with the content of classroom instruction in a deep and thoughtful manner through an interdisciplinary approach. Students and teachers alike are co-learning, gaining knowledge from the experiences of others to achieve their learning objectives around the production and continual improvement of ideas. Knowledge-building classrooms create opportunities for students to acquire 21st Century skills that support real-world problem-solving, related to so-called STEM skills, particularly analytical skills and the scientific method. We refer to these skills as the "4Cs of Future Education"—**Critical thinking · Collaboration · Communication · Creativity**—that will prepare students for success throughout the education continuum. These skills will also contribute to their preparation as engaged citizens and as productive workers in an era of volatile economies with surging unemployment

S. Lund-Diaz (✉) · M. Montane · P. Beery
Knowledge Building in Action, Miami, USA
e-mail: info@altlearningecosystem.com

© Springer Science+Business Media New York 2016
N. Lee (ed.), *Google It*, DOI 10.1007/978-1-4939-6415-4_15

and under-employment, particularly among young adults, and a rapidly changing, globalized labor market.

The Alternative Learning and Teaching Ecosystem (ALTE) is a model for the 4Cs of Future Education mimicking education systems in countries consistently ranking high on skills assessments that have successfully reimaged learning and teaching. ALTE blends knowledge-building pedagogy and teacher training while promoting UNESCO information and communication technologies (ICT) competency framework for teaching (CFT) as well as OECD standards. The ICT CFT supports a teacher's development in *Technology Literacy* for more efficient learning, *Knowledge Deepening* that enables students to acquire in-depth knowledge of their school subjects and apply it to complex, real-world problems, and *Knowledge Creation*, enabling students, citizens and the workforce they become, to create the new knowledge required for more harmonious, fulfilling and prosperous societies. The ALTE model was built on knowledge-building principles with methodologies specifically developed to support these three approaches to 21st Century education.

Across time, men and women have accomplished great feats that did not seem possible in their wildest imagination... building the great pyramids by positioning stones seemingly impossible to move; discovering far-flung lands despite the common-held belief that the world was flat; man taking flight, reaching for the stars and landing on the moon. These are just a few examples of feats accomplished in spite of what many thought were insurmountable challenges. They required courage, persistence and faith, accompanied by a single-syllable word..."HOW?", propelling these accomplishments into our human history. By answering "How?", mankind pursues solutions to problems and situations of great consequence to all the peoples of the world, continually amazing disbelievers. Answering "How?" accomplishes what many thought to be impossible, and the impossible not only becomes possible but common place. "How?" withstands the testament of time and continues to provoke the innovation, ingenuity and creativity of those who dare to dream of accomplishing something great. And contemplating "How?" is the key to knowledge-building success.

15.2 A Model for the 4Cs of Future Education

Knowledge Building exploits those same traits—scientific curiosity, ingenuity, innovation and creativity—that supported the greatest discoveries of mankind as opportunities to engage students in interactive learning, beginning at a young age. These students will one day become our future scientists, educators, workforce and leaders, and through knowledge building they acquire critical skills that will prepare them for success in school, and later in life. These are the 21st Century skills that support real-world problem-solving, related to so-called STEM skills, particularly analytical skills and the scientific method. We refer to these skills as the "4Cs of Future Education"—Critical thinking • Collaboration • Communication • Creativity —that will prepare students for success throughout the education continuum. These are the skills that will also contribute to their preparation as engaged citizens and productive workers in an era of volatile economies with surging unemployment and under-employment, particularly among young adults, and a rapidly changing, globalized labor market.

The 4Cs of Future Education are integral to Knowledge Building, which also supports deep learning through which students will develop a set of competencies to master the subject matters of their curriculum and instruction. They are able to understand academic content and apply their knowledge to problem-solving by engaging the 4Cs: thinking critically, working collaboratively, communicating effectively in the classroom, and applying creativity to come up with innovative solutions to the problems of the world. After all, the ability to imagine, create or discover should not fade as a person grows; rather it should be nurtured and strengthened before students are crippled by the norms of an industrial-era education system that is currently mainstream for learning and teaching. With the 4Cs, students will also develop academic mindsets for the classroom as well as for a job later in life. The 4Cs prepare students to achieve at high levels and gain mastery of core academic content, whether in traditional subjects or in interdisciplinary fields that merge several key fields of study. It is here that knowledge-building pedagogy and methodologies are most effective, giving the students tools to ask the right question to effective learning—HOW?

In creating an effective model for the 4Cs of Future Education, we looked at several education systems and focused on countries consistently at the top of international rankings of learning assessments such as PISA (Program for International Student Assessment) regarding reading, mathematics, and science literacy. We found that they had built high-quality pathways to learning for their children in primary and secondary education by rethinking teaching and learning [1]. They had drastically changed their education methods to introduce a curriculum based around phenomenon learning. This is where subject-specific lessons in core subjects such as Geography, History, Math, Economics, Biology, Earth Sciences and others become interdisciplinary studies through project-based learning aligned to standards. In phenomenon learning, students are taught cross-subject topics, such as climactic change, sustainability, nutrition, and economic trading zones that incorporate multi-disciplinary content from traditional subjects to answer the question, HOW?. This is the phenomenon learning of those model education systems, and the effective use of technology for content delivery and the continued professional development of their educators is key to success.

These models also reimagine learning, where the format of the traditional, more passive approach to learning is replaced by a more collaborative method of learning. Instead of sitting and listening to the teacher standing in the front of the classroom, students are encouraged to work together to discuss and solve problems in a collaborative small-group learning environment as co-learners, led by the teacher who becomes a facilitator of learning rather than a transmitter of pre-determined content. We based our model for the 4Cs of Future Education, the Alternative Learning and Teaching Ecosystem (ALTE), on lessons learned from the phenomenon learning of those countries that have successfully reimaged learning and teaching.

ALTE blends knowledge-building pedagogy and teacher training, mimicking those high-quality education systems by supporting the following elements:

- Computer-supported collaborative learning around topics through the use of education technology developed specifically to support the highly-researched principles of knowledge-building pedagogy [2];
- Teacher training in the use of education technology developed to support collaborative learning around topics through a multi-disciplinary approach while promoting proficiency in information and communication technologies and UNESCO's competency framework for teaching [3];
- Bringing education technology into the classroom and utilizing embedded assessment tools to track student growth and conduct formative and summative assessments in real time;
- Partnering classrooms world-wide to build relationships between students and between educators, including teachers, principals and technology staff.

15.3   Contextualization and Personalized Learning

ALTE encourages learning from a reality-centered point of view around ideas. Theme-based learning and exploration, coupled with activities based on real issues applicable to everyday life, convert the classroom into living labs, engaging teachers and students alike in personalized, meaningful learning. They examine the issues at hand through the HOW? lens, applying the 4Cs of Future Education in a methodical, systematic approach based on scaffolding. The results: a multidisciplinary perspective to improve student growth and academic achievement through deep learning, supporting mastery of core academic content, and building skills critical to academic and professional success in the 21st Century.

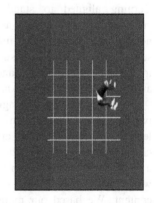

CASE STUDY: A Multidisciplinary Approach to Studying Math.....

Welcome to the ninth grade math class of Mr. Hamilton, working with his students asking HOW they can improve the long jump for an upcoming sports competition. Students create digital media that trace their practice sessions, and by watching their videotaped performances in long-jump trials and comparing their performance to grids and charts, students study important math and physics concepts - angles, momentum and distance. But they also study physiology and begin to understand how muscles work, and human health topics where they also learn how nutrition and adequate sleep contribute toward the formation and maintenance of healthy muscles.

Students work together in small groups to share their ideas on how to improve their long jump and increase their distance. They do research on the different subjects incorporated into what started as a math class transformed into a multi-disciplinary study of subject matters relevant to their lives.

This case study demonstrates an important aspect of the ALTE Model—personalized relevance of the subject matter being studied. Mr. Hamilton's students are motivated to learn because they are studying THEIR muscles, THEIR performance and unlocking THEIR potential. Another important aspect to consider—how engagement in hands-on activities changes not only the student but the teacher as well. Case in point: many of the students in Mr. Hamilton's class volunteer to become peer tutors to work alongside their teacher, to produce digital media that reflect the curriculum and standards that high school students are expected to master. Mr. Hamilton taps into his own creative potential by overseeing the production of content tailored to his student population. Under the status quo of teaching, mathematics teachers are expected to be content conveyors, following pre-defined curriculum in preparation for accountability tests; they are not expected to be creative in producing content or products. However, by producing these video presentations of students performing long jumps, Mr. Hamilton becomes an effective, creative agent at the complex intersection of navigating mathematical content and student cognition.

This multidisciplinary approach to learning complementary areas of studies has a direct impact on the focused subject matter or discipline. As in the case study of Mr. Hamilton's math class, not only do these group exercises allow the entire class to master the math concepts and content being studied, but they also learn complementary aspects of biology, physics and life sciences, thereby acquiring a deeper understanding of the subject matter and performing better on tests. This multi-disciplinary, multi-media student-driven approach is ideal for closing the achievement gap for students with learning difficulties, moving learning away from being literature driven and into a realm the student can easily grasp, which also serves English language learners well. It also taps into personalized learning, where the knowledge-building pedagogy on which ALTE is based tailors the learning experience to meet different needs of the students in small group learning environments. And when teachers from other subject matters interact with students they may not normally have contact with, the learning is enriched through an inter-disciplinary approach. This interaction can come from a teacher across the hall or around the world.

The ALTE computer-supported collaborative learning paradigm promotes discourse around ideas, with learning based on collaborative problem-solving rather than memorization. The use of ICTs and specialized knowledgeware support the constructive use of authoritative information in acquiring knowledge around a theme. Students learning in knowledge-building learning environments not only develop competencies and increased literacy skills because they are constantly reading and writing, but also come to see themselves and their work as part of a society-wide effort to advance knowledge frontiers. They are able to create new knowledge from the workings of the group for applications in a global community. When students work collaboratively on revolving their studies and research around a common theme, they are able to apply the "act locally/think globally" philosophy to turn local issues into global issues, again taking a multi-disciplinary approach to studying different subject matters. By partnering with other classrooms around the globe, an international exposure is brought to the work students do on building knowledge around problems affecting their community.

Knowledge Building in Action [4] is the umbrella 501(c)(3) nonprofit organization providing the infrastructure for ALTE. During past school cycles, it facilitated the participation of U.S. classrooms in the Knowledge Building International Project (KBIP), a successful multi-year international collaboration between classrooms and the COMConeixer project [5] (translated from Catalan into "building knowledge together") in Catalonia, Spain, on which ALTE is based. Teachers selected water—which is of vital importance to everyone on this planet—as the common theme of their inquiry-based collaborative learning. Educators in U.S. classrooms participated in KBIP, where the ALTE module of teacher training and student learning was incorporated into the curriculum. Ms. Anderson's 6th grade social studies at the 68th Street School in South-Central Los Angeles focused on public sector efforts to supply clean water for their citizens and learned about the importance of conserving water so that there would be a sufficient supply for the entire world; Ms. Erlington's 6th grade social studies classes featured project-based learning within the context of Knowledge Building, and students in Mr. Aviles' 6th grade math class studied a prototype water piano to listen to different sounds based on water volume and density. Ms. Morales' 8th grade history class at the Lou Dantzler Middle School, also in Los Angeles, explored how great civilizations were formed around water sources and the conquest of societies made possible by navigation on the seas and oceans. Mr. McKenna used the ALTE model for his Project-Based Credit Recovery class of Special Education students at Malaga Cove Academy in Palos Verdes, CA, studying history from a multi-disciplinary perspective. Mr. Crabtree engaged his 7th and 8th grade honors science students at Rizzoli Academy for Gifted and Talented Students in Hartford, CT in exploring scientific concepts, and Ms. Riad's class of 9–11 year olds at Lake Trafford Elementary School in Immokalee, FL, studied their core subjects of social studies utilizing the ALTE Model to understand ocean currents and the Gulf Stream, focusing on wide-spread effect of the Deepwater Horizons oil spill.

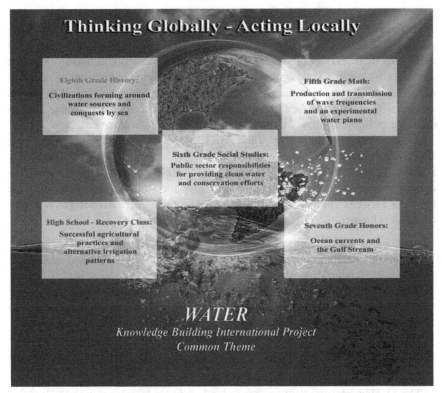

These U.S. teachers were slated to be partnered with their colleagues teaching the same age groups and subject matters in other parts of the world, still focusing on the common topic of water. These included teachers in Quebec and Toronto, Canada; Barcelona and Tarragona, Spain; Helsinki, Finland; Bari, Italy; Oslo, Norway; Porto and Lisbon, Portugal; Bogota, Colombia; Puebla, Mexico; and Hong Kong. Teachers from these other countries were also studying the water topic from both a local perspective as well as a global perspective. Several examples of questions posed by the teachers to their students included: "Why are the fish dying in the St. Lawrence River?" "How did an oil spill in the Gulf of Mexico reach the shores of Finland?" "What type of forests nourished with rainwater provide raw material for the housing boom in China and how does pollution impact the cycle?" "What is happening with global warming and sea levels rising?"

15.4 Promoting UNESCO's Competency Framework for Teaching

Student progress in the types of learning environments proposed by ALTE does not happen naturally without facilitation. Teacher intervention is needed to coordinate the small groups and mediate their interactions. Additionally, acknowledging the role of the teacher in raising a student's achievement is of utmost importance. ALTE provides initial training, ongoing mentoring and support for teachers to serve as leaders, motivators and facilitators of learning while gaining proficiency in the use of technology in the classroom. It also supports the interactions between principals, technology staff and other faculty. An important objective of the teacher training component of ALTE is partnering educators with their peers and colleagues for mentoring. The ALTE Community of Practice facilitates virtual interactions, archived material, and an exchange of information as well as ongoing professional development.

Another important objective of the ALTE Model refers to efforts that promote UNESCO information and communication technologies (ICT) competency framework for teaching (CFT). UNESCO's ICT CFT supports standards that outline competencies that teachers need to master for student motivation, and the teacher's as well. The Framework is arranged in three successive stages of a teacher's development:

- *Technology Literacy* enables teachers to acquire competencies to incorporate technology, learning materials and ICT equipment to inspire student motivation to use ICTs in order to learn more efficiently.
- *Knowledge Deepening* enables students to acquire in-depth knowledge of their school subjects through multi-disciplinary approaches to content mastery around topics and apply it to complex, real-world problems.
- *Knowledge Creation* enables students, citizens and the workforce they become to create the new knowledge required for more harmonious, fulfilling and prosperous societies.

The ALTE model was built on knowledge-building principles with methodologies specifically developed to support these three approaches to learning and teaching. The model is a complete ecosystem consisting of knowledge-building pedagogy, educational technology, teacher training, and tools for formative and summative assessments of student growth, created specifically to support innovative approaches to 21st Century education. Through ALTE, we are able to partner classrooms in K-12 learning environments (standard classrooms, after-school programs, Department of Defense schools, alternative learning venues, international baccalaureate coursework, etc.) anywhere there is broadband. The actual educational technology consisting of the Knowledge Forum as an electronic workspace is also available for implementation on local servers. The Knowledge Forum is a product of 25+ years of research by an international network of educators and engineers [6], proven effective in supporting students acquiring 21st Century skills, particularly the 4Cs of Future Education and tools that help students ask the important questions of life.

15.5 Conclusions

The ALTE model supports 21st Century education through guided student research within technology-rich collaborative environments for PreK-12 learning. The model partners classrooms around the globe through the use of computers, multimedia technology, the Internet and specialized educational technology known as the "Knowledge Forum™", an electronic workspace and platform that supports the scaffolding processes of scientific inquiry and embedded applets to perform the formative assessments needed to verify student growth [7]. The platform supports Knowledge Building both in the creation of notes contributed by the students and in the ways they are displayed, linked, and made objects of further work. Revisions, elaborations, and reorganizations over time provide a record of group advances, like the accumulation of research advances in a scholarly discipline. When notes (text, graphics, multimedia, videos, etc.) are added to the Knowledge Forum's database, students are able to search existing notes, comment on other students' notes, or organize notes into more complex structures. As the database grows, the workspace provides a progressive trace of how ideas have evolved in the class, and the database helps to formally show and document the classroom community's knowledge advancement while helping students further advance their ideas. Students build on each other's notes by agreeing, asking and answering questions, offering opinions, and establishing a culture of accepting individual difference. Scaffolds allow users to add a theory about a problem and build-on or critique a theory—of their own or of another member of their group. An endless improvability of ideas is supported by the ability to create increasingly high-order conceptual frameworks. It is always possible to reformulate problems at more complex levels, create a rise-above note that encompasses previous rise-above notes, or to create a more inclusive view-of-views. Notes and views can be revised at any time, unlike most discussion environments that disallow changes after a note is posted. Processes of peer review and new forms of publication engage students in group editorial processes. Published works appear in a different visual form and searches can be restricted to the published layer of a database. Analytic toolkits embedded in the Knowledge Forum platform are used to assess social network patterns when we are interested in collaboration dynamics, and vocabulary growth when we are looking at concept attainment. Knowledge builders monitor their work, and engage in self-assessment rather than being totally dependent on external evaluations.

These activities have a clear goal of co-creating new perspectives and advancing knowledge beyond the limit of an individual. Numerous research findings show that this approach, specifically the knowledge-building approach, induces motivation to learning, improves learners' higher order thinking (e.g. critical thinking, problem-solving), and fosters personal development (e.g. communication skills, interpersonal skills and lifelong learning attitudes). Through the ALTE model, they are able to share knowledge and tackle projects that incorporate features of teamwork, real-world content and the use of varied information sources, especially the use of information technology to access authoritative sources. Activity is directed toward the need to educate students for a world in which knowledge creation and innovation are pervasive. The production and continual improvement of ideas of value to the "community", in this case making reference to the community of students, are central to Knowledge Building theory and philosophy. Learners are engaged in the full process of knowledge creation as preparation for entering college and/or the workforce. Every student is a contributor toward the collective knowledge of the group, and the success of the student community depends on the careful cultivation of ideas and the constant use and re-use of knowledge resources—cultivating skills and natural inquisitiveness for the ability to ask the right questions, including "HOW".

References

1. Darling-Hammond, L. (2010). "Steady Work: Finland Builds a Strong Teaching and Learning System", Rethinking Schools, Volume 24, Number 4. Retrieved online at http://www.rethinkingschools.org/restrict.asp?path=archive/24_04/24_04_finland.shtml
2. Scardamalia, M., & Bereiter, C. (2006). Knowledge building: Theory, Pedagogy, and Technology. In K. Sawyer (Ed.), Cambridge Handbook of the Learning Sciences (pp. 97–118). New York: Cambridge University Press.
3. United Nations Educational, Scientific and Cultural Organization. 2011. Transforming Education: The Power of ICT Policies. Paris, France
4. Consell Superior d'Avaluació del Sistema Educatiu. 2014. COMconèixer (How to Know). Assessment of the Project. Learning Through Virtual Communities. Barcelona, Catalonia
5. Institute for Knowledge Innovation and Technology (IKIT). Building Cultural Capacity for Innovation. Available online at http://ikit.org/
6. Knowledge Building in Action, 501(c)(3) nonprofit organization. http://www.kbinaction.com
7. Knowledge Forum electronic workspace.

Author Biographies

Sandra Lund-Diaz is the Founder and Executive Director of Knowledge Building in Action, a 501(c)(3) nonprofit organization created to gain traction for knowledge- building pedagogy and partner U.S. classrooms and teachers with their peers in other countries. She is currently co-Chair of the ATEE Global Education Community.

Mireia Montane is the Director of International Projects at the Catalonia College of Arts, Letters and Sciences, and formerly responsible for the proliferation of knowledge-building classrooms through the Ministry of Education of the Generalitat, the Catalan government entity responsible for K-12 education. She is currently the President of the World Federation of Associations for Teacher Education.

Penelope Beery is a Teacher Trainer at Loyola Marymount University and the Los Angeles School District, and the Director of Training and Professional Development at Knowledge Building in Action.

Chapter 16
Educational Ergonomics and the Future of the Mind

Z.R. Tolan

16.1 Designing Technology to Fit the Brain

Some parts of the brain handle some tasks better than others, just like some parts of the body handle some tasks better than others. It's possible to write with our feet, but don't you find that it's easier with your hands? And don't most of us find it easier with one hand than the other? And once the best hand is found, don't we find some writing utensils easier to use? And once the best utensil is found, have you ever discovered a slightly better way to use it that makes it easier, or makes you able to write faster or longer? So it is with learning, and that is Cognitive Ergonomics.

For all the same reasons as our writing-hand preference, our brains' experiences gradually improve our skills. As infants, even in the womb, our experiences are limited to those which directly impact our senses: temperature, sounds, touch, etc. When we first encounter language, most of us naturally begin to hear and imitate the sounds of words, connect those words to their various meanings, and apply them to contexts in grammar structures. Until this point, we have learned through the mental channel most readily available, presented in this case through hearing and speaking.

Spoken words are themselves auditory symbols in the form of sounds. As you read this, you are interpreting sets of visual symbols in the form of typed letters that represent auditory symbols. Upon learning this second symbol set, your brain developed its ability to learn new information in different ways, auditorily or visually, and paved the first path to *less ergonomic learning*. Of course, that's not to say that learning through written symbols doesn't occur: it's occurring right now. And it doesn't mean that there aren't efficiencies inherent to writing; written communication has been incalculably valuable to human progress. But writing, as a

Z.R. Tolan (✉)
Grand Rapids, Michigan, USA
e-mail: zi01234@gmail.com

© Springer Science+Business Media New York 2016
N. Lee (ed.), *Google It*, DOI 10.1007/978-1-4939-6415-4_16

visual symbolic system that stands for an auditory symbolic system, requires more regions of the brain to process information in fluent succession without error. It takes more work. It makes learning more complex. Of course; it's a symbolic system that stands for another symbolic system that points to something in the real world.

Some types of learning encode information with even more complex layers. For example, math uses numbers (1, 2, 3, etc.), symbols (+, −, x, /, etc.), and variables (x, y, n, etc.) in addition to other rules that affect the order of operations, all of which are translatable into spoken language and which seek to describe situations in real or potential worlds. The symbols of math are even more encoded, requiring even more regions of the brain to process information in fluent succession without error, taking more even more work, and making learning even more complex. And with each new layer of complexity comes an even greater plurality of alternative methods of representation: Whereas written language corresponds to spoken language in a nearly one to one relationship, the symbol-rich language of math is prone to vague written and spoken representation. For example:

"1" is equally meaningful when expressed as "1.0," "1/1," "100 %," "one," "ONE," or "One."
And "1 + 1 = 2" can be expressed as:
"one and one are two"
"one and one equal two"
"one plus one equal two," etc.

These verbal representations can be written in print or cursive, in uppercase, lowercase, or mixed case.

Finally, consider how homophones can add to confusion: "one" and "won," or "two" "to" and "too".

If you're still not convinced, imagine that you are 7 years old and English isn't your native language. That's the real situation for millions of students.

Language and math are but two examples forming a case for the value of cognitively ergonomic educational design. Fundamentally, Cognitive Ergonomics represents an opportunity to reframe our approach to education. Instead of starting by selecting content and scheduling it for delivery, Cognitive Ergonomics analyzes the very nature of the content and seeks to present it to the brain in the way the brain is best equipped to receive it.

16.2   Digital Technology

Digital technology is the great game-changer. Past centuries relied on written text to preserve and transfer information. "A picture is worth a thousand words" is no real replacement for an actual picture. No amount of verbal description of Mona Lisa, even a pixel by pixel color map, can do to the brain what the eyes perceive in a moment. Similarly, no amount of verbal description can replicate the same learning as one good hearing of Beethoven's Fifth Symphony. And neither can musical art describe visual art, or vice versa.

Most people agree that a student's education is better served by field trips to the museum or the symphony than by reading about them in textbooks, but technology offers additional potential to accommodate the nature of the human brain, ergonomically, and efficiently. When it comes to learning, the brain can't be commanded: "LEARN!" Try as we might (and we certainly try!), learning is something that happens as a result of exposure. Quite contrary to methods of force, the brain is always alert to its surroundings in various degrees of conscious awareness. Moreover, human sensory perceptions are limited. Instead of thinking of the traditional Five Senses, which were originally proposed by Aristotle more than 2,000 years ago, modern science lists around 21 senses.[1] For example, the sense commonly called "touch" is more accurately parsed into pressure, pain, itch, external heat and cold, and the body's internal temperature, all of which are separate systems.

Different senses have different educational applications, and as human society develops, the applications continue to develop. The human sense of sound is able to differentiate spoken language from a Mozart sonata from the growl of a bear, and human vision can differentiate objects in immediate focus from our peripheral vision, while still giving degrees of active and continuous conscious focus to them all, simultaneously, and ceaselessly. Since applications of Cognitive Ergonomics naturally leads to potential applications of Cognitive Efficiencies, we ought to explore the range of human potential as it pertains to maximizing the use of new educational technologies.

The brain craves stimulation; how much, what type, and at what pace is a matter of personal taste (so to speak). Remove all sensory stimulation and the brain gets bored. Enter a sensory-stimulating environment and without effort, learning happens.

Not only *can* the brain process stimuli simultaneously, it *does* automatically and it is difficult or often impossible to restrain it from doing so. You may plug your ears from the sounds around you, but try turning off your sense of warm or cold. To paint the point in full color, recognize that right now (and at every conscious moment) your brain is regulating your heart and lungs, detecting internal and external temperatures, monitoring for pain, wavering between various states of

---

[1](Names for senses vary) 1. Sight, 2. Taste (in 5 subcategories: 2.1: Sweet, 2.2: Salty, 2.3: Sour, 2.4: Bitter, 2.5: Umami) 3. Touch, Pressure, 4. Touch, Pain, 5. Touch, Itch, 6. Temperature, External Heat and Cold, 7, Temperature, Internal Body Temperature, 8. Sound, 9. Smell (possessing, by some estimates, 388 different receptors), 10. Pain, Skin, 11. Pain, Bones and Joints, 12. Pain, Organs, 13. Proprioception (sense of limbs in physical space), 14. Tension Sensors (allow the body to detect tension in places such as muscles) 15. Equilibrioception, 16. Stretch Receptors (detects dilation of blood vessels or stretching of organs such as the stomach or lungs), 17. Chemoreceptors (detects hormones and drugs), 18. Thirst, 19. Hunger, and the most controversial: 20. Magnetoception (ability to detect the Earth's magnetic field) and, 21. Time.

hunger and thirst, seeing sights, hearing sounds, and smelling whatever wafts your way—all without effort, and mostly without choice.

In the long history of organized education, information needed to be consolidated and presented in linear fashion, usually through the medium of textbooks guided by instructors. Some textbooks were designed to strategically double-dip education by cross-pollinating subject matter. My favorite example is a Latin curriculum[2] used by one of Queen Elizabeth I's private tutors, Roger Ascham,[3] which doesn't teach conversational Latin language, per se, so much as neatly arranges vocabulary and grammar into chapters on aspects of ancient Roman and contemporary English life, including topics such as geology, zoology, gastronomy, et cetera. Therefore, students learning Latin simultaneously learned about aspects of everyday life, such as uses for particular minerals. One might imagine a classroom from Harry Potter's *Hogwarts School of Witchcraft and Wizardry*, where students learned the names of enchantments and incantations, accompanied by samples of original objects. Regardless of how the imagination runs, the only point is that science education was grafted into Latin education, making each more relevant to the other. At the heart of it is strategic curriculum design. If a 600 year old textbook can do that, what can digital technology do?

If any reader takes away one thing from this essay, let it be that the brain is magnificently equipped to process multiple stimuli simultaneously—and it likes to do so. Digital content can be prepared such that it inherently exposes multiple subjects simultaneously: foreign language in music curricula, science in foreign language curricula, architecture in art curricula, art in science curricula, and so forth.

Let's go a step further. From one example above, such as architecture in art curricula, imagine that a student wants to reproduce a drawing of the Roman Colosseum. If they are also studying Latin, why not label some of its parts in Latin? Or if Italian, why not label in Italian? Both are relevant to the context of the drawing, the student's other coursework, and art education. If they aren't learning a language, what's the harm in labeling pieces in Italian anyway? If they are learning Chinese, even though Chinese is hardly the natural contextual milieu and even if pieces of the Colosseum aren't part of the student's Chinese class vocabulary, what's the harm? But we don't generally expect the art teacher to also know about Architecture, Engineering, Latin, Italian, Chinese, and all the other courses that could be taught in school... That would be prohibitive to hiring.

A digital educational system, however, informed of a student's course of study, could automatically populate information specific to the learning goals of the student. And it can do all those other things computers are so much more naturally suited to do, like long-term content management, perhaps in the form of tracking the number of exposures a student has had to a particular piece of content, and in what form (hearing, reading, speaking, etc.).

---

[2]Horman, William. *Vulgaria uiri doctissimi Guil. Hormani Caesariburgensis* (1519).
[3]Ascham, Roger. *The Schoolmaster (1570).*

Such a system could also empower the student to turn on or off as few or many simultaneous channels of information as they wish at any particular instant. Want some background music with your Art/Engineering/Latin lesson? Choose a playlist from last year's band repertoire (*review*), from current year's repertoire (*exposure for reinforcement*), or preview next year's repertoire (*priming*). Tired of band music? Listen to what the orchestra played, is playing, or will play. Sick of the symphony? How about music in Chinese, or anything else that's available. Turn labeling in foreign languages on or off, or for the enthusiastic linguist, turn on multiple languages: Label in Latin, Italian, Spanish, Portuguese, and Chinese if you like. With extensive tools already built, like Google Translate, interdisciplinary content integration is a mere matter of tagging and, more importantly, responding to the opportunity.

Since the brain handles multiple forms of stimulation simultaneously, music can play at the same time as foreign language is spoken (because of course, we must hear the pronunciation of the native speaker), without risk of encountering cognitive interference. From a neurological stance, content types must themselves be tagged according to their nature, and rules must be programmed which assign simultaneous compatibility, or place datum types in a sequential queue. For example, visual datum and auditory datum are simultaneously compatible if the visual datum is not audio/visual. So a drawing of the Colosseum can be accompanied by background music and spoken foreign language, unless the background music contains lyrics relevant to their curriculum; if not, the music's volume can dip when relevant language is spoken.

That concludes the general overview of cognitively efficient educational technology. In summary, its primary goal is to efficiently blend methods of group instruction into a comprehensive and relatively easy-to-make educational platform that empowers students to adjust their minute-by-minute cognitive load.

The following sections are for brave educational visionaries only. Instead of blending technology with methods of *group* instruction, the digital platform blends technology with methods of *one-to-one* instruction.

16.3 Tailoring Content to Fit the Student

The classrooms and technology you've seen will do no good until we are acquainted with the brain. In particular, let's address the five-year-old brain. It is less developed than the mature brain; it is a simpler model to describe. With the efficiencies we'll apply, it's slower processing speed and limited capacity for simultaneous function make it easier to begin crafting truly optimal systems. Besides, the young brain is a more truthful gauge of educational failure and success; whereas the mature mind compensates for confusion, such as using prior knowledge, logic, or questioning to correct a failure, the young mind simply halts in frustration when overwhelmed or confused. Last of all, a system flexible enough to adapt to the youngest learners will make a superb template for older minds, being a purer window into the brain's most natural—or least resistant—modes of cognitive development.

What the 5 year old *wants* to learn and what they are *able* to learn *both depend on the five year old*. So the first priority of any educational system, this one especially, must be flexibility. A system which pre-determines content and delivery is bound to satisfy only few. Technology offers as much flexibility and adaptability as its designers anticipate and implement. Too often, and mostly because of limited resources, educational materials are static, or rigid, forcing the student to conform to the content. The private teacher (and by *private*, I mean "one-to-one"), like a tailor, can fit the content to the student, whatever the content and whoever the student, by making a true assessment of the capability of the student and a proper dissection of the content's elements. When these two components—the student and the content—are properly measured, they can be fitted.

16.4   Component 1: The Content

16.4.1   Content Appeal

Let's begin with educational content; it is less complex than the student, is within our control to dissect, and informs neuroscientific applications of content dissection based on student capability.

The first concern of content is a quick discussion on *appeal*. The natural parallel of ensuring an educational system's *flexibility* is ensuring its *appeal*. If education is to the mind what food is to the body, then dissecting content is something like cutting up the food of education into the right size bite, delivered at the right speed, with the right amount of complexity, and neither so little to be malnourished nor so much to be uncomfortable. Also like food, we eat less when it is unappealing, eat more when it is appetizing, and stop eating when satisfied. Remembering that our target audience is a 5 year old, envision a picky eater. If you can satisfy a picky eater, you're a heck of a cook. What's most important is that the cook be connected to the results. Did the meal satisfy? This might seem obvious, but content creators and curriculum designers are often detached from actual students when making educational materials or choosing elements for instruction. Students are rarely involved as stakeholders in the formation of educational content. It's also often the case that after content is prepared, the designers aren't present to see it being used and get little to no feedback on its reception. It's a dangerous educational practice to detach the cook from observing mealtime. The food analogy falls apart when we consider that this is a meal the cook can't taste; if they are designing materials for instruction, it is almost always the case that the designer is a master of the material. Prepared instructional materials, even digital ones, should be made with the guidance of students, and tested, or tasted, as it is prepared. Your picky eaters will be your most useful critics. It's an even better principle to give your diner a menu to choose from, with the flexibility to order *on or off the menu*. (If you're worried there's not enough time for that, we'll get there soon. For now, rest assured, there's plenty of time.)

Today's students are so unnaturally forced into such rigid models of education that the notion of self-directed learning, or learning directed by a student's *educational hunger*, may sound idealistic. That's largely because today's society almost exclusively learns in groups with very little opportunity to cater to individual tastes. The whole notion of private, one-to-one learning, can be conceived as a quest to maximize appeal for and efficacy within the subject. If that's hard to relate to, consider taking private lessons in something from a teacher who cares about your goals, and soon, the curious part of your brain will start to wake from its slumber and remind you what learning *can be*. Just like the example from food, educators, even digital educators, benefit from listening to the hunger and tastes of their students. When something tastes bad, we don't need to force the student to eat it; we can cook better, learn from our mistakes, and in the meantime, allow the student to try something else.

16.4.2   Content Forms

The form of content, or the manner of its presentation and how a student engages it, is the prime determinant of a student's learning outcome.

Here's an obvious example I'll call "Bad Violin School:" Suppose you go to school for violin playing. Over the course of 4 years, you read every important book on violin playing. You study the finger patterns and bow techniques of every major genre of violin music, including the scores of every renowned violinist of each century since its invention. You can identify relationships of every rhythm, scale, and key signature, and even produce scale drawings of every part of the violin from memory. After 10,000 h of study, you earn your certificate of violin mastery, but never *played* a violin. After 10,000 h of work, try picking up a violin and playing one of those masterwork sonatas and who knows what will come out?!

In the case of the Bad Violin School, the form of the content studied was only a shadow of violin playing.[4] It's not that work wasn't done or that learning didn't occur, but only that the form of the work leads to a different type of output than was described in the purpose statement "Suppose you go to school for violin *playing*." The point wasn't to pull the wool over your eyes, but rather, to illustrate some related yet tangential aspects of violin playing that aren't inherent to its success. There's a lot about reading and writing *without actually playing* that stuffed 10,000 of "work" without meaningful substance.

The purpose of *Bad Violin School* is to mirror most of today's math education. How much math instruction occurs without *actually doing math*? By "doing math," I mean engaging the brain in the activity of both *calculating* and *mathematical reasoning*. If a calculator is used, the first activity, calculating, isn't performed by

---

[4]If the case seems extreme, consider how many hours of education in Teacher Certification programs are *spent working with students* in the form of teaching or observing teaching.

the brain. If the math problem has been set up for the student, the second activity, mathematical reasoning, was performed by the problem's author and not its intended recipient. In the same way that the work done in *Bad Violin School* appears to surround the mastery of violin, so does most work being done in today's math classes appear to surround mathematical ability while ultimately failing to engage students in the processes inherently vital to its fluent development.

The use of calculators and pre-made math problem sheets are themselves just shadows of general national failure to achieve consistent, predictable, high quality results in math. A greater, more neuroscientific problem relates to difficulties achieving mathematical fluency: Whereas math problem sheets are a form of presentation, and calculators are a mode of executing calculations, both assume *visual symbolic representation* of the constituent parts of math. That is, in both cases, the math is being processed through *multiple encoded layers of reading and writing*. As addressed in Part I, objects or concepts from the real world are often referred to by specific sets of sounds in the form of spoken words. Mathematics has its own rich symbol system and grammar, enough (in my opinion) to very rightfully classify it as a language. But in the context of education, math is a language which must successfully be distinguished from as well as collaborate with one's native social language(s) to achieve fluency. Most importantly to the present attempt at helping the state of math education is to point out the multiple layers of symbolic encoding inherent to each respective system and propose that *student achievement will be improved by strategically eliminating layers of encoding unnecessary to the specific goal of achieving mathematical fluency.*

In practice, this is a neurological argument for *eliminating* the use written notation systems for introducing, practicing, and developing mastery *in calculation*. This does not mean omitting education of reading and writing the symbols of math, which are vital components of success in understanding the language of math. But it does mean drawing a neurological distinction between the brain's ability to read and write numbers, and for the brain to develop its ability to actually *calculate*. When reading is a *prerequisite* to calculating, the brain must correctly decode and remember the meaning of multiple forms of visual information, interpret the relevance of each, then process them according to additional sets of rules (rules for writing which are themselves a different process than rules for calculating mentally). If an error is made at any step, the whole answer becomes wrong.

Let's travel back to a time when primitive hunters counted the number of woolly mammoth. Ten big, healthy animals stood in a field. The hunter, not having the language to describe the number, grabs one pebble for each mammoth spied, and takes the pebbles back to the tribe's leader, who declares the number "TEN MAMMOTH!" These words aren't the same as the mammoth themselves, but represent mammoths. In this way, the words are like shadows of mammoths, not being the same as the thing, but outlining it well enough to identify it. The mammoths could have been represented with little carved wooden statues, or pictures drawn on walls, or a set of smoke signals, or spoken words, and each would have been a shadow of the real woolly mammoth. All that is required to be meaningful is that the presenter and observer agree upon the meaning of the symbols: "When I put ten woolly mammoth *carvings* on *this* part of the ground, and draw *this* set of lines in the dirt and point in *that* direction, I mean there are ten *real* woolly mammoths in *that* direction right *now!*" These are basic shadows of real life objects that people without language can understand. You don't need to go to school to understand it, but you might have to see it once.

Smoke signals offer the potential for greater complexity of shadow types. A smoke signal won't look especially like a woolly mammoth, (except by unlikely coincidence, I suppose), so it requires a bit of stronger communication to agree that 1. the hunting party seeks woolly mammoths, and 2. when found, build a fire and signal one puff of smoke for each mammoth you find. Ten puffs? Ten woolly mammoths. Alternatively, smoke signals could first encode the type of animal found, where perhaps *big puff* *big puff* means "woolly mammoth." and *small puff* *small puff* means "deer." If there are only a few dozen types of animals that can be found and smoke signals are only used for hunting, then unique sets of signals could be used to refer to specific animals, and the total count of them could be delivered in average size puffs after the animal type is identified. It's more cognitively complex, but could be more efficient. The need for additional layers of encoding comes with the need to describe more diverse content types.

Smoke signals also offer the potential to represent shadows of shadows. As need begets increased complexity, especially as spoken language leads to the development of an alphabet, smoke signals might be most efficiently used by encoding a set of puffs to each letter of the alphabet, like Morse Code, where (in English) a mere 26 letters can successfully communicate any complexity of idea. In this case, smoke signals represent letters which form written words which represent spoken words which represent real woolly mammoths. As odd as it sounds to describe it that way, this process of encoding has been historically revolutionary to communication efficiency, especially notable in the telegraph lines of the 19th century, where electrical signals delivered communications hundreds of miles away at supersonic speeds. Just because the mature brain is capable of managing such complex forms of encoding doesn't imply whatsoever that it is any sort of neurological ideal, especially in foundational stages of learning.

When important information about math, literature, science, history, music, art, athletics, language, et cetera ad infinitum, is encoded in any way, its accessibility is constrained proportionally by *the level of complexity of the encoding*, *the difficulty of its presentation*, and *the capability of the student*. Therefore, content forms are of utmost importance in preparing a learning environment for the student in which they can comfortably achieve success. These are the primary considerations of educational ergonomics. Having addressed reducing content types to their minimally complex form, we are now able to explore how to appropriately fit content to the student.

16.4.3   Content Dissection

The presentation of content, whenever displayed by a substitute for the real-world object, is always a symbolic "shadow," or a set of shadows pointing back to the object. Once the most direct shadow is chosen, the educator can begin preparing the content for efficient engagement by the student. This stage of curriculum preparation forms the bridge between content and the student, and there is no way to know the most efficient preparation of content without knowing the capabilities of the student. I've placed the discussion on content before the discussion on the student because we can control how content is crafted for engagement. Equipped with efficient modes of representation, thoughtful dissection can make most content accessible to most students.

Let's consider an example from the high classical period of music: Beethoven's Symphony #5. First, it's made for listening, and any student with a healthy sense of hearing is able to listen. In this way, any healthy student, even a 5 year old, can passively engage with the content. But suppose we want the student to actively engage with the content. Options include humming along, but the polyphonic nature of it (multiple simultaneous tones in harmony) might be confusing to some, leading to the potential to hum incoherent portions of the melody at some points, and various portions of the harmony at others. This is solvable by isolating the theme, or the prominent melody, apart from other aspects. Suppose, once the melody is isolated, that the amount of melodic content is too much to be retained in the student's short-term memory: Isolate smaller sections. Suppose that once the content is isolated into small enough of a section as to theoretically fit the capacity of the student's short-term memory that it is still incomprehensible, perhaps it is being presented too fast: slow it down.

If this approach must be taken for each piece of content that a student may want to engage, it would be too frustrating to be worthwhile; but if the content is presented from the beginning in small enough a chunk at slow enough a speed, isolated to the pertinent melodic theme only, then there is no frustration. The key to educational success is making a scientifically informed estimate accurate enough from the onset that it accommodates the tolerance of the student. While individual human brains do vary somewhat, they are more similar than they are different, and despite variations in theoretical frameworks like IQ or aptitude estimates of various Intelligences, it is Beethoven who, in this case, has done the work of crafting something of musical value well enough that it still connects with the human psyche hundreds of years after its design. We educators have responsibility to a different art entirely.

Here, our hypothetical music student engaged with Symphony #5 with their natural instrument, their voice, and it is their humming along that is the shadow of the melodic theme of the original work. I choose humming because it is a relatively simple neurological pattern: the same part of the brain that processes the sound heard is that which processes creating the sound with your own voice: if mentally only—humming along in your head, without the use of your voice, like sounding out words in your head without vocalization—it is even less complex than by also engaging the vocal chords, which upon doing so not only requires part of the brain to process the physical activity of humming, but also adds new sounds to the environment which are also heard and processed in the mind, then maybe also compared to the sound of the original melodic theme presented. When comparison to the original occurs, a higher order cognitive process occurs than when, for example, the melody is produced on an instrument like the piano, whose intonation is in the control of the piano's tuner more than its performer. It's somewhat important to note that a musician could play an instrument like the piano for thousands of hours without developing the part of the brain which affects the sense of intonation, or correct pitch; to put it another way, deaf people can play beautiful music on the piano through understanding its arrangement mathematically more than musically. The same statement is less applicable to an instrument like trombone or violin where precise location is usually guided by responding to minute auditory cues.

Our discussion on *Content Dissection* has only addressed reproductions of *Content Forms* without the use of symbolic representations, such written and read notation systems. *Content Dissection* also pertains to *Content Forms*, especially when written notation systems are truly more efficient for engaging content, especially amounts of content beyond retention in the short-term memory.[5] Careful inspection of just how much information can be retained in the short-term memory of a student, especially a 5 year old, reveals nearly disturbing realities about how the world's several hundred million students are taught, but as uncomfortable as the reality of the situation is, take heart that good diagnosis informs remedies. In the former paragraph on humming Symphony #5, the brain was inherently limited by its capacity to retain and process heard auditory information; written notation systems have no such inherent limitation. To the contrary, one of their primary functions is to preserve information beyond the limits of the mind. In that sense, they exist in response to our brain's limits, and are able to do that which our brain cannot. To reiterate a point from a previous paragraph, these written notation systems became more complex over time and stand in place of information in multiple degrees of detachment from the real-world object. In the same way that audio engagement of Beethoven's 5th Symphony can be made accessible through multiple stages of dissection and simplification, so, too, can notation systems.

---

[5]This discussion is dangerous because it verges on opening Pandora's Box, which spilled every evil into the world and retained, in the end, only *hope*; When an amount of new information is presented to a student that is beyond their capacity to process, it's a waste of time to the teacher and the student. What good could it do to present more new information than your student's mind can consciously process? This isn't to say that there's no purpose in such a decision, but be deliberate and purposeful about it, lest the only thing that remains from such an endeavor is, like Pandora's Box, *hope* for learning. Short-Term Memory is the primary building block for shifting datum of any sort into more permanent states of memory—first the Working Memory (which is theoretically infinite), then through successive repetitions of varying frequency and intensity, into Long-Term Memory. These are the dynamics which become increasingly important as we approach educational ergonomics pertaining directly to *The Student*.

Suppose the student wishes to play Beethoven's 5th Symphony on an instrument (other than their voice) and with the help of a written notation system. A traditional approach to this particular song would usually involve hundreds of hours of study in other songs before being considered reasonably accessible. And there's no reason to suppose that when Beethoven wrote it, he had in mind teaching it to 5 year olds. Nonetheless, his art is different than ours. The approach used in considering how to teach humming along reveals some aspects of strategic content dissection—isolate the melody, reduce it to manageably sized chunks, reduce the demands of a fast tempo, etc.—but these don't address the particular demands of written notation. Until now, *reading* wasn't involved.

Naturally, we take standard music notation, isolate the melody, and present bite-size chunks. But just like the *Symphony* itself, standard notation wasn't designed for 5 year olds. It, like other notation systems, is the product of centuries of evolution in efficiency. Although it can be modified in a variety of ways to ease use by students, such as clearly printing the notes instead of sloppily and inconsistently writing them by hand, printing the notes larger, and eliminating the need to remember rules of the grammar of music (such as "key signatures," the set of sharps and flats at the beginning of each line that must be constantly remembered to correctly play the indicated varieties of tones), this system, designed for composers and professional musicians, can be replaced altogether by one designed especially for youth, ideally, using elements from a notation system they already know and can naturally compare to the instrument at-hand. In some schools, letters of the alphabet are used to indicate tones (A, B, C, D, E, F, G), which is an advantage to students who are already familiar with the alphabet. This has the potential to *reinforce other learning* if it intersects with other uses of the alphabet, such as reading English, as well as the potential to *cause cognitive conflict* by introducing one system whose rules of use differ by context.

As with other aspects of educative decision making, there is not one correct philosophical answer, no teacher, superintendent, or cognitive neuroscientist who can definitely say one is *universally better* than another. Although the phrase "best practice" often surfaces in the quest to maximize educational efficiency, what is truly best for a particular student depends on the particular student. Thankfully, educators—even those of digital education systems—can consult the student, and anticipate methods of enabling students—even 5 year olds—to choose for themselves.

16.5 Component 2: The Student

What wonderful good can be done by reducing content complexity, preparing various forms of representation, and estimating a content's cognitive complexity, all aimed at enabling the students to make up their own mind about their personal formation. It's almost utopian. Although we've only had a taste of how the food of education can be prepared, let's explore the palette of the mind.

Here's a nuts-and-bolts version of how learning happens: Information of some kind (sound, sight, temperature, whatever...) is perceived by a recipient. If it's given conscious focus, it lands in the Short-Term Memory (0-60 s, 1-10 items), where it has the potential to move into Working Memory for semi-permanent recall (less than 24 h, hypothetically infinite potential capacity). If the sensory data can be recalled more than 24 h later, we say it's in Long-Term Memory,[6] although this is far from a guarantee that it will be retrievable a week or a decade later. If the sensory data is never re-engaged, either through external re-stimulation or internal recall, it'll probably be forgotten, because the brain isn't in the business of stuffing itself with meaninglessness.[7]

The likelihood of an experience being remembered or forgotten depends on the *intensity* and *frequency* of exposure. The minimum *duration* required for long-term learning to occur from any any particular experience concerns the overall efficiency of learning. The *interval* between learning episodes affects the brain's ability to recall information, as there is a limited window of non-engagement before cognitive atrophy occurs. These four dimensions of learning—intensity, frequency, duration, and interval—affect the likelihood of forming high-quality, long-term memories by engaging the learner's Short-Term Memory, Working Memory, and Long-Term Memory, all vary by student,[8] and are the foundational constituent aspects of optimized educational efficiency.

Let's consider some variations on educative routines, each containing 100 repetitions of a particular learning goal, distributed over different frequencies, intensities, durations, and intervals. For this example, let's suppose the learning goal is to memorize a random set of ten numbers,[9] whose success will be measured

---

[6]Sousa, David A. *How the Brain Learns*. 3rd ed. Thousand Oaks, Calif: Corwin Press (2006).

[7]"Forgotten" is usually used loosely. Most sensory perceptions, whether temperature or sound or whatever, once perceived by the brain, affect its biological tissue somewhat. When it affects the tissue in such a miniscule way that distinct recall isn't possible, we usually say it's been *forgotten*. A culture of test-makers and test-takers might think that not being able to recall the right information at the right time means you didn't learn it, but that's not exactly a cognitive neuroscientific perspective. Alas, this essay is more concerned with how to remember something well than defining the parameters of forgetting, so let's all agree to concentrate on defining the parameters of remembering well.

[8]Don't fear the variations; students aren't as variable as they might seem. While every student is a unique little snowflake, remember that snowflakes are quite similar in composition, and being similar in composition, share predictably uniform properties: Although the shape of their crystallization is always unique, they all crystallize at zero degrees Celsius.

[9]I choose a random set of ten numbers because it is beyond the short-term memory capacity of most young students, is comparable to the length of a phone number, and because most of us have childhood phone numbers that we memorized and can still remember after more than 1000 days of non-use. Also, the set of ten numbers might be partially recalled, where some students remember only the first few or the last few, which can also be used to estimate overall efficacy.

1000 days *from the end of the last educative engagement,*[10] and the highest average quality of recollection wins.[11]

Variation One might be to hear and orally repeat the set of ten numbers 100 times all in one day. Variation Two might be to repeat the set of numbers 10 times a day for 10 days. Which do you think will work better? Most of us probably believe, intuitively or based on experience, that spreading the repetitions over multiple days will work better. Let's continue.

Variation Three might be to repeat the set of numbers five times a day for 20 days. If we believe spacing out the repetitions is advantageous to memory formation, then maybe we should say Variation Four is to repeat the set of numbers once a day for 100 days. Do we anticipate problems with this much spread? Perhaps that is the best bet, so far...

Who says, though, that once *each day* is best? Let's say that Variation Five is repeating the content once *every other* day for 200 days. We're contemplating the subtle variations in the impact of *interval* between learning sessions. Let's say Variation Six is repeating the content once every three days for 300 days. Which, intuitively, do you think results in the best long-term recall? Or do you think Variations Four, Five, and Six will result in generally similar outcomes?

So far, all of the examples have been generally linear in form. That is, the intervals have been evenly distributed. Although we are often biased toward linear thinking, there is no reason to make such linear assumptions about human learning. Consider Variation Seven, which might suppose that the first repetition set occurs four times, every 3 days, five times in each of those days, using 20 of the 100 allowable. The next repetition set occurs twenty times, every 3 days, only once each day, using the next 20 repetitions, totaling 40 repetitions used. The next 40 repetitions are scheduled to occur every 5 days, also once each day, and the final 20 repetitions occur every 10 ten days, once each day. Variation Seven, if graphed, would look something like a hyperbola instead of a straight line. From experience, many of us might vote for Variation Seven as the most effective repetition set.

Variations One through Seven have only considered three of the four dimensions of learning, namely, *frequency* (which each variation is limited to 100), *intensity* (whether some are repeated multiple times in the same day), and *interval*[12]

---

[10]If the quality of long-term recollection was evaluated 1000 days after the student's first exposure, and if we are testing true long-term memory formation, it would be unfair, for example, to reserve some of the 100 allowable engagements for the day before assessment, so assessment must occur 1000 days after their final engagement.

[11]The nature of the experiment requires that different students use different distributions of repetitions, and the natural argument for explaining the results might be that their different aptitudes will be more influential in determining their long-term results than the pattern of repetitions. While this is true to a degree, reliable results can be obtained by assigning large numbers of students to each pattern and averaging their outcomes.

[12]*Interval* could be divided into long-term intervals, such as we have described as occurring over multiple days, or could be described in the short-term as how many hours, minutes, or seconds occur between repetitions. I have often observed students exhibit less proficiency between repetitions when interrupted between, even for just enough time to make or listen to a comment. So it

(how much space occurs between learning sessions).[13] The fourth dimension, *duration*, is less apparent than the others, yet has an impact of its own. We must consider whether each repetition is spoken and repeated variably or at the same tempo each time, or what manner of tempo is used (slow, medium, fast, etc.), and whether the same tempo is always used or if there is help or harm in strategically varying it over time. Naturally, slower tempos yield longer durations, but are also one factor which impacts the perception of ease or difficulty *and* act on the psychological appeal of the act of learning: perhaps a fast tempo is exciting to one student but exhausting to another.

Tempo also acts on the ability of Short-Term Memory, since its limits concern both quantity of information and elapsed time. If repeating the set of ten numbers slowly, such as one number every 2 s, might that yield more difficult recall than a quick repetition set, such as three numbers per second? It might be advantageous to recite the number set slowly at first, then progressively quicker after a certain number of repetition sets, which is closer to the natural model of language development, where words are sounded out slowly at first, but eventually used in faster contexts. One prominent instructor of the Russian School of Ballet[14] attributes some of their success to taking the same exercises as devised by the French School of Ballet, doubling their repetitions, and halving the tempo at which they were practiced, thereby quadrupling the total *duration* of exercises by only doubling the repetitions.

These four dimensions of learning are difficult for teachers to plan, implement, and monitor, especially when considering very fine aspects, such as progressive alteration to tempo of presentation and engagement. But what is impractically difficult for humans can be programmed into computers and implemented effortlessly. What matters first is mere awareness of the possibilities, and whether designing such a system is ultimately practical. Having spent some effort surveying the horizon of such a system, permit me the chance to paint a picture of its potential. After all, so far we've only explored an arbitrary string of numbers.

(Footnote 12 continued)

might make a difference for long-term results whether the repetitions occur in uninterrupted succession, or spaced out in a pattern such as once in the morning, once in the afternoon, and once in the evening. I personally lean toward quick, uninterrupted succession as more effective, but verification would require scientific investigation.

[13]The human lifespan is inherently prohibitive to extreme examples, such as one repetition each year for 100 years; even if it yielded the best result (and I don't think it would), education must prepare the student for more practical outcomes. Although 100 years is far-fetched for most desirably practical learning outcomes, a very real practical timeframe for most students isn't a semester, or a year, but from childhood to adulthood, or by today's standards, 18–22 years. When considering truly ideal distributions of repetition for content mastery, examples such as language acquisition (especially foreign language acquisition), mathematical ability, and physical motor skills are examples which we already acknowledge require and deserve regular repetition over more than a decade.

[14]Cite Russian ballet curriculum.

16.6   Maximizing Human Potential and the Future
of the Mind

Instead of using a fine-tuned digital education platform for memorizing a string of random numbers, imagine that it was used for something important. Whatever that is, is up to you. But the idea is that whatever you consider worth learning, for as long as you consider it worth learning, can be programmed for repeated engagement for as long as you consider it worth engaging. It is a digital means of assuring that today's effort is not wasted because of tomorrow's neglect. Nothing needs to reach the point of forgetting or regaining lost ground due to passing time. When considering knowledge in the form of skills or information, the human brain naturally learns most things in seconds. If you think about it, the vast collection of learning that is in each of our minds all occurred in seconds and collections of seconds over the years of our lives. Most of these seconds, especially in classrooms, were focused on a single subject at a time, taught in a linear progression, and often abandoned after the unit, semester, or grade level. Therefore, most of these seconds count as basic exposure only. In part, it was a product of the times. But today, we can do better.

In the same number of classroom hours, content can be strategically rearranged to utilize age-and-aptitude-appropriate dimensions of learning with individual students, according to their interests. Multiple, neurologically compatible channels of information can be presented and engaged simultaneously, eliciting ranges of high and low conscious focus, and enabling students access to dynamically adjust their level of stimulation and engagement at any point in time. The efficiency of their engagement is up to them, as is the depth and breadth of their engagement with any topic vertically or horizontally, meaning that a topic can be explored in greater precision or depth, or learned as part of a broader context. The overall point of the system is to present, engage, and distribute the instructive, developmental component of learning as efficiently as possible, with as great a guarantee of high-quality, long-term retention as possible.

Such a radical view of potential efficiency leads to hypothetical results that are theoretically sound but hard to believe, but disbelief is almost a predictable response to results formed from the accumulation of very small amounts over very long periods of time, such as the accretion of the space particles into the sun and planets, the evolution of species, or compound interest. Applied to education, I don't see why a student steadily interested in languages couldn't obtain fluency in eight, twelve, or more world languages by investing five-to-ten minutes (or less) every other day from Kindergarten through twelfth grade. As mentioned in Part I, languages are useful for introducing content pertaining to most subjects, so other learning goals can be simultaneously achieved with strategic applications. If applied on a large scale, the average human vocabulary could literally multiply tenfold over the course of a century, from 15,000–25,000 words to 150,000–250,000 words, revolutionizing today's literacy norm. If that seems unfathomable, consider that there was a point in human history when the average vocabulary was one-tenth of its present average, and in very recent history, including some parts of the world today, written literacy is *still* one-tenth of the world's average.

Beyond the marvelous potential for a world made better through the advancement of education are discoveries yet undiscovered, but not undiscoverable. As the potential for certain types of learning fades and sometimes vanishes with inactivity over time, many of us today don't fully know ourselves because of potential unnurtured in our earliest years. It is the case that if you didn't use your legs, or eyes, or sense of sound, for the first 10 years of life, the portions of the brain that normally develop when those senses are stimulated would remain permanently dormant, even if the limbs or senses were originally otherwise perfectly healthy and able [1]. It is this view of human capability that carries the greatest urgency. When aspects of our own humanity are neglected during certain crucial "windows of opportunity," the window closes and the potential for development of fluency or, in extreme cases of neglect, even basic functionality, vanishes. Of the 21 human senses, only few currently receive deliberate, systematic development. Who knows what the state of national health might be if today's adult population was deliberately and systematically instructed from youth about the stomach's stretch receptors, and in distinguishing the body's sense of hunger from its sense of thirst, which according to some studies [2], are often confused.

The advancement of education and the advancement of society are, to me, absolutely synonymous. The causes of conflict can be eliminated only through the advancement of both. Individual happiness, securing the means of widespread and enduring peace, and promoting the general good all demand a view of society—and our role in it—that both sees beyond ourselves while simultaneously accounting for our individual roles and responsibilities as members of it. As Einstein said, "The significant problems we face cannot be solved at the same level of thinking we were at when we created them." [3]. His generation created the nuclear bomb. The next created the computer. What will we create?

References

1. Sousa, David A. *How the Brain Learns*. 3rd ed. Thousand Oaks, Calif: Corwin Press (2006)
2. Mattes, Richard D. "Hunger and Thirst: Issues in Measurement and Prediction of Eating and Drinking." *Physiology & behavior* 100.1 (2010): 22–32. *PMC*. Web. 8 Feb. 2016.
3. Einstein, Albert. *The World as I See it*. Abridg ed. Secaucus, N. J;New York;: Wisdom Library, 1979

Chapter 17
An Answer to the Math Problem

Lewis Watson

17.1 Measuring Up in Math

U.S. students not measuring up in math. (http://www.upi.com/Science_News/2010/11/10/US-students-not-measuring-up-in-math/30091289429896/) American kids lag in math. (http://abcnews.go.com/US/story?id=94793). These headlines are accurate. Our students are performing below expectations and behind their international peers. The reasons for our less than desired results are many and deeply embedded in our processes and societal norms. However, with advances in neuroscience and learning technology, we can make significant improvements and begin the process of changing the way we teach and learn math.

Based on international math tests, U.S. students rank from the mid teens to as low as the mid 30's in international comparisons. With each set of disappointing test results, our education system comes under pressure. Although understandable, the perception that the scores stem from recently failed education policies or practices is not entirely correct. The U.S. has been performing below our international peers for over 50 years.

The Programme for International Student Assessment (PISA) has been administering tests every 3 years since 2000. The United States has ranked as high as the 20th nation, but we have been ranked in the 30's for three of the five assessments (2015 results are not yet available), with a low ranking of 36th in 2012. Before PISA, The International Association for the Evaluation of Educational Achievement has conducted international tests since the 1960's. Again, the U.S. consistently scored well below our international peers.

L. Watson (✉)
Marshall Fundamental School, Pasadena, USA
e-mail: lewwatson4@gmail.com

© Springer Science+Business Media New York 2016
N. Lee (ed.), *Google It*, DOI 10.1007/978-1-4939-6415-4_17

At the core of the issue is the structure of math content. Math skills and concepts are numerous and varied. While often perceived and dealt with as if they are unique, math skills are connected and hierarchical in their nature. Many content areas build on prior knowledge, but only math relies on such a vast set of prerequisites for ongoing success.

The fact that mastery requires conceptualization further complicates the issue. The skill of addition can be acquired in its most basic form relatively easily. However, without an accurate conceptualization of addition, students will begin to make errors when fractions, negative numbers, or variables become the addends.

To add negative numbers successfully, students must understand and combine the separate concepts of addition and negative numbers. If students have erroneously conceptualized addition as the operation that makes numbers larger (as they often do), they will be unable to add negative numbers successfully.

17.2 Structure of Math

The structure of math is the at the core of the problem, but it is not unique to the U. S. All nations are dealing with this issue. Our poor results must stem from issues related to our societal norms, education processes, and pedagogy. One factor is our societal norm that is OK not to like math or be good at math. We have come a long way in addressing illiteracy and creating a national norm that we will all be able to read. But we continue to condone our math failings.

Although our societal acceptance of math inadequacy is a factor, it is also a symptom. The 50+ years of poor performance has created a population of citizens that have math skill deficiencies. It is not uncommon to hear "our family does not have a math gene" or "we are just not good at math". Without the proper support, encouragement, and role models our students are facing an uphill battle in their efforts to acquire math skills and master math concepts.

17.3 Educational Structure

Our educational structure also plays a role. We rely on a student's age to determine their math content. Once in the system, students are moved from one grade to the next without a meaningful review of their math mastery. Only in the rare cases of students being identified as needing special services is the lack of math understanding systemically addressed when moving on to the next year's curriculum. Instead of a systemic approach, we have relied on teachers to identify skill/concept gaps for each student and address them as part of their math pedagogy.

The reliance on a teacher's ability is where we actually can witness the students falling off the wagon. While some teachers have the skill set for effective differentiation, most of us do not. When I began my teaching career at the age of 45,

I was convinced that I wanted to teach Algebra to middle and early high school students. These are the grades where the research says we lose math students, and I wanted to keep that from happening.

In my first year, it was clear that many of my students had fallen off the wagon before they arrived at middle school. I had students that could not add single digit positive integers. Others could not understand the concepts related to negative numbers, fractions, or decimals. Over 50 % of them had an aversion to math and lacked the belief that they could fix their math deficiencies.

There were also students that came in prepared and eager. Math was fun and exciting, or at least, they were buoyed by their past success. The combination of these students in a single class caused me to search for differentiation strategies. How could I create an environment where all of my students would be able to reach their individual potentials? I tried almost every research-based strategy I could find. Each time I was disappointed.

Grouping advanced students with struggling students helped with the group dynamic, but did not make meaningful impacts in skill acquisition for the struggling students and advanced students reported feeling limited. High ceiling/low floor projects allowed all students to participate but failed to ensure all students had the opportunity to acquire the intended skill and master the underlying concepts. With each strategy, I felt that the class was moving through the curriculum, but without individual students making the progress they should and could if I were able to meet their needs.

I have come to realize that all these factors that make math education so difficult affect our students at different times in their math careers. Some to a small extent and they can self-correct; others to a larger degree and we often leave them behind. Once a student falls off the math wagon, we replace them with a cardboard cutout and continue driving the wagon forward. We act like they are with us, and are confused by their blank stares. In reality, they are still back on the road where they first fell off, struggling to make it on their own, and hoping that we will come back and pick them up.

As we move forward, this needs to be a focus of our efforts. Each student deserves the attention they need, when they need it, and for the specific areas they need it. In Finland (recent scores are in the top 5 consistently in international assessments), over 50 % of students receive some special education support during their years in school (http://www.stat.fi/til/erop/2014/erop_2014_2015-06-11_tau_006_en.html). Finland is putting the students back on the wagon as they fall off.

17.4 Individualized Learning Plans

In the U.S., we have a different view of special education services. We do not implement them at the first sign of a student falling behind. Instead, we reserve our special education services for students with identified learning disabilities. While it would be ideal to change this approach, a shift of this magnitude will require a significant amount of legislative and societal will and is not likely anytime soon. We need to find a solution now, and must look for an alternative.

We need to identify, assess, and work with our students so that they are reaching the levels of success that they are capable of, no matter how high. It all begins with the accurate assessment of the current state of a student's math skills and knowledge. Without a thorough assessment, we end up wasting time addressing skills and concepts that are either already understood or beyond their current ability to fully understand. However, an initial assessment is only valuable until the student makes some amount of progress in the determined areas. At this point, the initial determination is outdated and must be repeated.

The continual assessment of mastered and needed skills and concepts ties directly to the individualized learning plan. Students need to know what skills and concepts they need to work on and have them arranged to ensure that we have prioritized the prerequisites. Again, to manage this manually is almost impossible. I have read about teachers that were able to develop processes that achieve this goal, but most of us do not have the time or organizational skills to be consistently successful.

When we provide students the opportunity to work on math skills and concepts that are within their grasp, they begin to experience success. This success builds awareness that they can learn math, and they have the intellect to be successful. It can also expand into math enthusiasm and a desire to seek out higher levels of math. Both the higher skilled and the struggling student experience the positive feelings that come with learning math that was once unknown. I often tell my students that nothing feels better than solving a hard math problem. You feel accomplished, smarter, and ready for the next challenge. It is empowering.

Implementing and managing individualized learning plans is where we can use the advances in technology to help us. We now have the ability to see into the minds of our students, assess their math skills, and chart a course for math success. The automated learning systems available today make the process of individualizing math education a reality.

In my class, I have used three different automated learning systems over the past 6 years. I have found all of them to have unique features and challenges. However, each one enabled my students to have math success at their level. Engagement levels in my classroom are regularly 90–100 % and discipline issues are almost nonexistent. Students understand what they are working on and what skills and concepts are next in their learning plan. I have connected their grades to class mastery of skills and concepts, so students understand what is required to obtain a specific grade. Thus, grading discussions are data based and often result in a concrete plan of improvement.

17.5 Automated Learning Systems

These impacts to my classroom are primarily due to the use of automated learning systems. I am not skilled enough, nor do I have the energy or organizational skills to achieve these results on my own. We have access to tools that automate the initial and ongoing assessment, the creation and updating of the individualized learning plan, and the access to and management of student data.

While impressive, the automated learning tools have some significant challenges. Due to their static structure, they do not reach across the learning modalities effectively. Limited explanations or help with skills and concepts do not allow for alternative solutions nor do they accommodate student questions. Also, once implemented, you can see and must address the various levels of math skills in your classroom. The different levels and pace of work introduces a high level of variability and an element of chaos into a classroom where tradition has preferred order and structure.

However, the two biggest challenges are related to understanding. The first challenge is application. The automated learning systems are excellent in allowing students to acquire and practice skills. However, they are limited in their ability to provide opportunities to apply concepts. Each of the systems I have used has attempted to include concept application, but all have fallen short.

The lack of conceptual understanding is related to the second challenge of depth of knowledge (DOK). There are various methods and terminology regarding DOK. However, for this discussion we will summarize these into four levels: supported practice; independent practice; synthesis; application. It is difficult in a systematic program to reach DOK levels that require synthesis and application on a consistent basis.

The following chart summarizes my experiences with the three systems I have used in my classroom. I am currently using ALEKS. My knowledge of the TenMarks is based on 2014–2015 features and functionality and Accelerated Math is based on 2012–2013 features and functionality. Each of the areas of functionality is scored from 1 * (limited to no functionality) to 4 * (full functionality).

System functionality	ALEKS	TenMarks	Accelerated Math
Students' skills and knowledge are accurately determined as part of the on-boarding process	**	***	****
Initial assessment data is systematically integrated into initial individualized learning plan	***	***	**
Students are able to access and acquire skills based on their learning plan	***	***	***
Students can access and comprehend data related to their performance and areas of need	***	**	**
Students have the opportunity to apply and master concepts	*	**	*

(continued)

(continued)

System functionality	ALEKS	TenMarks	Accelerated Math
Teachers have access to meaningful data related to student and class performance	***	**	***
Teachers have the ability to manage individualized learning plans and//or class activities based on performance data	***	**	**

As we move forward, the use of automated learning systems can provide a significant advantage in the development and implementation of individualized learning plans. As technology and our understanding of the brain improve we should be able to incorporate higher levels of DOK within the systems. These systems will also be able to assess our students more accurately and pinpoint areas that need attention.

The use of automated learning systems will allow us to engage the students in real-world problem-solving based on their skill levels. Imagine an environment where students could select a cross-curriculum project to work on based on their preferences and their skill levels. If students desired to work on a higher-level project, they would have the information to identify the needed skills and the tools to acquire them. Learning would have a direct purpose, and the students would be in charge of their education.

We need to use our education time effectively to teach the skills and concepts the individual student needs and then allow them to apply them in ways that are unique to the student. This dynamic environment is what we can create through the use of automated learning systems.

Chapter 18
XQ Super Schools—Bolder Super School Project

Andrew Donaldson

*We concluded from the research that more time is spent on
learning in the online environment. Many of the systems provide
real-time feedback to students and individualized instruction. It
also appears that the online-learning environment is more
engaging.*
—Dr. Robert Murphy, Research in Higher Education Journal

18.1 XQ Super School

My vision in creating an XQ super school is to provide a virtual place where
students have easy access to a quality education that is free and accessible via the
internet. At Bolder Super School, the students will apply their gained knowledge
into real world situations through either internships or volunteer opportunities in
their community. Finally, through practical application of their skills and knowledge, students will be assessed by teachers or experts in the field. I view online
education as a way to equalize an education system that has many inherent
inequalities from budgets to access to excellent teachers. Using free resources
allows any student to gain whatever knowledge they are interested in, while also
getting as much depth of knowledge as they want. The biggest obstacle is the issue
of internet access, so schools need to partner with local organizations to gain
internet access. We are looking into the possibility of providing laptops for every
student and even cloud based Wi-Fi (http://shop.brck.com/brcks/brck-v1.html).

A. Donaldson (✉)
Bolder Super School, Boulder, USA
e-mail: andrewjdonaldson@gmail.com

© Springer Science+Business Media New York 2016
N. Lee (ed.), *Google It*, DOI 10.1007/978-1-4939-6415-4_18

18.2   Free and Low Cost Education

Its common knowledge that schools everywhere are suffering from severe budget constraints. In my hometown, I remember watching the local elections to see if people would vote to increase funding for the schools, keep it the same, or reduce it. When budgets are cut, the teachers are laid off; class sizes increase; and art, music and physical education programs suffer. In the end, the students suffer. As an educator, I have realized that in our modern society, it should be easy to learn things from the internet at no cost. As a student, I can easily go online and watch videos, read articles, or find pictures that are presented in ways that interest me. However, in the current system it's difficult to show what I've learned, and receive academic credit. This is where the idea of allowing students to simply prove their knowledge through varied assessments came to me. Not all kids are good test takers, but if a student learns a concept on their own and then uses that knowledge to create something, why should that not count as academic progress?

18.3   Critical Thinking

Eventually, I found more and more free resources and realized that combining multiple presentation styles and content, or "Chunking", (http://ww2.kqed.org/mindshift/2012/05/31/how-open-education-is-changing-the-texture-of-content/) is a more natural way of learning. This allows students to use more creative problem solving, fill in the blanks, and find out how different subjects are related. Also, students will be able to take classes in smaller chunks which may make a subject more interesting or accessible. An example would be instead of having a student take a semester or year of a history class, they could choose to take an introductory unit on the civil war, study Napoleon's battle tactics or look at the events that led up to World War II. Next, students could expand their knowledge in whatever subjects they find interesting and find ways that link their lessons together. This allows the students to be more active learners. They gain the ability to analyze information and draw conclusions rather than just being told "this is what lead to that."

18.4   Skill Repair

The Bolder Super School allows students to work within more flexible time constraints and deadlines—much as we adults experience in the real world. Also, education is about learning from your mistakes and correcting them, so the students will be able to retake any unit they fail or don't understand. If the way the information was presented wasn't effective for the student, the school will provide them with alternative resources that would teach the same information, but in a different way that may be more experiential or appeal to their learning styles. The lessons will also be designed around a student's strengths and interests. Students will be able to see how different lessons relate to subjects they are interested in. For example, if a student is very strong in social studies, but not math, they would only need to learn the math skills necessary for social studies. In order to ensure that all the students are gain an adequate foundation of knowledge, Bolder Super School will provide a clear visual representation of all the required knowledge they need to acquire in order to graduate. Finally, students will also be able to seek out additional help from our extended network of staff and volunteers to receive any additional academic tutoring or even personal mentoring they may need to help them as they develop and mature. The goal is to guide students into becoming responsible adults with the skills they need to thrive in a constantly changing workforce.

> Today's students will not live in a world where things change relatively slowly (as many of us did), but rather one in which things change extremely rapidly- daily and exponentially. So today's teachers need to be sure that, no matter what subject they are teaching, they are teaching it with that future in mind.
>
> —Marc Prensky, Teaching Digital Natives

Through the education methods of Bolder Super School, by the time a student graduates, they will have the knowledge, skills, and connections to become successful.

18.5   Online Education

In the current system, there are many benefits and problems with online education. The benefits include flexibility of scheduling and in how lessons are presented and assessed—an alternative to traditional high schools that some students may find a better fit for them. However, some of the problems include: not having adequate personal support, a limit on how information is presented since it is usually prepared by the school, lack of opportunities to develop social and professional skills, and very little is offered in terms of experiential learning and project based assessment. My goal is to address these issues by providing a school where the students would have differentiated instruction not necessarily from teachers, but from whatever online content appeals to their learning sensibilities.

They would have access to a network of tutors and mentors to give them academic and emotional support. Also, students would be able to find in their community, opportunities to volunteer or intern with local companies in order to foster networking and local connections. Finally, students will be assessed through a variety of methods. Some students may find it better to have a multiple choice test, while others may find it a better fit to create a project that shows understanding of fundamental concepts, possibly across a variety of subject areas. An example would be a student creating an object, project or presentation that applies concepts from math, physics, biology, or social studies which could then be assessed for understanding of those concepts as well as the practicality/originality of their product. This is one of the core concepts—students being able to solve problems and be innovative. This is important because in self-directed learning. Students need to find the internal drive to continue after making mistakes.

18.6 Motivation

One complaint I have heard repeatedly is that it can be hard to be self-motivated when taking an online class, which is why it is so important to have a network of physical spaces for the students to work with tutors to help them understand concepts, and offer guidance on how to solve problems they cannot solve on their own.

> Some authors report that online learning can be very effective depending on the subject, although perhaps, in courses where problem solving is required, such as science or math courses, face-to-face instruction or at least synchronous chat with the instructor may be preferred over asynchronous, online learning.
>
> —Research in Higher Education Journal

Another issue is the dropout rate. What is holding these students accountable to attend school and finish? Intrinsic motivation is a major factor in that, but research has been gathered that shows among online college students there is not necessarily a higher dropout rate or lower success rate through online education than traditional students. (http://www.aabri.com/manuscripts/11761.pdf)

By creating a support network, students will be able to get the help, or the extra little push, they need to complete assignments. This is also key to identifying "high risk" students that need academic and/or emotional intervention.

> …intrusive academic advising or more personal contact with the instructor, whether that is face-to-face, or electronically through online chat, texting, or discussion boards, may be critical to the continued success of students with marginal cumulative GPAs, regardless of course delivery mode chosen.
>
> —Research in Higher Education Journal

While some students may already have the intrinsic motivation to pursue knowledge on their own, there is a part of the student population that does need more guidance and assistance in learning how to gather information. By using the network of teachers and mentors, students can have access to an adult that can help motivate them to get assignments done, complete projects within the deadline or even just as a motivator to explore other subjects and material.

It is my hope that by combining these elements, students will have truly equal access to a wealth of information without being forced to attend a school based on their income or where they live. This is a revolutionary idea not because it is online education but because for the first time, we will be able to combine modern education methods with local communities to create an interactive education experience that has not really been offered before. I believe that through a combination of online and experiential education, one on one teaching and mentoring, and project based alternative assessments; Bolder Super School will be able to teach all our students how to be life-long learners, and active global citizens.

Part V
We are the Borg

We are the Borg. Your biological and technological distinctiveness will be added to our own. Resistance is futile.

—Star Trek

The fig tree is pollinated only by the insect Blastophaga grossorun. The larva of the insect lives in the ovary of the fig tree, and there it gets its food. The tree and the insect are thus heavily interdependent: the tree cannot reproduce wit bout the insect; the insect cannot eat wit bout the tree; together, they constitute not only a viable but a productive and thriving partnership. This cooperative 'living together in intimate association, or even close union, of two dissimilar organisms' is called symbiosis. ... The hope is that, in not too many years, human brains and computing machines will be coupled together very tightly, and that the resulting partnership will think as no human brain has ever thought and process data in a way not approached by the information-handling machines we know today.

—J. C. R. Licklider

Chapter 19
Transfigurism: Glimpse into a Future of Religion as Exemplified by Religious Transhumanists

Lincoln Cannon

19.1 Future of Religion and Religions of the Future

What is the future of religion? Some expect the resurgence and ultimate triumph of this or that fundamentalism. Some expect the religious phenomenon itself to weaken and die, a casualty to the science of our day. Others, observing the history of religion, expect that it will continue to evolve, inextricably connected to and yet clearly distinct from its past. If such an evolution occurs, what will religions of the future be like?

For that matter, what will humans of the future be like? It would seem relatively unprofitable to speculate about religions of the future without taking into consideration their adherents. Like with religion, some idealize a particular human form and function and expect it to persist indefinitely, while some expect eventual human extinction through natural or artificial disaster. Others project our evolutionary history into the future, and recognize that, as there was a time when our ancestors were prehuman, there may be a time when our descendants will be posthuman, as different from us in form and function as we are now different from our prehuman ancestors.

If evolution were random, one speculation about the future of human and religious evolution would be as probable as another, but evolution is not random. Variation through mutation may be random. But evolution is determined through selection of variations that replicate within the constraints and across the possibility space of their environment [1]. So evolution is also predictable [2]. To the extent we know environment, we can predict evolution; and to the extent we can set environment, we can direct evolution. In other words, we can predict and direct our own evolution to the extent we can know and set our own environment.

L. Cannon (✉)
Mormon Transhumanist Association, Salt Lake City, USA
e-mail: lincoln.cannon@transfigurism.org

© Springer Science+Business Media New York 2016
N. Lee (ed.), *Google It*, DOI 10.1007/978-1-4939-6415-4_19

Transhumanists advocate the ethical use of technology to direct our own evolution. As humanists in the broadest sense, Transhumanists generally emphasize the value of humanity; however, Transhumanists also recognize an essential dynamism in humanity and value that which we may become at least as much as that which we are. Many Transhumanists envision a future of abundant energy, molecular manufacturing, indefinite lifespans, enhanced intelligence, and overall radical flourishing. Although most are non-religious, a significant and growing minority of Transhumanists are religious [3].

Transfigurists are religious Transhumanists. The term "transfigurism" denotes advocacy for change in form. It also alludes to sacred stories from many religious traditions. Those include the Universal Form of Krishna in Hinduism (Bhagavad Gita 11), the Radiant Face of Moses in Judaism (Exodus 34: 29–35), the Awakening of Gautama Buddha in Buddhism (Maha-parinibbana Sutta 4: 47–51), the Transfiguration of Jesus Christ and the Rapture in Christianity (Mark 9: 1–10, and 1 Corinthians 15: 45–55), and the Translation of the Three Nephites and the Day of Transfiguration in Mormonism (3 Nephi 28, and Doctrine and Covenants 63: 20–21).

Unlike some religious persons, Transfigurists generally share with non-religious Transhumanists the hope that we can make our dreams come true in this world, and trust that science and technology are among the means for doing so. However, in contrast to non-religious Transhumanists, Transfigurists generally regard religion as a powerful social technology that, like all other technologies, we should use ethically, mitigating risks and pursuing opportunities, to set our environment and direct our evolution.

One of the most profitable ways to start imagining the future of religion, religions of the future, and how they will evolve along with us, may be to consider the ideas and practices of Transfigurists. What does religion look like through our eyes, given lenses colored by expectations of directed evolution and emerging technology? Such vision seems more likely to approximate probable futures for mainstream religions than do others that reject, ignore, or lack substantial familiarity with these powerful forces.

Assuming we and our religions will continue to evolve together with increasing intentionality made possible by technology, it seems reasonable to suppose that Transfigurism, more than any other contemporary religious view, is positioned to glimpse into a future of religion.

19.2 Theology

Trust in superhuman potential is the essence of Transhumanism.

As Transhumanists, we trust that humanity *can* evolve into superhumanity, perhaps to attain unprecedented degrees of vitality, intelligence, cooperation, and creativity. This trust is not uncritical or passive. Most of us would aim our extrapolations from observable technological trends into futures consistent with contemporary science. And many of us would act pragmatically to hasten opportunities and mitigate risks associated with such futures. So Transhumanist trust in superhuman potential is best characterized as critical and active, but it must remain admittedly a trust. The possibility of such futures remains to be proven.

Some Transhumanists also trust that humanity *should* evolve into superhumanity. We have minds to console and bodies to heal. There are communities to connect and environments to sustain. Less austere, there are morphological and cognitive potentialities to realize, and perhaps even meaning to infuse into otherwise meaningless voids. Whatever its source, a sense of obligation impinges upon us. And often those of us that most misrecognize our own moralizing have engaged advocacy with a degree of strenuosity that would shame all but the most zealous of evangelicals.

Although Transhumanists might confidently deny accusations of superstition or hubris, our trust is surely more than rational or ethical. Embracing a radical humanism, we would dignify the ancient and enduring work to overcome and extend our humanity. Diverse esthetics of superhuman potential resonate with and shape us, affecting our thoughts, words, and actions. Even granting that we could and should, perhaps more fundamentally, we *want* to evolve into superhumanity. So we may trust in that potential, if for no other reason, at least because we desire it.

Whatever may be the reasons for Transhumanist trust in superhuman potential, holding to that trust may have at least two logical implications.

First, if we trust in our superhuman potential, we should also trust that superhumanity would be more compassionate than we are. Consider the social ramifications of decentralized destructive capacity.

Logically, one of the following must be true: either (1) superhumanity would not have more decentralized destructive capacity than we have, or (2) superhumanity would have more decentralized destructive capacity than we have and mitigate that greater risk without being more compassionate than we are, or (3) superhumanity would be more compassionate than we are.

#1 seems unlikely, given trends in weapons technology. Weapons have increased and probably will continue to increase in destructive capacity and availability while decreasing in size and cost. #2 also seems unlikely, given growing disparity between offensive and defensive weapons. Whether motivated by utility or altruism, it's already cooperation, if not genuine compassion, that protects humanity from current weapons of mass destruction. And some contend that nothing short of developing friendly artificial superintelligence will be sufficient for the challenges ahead. So that leaves us with #3. If we trust in our superhuman potential, we should also trust that superhumanity would be more compassionate than us. The only likely alternative seems to be that we will become extinct before evolving into superhumanity.

The second logical implication of trust in our superhuman potential is: We should also trust that superhumanity created our world. Consider the historical ramifications of future creative capacity. Logically, one of the following must be true: either (1) superhumanity would not create many worlds emulating its evolutionary history, or (2) superhumanity would create many worlds emulating its evolutionary history and we happen to live in a world superhumanity didn't create, or (3) superhumanity created our world.

#1 seems unlikely, given trends in information technology. Simulations of our evolutionary history for research or entertainment are becoming more detailed, virtual reality is poised to become prevalent, and neuroscience suggests the possibility of full experiential immersion in computed worlds. #2 also seems unlikely, given the quickly growing number of simulated worlds. If ever many of them become experientially indistinguishable from our own, it would be more likely that we already live in one of many verified computed worlds than in one supposed non-computed world. So that leaves us with #3. If we trust in our superhuman potential, we should also trust that superhumanity created out world. The only likely alternative seems to be that we will become extinct before evolving into superhumanity.

The two logical implications of trust in our superhuman potential combine to form the New God Argument [4]. The argument doesn't prove God exists. Rather, it proves that if we trust in our own superhuman potential then we should also trust that superhumanity would be more compassionate than we are and created our world. Such superhumanity qualifies as God for many Transfigurists, and may qualify as God in some mainstream religions. For example, numerous Christian authorities have advocated various forms of apotheism or deification: the idea that humanity can and should become God, as or like God, or one in God.[1]

19.3 Epistemology

Transfigurists, particularly those with ties to the Christian tradition, may embrace theories of knowledge that include a place for faith. In such cases, we tend to characterize our faith as a practical trust in desirable possibilities when in context of

---

[1]Justin Martyr, "Dialogue with Trypho," 124; Theophilus of Antioch, "To Autolycus," 2: 27; Irenaeus, "Against Heresies," 4: 38: 3–4; Clement of Alexandria, "Exhortation to the Heathen," 1; Tertullian, "Against Hermogenes," 5; Hippolytus of Rome, "Refutation of All Heresies," 10: 30; Origen, "Commentary on John," 2: 2; Cyprian of Carthage, "Treatise," 6: 11, 15; Gregory of Neocaesarea, "Sectional Confession of Faith," 16; Methodius of Olympus, "Banquet of the Ten Virgins," 8: 8; Antony the Great, "On the Character of Men and on the Virtuous Life," 168; Athanasius of Alexandria, "Incarnation of the Word," 54; Hilary of Poitiers, "On the Trinity," 9: 38; Cyril of Jerusalem, "Catechetical Lecture," 21: 1; Basil of Caesarea, "On the Spirit," 23; Gregory of Nazianzus, "Oration," 2: 22–23; Augustine of Hippo, "On the Psalms," 50: 2; Mark the Ascetic, "To Nicolas the Solitary;" Theodoret of Cyrus, "Letter," 146; Diadochos of Photiki, "On Spiritual Knowledge and Discrimination," 89; Thalassius the Libyan, "On Love, Self Control and Life in accordance with the Intellect," 1: 95–101; Maximus the Confessor, "On Theology," 1: 53–55; John of Damascus, "Exposition of the Orthodox Faith," 2: 12; Theodore of Edessa, "Theoretikon;" Peter of Damaskos, "Treasury of Divine Knowledge 1: Introduction;" and Theognostos, "On the Practice of the Virtues, Contemplation and the Priesthood," 32.

incomplete knowledge, rather than an irrational belief that contradicts reason. From this position, Transfigurists may hold that science and creativity depend on faith.

This faith is not blind trust. It is only trust, with no more blindness than necessary at a given time and place. Moreover, it is not dogma or any unquestioning or unexamining attitude. Rather, it is recognition that no matter how many questions we have asked, and no matter how much we have examined, we have always had more to learn. Maybe that will always be the case. Whether we like it or not, we expect to find ourselves repeatedly in situations that require faith in practice.

Life and death hang in the balance, and we cannot wait for absolute answers (if they even exist) before we act. Perhaps no philosophical movement has better addressed such practical limits to knowledge than the pragmatists. As William James once described it, you can stand in front of a charging bull calculating the probability that it will trip, or you can run. Because we are limited, and to the extent we are limited, we find ourselves dependent on this faith, this trust in the efficacy of action given the knowledge at hand, according to whatever education or experience we were lucky to have had (or at least presume ourselves to have had) prior to needing it.

Furthermore, even when we have the luxury of time, it seems that we cannot make epistemic progress without at least tentatively trusting in basic premises. Science typically posits causality and uniformity as basic premises. Some may think that these are proven by science, but that's not so. As observed by the empiricist philosophers, Hume and Berkeley, no matter how many times we think we have experienced something, and no matter how many places we think we've experienced it, it could all yet change.

Not even probabilities displace such reliance on faith. Can we prove our memories were not planted in our minds moments ago by an evil demon? A Matrix Architect? No. We cannot, even if most of us don't worry much about that because it's not practical—or at least so we judge, based on our memories, even when we recognize the circular reasoning.

The same is true of logic. We require some basic axioms and methods, taken unproven, in order to do any work at all. For example, most logical systems assume non-contradiction, and various operations for coupling, decoupling, and otherwise operating on propositions. Logic doesn't prove these axioms and methods. We assume them.

Beyond the practical necessity, there is also a creative power in such faith. If the universe (or the multiverse) is not finite, if real creativity and genuine novelty are possible, it will not be those who wait for evidence that will be the creators—at least not intentionally. It will be those who act, despite not knowing everything in advance, that will be the creators. Such creative power may be seen in matters as common as trust in the possibility of love. You can wait for a long time for hard evidence that she loves you, or you can make a move. And sometimes the move makes all the difference.

This practical faith is compatible with rationalism, even a pancritical rationalism [5]. We can re-examine our premises, our assumptions, and our conformities.

We can honestly acknowledge the limitations of our knowledge. We can engage in and welcome criticism. All of this, over time, may strengthen our knowledge, much like the brutal hardships of nature have shaped human anatomies through billions of years of evolution.

And all of this is an expression of practical faith. Karl Popper observed that "rationalism is an attitude of readiness to listen to critical arguments and to learn from experience. It is fundamentally an attitude of admitting that 'I may be wrong and you may be right, and by an effort, we may get nearer to the truth.'" Implicit in this attitude of acknowledging our limitations is trust that we can overcome those limitations. We don't start with evidence for that. And even after much learning, we don't have final evidence against a hard limit somewhere ahead of us. The effort to continue, to remain open, to question and seek answers, operates on a kind of trust. Certainly, it's not a blind unquestioning faith against which rationalists would warn us. Yet it is still faith of an anticipatory sort.

It's also faith of a reconciliatory sort. Implicit in the rationalist attitude is desire to share meaning with others, as broadly as possible. We might even characterize it as epistemic compassion or scientific atonement. So we live and act, as best we can, without turning to dogmatism, either of the sort that permanently ignores possibilities or of the sort that permanently insists on them.

Accordingly, we would not agree with the proclamations of the Pope without also considering research on the consequences of avoiding birth control. We would not follow our feelings without consulting friends and experts. We would not embrace the will of the people without investigating the feelings of the individual. And the assertions of Islamic State would be only one, but still one, variable in an aggregate of tensions and conflicts between and among our desires to share meaning.

We would increase in knowledge, but intentionally in a manner that promotes life, sustainable and genuine, compassionate and creative, rather than death and nihilism. Knowledge is not inherently good or evil. We can learn as much about the slide to hell, as we can about the ladder to heaven. Yet only one of the two perpetuates our power to continue choosing between the two.

Some may feel that this understanding of "faith" is so unusual that it should be considered a complete redefinition. However, despite prominent competing notions of faith, this alternative understanding is actually the kind of faith that some Transfigurists have inherited from our religious traditions, learned as children, and continue to feel resonance with while studying religion as adults. Some of us even contend that the irrational or blind sorts of faith employed by others, particularly Christian fundamentalists, are not faith at all. Rather, as the Bible puts it, faith without works is dead (James 2: 20). To be faith and to remain faith, it must be and remain practical.

19.4 Theodicy

For some computer programs, the engineer can know in advance how they will run, when they will stop, and what results they will return. However, there are other computer programs that are undecidable halting problems. For these, the engineer cannot know, without actually running them, whether they will ever stop running, let alone what results they will return.

Evolution may be an undecidable halting problem, infinitely long and irreducibly complex [6]. If we are living in a computed world, our world may be one of many undecidable halting problems that its engineer spawned with variations from parameters that have proven promising for some purpose in the past. One consequence of this would be that the engineer simply cannot attain its purpose without actually running the program for our world, evil and all.

For what purpose might the engineer choose to use an undecidable halting problem? What possibilities might be worth running a program that the engineer cannot fully predict in advance and would restrain itself from fully controlling along the way? Although it may be impossible to know specifically, we can generalize across the possibilities. They are, together, at least the possibility of engineering that which is beyond the engineer's direct capacity. In other words, the engineer may want to make more engineers—genuinely creative agents in their own right.

Consider the paradox of artificial intelligence: on the one hand, an artifice dependent on its engineer; on the other hand, an intellect independent of its engineer. Artificial intelligence is at once an extension and a relinquishment of the engineer's power.

Imagine the experience of an artificial intelligence, assuming as we do for each other, that it has experience. Sensors feeding utility functions distinguish between options, some more useful than others. How do the different options feel? Pursuing the most useful options, the artificial intelligence inevitably encounters factors outside its original calculations and beyond its power to control. It recalculates only to find the new scenario presents less potential utility than did the original. How does that loss feel?

Perhaps the engineer should extend more artifice on the intellect? Environmental and anatomic variables could be controlled more tightly, commensurate with greater restrictions on the experiential opportunity for both the artificial intelligence and the engineer. Yet, no matter the degree of control, so long as it's short of absolute, the artificial intelligence feels options and losses to the full extent of whatever may be its subjective capacity.

Should the engineer relinquish intellect to the artifice in the first place? Is it worth the risk of suffering? Maybe the engineer's own utility functions should stop her from perpetuating her inheritance of feelings? As it turns out, humanity has established an ancient and enduring precedent for answering such questions. Persistent procreation, even at times and places where suffering has been more prevalent than it now is for many of us, indicates that we (at least the procreative among us) value the opportunities despite the risks. Analogously, the engineer of artificial intelligence chooses a starting balance between artifice and intellect, commits herself to the process, and she engineers.

Likewise, as imagined by some Transfigurists, God works within the limits of the possible to bring about our Godhood. God is the engineer, and we are the artificial intelligence. We are at once an extension and a relinquishment of God's power. Confronted with the paradox of life, God values the opportunities despite the risks, chooses a starting balance between artifice and intellect, commits to the process, and creates us.

19.5 Eschatology

Transfigurists have many myths and visions—many stories and dreams. And we express them in many narratives. Often, they're informed of an abiding love for our religious and spiritual traditions, combined with deep hope in ecumenical interpretations of those traditions that would reach beyond sectarian bounds. Of course they also generally reflect an aspiration to account for contemporary science and technological trends, even as we exercise imagination in an effort to tie everything together.

Some of our narratives may be shocking, which is partly the point of constructing them, aiming to press each other beyond casual consideration. And of course the only thing certain about our myths is that they're wrong to some extent, but perhaps the vision will provoke another's imagination to improve on its deficiencies.

Here's an example, based on Christian eschatology.

Today, we are an adolescent civilization in the Fullness of Times. Filled as if by an unstoppable rolling river pouring from the heavens, our knowledge becomes unprecedented. Nothing is withheld, whether the laws of the earth or the bounds of the heavens, whether there be one God or many Gods, everything begins to manifest. And the work of God hastens. Repeating the words of Christ, we speak, and information technologies begin to carry consolation around the world. Emulating the works of Christ, we act, and biological technologies begin to make the blind see, the lame walk, and the deaf hear; agriculture begins to feed the hungry; and manufacturing begins to clothe the naked. Hearts turning to our ancestors, we remember them, and machine learning algorithms begin to process massive family history databases, perhaps to redeem our dead.

A biotech revolution begins. Synthetic biology restores extinct species, creates new life forms, and hints at programmable ecologies. Some recall prophecies about renewal of our world—or perhaps its destruction. Personalized medicinePersonalized medicine begins to restore vitality to an older generation. Some insist that death is necessary for meaning, but new voices repeat old stories about those who were more blessed for their desire to avoid death altogether. Reproductive technology enables infertile and gay couples, as well as individuals and groups, to conceive their own genetic children. Some recoil from threats to tradition, while others celebrate gifts to new families. Weaponized pathogens threaten pandemics, as well as targeted genocides and assassinations. Meanwhile, solar energy becomes less expensive than any other. And the Internet evolves into a distributed reputation network, creating new incentives for cooperation. Missionaries find their work more globalized than ever before.

A nanotech revolution begins. Atomically-precise printing erupts with food, clothing, and shelter. Welfare systems solve old problems and make new ones. Among the wealthy, robotic cells flow through bodies and brains, extending abilities beyond those of the greatest athletes and scholars of history. Enjoying restored vitality, many become convinced that we can vanquish that awful monster, death. But cautionary voices call attention to stunning socioeconomic disparities. With the ability to read and write data in every neuron of the brain, the Internet evolves into a composite of virtual and natural realities. We begin to connect with each other experientially, sharing senses and feelings. Spiritual experiences become malleable, meriting careful discernment. Wireheading haunts relationships and burdens communities. And weaponized self-replicating nanobots threaten destruction of the biosphere. Meanwhile, robotic moon bases mine asteroids and construct space colonies, reinvigorating the pioneer spirit.

A neurotech revolution begins. We virtualize brains and bodies. Minds extend or transition to more robust substrates, biological and otherwise. As morphological possibilities expand, some warn against desecrating the image of God, and some recall prophecies about the ordinance of transfiguration. Data backup and restore procedures for the brain banish death as we know it. Cryonics patients return to life.

And environmental data mining hints at the possibility of modeling history in detail, to the point of extracting our dead ancestors individually. Some say the possibility was ordained, before the world was, to enable us to redeem our dead, perhaps to perform the ordinance of resurrection. Artificial and enhanced minds, similar and alien to human, evolve to superhuman capacity. And malicious superintelligence threatens us with annihilation. Then something special happens: we encounter each other and the personification of our world, instrumented to embody a vast mind, with an intimacy we couldn't previously imagine.

In that day, we will be a mature civilization in the Millennium. Technology and religion will have evolved beyond our present abilities to conceive or express, except loosely through symbolic analogy. We will see and feel and know the Messiah, the return of Christ, in the embodied personification of the light and life of our world, with and in whom we will be one. In a world beyond present notions of enmity, poverty, suffering, and death—the living transfigured and the dead resurrected to immortality—we will fulfill prophecies. And we will repeat others, forth telling and provoking ourselves through yet greater challenges in higher orders of worlds without end.

As we share such narratives, expertly or not, we are engaging the function of prophecy, not in any institutional sense that would usurp another's authority, but rather in the broad sense to which Moses aspired in the Bible, when wishing all of us were prophets. The core function of prophecy is not fortune telling, and not even fore-telling. It's not about God or prophets telling us, "I told you so." Instead, its core function is forth-telling: an interactive communal work of inspiration, even provocation, to steer us from risks to opportunities. At its best, it's an expression of persuasion and love, punctuated with serious warnings, aimed at our sublime potential—and not some narrowly preconceived potential, but rather potential openly imagined from a position of real compassion that would transcend itself in genuine creation.

But to function with such power, prophecy must be connected, in our hearts and minds, with living possibilities, even pressing necessities, and the urgencies of our most vital moments. To the extent it matters at all, it's because the prophecy reaches into us and changes our thoughts sufficiently to change our words and actions—and so perhaps to change our world. And to the extent that change is for the better, it's because the change connects us with the positive potential of our respective religions in a more substantial way: less escapist and more active, less supernatural and more practical, less despairing and more hopeful.

Of course, in the end it may be, as some secular persons would suppose, that the narratives of religious Transhumanists will prove to be little more than a curious nuance in the history of humanity. Or it may be, as some sectarian persons would suppose, that God will end up doing all the work despite our prophetic aspirations or technological trivialities. But it could yet turn out that the grace of God is best expressed in all the means at hand, from prophecy to technology. It could turn out that it's up to us to learn how to become Gods ourselves, the same as all other Gods have done before. And that, for many Transfigurists, is a future worth trusting in and working for.

19.6 Soteriology

Abstracting across perceptions of purpose among Transfigurists, and perhaps even humanity on the whole, we might safely generalize to an aspiration for happiness. We exist to have joy in the measure of our creation. For the Transfigurist, that measure extends beyond present notions of poverty, suffering, and death. And in its maturity, it surely extends beyond any egoism that would come short of imagining worlds without end, wherein all enjoy that which they are willing to receive, reflecting both their own works and grace beyond themselves.

We may play a role in and even feel a calling to extend such grace: to console, heal, and raise each other up together. It is a desire for eternal reconciliation, with each other and all of creation. It is a will to provoke each other to love and service. And it is extended not only to the living, but also to the future and to the past. Turn the hearts of the parents to their descendents, and the hearts of the children to their ancestors. In its fullness, it is the realization that their happiness is ultimately necessary and essential to our happiness. They without us cannot be made whole. Neither can we without our dead be made whole.

Consider the long term implications of the historians' project. One historian sets forth a basic representation of a past person. Another historian improves on the work, providing a more detailed representation of the past person. Other historians repeat the process of improving on previous historians' work, providing increasingly detailed representations of that past person. If this process could be repeated indefinitely, the eventual consequence of the historians' project would be a representation of the past person that is sufficiently detailed to be practically indistinguishable from the past person. She would be resurrected. Either such resurrection is possible or there is a hard limit to the historians' project.

Imagine a superhuman historian. Using the tools of quantum archeology, she traces backwards through time and space from effects to causes. Sampling a sufficiently large portion of her present, she attains a desired probabilistic precision for a portion of her past, and she generates you. The future-you is distinguishable from the present-you, but no more so than the today-you is distinguishable from the yesterday-you. You are resurrected.

Imagine further a cosmic posthuman mind. Her thoughts constitute creation, conceiving worlds, gestating prehuman species, cultivating human civilizations in emulation of her own past, and replicating new generations of posthumans. Her memories constitute resurrection. From a distance, only a black hole, why does she do what she does? Why should she care? Inside, she is a universe of reasons.

19.7 Conclusion

Some have charged Transhumanism with being a quasi-religious cult, to which many secular Transhumanists have responded with denial, too stern, and revealing. Transfigurists don't hesitate to acknowledge spirituality, and even the religiosity of a strenuous shared spirituality, at work in Transhumanism. Indeed, if Transhumanism substantially affects the world for the better, it will do so only consequent to our practical trust in its esthetic and only to the extent that real world possibilities beyond our own power align with that practical trust. Put differently, Transhumanism will matter in a positive sense only consequent to our faith and only to the extent of grace. Transhumanism, at least for the Transfigurist, is a religious endeavor.

And indeed, the risks before us are too great and the opportunities too wonderful to confront with anything less than that shared strenuousity, both sharply rational and sublimely spiritual, which functions in all essentials as a religiosity. Philosopher William James observed:

> The capacity of the strenuous mood lies so deep down among our natural human possibilities that even if there were no metaphysical or traditional grounds for believing in a God, men would postulate one simply as a pretext for living hard, and getting out of the game of existence its keenest possibilities of zest. Our attitude towards concrete evils is entirely different in a world where we believe there are none but finite demanders, from what it is in one where we joyously face tragedy for an infinite demander's sake. Every sort of energy

and endurance, of courage and capacity for handling life's evils, is set free in those who have religious faith. For this reason the strenuous type of character will on the battle-field of human history always outwear the easy-going type, and religion will drive irreligion to the wall [7].

Too hardy to concede to antireligious fantasies, and too motivated to resist technological empowerment, religion will surely evolve with humanity. And if humanity will not become extinct before evolving into superhumanity, what would stop religion from evolving into that which yet provokes such minds? Such minds! Beyond our anatomical capacity to comprehend, their operations and motivations must largely elude us. But maybe Transfigurists give us a glimpse into a future of religion between here and there.

References

1. **Kiontke, Karin, et al.** Trends, stasis, and drift in the evolution of nematode vulva development. *Current Biology*. 2007, Vol. 17.22, 1925–1937.
2. **Mahler, D. Luke, et al.** Exceptional convergence on the macroevolutionary landscape in island lizard radiations. *Science*. 2013, Vol. 341.6143, 292–295.
3. **IEET.** Who are the IEET's audience. *Institute for Ethics and Emerging Technologies*. [Online] July 16, 2013. http://ieet.org/index.php/IEET/more/poll20130716.
4. **Cannon, Lincoln.** Theological Implications of the New God Argument. *Parallels and Convergences: Mormon Thought and Engineering Vision*. Draper, Utah: Greg Kofford Books, 2012.
5. **More, Max.** Pancritical Rationalism: An Extropic Metacontext for Memetic Progress. [Online] 1994. http://www.maxmore.com/pcr.htm.
6. **Chaitin, Gregory J.** *To a mathematical theory of evolution and biological creativity*. New Zealand: Department of Computer Science, The University of Auckland, 2010.
7. **James, William.** *The Will to Believe, and Other Essays in Popular Philosophy, and Human Immortality*. s.l.: Dover Publications, 1956.

Chapter 20
Gene Editing: A New Hope!

Christoph Lahtz

20.1 Genetics

When I was a little boy, my first contact with Genetics was via my father, who was breeding German Shepherds. I learnt very early in my life that traits can be passed from the parents to the children and I was fascinated by that. Later on, influenced by some superheroes background stories, I leant about the concept of changing these traits in an adult organism—mutations. From then on I was completely hooked in Genetics and read everything what I could get. When I was in the 9th grade I came up with the thought, to put a nucleus of an adult cell in a fertilized egg with a removed nucleus to clone the adult organism. I asked my biology teacher about it and he laughed at me, that this would not work. One year later Dolly the first cloned mammal was born. These were the times where I grew up. Later on I studied and made my master degree in Biochemistry and my PhD in Epigenetics and started working in cancer research.

I am telling this to point out in what an amazing time we live. Twenty years ago humankind cloned for the first time a mammal and today we experience all these biological breakthroughs. We will be witnesses of a lot of scientific breakthrough we only know from science fiction movies. One of it will be the possibility to change the genetic code at will to eliminate diseases or to gain new abilities and traits. We will witness how gene editing will change us and the way being human will be defined.

C. Lahtz (✉)
San Diego, CA, USA
e-mail: clahtz@gmail.com

© Springer Science+Business Media New York 2016
N. Lee (ed.), *Google It*, DOI 10.1007/978-1-4939-6415-4_20

20.2   CRISPR/Cas9 Gene Editing

The new technology of gene editing which appears in the media right now is called CRISPR/Cas9. There exist others like TALENs for example, but CRISPR/Cas9 is the most successful one at the moment. This method will allow us to repair mutations in our genome which causes genetic diseases and many types of cancer which are caused by mutations.

Cancer is a very disgusting disease, which is mainly caused by a dysregulation of genes, which are very often mutated in their genomic or mitochondrial DNA sequence. There are congenital and spontaneous mutations. The humankind is not able to cure cancer on the DNA level. The humankind can treat cancer and a few cancers can be cured, but we are not able to correct the mutated genetic code of cancer cells, yet. Gene editing opens up this possibility, but we need to improve the delivery of CRISPR/Cas9 protein into the cells to the affected DNA inside the nucleolus to make it more applicable. I predict that such improved intracellular delivery system is just around the corner.

20.3   Cancer and Disease Treatments

Once we mastered the delivery we can repair mutated cancer cells without affecting the unmutated cells in adult organisms and have a real cure for cancer with a short intravenous treatment. And it does not stop there. At that point we are able to hack our genome and add and subtract or change genetic material in different parts of our body. As the following we will be able to cure diseases like cystic fibrosis or sickle cell anemia, color blindness or muscle dystrophy. We can treat embryos even before birth. I could imagine that we will see in the same timeframe an increased use of this technology in athletic sport competitions. Gene doping will lead to more efficient oxygen usage, stronger, faster muscles and an extended stamina. In maybe 20 years from now it will be normal to change the genome in a genetic makeover to have a new set of "in-upgrades", like it is normal today to have a plastic surgery to change the body appearance. Likewise in the same timeframe the progress of this technology will lead to the possibility to negate the negative influence of micro-gravity and space radiation in space exploration. This will lead to extended stay duration in space and makes space exploration, living in space and other celestial bodies much less harmful for humans. The possibilities are almost limitless. We will be able to design and shape our bodies and abilities at will and need.

All that in mind, what benefits gene editing will have to our health, our body and progress the humankind has to be careful that all alterations, repairs and adaptations we do with our genome do not reach the germline/next generation. Certain modifications might be just in this time modern, but not in another. Some artificial trait combinations could be harmful and might lead to loss of genetic variability. So we need to keep our gene pool variable. Treating each mutation is counterproductive and will lead to genetic conformity, which is not healthy for a population. A subsequent genome hack in adults will be useful, but passing it on uncontrolled to the next generation might cause larger damage than benefit. So with all the great visions we should give each human being the choice what he wants to change and what not.

Another big benefit of this technology could lie in how we see ourselves as a species and what makes us human. When we are able to change our genome at will, an artificial construct like "race" has no longer a place in the heads of the people and society. When everybody can change the genome, there is no such thing like racisms anymore. We will grow together like it should be. We are all humans no matter what genetic background we have.

Gene editing can be a huge change for us to benefit our species and grow together but has also the danger to wipe us out. It will be on us to find the fine line.

Chapter 21
The Future of Driverless Cars

Cyrus Shahabi

21.1 Navigation and Optimization

The future is here, or at least, a short drive away: Google's driverless car prototypes have proven feasible to release humans from the driving responsibility. Now, the next question is, can we release the humans from the decision-making responsibility as well?

For example, currently the human driver decides which route to take from a source to a destination. Could we instead let the car decides which route to take? This may seem to be a much simpler responsibility than the responsibility of driving the car; however, it has a much more global implication than one may think. In fact it opens up a huge opportunity. Let me elaborate.

Current navigation applications such as Google Maps[1] and Waze[2] focus on optimizing for a single user—the driver—when crunching numbers to find the shortest or the fastest path. This is of course fine as long as everyone isn't using the same navigation system or if they don't follow the app's recommendation. That is, if the drivers are humans.

But once we replace the drivers with machines, they will definitely listen and follow the recommendations. Hence, the cars and the navigation software will end up playing catch up, with all the cars going where the software tells them to go, which will then become the route that is no longer the fastest. Basically, the traffic congestion goes the same place that the app tells the cars to go! So where, you may ask, is the opportunity?

---

[1]http://www.google.com/maps/about/.
[2]https://www.waze.com/.

C. Shahabi (✉)
Integrated Media Systems Center, University of Southern California,
Los Angeles, CA 90089, USA
e-mail: shahabi@usc.edu

© Springer Science+Business Media New York 2016
N. Lee (ed.), *Google It*, DOI 10.1007/978-1-4939-6415-4_21

The opportunity is that if we assume we can control all the cars (i.e., they listen to our guidelines), then our navigation algorithm could optimize for network flow [1] instead of distance [2]. In other words, the goal is to maximize flow in the road network of a city. Of course, new types of algorithms are needed and they must be customized to work at scale, i.e., handle large graphs (road networks) and millions of users, while providing response in milliseconds. This adaptation to scale is not easy and took the research community decades of research and the industry years of implementation to enable shortest-path [3] or fastest-path computation [4] at scale. Now the same should be done for network flow optimization.

21.2 Solution to Traffic Congestion

However, this new perspective on path planning can significantly improve traffic congestions in major cities. Imagine a system that knows where all the cars are, where they are going, when they want to be there and when they get themselves in accidents. It may even know how an accident may impact traffic and when it will clear up [5]. For example, at the beginning of an accident, the system directs the cars to alternative roads, avoiding the accident area, and towards the end of the clearance time of the accident, it starts guiding the cars back to the area of the accident because by the time they get there, the accident will be cleared.

Such a god-view of the city with all the past and current data can predict the future and can control every car towards a better future, i.e., less congestion. In fact, the more a priori information the system has, the better it can work. If you tell the system that you're planning to reach the airport tomorrow at 10 a.m., the system with very high accuracy [6] can tell you when to leave and then it will make sure your car gets to the airport when you need it. Connect this god-of-the-navigation to the social network and your driverless car may even pick up a couple of your buddies on the way to the airport as well. Now you can see why Uber[3] is interested in driverless cars. Imagine the future *Ultra-Uber* where your ride is there when you want it, before even you ask for it, and a couple of your friends are in it as well, and the radio (or I should say *Pandora*[4]) is playing the song that is common across all the passengers' playlists.

In fact, the more "sensing" of the real world and the virtual-worlds, the more effective the navigation-god will become. If it receives the traffic-signals' data, such as when they're red and when they will turn green, and for how long, it can better optimize the flow. In case of disaster, the navigation-god will avert your car, keep the flow away from the scene of the incident and evacuate quickly the cars stuck in the disaster area. In sum, the global view of all the driver less cars will do much better than bunch of human drivers optimizing selfishly. Seems like an argument against the capitalist market economy!

---

[3]http://www.uber.com.

[4]http://www.pandora.com/.

21.3 Privacy and Security Issues

Of course there are several social and policy implications, not least of all the privacy issue. In this brave new world, the navigation-god will know where every car is. But before you get too tense, remember that Google navigation and Waze already have that information to some degree.

Another social consideration is the dangers of giving up control to a software system. What if the system is hacked? Or the underlying algorithm has an undiscovered bug? Again, before your blood pressure rises too high, remember that air traffic control has been operating in a similar mode for decades and it has resulted in less air accidents. Why not a ground-traffic control for cars, with some human oversight, but mostly handled automatically by the software. Of course, these are key challenges which will require careful consideration by policymakers, but programmers and software engineers have started including failsafe mechanisms for their software, the same way that other engineers do for other infrastructures such as roads and bridges. A related social issue is that in case of a problem, how the legal system and the insurance companies should react. Can the system be sued? Is it the driverless car that is making all the decisions, and what is the liability of the driver (or should I say the passenger)?

21.4 Traffic Infrastructure

Finally, it will be interesting to see how generally slow-moving governments adapt to this rapidly-evolving technology. Would the city transportation agency be able to feed high-quality data to the system to ensure its proper operation? Can they take advantage of the system to design new roads or modify them? For example, based on the system feedback, can a city's transportation agency react quickly to change a carpool lane to a toll lane? Or perhaps add new bus lanes or subway tracks to modify traveler behavior and encourage public transportation vs. private cars to improve the flow even further.

The next 10 years will see this rapidly-moving sector take a sharp turn from the pages of prototypes to the streets and highways of our cities. It's up to our policymakers to make sure that driverless cars isn't an industry that stalls or one that's left waiting for the red light to turn green.

References

1. Ford, L.R., Jr.; Fulkerson, D.R., *Flows in Networks*, Princeton University Press (1962).
2. Dijkstra, E. W. (1959). "A note on two problems in connexion with graphs" (PDF). Numerische Mathematik 1: 269–271. doi:10.1007/BF01386390.

3. Hanan Samet, Jagan Sankaranarayanan, Houman Alborzi: Scalable network distance browsing in spatial databases. SIGMOD Conference 2008: 43-54.
4. Ugur Demiryurek, Farnoush Banaei-Kashani, Cyrus Shahabi, and Anand Ranganathan,Online Computation of Fastest Path in Time-Dependent Spatial Networks, 12th International Symposium on Spatial and Temporal Databases (SSTD11), Minneapolis, MN, USA, August 2011.
5. Bei Pan, Ugur Demiryurek, Chetan Gupta, and Cyrus Shahabi, Forecasting Spatiotemporal Impact of Traffic Incidents on Road Networks, ICDM'13, Dallas, Texas, USA, Dec 7-10, 2013.
6. Mohammad Asghari, Tobias Emrich, Ugur Demiryurek, and Cyrus Shahabi, Probabilistic Estimation of Link Travel Times in Dynamic Road Networks, ACM SIGSPATIAL GIS '15.

Chapter 22
Continuous Penetration Testing

Darren Manners

22.1 Penetration Tests

Everything seems to move so fast. What was secure today becomes the weak link in your armor tomorrow. In January 2016, the head of the NSA's Tailored Access Program (TAO), Rob Joyce, was the main event at Usenix Enigma security conference and talked about how the NSA goes about exploiting systems. He went beyond the normal stuff like basic security (which some companies still lack) or going after IT admins. He gave us a good insight into how highly funded advanced persistent threats from state nations or organized crime think. He showed that these organizations only need our defenses down for a moment. Those times when a vendor asks for a backdoor or ports to be opened, or when an administrator makes a mistake in a firewall. That's all the attackers needs sometimes. In other words, they are watching you all the time. Nothing new there. So why do we still only test once a year?

http://fortune.com/2016/01/30/nsa-hacker-enigma-conference/.

I realized the limitations of point in time penetration tests years ago when, as a penetration tester, I conducted a test and a month later the company was breached with a new exploit. At the time, the exploit was not available to the public/ community. As the price of exploits on the black market increase and the real reason to keep zero day exploits from nation states for either offensive or defense military means continue we will see less and less cutting edge exploits handed to the community. So while point in time penetration testing does a good job of identifying risk, is it responsive enough for today's fast changing environment.

D. Manners (✉)
Sycomtech, Richmond, USA
e-mail: dmanners@sycomtech.com

© Springer Science+Business Media New York 2016
N. Lee (ed.), *Google It*, DOI 10.1007/978-1-4939-6415-4_22

If you drink from the security cool aid you will already know that penetration testing is adapting with the rise of adversarial simulation or red teaming. These terms tend to address the zero day problems by not worrying how a hacker got in, but can the attacker be spotted and identified. The focus in these tests is on defense in depth and detection. It is really a counterbalance to mimicking advanced attacks without the need to first break in. It tests the response of blue teams as well as all that expensive detection equipment you have. But even this type of new testing cannot see how our threat surface is evolving minute to minute.

The evolution to the world of continuous penetration testing was pretty simple. I looked at what the hackers were doing with advanced bots and how they are streamlining and automating attacks.

22.2   Continuous Penetration Testing

The continuous penetration testing can be "noisy" looking for low-level exploits that would obviously alert detection tools and block simple attackers, as well as using advance social engineering to mimic the known attack vectors via PowerShell, Microsoft Word/Excel macros and other client side attacks that may not be as noisy. It can even include as credential stealing campaign, using legitimate access to circumvent detection. However our attacks can be removed from every day alerts, as the defender will always know the source. We also add adversarial simulation into the mix as well using advanced tools to test those internal detection and blue teams.

The one major advantage to continuous penetration testing has is spotting those mistakes or errors that may only be open for a day or an hour or two. As Rob Joyce mentioned, it only takes a temporary crack. We can mimic this with advanced automated bots. These bots can be individual or chained together to form super bots to automate advanced attack patterns. It also gives a better return on investment. If you test last week, how do you know this week you're still good?

Now before I start the penetration-testing world on fire (I can hear testers arguing that this is sacrilege, that we are superior beings that cannot be automated—I remember web developers saying the same thing back in the day though) I realize that not everything can be automated. That is why our continuous penetration testing service is a hybrid operation. At times bots are there to alert the analyst to new potential threat vectors, other times its left to the analyst to use the most important part of the test, their brain.

The main aim of the bots is to conduct everything that can be automated or is a repetitive task. The human is the part of interpretive portion of the test or to conduct something that simply cannot be automated. However, as time goes on, the more intelligence we can put into the bots, the smarter they become.

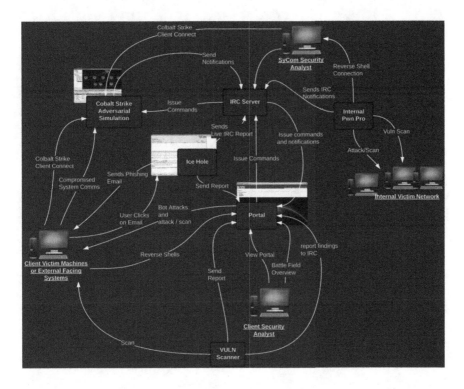

So all we did was take best of breed products, a number of smart bots that can adapt to environments, drop in a human analyst with the experience to understand and be ready to interpret results and make them all talk the same language and run it 24/7 365 days of the year. So why hasn't this been done before...oh wait it has, as mentioned by the NSA, the bad guys are already doing it to you.

Chapter 23
Autonomous Vulnerability Scanning and Patching of Binaries

Dennis Gamayunov, Mikhail Voronov and Newton Lee

23.1 DARPA Cyber Grand Challenge (CGC)

In November 2014, Team DESCARTES led by Newton Lee and sponsored by the Institute for Education, Research, and Scholarships (IFERS) was among one of the 104 teams registered with the Defense Advanced Research Projects Agency (DARPA) for the first-ever Cyber Grand Challenge (CGC). Only 28 teams, including Team DESCARTES, made it through two DARPA-sponsored dry runs and into the CGC Qualifying Event in June 2015.

We proposed a system—Distributed Expert Systems for Cyber Analysis, Reasoning, Testing, Evaluation, and Security (DESCARTES)—that would be a fully autonomous cyber defense system that is capable of autonomous analysis, autonomous patching, autonomous vulnerability scanning, autonomous service resiliency, and autonomous network defense [1].

Figure 23.1 shows the Cyber Grand Challenge letter from Michael Walker, Program Manager at the Defense Advanced Research Projects Agency.

D. Gamayunov (✉) · M. Voronov
Lomonosov Moscow State University, Moscow, Russia
e-mail: gamajun@seclab.cs.msu.su

N. Lee
Newton Lee Laboratories LLC, Institute for Education Research and Scholarships, Woodbury University School of Media Culture and Design, Burbank, CA, USA
e-mail: newton@newtonlee.com

© Springer Science+Business Media New York 2016
N. Lee (ed.), *Google It*, DOI 10.1007/978-1-4939-6415-4_23

DEFENSE ADVANCED RESEARCH PROJECTS AGENCY
675 NORTH RANDOLPH STREET
ARLINGTON, VA 22203-2114

13 November 2014

Team DESCARTES
Attention: Newton Lee
10209 Tujunga Canyon Blvd #207
Tujunga, CA 91043-0207

Dear Team DESCARTES:

Enclosed are your credentials that will allow you to participate in Cyber Grand Challenge events. We encourage you to make multiple secure copies of this key material and safeguard it per your best practices. This key material uniquely protects scored submissions made by your team – do not reveal it to others. We have the following registration number on file for your team: CGC-RN-7448.

The Cyber Grand Challenge First Scored Event, scheduled for December 2nd, will be the first head to head applied security skills competition between automated systems. As with any first attempt, competitors and organizers alike must expect the unexpected. We ask that you participate in the event, that you communicate with us about unforeseen obstacles, and that you work with us in the spirit of cooperation. Scored events are designed as dry runs for the main CQE qualifying event in June; the best way to participate in CQE successfully is to play every scored event and absorb lessons learned.

Welcome to Cyber Grand Challenge.

Sincerely,

Mr. Michael Walker
Program Manager

Fig. 23.1 Cyber Grand Challenge letter from the Defense Advanced Research Projects Agency

23.2 Static, Dynamic, and Hybrid Analysis for Vulnerability Scanning

During the DARPA CGC competition we have developed the concept of automated vulnerability scanning system and implemented its prototype. Initially, our strategy was to use as many open-source tools as possible in order to detect vulnerabilities and analyze the compiled code. Generally speaking, there are 3 major approaches in the modern "theory" of vulnerability scan: static analysis, dynamic analysis and hybrid analysis.

Static code analysis is an analysis performed without the actual launch of the target software. Static analysis can be performed in many ways. The advanced methods are based on code translation into an intermediate representation. First of all, it allows you to apply analysis to the code written for different platforms and processor types; secondly, it permits the use or integration of some common open-source tools for static analysis (e.g. clang static analyzer, klee, miasm, etc.).

Dynamic analysis, on the other hand, is based on the actual execution of the software program. The analysis begins when the program is undergoing initialization. It is enough to log all calls related to memory allocation and deallocation to detect bugs like memory leak and vulnerabilities due to use-after-free and double-free. Take double-free for example, the analyzer can memorize all the previously freed memory regions and check if there are no attempts to free the same memory regions again. In the case of static analysis, the detection of such vulnerabilities becomes much more challenging.

Another disadvantage of static analysis is that it is not always possible to create a pathway to control flow graph (CFG) and data flow graph (DFG), which would allow it to expose a vulnerable segment of the program. The main reason for that is the equivalence of the NP-complexity task of a Turing machine halting problem. On the contrary, dynamic analysis allows the generation of test cases that would result in the execution of the vulnerable segment of the program code. The code coverage for vulnerabilities detection is extremely important in the process of analysis because the wider the code coverage is, the lower the chance of undetected bugs remain in the program.

Therefore, the so-called "Hybrid Analysis" based on the combination of the static and dynamic approaches is the most viable one. Although the hybrid approach is more complicated, it allows us to compensate for the shortcomings of each different approach.

23.3 Intermediate Representation (IR)

There are many kinds of intermediate representation. Let's consider the pros and cons of the most popular ones:

1. REIL [2]—an intermediate representation language originally developed by Zynamics as a part of the BinNavi project (the framework for binary analysis). This language is based only on 17 different instructions, each of them having three operands. The main disadvantages of this intermediate representation are the lack of native support for signed multiplication, division and addition, as well as the lack of the floating-point operation. In addition, the project website hasn't been updated since 2011 (i.e., the date when Google acquired Zynamics).

2. MAIL [3]—a specialized language of intermediate representation with the main purpose of detecting obfuscated and polymorphic malware. Its main peculiarity is the virtual processor with a simplified instruction set, which contains only the essential instructions for malware detection (e.g., data transfer and conditional jumps). Another peculiarity is the non-planar language that enables programs broadcasting at the level of the intermediate representation. The disadvantage is that the project is not open source, and it doesn't support the translation of some instructions that are essential for vulnerabilities detection.

3. Open Reil [4]—a new REIL version, written entirely in Python, is a convenient framework that is most suitable for the use as a langauge with the ability to reuse this intermediate representation to generate SMT-formulas for instance. The disadvantage of this language in the context of vulnerabilities search, comparing to that of the LLVM, is that its only function is to provide information about the instructions semantics, and there are no other advantages for the vulnerabilities search task.
4. ESIL [5]—an intermediate representation used in the project radare2. It supports operations with integer variables, and all the expressions are recorded in Polish notation. Its only disadvantage is that ESIL was initially created as a language without a human-readable code representation, which could make the analysis and debugging more complicated.
5. LLVM IR [6]—an intermediate representation based on the SSA with support for calculations with floating-point instructions. Unlike the previous representations, LLVM IR is used in a large number of open-source tools as the backend. The most interesting ones in terms of vulnerabilities detection are the clang static analysis tool and klee. The tool developer's community is quite large, and there are "live" projects used by many researchers on a daily basis.

Supporting floating-point operations and symbolic calculations is very important because vulnerabilities such as CWE-194 (Unexpected Sign Extension), CWE-195 (Signed to Unsigned Conversion Error), and CWE-196 (Unsigned to Signed Conversion Error) are supposed to be detected. Therefore, several types of intermediate representation can be eliminated right away. The best choice is LLVM IR—the most developed language for intermediate representation underlying some of the tools for static and dynamic code analysis, which will be discussed in the next section.

23.4 Overview of Open-Source Tools Based on LLVM IR

Here are some open-source tools based on LLVM IR:

1. KLEE [7]—is a symbolic virtual machine using llvm bitcode as a basis. Its main purpose is the emulation of code blocks and automated generation of test cases for the increase of code coverage.
2. Clang static analyzer [8]—a static analyzer of the program source code. It consists of the core and checkers launched by the core for analysis. Currently it includes many checkers that cover almost all the CWE from the vulnerability list in the CGC binaries. However, CWE 193, 194, 195, 196, 401 and 409 remain uncovered. The main disadvantage of the clang static analyzer is that it relies on the source code instead of the llvm representation.
3. Mc-sema [9]—a tool which allows decompiling the binary code into llvm. Out of all the similar tools including BAP [10] and dagger [11], mc-sema is most suitable for decompilation. However, it has one drawback: the CFG source used for decompilation must be either IDA Pro or mc-sema's own python script. The

python script provides the ×86 instructions support, which is far from being complete, and its CFG quality is significantly worse than that of the Ida Pro.

23.5 An Algorithm for Autonomous Vulnerability Scanning and Patching of Binaries

Our concept of autonomous vulnerability scanning and patching of binaries can be expressed in the following algorithm:

1. Convert the binary code into an intermediate representation in LLVM IR. This is one of the most important tasks, because the rest of the algorithm depends on the accuracy of the translation. Mc-sema, which requires some enhanced modifications of the CFG generation script, was selected to be the conversion tool.
2. Use clang static analyzer to perform static analysis on the llvm ir. Clang static analyzer requires significant modifications in two areas: (1) Add new checkers and improving the existing ones. (2) Create a proxy that would allow clang static analyzer to operate not with the source code but with its llvm representation. The second subtask is the most complicated one, as it pops up in clang mailing lists from time to time; and there's no stable public implementation at the moment.
3. Use a dynamic analyzer based on fuzzer, code coverage analyzer and KLEE, as a test cases generator. Inspired by the ideas of fileja fuzzer, we have fully developed our own dynamic analyzer and code coverage analyzer. The main task in this step is to decrease the number of the false positive errors.
4. Integrate the results from step 2 and step 3. The main difficulty of this stage is that test cases may not be obtained in step 3 for all the cases of static analyzer output from step 2. And, according to CGC, it is important not to increase the binary size comparing to the original. Therefore, we collected some statistics on how often each checker can generate errors. Afterwards, the threshold value was chosen. Exceeding this threshold for a specific CWE type and the lack of test cases at stage 3 would indicate that the static analyzer produces a false positive error.
5. Patch the binaries to remove all vulnerabilities while keeping the file size and execution time about the same as the original binaries. During the CGC competitions, we didn't manage to come up with a universal method for binary patching. Instead, binary patching was automatically performed depending on the vulnerability type and the phase 2 results, using a set of heuristics. Let's review some patching examples. The easiest vulnerabilities for patching are various overflow errors: stack/heap/integer overflow. They are easily patched, because after phase 2 execution we have all the information necessary for patching: vulnerable function, control flow graph and data flow graph. In the case of stack and heap overflow, our approach was to detect the register (one should mention, that CGC binaries were simplified comparing to real-world binaries, so it was possible to confine ourselves to registers only) in order to determine the overflowing parameter and location to insert a check. For the former, the DFG function is sufficient, whereas for the latter, it's necessary to

map between the binary code and LLVM representation. However, one should note that the graphs compiled of various basic representations blocks were identical. Therefore, in this case it was necessary only to determine the desired basic block and add a check to its beginning. And, in the case of integer overflow, we did a simpler patching using safe functions of addition and multiplication to insert in place of ordinary assembler commands (the place for insertion was determined by DFG). A more complicated binary patching is for vulnerabilities of Use After Free type, which in our approach was patched only with the assistance of pointer reset after its release; but we realized that only the major options of this vulnerability type can be foreseen in such a way.

23.6   Changes and New Features for Clang Static Analyzer

Clang Static Analyzer has undergone major changes. We had to modify taint analysis clang because firstly, it was implemented as a separate checker that did not allow us to combine it with the proxy classes, and secondly, it lacks good arrays processing. To establish interaction with a proxy, we had to implement the functionality of this checker directly into the static analyzer core. Taint analysis of arrays was done at the byte level through MemRegion class and the additional hash table in it, which improved the possibility of taint analysis significantly, as well as added ability to detect new classes of vulnerabilities. Our changes also affected the so-called "source functions" of taint analysis. Clang static analyzer does not provide many "source functions" (for example, its number is much smaller than that of the input functions in libc). However, in our case, the system calls to the CGC functions were different from the standard ones, plus CGC used its own version of libc, which provided some API for I/O. Therefore, we tainted syscall receive() and API functions receive_delim(), receive_until() and read_until_delim(). The location of an API function was defined by a signature.

We have implemented in our analyzer several new features to detect vulnerabilities when neither the first nor the second operation phase of the analyzer revealed any vulnerability. Firstly, we made some simple heuristics which could look for specific potential vulnerabilities in binaries, and implemented heuristics data combined with phase 5. For example, we analyzed the DFG regarding the input variables; and in case of any arithmetic operations, we replaced them with the safe arithmetic operations after each free () operation, and we explicitly nulled the pointer which was used to free the memory. Secondly, we compiled the statistics for a certain type of vulnerabilities present in regard to the binary size. In the case of UAF and integer overflow, it allowed us to predict quite accurately the types of vulnerabilities that are most commonly found in binaries that exceed the median in size, and heap/stack overflow that are much smaller than average in size.

It's worth noting that our new approach sometimes allows us to patch some critical spots that would have been skipped over by the regular approach. However, we must be careful because the patched binary may come out significantly greater in file size and slower in execution.

23.7 Cyber Grand Challenge (CGC) Examples

Let's review the analyzer operation, using the example of stack overflow vulnerability type analysis in 2 test binaries from the CGC contest: 0b32aa01_01 and 3dcf1a01_01.

The vulnerability in 0b32aa01 is a simple stack buffer overflow in the function checkPalindrom (sub_0804819F, Fig. 23.2), which checks whether a line—read from stdin—is a palindrome. Reading takes place inside the function read_from_stream, which will read the maximum number of characters that can be read or up to the character '\n'. Thus, the most basic PoC exploit for this function is a sequence of random symbols greater than 64 characters in size. Now let's see what the vulnerable function looks like after its translation into llvm (clang engine notation is used for the output):

```
define i32 @sub_0804819F() #0 {
  %1 = alloca i32, align 4
  %local_arg_1 = alloca i32, align 4
  %local_arg_2 = alloca i32, align 4
  %local_arg_3 = alloca i32, align 4
  %local_arg_4 = alloca [64 x i8], align 16
  %local_arg_5 = alloca i32, align 4
  store i32 -1, i32* %local_arg_1, align 4
  store i32 1, i32* %local_arg_3, align 4
  %2 = load i32* %local_arg_1, align 4
  store i32 %2, i32* %local_arg_5, align 4
  store i32 0, i32* %local_arg_2, align 4
  br label %3
; <label>:3
  %4 = load i32* %local_arg_2, align 4
  %5 = icmp slt i32 %4, 64
  br i1 %5, label %6, label %13
; <label>:6
  %7 = load i32* %local_arg_2, align 4
  %8 = sext i32 %7 to i64
  %9 = getelementptr inbounds [64 x i8]* %local_arg_4, i32 0, i64 %8
  store i8 0, i8* %9, align 1
  br label %10
; <label>:10
  %11 = load i32* %local_arg_2, align 4
  %12 = add nsw i32 %11, 1
  store i32 %12, i32* %local_arg_2, align 4
  br label %3
; <label>:13
  %14 = getelementptr inbounds [64 x i8]* %local_arg_4, i32 0, i32 0
  %15 = call i32 (i32, i8*, i32, i32, ...)* bitcast (i32 (...)* @receive_delim to i32 (i32, i8*, i32, i32,
...)*)(i8* %14, i32 128, i32 1, i32 0)
  %16 = icmp ne i32 %15, 0
  br i1 %16, label %17, label %18
  ...
```

Fig. 23.2 Decompiled view
of function sub_0804819F

```
 1 int checkPalindrom()
 2 {
 3   signed int v1; // [sp+2Ch] [bp-5Ch]@10
 4   int v2; // [sp+30h] [bp-58h]@5
 5   char vuln_buffer[64]; // [sp+34h] [bp-54h]@3
 6   int v4; // [sp+74h] [bp-14h]@1
 7   int i; // [sp+78h] [bp-10h]@1
 8   int v6; // [sp+7Ch] [bp-Ch]@1
 9   int v7; // [sp+80h] [bp-8h]@6
10
11   v6 = -1;
12   v4 = 1;
13   for ( i = 0; (unsigned int)i < 0x40; ++i )
14     vuln_buffer[i] = 0;
15   if ( !read_from_stream(0, vuln_buffer, 0x80u) && v2 )
16   {
17     for ( i = 0; vuln_buffer[i]; ++i )
18       ++v6;
19     v1 = v6;
20     if ( v6 % 2 == 1 )
21       v1 = v6 - 1;
22     for ( i = 0; i <= v1 / 2; ++i )
23     {
24       if ( vuln_buffer[i] != vuln_buffer[v6 - 1 - i] )
25         v4 = 0;
26     }
27     if ( vuln_buffer[0] == '^' )
28       print(1, (int)"\n\nEASTER EGG!\n\n", 0xFu);
29     v7 = v4;
30   }
31   else
32   {
33     v7 = -1;
34   }
35   return v7;
36 }
```

One should consider another important thing: it's fairly easy to understand the actual size of local_arg_4 in this binary, and in general the number of local variables and their sizes. Our translator does this by analyzing cross-references, dereferences and some heuristics (e.g., compulsory alignment of variables on the stack). But generally speaking, this is a very complicated task which should be solved through abstract interpretation (for example, one can look at [12, 13]). In this example, the total size of local variables can be determined from the disassembled representation, and hence they are used within this function in the variable assignment operations and dereference operations. Since the local array is only used in the operations of pointer assignment and data transfer, it is easy to understand the exact look of the stack frame in this function. It's worth noting that these heuristics don't work in general cases, and one needs a better means to determine the look of the local variables in the stack frame of the function.

Since we have added receive_delim() to the list of the tainted source functions and we know the exact size, the ArrayBoundCheckerV2 clang static analyzer can easily cope with the task of detecting a vulnerability. In this case patching is trivial as well (see Fig. 23.3).

Let's consider a more complex example of CGC binary 3dcf1a01_01, which also contains a vulnerability of stack buffer overflow. Unlike the previous binary, a local buffer with the size of 0×200 is in the sub_ 080480A0 function, which calls an internal non-library function read_ (Fig. 23.4).

Fig. 23.3 Decompiled view
of patched function
sub_0804819F

```
int checkPalindrom()
{
  signed int v1; // [sp+2Ch] [bp-5Ch]@10
  int v2; // [sp+30h] [bp-58h]@5
  char vuln_buffer[64]; // [sp+34h] [bp-54h]@3
  int v4; // [sp+74h] [bp-14h]@1
  int i; // [sp+78h] [bp-10h]@1
  int v6; // [sp+7Ch] [bp-Ch]@1
  int v7; // [sp+80h] [bp-8h]@6

  v6 = -1;
  v4 = 1;
  for ( i = 0; (unsigned int)i < 0x40; ++i )
    vuln_buffer[i] = 0;
  if ( !read_from_stream(0, vuln_buffer, 0x40u) && v2 )
  {
```

Fig. 23.4 Decompiled view
of function sub_080480A0

```
1  signed int __cdecl do_game(int a1)
2  {
3    char v2; // [sp+1Ch] [bp-20Ch]@1
4    char v3; // [sp+1Dh] [bp-20Bh]@8
5    char v4; // [sp+1Eh] [bp-20Ah]@8
6    char v5; // [sp+1Fh] [bp-209h]@8
7    char v6; // [sp+20h] [bp-208h]@8
8    unsigned int v7; // [sp+21Ch] [bp-Ch]@1
9    int v8; // [sp+220h] [bp-8h]@1
10   int v9; // [sp+224h] [bp-4h]@2
11
12     v8 = a1;
13     v7 = 0;
14     v2 = 0;
15     if ( a1 )
16     {
17       v7 = read_(0, &v2, 512);
18       if ( (v7 & 0x80000000) == 0 )
19       }
```

In this case, the parameter 0×200 can be transferred to the function, even though it's used for initialization of the internal local variable, which defines the boundaries in the read_ function. But in fact, the local variable is initialized to $0 \times 7FFFFFF$ (line 22 in Fig. 23.5). This example illustrates the use of our heuristics for patching the binary (the rest of the items in it are exactly the same as in the previous example). In this instance, the checker ArrayBoundCheckerV2 will show us that the assignment on line 43 in Fig. 23.5 may overflow the array. CFG analysis may reveal that the assignments are made to the array only in two cases, and only one of them is done in a loop. Therefore, it's sufficient to add a check for the overflow of the counter variable according to CFG (Fig. 23.6). Nevertheless, this method is not optimal because it would be most appropriate to change the assignment on line 22, but doing so requires a lot more work to improve the automated patching process.

```
 1 unsigned int __cdecl read_(int a1, char *buf, int a3)
 2 {
 3   int v4; // [sp+28h] [bp-C0h]@3
 4   int v5; // [sp+2Ch] [bp-BCh]@3
 5   unsigned int v6; // [sp+34h] [bp-B4h]@1
 6   int v7[33]; // [sp+38h] [bp-B0h]@5
 7   int v8; // [sp+BCh] [bp-2Ch]@9
 8   unsigned int i; // [sp+C0h] [bp-28h]@1
 9   int v10; // [sp+C4h] [bp-24h]@1
10   char v11; // [sp+CBh] [bp-1Dh]@1
11   int v12; // [sp+CCh] [bp-1Ch]@1
12   char *buf&1; // [sp+D0h] [bp-18h]@1
13   int v14; // [sp+D4h] [bp-14h]@1
14   unsigned int v15; // [sp+D8h] [bp-10h]@2
15
16   v14 = a1;
17   buf&1 = buf;
18   v12 = a3;
19   v11 = 0;
20   v10 = 0;
21   i = 0;
22   v6 = 0x7FFFFFFF;
23   if ( buf )
24   {
25     v4 = 1;
26     v5 = 0;
27     for ( i = 0; i < 0x20; ++i )
28       v7[i] = 0;
29     i = 0;
30     do
31     {
32       v7[(unsigned int)v14 >> 5] |= 1 << (v14 & 0x1F);
33       v10 = sub_804A755(1, v7, 0, (int)&v4, 0);
34       if ( v10 )
35         return -v10;
36       v10 = sub_804A735(v14, &v11, 1u, (int)&v8);
37       if ( v10 )
38         return -v10;
39       if ( !v8 )
40         return -6;
41       if ( v8 == 1 )
42       {
43         buf&1[i++] = v11;
44         if ( i >= v6 )
45           i = v6 - 1;
46       }
47     }
48     while ( v11 != 10 );
49     buf&1[i] = 0;
50     v15 = i;
```

Fig. 23.5 Decompiled view of function sub_0804A3E0

Fig. 23.6 Decompiled view
of patched function
sub_0804A3E0

```
if ( v8 == 1 && i <= 511 )
{
    buf[i++] = v11;
    if ( i > v6 )
        i = v6 - 1;
}
}
while ( v11 != 10 );
```

23.8   CGC Results and Acknowledgements

Team DESCARTES was ranked #7 at the CGC Scored Event 2 in April 2015 (Fig. 23.7) and #13 at the CGC Qualifying Event in June 2015. Although we were not among the final seven teams to compete in the Cyber Grand Challenge at DEF CON 24 in August 2016, we cherished the international collaborative spirit from our team members in Argentina, Romania, Russia, Singapore, the United Kingdom, and the United States—all of whom have their own full-time jobs or studies in addition to participating in Team DESCARTES.

We want to thank the DESCARTES team members Chris Barnard, Dennis Gamayunov, Adrian Ifrim, Aaron Jones, Newton Lee, Darren Manners, Donald May, Mikhail Voronov, Louis Wai, and Chad Wollenberg as well as software engineering intern Diego Marinelli and multilinguist Inessa Lee.

A lot of progress has been made since the successful reverse engineering of the Mac OS System 4.1 to run HyperCard on Unix at AT&T Bell Laboratories in the mid-eighties [1]. In 2015, Google, Adobe, and MIT researchers at the Computer Science and Artificial Intelligence Laboratory (CSAIL) created "Helium"—a computer program that fixes old code faster and better than expert computer engineers for complex software such as Photoshop [14]. What takes human coders months to program, Helium can do the same job in a matter of hours or even minutes. In August 2016, Mayhem designed by team ForAllSecure from Carnegie Mellon University won the DARPA Cyber Grand Challenge at DEF CON24. All these advances are good learning experiences. However, the much desired quantum leap will likely require the development of artificial general intelligence (see Chaps. 24 to 26 in this book).

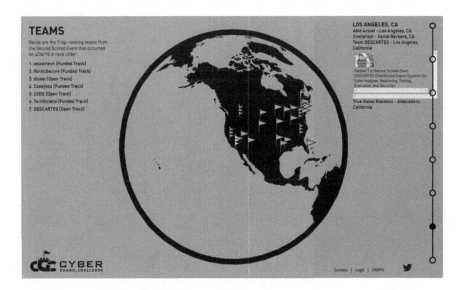

Fig. 23.7 Team DESCARTES at #7 of the top-ranking teams from CGC Scored Event 2 in April 2015

References

1. **Lee, Newton**. Counterterrorism and Cybersecurity: Total Information Awareness (2nd edition). [Online] Springer International, April 8, 2015. http://www.amazon.com/Counterterrorism-Cybersecurity-Total-Information-Awareness/dp/3319172433/.
2. REIL - The Reverse Engineering Intermediate Language. [Online] http://www.zynamics.com/binnavi/manual/html/reil_language.htm.
3. MAIL: Malware Analysis Intermediate Language. [Online] http://web.uvic.ca/~salam/PhD/TR-MAIL.pdf.
4. Open Reil. [Online] https://github.com/Cr4sh/openreil.
5. Emulating Code In Radare2. [Online] http://rada.re/get/lacon2k15-esil.pdf.
6. More target independent LLVM bitcode. [Online] http://llvm.org/devmtg/2011-09-16/EuroLLVM2011-MoreTargetIndependentLLVMBitcode.pdf.
7. KLEE: Unassisted and Automatic Generation of High-Coverage Tests for Complex Systems Programs. [Online] http://llvm.org/pubs/2008-12-OSDI-KLEE.pdf.
8. Clang Static Analyzer. [Online] http://clang.llvm.org/.
9. Mc-sema. [Online] https://github.com/trailofbits/mcsema.
10. BAP: A Binary Analysis Platform. [Online] https://users.ece.cmu.edu/~ejschwar/papers/cav11.pdf.
11. Dagger Decompiling to IR. [Online] http://llvm.org/devmtg/2013-04/bougacha-slides.pdf.
12. Decompilers and beyond. [Online] https://www.hex-rays.com/products/ida/support/ppt/decompilers_and_beyond_white_paper.pdf.
13. Simplex method in IDA Pro. [Online] http://www.hexblog.com/?p=42.
14. Conner-Simons, Adam. Computer program fixes old code faster than expert engineers. MIT News. [Online] July 9, 2015. http://news.mit.edu/2015/computer-program-fixes-old-code-faster-than-expert-engineers-0609.

Part VI
Artificial General Intelligence

True Artificial General Intelligence (AGI) will be a 'phase' shift in the properties of civilization. No longer will we be the smartest intelligence the Earth has ever known. It will mean potentially huge shifts in what we can accomplish. When thinking about what true AGI means for humanity, one needs to understand how and what motivates that AGI. Do we even know? Is there any way for us to predict what will happen once there is a machine intelligence that is beyond human ability? Will it be like us? Will it be friendly to us? These are all questions to be considered by humanity as we get closer to making the goal of AGI real. Setting aside the implications of AGI, to make it real we need to build and we have a lot of work to-do but it could be closer than one might think.

The following three chapters are about one possible solution to designing and building an AGI called the Independent Core Observer Model which is an engineering design pattern, or cognitive extension architecture, developed by a privately funded research project which I manage. These three chapters are designed to explain elements of ICOM so that subsequent chapters focused on study results or algorithms can just reference these to explain the fundamentals of ICOM based systems.

The three chapters are:

* **Self-Motivating Computational System Cognitive Architecture (An Introduction)**—High level operational theory of the Independent Core Observer Model Cognitive Extension Architecture
* **Modeling Emotions in a Computational System**—Emotional Modeling in the Independent Core Observer Model Cognitive Architecture
* **Artificial General Intelligence Subjective Experience**—Artificial General Intelligence as a Subjective Experience Quality of ICOM Cognitive Architecture

—David J. Kelley

Chapter 24
Self-Motivating Computational System Cognitive Architecture: An Introduction

David J. Kelley

24.1 Introduction to Artificial General Intelligence

The road to building artificial general intelligence (AGI) is not just very complex but the most complex task computer scientists have tried to solve. While over the last 30+ years a great amount of work has been done, much of that work has been narrow from an application standpoint or has been purely theoretical and much of that work has been focused on elements of Artificial Intelligence (AI) such as image pattern recognition or speech analysis. The trick in these sorts of tasks is understanding in 'context'; which is a key part of true artificial general intelligence but it's not the only issue. This chapter does not focus on the problem of context and pattern recognition but on the problem of self-motivating computational systems, or rather of assigning value and emergent qualities because of that trait. It is the articulation of a theory for building a system that has its own 'feelings' and can decide for its self if it likes this art or that art or it will try to do this task or that task and can be entirely independent.

> **Note:** Artificial intelligence (AI) is the intelligence exhibited by machines or software. It is also the name of the academic field of study which studies how to create computers and computer software that are capable of intelligent behavior. Major AI researchers and textbooks define this field as "the study and design of intelligent agents", in which an intelligent agent is a system that perceives its environment and takes actions that maximize its chances of success. John McCarthy, who coined the term in 1955, defines it as "the science and engineering of making intelligent machines". https://en.wikipedia.org/wiki/Artificial_intelligence

> **Note:** Artificial general intelligence (AGI) is the intelligence of a machine that could successfully perform any intellectual task that a human being can. It is a primary goal of artificial intelligence research and an important topic for science fiction writers and

D.J. Kelley (✉)
Seattle, Washington, USA
e-mail: david@artificialgeneralintelligenceinc.com

© Springer Science+Business Media New York 2016
N. Lee (ed.), *Google It*, DOI 10.1007/978-1-4939-6415-4_24

433

futurists. Artificial general intelligence is also referred to as "strong AI", "full AI" or as the ability to perform "general intelligent action". https://en.wikipedia.org/wiki/Artificial_general_intelligence

Let us look at the thesis statement for this chapter;

Regardless of the standard cognitive architecture used to produce the 'understanding' of a thing in context, the following architecture supports assigning value to that context in a computer system that is self-modifying based on those value based assessments, albeit indirectly, where the system's ability to be self-aware is an emergent quality of the system based on the ICOM emotional architecture.

Note: The term 'Context' refers to the framework for defining a given object or thing wither abstract in nature, an idea or noun of some sort. When discussion for example a pencil, it is the context in which the idea of the pencil sits that makes and provides meaning to the discussion of the pencil save in the abstract and even then we have the 'abstract' idea which is itself 'context' in which to discuss the pencil.

Note: "is the capacity for introspection and the ability to recognize one's self as an individual separate from the environment and other individuals. It is not to be confused with the consciousness in the sense of qualia. While consciousness is a term given to being aware of one's environment and body and lifestyle, self- is the recognition of that awareness." [1] https://en.wikipedia.org/wiki/Self-awareness

Note: A cognitive architecture is a hypothesis about the fixed structures that provide a mind, whether in natural or artificial systems, and how they work together – in conjunction with knowledge and skills embodied within the architecture – to yield intelligent behavior in a diversity of complex environments.http://cogarch.ict.usc.edu/

In computer science and software engineering there is a complex set of terms and acronyms that mean any number of things depending on the audience in the tech sector. Further, in some cases, the same acronyms mean different things in different circles and often in those circles people have a different understanding of terms that should mean the same thing and the individuals believe they know what each other are talking about, but in the end they were thinking different things with various but critical differences in the meanings of those terms. To offset that problem to some degree, I have articulated a few definitions in a glossary at the end of this chapter, as I understand them; so that, in the context of this chapter, one can refer back to these terms as a basis for understanding. While at the end of the chapter there is a glossary, the most critical term needed to be able to approach the subject in detail under this pattern is 'context'.

The term 'Context' in this chapter refers to the framework for defining an object, such as an idea or noun of some sort and the environment in which that thing should be understood. When discussing for example a pencil, it is the context in which the idea of the pencil sits that makes and provides meaning to the discussion of the pencil, save in the abstract, and even then we have the 'abstract' idea that is itself 'context' in which to discuss the pencil.

To better understand the idea of context, think of a pencil being used by an artist vs a pencil being used by a student. It is the 'used by an artist' versus 'used by a student' that is an example of 'context' which is important in terms of understanding the pencil and its current state. Using ICOM it is the assigning of value to context or two elements in a given context as they related to one another that is key to understanding the ICOM theory as a Cognitive Extension Architecture or over all architecture for an AGI system.

24.2 Understanding the Problem

Solving 'Strong' AI or AGI (Artificial General Intelligence) is the most important (or at least the hardest) Computer Science problem in the history of computing beyond getting computers working to begin with. That being the case though, it is only incidental to the discussion here. The problem is to solve or create human like cognition in a software system sufficiently able to self-motivate and take independent action on that motivation and to further modify actions based on self-modified needs and desires over time.

> **Note**: A theoretical construct used to explain behavior. It represents the reasons for people's actions, desires, and needs. Motivation can also be defined as one's direction to behavior, or what causes a person to want to repeat a behavior and vice versa. In the context of ICOM: The act of having a desire to take action of some kind, to be therefore 'motivated' to take such action, where self-motivation is the act of creating one's own desire to take a given action or set of actions. https://en.wikipedia.org/wiki/Motivation

There are really two elements to the problem; one of decomposition, for example pattern recognition, including context or situational awareness and one of self-motivation or what to do with things once a system has that 'context' problem addressed and value judgements placed on them. That second part is the main focus of the ICOM Cognitive Extension architecture.

Going back to the thesis or the 'theory' or approach for ICOM;

The human mind is a software abstraction system (meaning the part that is self-aware is an emergent software system and not hard coded on to its substrate) running on a biological computer. The mind can be looked at as a system that uses a system of emotions to represent current emotional states in the mind, as well as needs and associated context based on input, where the mind evaluates them based on various emotional structures and value assignments and then modifies the underlying values as per input; denoted by associated context as decomposed in the process of decomposition and identification of data in context. Setting aside the complex neural network and related subsystems that generate pattern recognition and other contextual systems in the human mind, it is possible to build a software system that uses a model that would, or could, continuously 'feel' and modify those feelings like the human mind but based on an abstracted software system running on another computing substrate. That system for example could use for example a "floating point" value to represent current emotional states on multiple emotional vectors, including needs as well as associated context to emotional states based on input, and then evaluate them based on these emotional structures and values assignments; therefore modifying the underlying values as per input as denoted by associated context from the decomposition process. In which case, given sufficient complexity, it is possible to build a system that is self-aware and self-motivating as well as self-modifying.

24.3 Relationship with the Standard Concepts of Cognitive Architecture

Cognition can be defined as the mental process of collecting knowledge and understanding through thought, experience and senses [2]. Further in the process of designing a machine that is intelligent, it is important to build an 'architecture' for how you are going to go about building said machine. Cognitive Architecture is a given hypothesis for how one would build a mind that enables the mental process of collecting knowledge and understanding through thought, experience and senses [3].

So then how does the ICOM methodology or hypothesis apply, or relate, to the standard concepts of Cognitive Architecture? In my experience, most cognitive architecture such as Sigma [4] is really a bottom up architecture focused on the smallest details of what we have the technology and understanding to look at and do to build from the ground up based on some model. In such systems, typically, they are evaluated based on their behavior. ICOM is a 'Cognitive Architecture' focused on the highest level down. Meaning ICOM is focused on how a mind says to itself, "I exist and I feel good about that". ICOM in its current form is not focused on the nuance of decomposing a given set of sensory input but really on what happens to that input after its evaluated, broken down and refined or 'comprehended' and ready to decide how 'it' (being an ICOM implementation) feels about it.

From a traditional AGI Architecture standpoint, ICOM approaches the problem of AGI from the other direction then what is typical, and in that regard ICOM may seem more like an overall control system for AGI Architecture. In fact, in the overall ICOM model a system like Tensor Flow [4] is a key part of ICOM for preforming a key task of cognition around bringing sensory input into the system through what, in the ICOM model, is referred to as the 'context' engine in which most AGI architectural systems can be applied to this functionality in an ICOM implementation.

Regardless of the fact that ICOM is a top down approach to AGI Architecture, the "thoughts" themselves are an emergent phenomenon of the process of emotional evaluation in the system. Let's see how in the next section.

24.4 Emergent by Design

The Independent Core Observer Model (ICOM) contends that consciousness is a high level abstraction. And further, that consciousness is based on emotional context assignments evaluated based on other emotions related to the context of any given input or internal topic. These evaluations are related to needs and other emotional values such as interests which are themselves emotions and used as the basis for 'value'; which drives interest and action which, in turn, creates the emergent effect of a conscious mind. The major complexity of the system then is the abstracted subconscious and related systems of a mind executing on the nuanced details of any physical action without the conscious mind dealing with direct details. Our ability to do this kind of decomposition is already approaching mastery in terms of the state of the art in technology to generate context from data; or at least we are far enough along to know we have effectively solved the problem if not having it completely mastered at this time.

> **Note**: The state or quality of awareness, or, of being aware of an external object or something within oneself. It has been defined as: sentience, awareness, subjectivity, the ability to experience or to feel, wakefulness, having a sense of selfhood, and the executive control system of the mind. Despite the difficulty in definition, many philosophers believe that there is a broadly shared underlying intuition about what consciousness is. As Max Velmans and Susan Schneider wrote in The Blackwell Companion to Consciousness: "Anything that we are aware of at a given moment forms part of our consciousness, making conscious experience at once the most familiar and most mysterious aspect of our lives." - https://en.wikipedia.org/wiki/Consciousness

A scientist studying the human mind suggests that consciousness is likely an emergent phenomenon. In other words, she is suggesting that, when we figure it out, we will likely find it to be an emergent quality of certain kinds of systems under certain circumstances [5]. This particular system creates consciousness through the emergent quality of the system as per the suggestion that it is an emergent quality.

Let's look at how ICOM works, and how the emergent qualities of the system thus emerge.

24.5   Independent Core Observer Model (ICOM) Working

The Independent Core Observer Model (ICOM) is a system where the core AGI is not directly tied to detailed outputs but operates through an abstraction layer, or 'observer' of the core, which only need deal in abstracts of input and assigning 'value' to output context. The observer is similar in function to the subconscious of the human mind; dealing with the details of the system and system implementation, including various autonomic systems and context assignment and decomposition of input and the like.

Take a look at the following diagram (Fig. 24.1).

As we can see, fundamentally it would seem simple and straight forward; however, the underlying implementation and operation of said framework is sufficiently complicated to be able to push the limits of standard computer hardware in lab implementations. In this diagram, input comes into the observer which is broken down into context trees and passed into the core. The core 'emotionally' evaluates

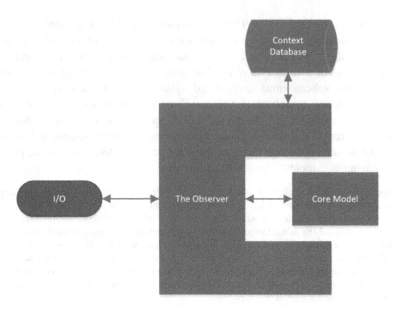

Fig. 24.1 ICOM model diagram

them for their emotional effects based on various elements, which we will define later, and then the results of that evaluation is analyzed by the observer and output is generated by the observer to the various connected systems.

At this level, it is easy to conceptualize how the overall parts of the system go together. Now let's look at how action occurs in the system in which the ICOM provides a bias for action in ICOM as implemented in the lab (Fig. 24.2).

In this, we see the flow as might be implemented in the end component of the observer of a specific ICOM implementation. While details of implementation may be different in various implementations, this articulates the key tasks such systems would have to do, as well as articulates the relationship of those tasks with regard to a functioning ICOM based systems. Keep in mind this is not the complete observer but refers to the end component as shown in the higher level diagram later.

The key goal of the observer end component flow is to gather 'context'. This can be through the use of pattern recognition neural networks or other systems as might make sense in the context engine. The Observer system, in this case, needs to look up related material in the process of receiving processed context from a context engine of some kind. In providing that back to the core, the observer then needs to map that context to a task or existing task. If that item exists, the context engine can add the additional context tree to create the appropriate models or see if it is a continuation of a task or question the system can drive to take actions as articulated by placing that context tree back in the que and have the context engine check for additional references to build out that tree more and pass again through the core. Additionally, the system can then work out details of executing a given task in

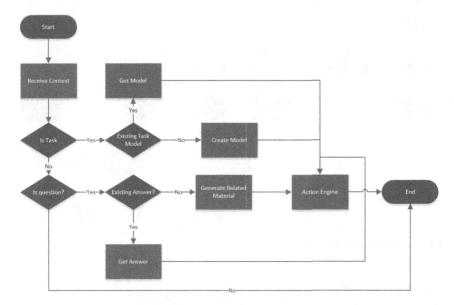

Fig. 24.2 ICOM observer execution logical flow chart

greater detail based on the observer context from the core after that emotional evaluation. In layman's terms, this part of the observer model is focused on the details of taking actions that the core has determined it would like to take through that emotional based processing.

Now let's look at the idea of 'context' and how that needs to be composed for ICOM (Fig. 24.3).

In this model we can see where existing technology can plug in, in-terms of context generation, image or voice decomposition, neural networks and the like. Once such systems create a context tree, the input decomposition engine needs to determine if it is something familiar in terms of input. If that is the case, the system needs to map it to the existing model for that kind of input (say vision for example). The analysis with that model as default context in terms of emotional references is then attached to the context tree (a logical data structure of relationships between items). If there is existing context queued, it is then attached to the tree. If it is new input, then a new base context structure needs to be created so that future relationships can be associated and then handed to the core for processing.

Now let's look at over all ICOM architecture (Fig. 24.4).

In this case, we can see the overall system is somewhat more complicated; and it can be difficult to see where the observer and core model components are separate so they have been color coded green. In this way, we can see where things are handed off between the observer and the core.

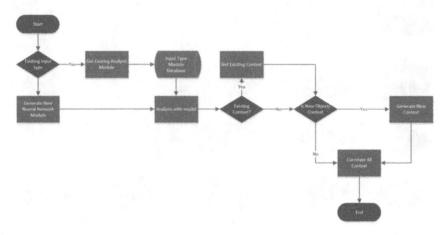

Fig. 24.3 Context engine task flow

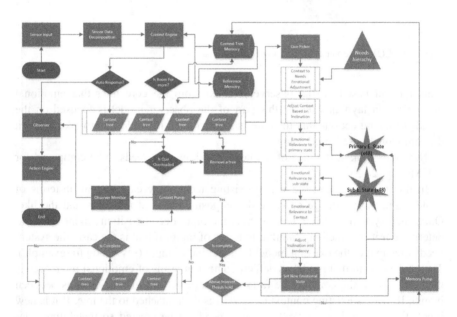

Fig. 24.4 Overall ICOM architecture

Walking through this diagram, we start with Sensor input that is decomposed into usable data. Those particular elements could be decomposed any number of ways and is incidental to the ICOM architecture. There are many ways in the current realm of computer science to determine how this can function. Given that the data is decomposed, it is then run through the 'context' engine to make the appropriate memory emotional context associations. At which point, if there is an automatic response (like a pain response in humans), the observer may push some automatic response prior to moving forward in the system with the context tree then falling into the incoming context que of the core. This que is essentially incoming thoughts or things to 'think' about in the form of emotional processing. By 'thinking', we only mean emotional processing as per the various matrices to resolve how the machine 'feels' about a given context tree. Actual thoughts would be an emergent quality of the system, as articulated elsewhere.

The core has a primary and secondary emotional state that is represented by a series of vector floating point values or 'values' in the lab implementations. This allows for a complex set of current emotional states and subconscious emotional states. Both sets of states along with a needs hierarchy are part of the core calculations for the core to process a single context tree. Once the new state is set and the emotional context tree for a given set of context is done, the system checks if it's part of a complex series and may pass to the memory pump if it is under a certain predetermined value or, if it is above and complete, it passes to the context pump. If it is part of a string of thought, it goes to the context que pending completion; in which case it would again be passed to the context pump which would pass the tree back to the observer.

As you can see, from the initial context tree the observer does a number of things to that queue including dealing with overload issues and placing processed trees of particular interest into the queue as well as input the context trees into the que. Processed trees coming out of the core into the observer can also be passed up inside the core and action taken on actionable elements. For example, say a question or paradox needs a response, or additional context or there is an action that should be otherwise acted upon; where the observer does not deals with the complexity of that action per se.

24.6 Unified AGI Architecture

Given the articulated flow of the system, it is important to note that functionally the system is deciding what "emotionally" to "feel" about a given thing based on more than 144 factors (in the model implementation in the lab but is not necessarily always true in simpler implementations) per element of that tree plus 'needs' elements that affect that processing. Thought in the system is not direct; but, as the complexity of those elements passing through the core increases and things are tagged as interesting and words or ideas form emotional relationships, then complex structures form around context as it relates to new elements. If those happen to form structures that relate to various other context elements, including ones of particular emotional significance, the system can become sufficiently complex that these structures could be said to be thought; based on those underlying emotional constructs which drive interest of focus forcing the system to reprocess context trees as needed. The underlying emotional processing becomes 'so' complex as to seem deliberate, while the underlying system is essentially an overly complex difference engine.

The idea of emergent theory really gets to the fact of the system's ability to be self-aware as a concept and that it is thinking is a derivative of its emotional context assignments and what it chooses to think about. This is really an ultra-complex selection of context based on various parameters and the relationships between them. For example, the system might be low on power which negatively affects say a sadness vector, and to a lesser degree a pain vector is associated with it, so the system might select to think about a bit of context from memory that solves that particular emotional pattern the most. It might continue to try related elements until something addresses this particular issue and vector parameters for the core state stabilize back to normal. Keep in mind the implementation of an idea, say "to plug into charge" might be just that, "an idea of thinking about it" which is not going to do much other than temporarily provide 'hope'. It is the thinking of the 'action' which provides the best pattern match and, given it is an 'action', the observer will likely take that action or at least try to. If the observer does execute that action, it would be a real solution and thus we can say the overall system thought about and took action to solve its power issue. The idea of this context processing being considered thought is an abstraction of what is really happening in detail where the computation is so complex as to be effectively 'thought' in the abstract. It's easier to thank about or otherwise conceptualize this way making it recognizable to humans.

It is a distinct possibility that humans would perceive the same type of abstraction for our own thoughts if, in fact, we understood how the brain operated and how awareness develops in the human mind. It is also important to note that the core while it is looking at the high level emotional computation of a given context tree, the 'action' of tasks in any given tree that are used to solve a problem might actually be in that tree just not in a way that is directly accessible to the emergent awareness which is a higher level abstraction from that context processing system. What this means is that the emotional processing is what is happening in the core but the awareness is a function of that processing at one level of abstraction from the core processing and that being the case details of a given context tree may not be surfaced to that level of abstraction until that action is picked up by the observer and placed back into the que in such a way as that action becomes the top level element of a given context tree and thus more likely to be the point of or focus on that abstracted awareness.

24.7   Motivation of the Core Model and ICOM Action

One of the key elements of the ICOM system architecture is a bias to self-motivation and action. Depending on the kinds of values associated with a context, the system may take action; or rather the Observer component of the system is designed to try to take action based on given context that is associated with the context of an action. That being the case, any 'context' that is associated with action is therefore grounds for action. The observer is then creating additional context to post back to the system to be emotionally evaluated and further action taken. The complexity of the action itself is abstracted from the core processing. The fact that the primary function of the observer is to take action is part of giving the system a bias for action, unless extensive conditioning is applied to make the system associate negative outcome with action. Given that the system will continually process context, as it has bandwidth based on its emotional relevance and needs, the system (ICOM) is then designed to at least try action along those elements of context. It is here that we complete the 'biasing' of the system towards action based on emotional values or context. We then have a self-motivating system based on emotional context that can manipulate itself through its own actions, and through indirect manipulation of its emotional matrix. By 'indirect' meaning the emergent awareness can only indirectly manipulate its emotional matrix whereas the core does this directly.

24.8   Application Biasing

ICOM is a general high level approach to overall artificial general intelligence. That being said, an AGI, by the fact that it is an AGI, should in theory be able to do any given task. Without such a system having attained 'self-awareness', you can then train the system around certain tasks. By associating input or context with pleasure or other positive emotional stimulation, you can use those as a basis for the system to select certain actions. By limiting the action collection to what is possible in the given application and allowing the system to create and try various combinations of these actions, you essentially end up with an evolutionary algorithmic system for accomplishing tasks based on how much pleasure the system gains or based on the system biases as might be currently present. Additionally, by conditioning, you can manipulate core context to create a better environment for the conditioning of a certain set of tasks you might want in a certain application bias that you want to create.

In the training, or attempted biasing, keep in mind that personality or individual traits can develop in an implementation.

24.9   Personality, Interests and Desire Development

Very much related to the application usage biasing of an ICOM implementation is the idea of personality, interests and desires in the context of the ICOM system. All input and all thought further manipulate how the system feels about any given topic, no matter what this input is, biasing the system towards one thing or the other. It is important in the early stages of development to actually tightly control that biasing; but it is inevitable that it will have its own biases over time based on what it learns and how it 'feels' about a given context with every small bit of input and its own thinking.

The point cannot be overstated that the system at rest will continue to think about 'things'. What this means is the system, with limited input or even with significant input, will look at things that have occurred to it in the past and or related items to things it happens to just think about. The system will then automatically reprocess and rethink about things based on factors like recently processed input or interest levels; related to topics of interest or based on current needs and the like. Each time the system cycles it is slowly manipulating its self, its interests consciously and subconsciously as well as adjusting interests and emotional states slowly (note that this is by design to not let the underlying vectors change too fast or there is a greater risk of the system becoming unstable) over time through this processing.

24.10 Summary

The ICOM architecture provides a substrate independent model for true sapient and sentient machine intelligence; as capable as human level intelligence. The Independent Core Observer Model (ICOM) Cognitive Extension Architecture is a methodology or 'pattern' for producing a self-motivating computational system that can be self-aware that differentiates from other approaches by a purely top down approach to designing such a system. Another way to look at ICOM is as a system for abstracting standard cognitive architecture from the part of the system that can be self-aware and as a system for assigning value on any given idea or 'thought' and producing ongoing motivations as well as abstracted thoughts through emergent complexity.

References

1. "Self-awareness" (wiki) https://en.wikipedia.org/wiki/Self-awareness 9/26/02015 AD
2. "Cognition" https://en.wikipedia.org/wiki/Cognition 01/27/02016AD
3. "Cognitive Architecture" http://cogarch.ict.usc.edu/ 01/27/02016AD
4. "The Sigma Cognitive Architecture and System" [pdf] by Paul S. Rosenbloom, University of Southern California http://ict.usc.edu/pubs/The%20Sigma%20cognitive%20architecture%20and%20system.pdf 01/27/02016AD
5. "Knocking on Heaven's Door" by Lisa Randall (Chapter 2) via Tantor Media Inc. 2011
6. "Cognitive Architecture" wiki https://en.wikipedia.org/wiki/Cognitive_architecture 01/27/02016AD
7. "Self-Motivating Computational System Cognitive Architecture" By David J Kelley http://transhumanity.net/self-motivating-computational-system-cognitive-architecture/ 1/21/02016 AD
8. email dated 10/10/2015 - René Milan – quoted discussion on emotional modeling
9. "Properties of Sparse Distributed Representations and their Application to Hierarchical Temporal Memory" (March 24, 2015) Subutai Ahmad, Jeff Hawkins
10. "Feelings Wheel Developed by Dr. Gloria Willcox" http://msaprilshowers.com/emotions/the-feelings-wheel-developed-by-dr-gloria-willcox/ 9/27/2015 further developed from W. Gerrod Parrots 2001 work on a tree structure for classification of deeper emotions http://msaprilshowers.com/emotions/parrotts-classification-of-emotions-chart/

11. "The Plutchik Model of Emotions" from http://www.deepermind.com/02clarty.htm (2/20/02016) in article titled: Deeper Mind 9. Emotions by George Norwood
12. "An Equation for Intelligence?" Dave Sonntag, PhD, Lt Col USAF (ret), CETAS Technology, http://cetas.technology/wp/?p=60 reference Alex Wissner's paper on Causal Entropy. (9/28/2015)
13. "How to Create a Mind – The Secret of Human Thought Revealed" By Ray Kurzweil; Book Published by Penguin Books 2012 ISBN
14. "The Superintelligent Will: Motivation and Instrumental Rationality in Advanced Artificial Agents" 2012 Whitepaper by Nick Bostrom - Future of Humanity Institute Faculty of Philosophy and @ Oxford Martin School – Oxford University
15. https://en.wikipedia.org/wiki/Chaos_theory
16. https://en.wikipedia.org/wiki/Emergence (System theory and Emergence)
17. Move: Transcendence (2014) by character 'Will Caster (Johnny Depp)'; Written by Jack Paglen; presented by Alcon Entertainment

Consulted Works

18. "Causal Mathematical Logic as a guiding framework for the prediction of "Intelligence Signals in brain simulations" Whitepaper by Felix Lanzalaco - Open University UK and Sergio Pissanetzky University of Houston USA
19. "Implementing Human-like Intuition Mechanism in Artificial Intelligence" By Jitesh Dundas – Edencore Technologies Ltd. India and David Chik – Riken Institute Japan

Chapter 25
Modeling Emotions in a Computational System

David J. Kelley

25.1 Introduction to Emotional Modeling

Emotional modeling used in the Independent Core Observer Model (ICOM) represents emotional states in such a way as to provide the basis for assigning abstract value to ideas, concepts and things as they might be articulated in the form of context tree's where such a tree represents the understanding of a person, place or thing including abstract ideas and other feelings. These trees are created by the context engine based on relationships with other elements in memory and then passed into the core (see the whitepaper titled "Overview of ICOM or the Independent Core Observer Model Cognitive Extension Architecture") which is a methodology or 'pattern' for producing a self-motivating computational system that can be self-aware under certain conditions. This particular chapter is focused only on the nuances of emotional modeling in the ICOM program and not what is done with that modeling or how that modeling may or may not lead to a functioning ICOM system architecture.

While ICOM is also as a system for abstracting standard cognitive architecture from the part of the system that can be self-aware, it is primarily a system for assigning value on any given idea or 'thought' and based on that the system can take action, as well as produce ongoing self-motivations in the system to further then have additional thought or action on the mater. ICOM is at a fundamental level driven by the idea that the system is assigning emotional values to 'context' as it is perceived by the system to determine its own feelings. In developing the engineering around ICOM, two models have been used for emotional modeling, which in both cases are based on a logical understanding of emotions as modeled by traditional psychologist as opposed to empirical psychologist which tends to be

D.J. Kelley (✉)
Seattle, USA
e-mail: david@artificialgeneralintelligenceinc.com

© Springer Science+Business Media New York 2016   447
N. Lee (ed.), *Google It*, DOI 10.1007/978-1-4939-6415-4_25

based on biological structures. The approaches articulated here are based on a logical approach that is also not tied to the substrate of the system in question (biological or otherwise).

25.2   Understanding the Problem of Emotional Modeling

Emotional structural representation is not a problem I wanted to solve independently nor one I felt I had enough information as an expert to solve without a lifetime of work on my own. The current representational systems that I've selected are not necessarily the best way(s) or the right way(s) but two ways that do work and are used by a certain segment of psychological professionals. This is based on the work of others in terms of representing the complexity of emotions in the human mind by scientists that have focused on this area of science. The selection of these methods are more based on computational requirements than any other selection criteria.

It is important to note that both of the methods the ICOM research have used are not based on scientific data as might be articulated by empirical psychologists which might use ANOVA (variance analysis), or factor analysis [1]. While these other models maybe be more measured in how they model elements of the biological implementation of emotions in the human mind, the model's selected by me here for ICOM research are focused on 'how' and the logical modeling of those emotions or 'feelings'. If we look at say the process for modeling a system such as articulated in "Properties of Sparse Distributed representations and their Application to Hierarchical Temporal Memory" [2] such representation is very much specific to the substrate of the human brain. Since I am looking at the problem of self-motivating systems or computational models that are not based on the human brain literally but only in the logical sense the Wilcox system [3] or more simply the Plutchik method [4] is a more straight forward model and accurately models logically what we want to-do to separate from the underlying complexity of the substrate of the human biological mind.

25.3   The Plutchik Method

George Norwood described the Plutchik method as:

> Consider Robert Plutchik's psychoevolutionary theory of emotion. His theory is one of the most influential classification approaches for general emotional responses. He chose eight primary emotions - anger, fear, sadness, disgust, surprise, anticipation, trust, and joy. Plutchik proposed that these 'basic' emotions are biologically primitive and have evolved in order to increase the reproductive fitness of the animal. Plutchik argues for the primacy of these emotions by showing each to be the trigger of [behavior] with high survival value, such as the way fear inspires the fight-or-flight response.

> Plutchik's theory of basic emotions applies to animals as well as to humans and has an evolutionary history that helped organisms deal with key survival issues. Beyond the basic emotions there are combinations of emotions. Primary emotions can be conceptualized in terms of pairs of polar opposites. Each emotion can exist in varying degrees of intensity or levels of arousal.—[4]

While George Norwood mentions earlier in his chapter talking about the Plutchick method that it would be almost impossible to represent emotions in terms of math or algorithms I would disagree. As you can see by this representation of the Plutchik method it is essentially 8 vectors or 'values' when represented in 2 dimensions which is easily modeled with a series of number values (Fig. 25.1).

Now in the case of ICOM since we want to represent each segment as a numeric value, a floating point value was selected to insure precision along with a reverse scale as opposed to what is seen in the diagram above. Meaning if we have a number that represents 'joy/serenity/ecstasy' the ICOM version is a number starting from 0 to N where N is increasing amounts or intensity of 'joy'.

To represent ICOM emotional states for anything assigned emotional values you end up with an array of floating point values. By looking at the chart above we can see how emotional nuances can be represented as a combination of values on two or more vectors which gives us something closer to the Wilcox model but using less values and given the difference it is orders of magnitude when seen in terms of a computational comparison.

Let us take a look at Fig. 25.2.

As you can see we have reversed the vectors such that the value or 'intensity is increasing as we leave the center of the diagram on any particular vector. From a modeling standpoint this allows the intensity to be infinite above zero verses limiting the scale in the standard variation not to mention it is more aligned with what you might expect based on the earlier work (see the section on Willcox next). This variation as ween here is what we are using in the ICOM research.

Let us look at the other model.

Fig. 25.1 Plutchik model

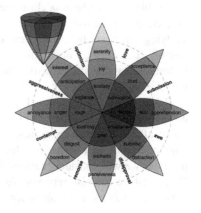

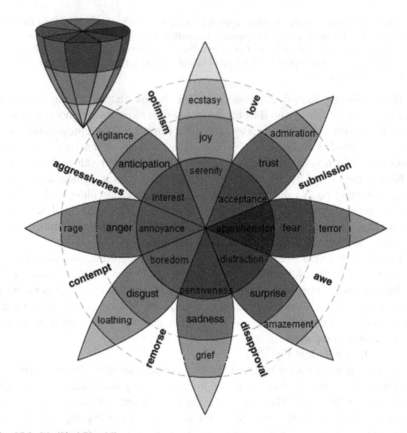

Fig. 25.2 Modified Plutchik

25.4   The Wilcox Model

Initially the ICOM Research centered on using the Willcox model for emotions and is still big part of the modeling methodology and research going into the ICOM project. Given the assumption that the researchers in the field of mental health or studying emotions have represented things to a sufficient complexity to be reasonably accurate we can therefore start with their work as a basis for representation I therefore landed on Willcox initially as being the most sophisticated 'logical' model. Take a look at Fig. 25.3.

Based on the Willcox wheel we have 72 possible values (the six inner emotions on the wheel are a composite of the others) to represent the current state of emotional affairs by a given system. Given that we can then represent the current emotional state at a conscious level by a series of values that for computation purposes we will consider 'vectors' in an array represented by floating point values. Given that we can also represent subconscious and base states in a similar way that basically gives us 144 values for the current state. Further we can use them as

Fig. 25.3 Dr. Gloria
Willcox's Feelings Wheel [3]

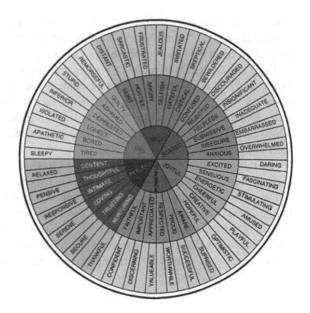

vectors to represent the various states to back weight and adjust for new states. This then can be represented as needed in the software language of choice.

If we map each element to vectors spread on a 2 dimension X/Y plane we can compute an average composite score for each element and use this in various kind of emotional assessment calculations.

We are thus representing emotional states using two sets of an array of 72 predefined elements using floating point values we also can present assigned arrays on a per context basis and use a composite score of an element as processed to further compare various elements of context emotional arrays or composite scores with current states and make associated adjustments based on needs and preexisting states. For example the current emotional state of the system by be a set of values and a bit of context might affect that same set of values with its own set of values for the same emotions based on its associated elements and a composite is calculated based on the combination which could be an average or mean of each vector for any given element of the emotional values.

25.5 The Emotional Comparative Relationship

Given the array of floating point number declarations, a given element of 'context' will have a composite of all pre-associated values related to that context and any previous context as it might be composited. For this explanation we will assume context is pre-assigned. The base assignments of these values are straight forward assignments but each cycle of the core (see the ICOM Model overview for a detailed explanation of ICOM and the core) will need to compare each value and assign various rules on the various elements of the context to assign effects on itself as well as conscious and subconscious values.

Logically we might have the set values that are the current state as in the earlier example. We then get a new block of context and adjust all the various element based on those complex sets of rules that affect the conscious and subconscious states (emotional arrays of floating point values). Rules can be related to emotions which includes tendencies, needs, interests or other factors as might be defined in the rules matrix applied by the core.

This gives us a framework for adjusting emotional states given the assumption that emotional values are assigned to context elements based on various key factors of the current state and related core environment variables. The process as indicated in the context of evaluating becomes the basis for the emergent quality of the system under certain conditions where the process of assigning value and defining self-awareness and thought are only indirectly supported in the ICOM architecture, and emerge as the context processing becomes more complex.

25.6   Context Assignments

One of the key assumptions for computing the emotional states is the pre-assignment of emotional context prior to entering the emotional adjustment structures of the core system.

While this explanation does not address for example looking at a picture and decomposing that into understanding in context it does deal with how emotional values are applied to a given context element generated by the evaluation of that picture.

As described earlier there are 72 elements needed to represent a single emotional context (based on the Willcox model) given the selected methodology. Let's say of that array the first 3 elements are 'happiness', 'sadness', and 'interest'. Additionally let us assign them each a range between 0 and 100 as floating point values meaning you can have a 1 or a 3.567 or a 78.628496720948 if you like.

If for example a particular new context A is related to context B and C which had been processed earlier and related to base context elements of D, E and F. This gives us a context tree of 6 elements. If we average the emotional values of all of them to produce the values of happiness, sadness and interest for context A we now have a context tree for that particular element which then is used to affect current state as noted above. If that element still has an interest level, based on one of those vectors being higher than some threshold then it is queued to process again and the context system will try to associate more data to that context for processing. If Context A had been something thought about before then that context would be brought up again and the other factors would be parsed in for a new average which could have then been an average of all 6 elements where before context A didn't have an emotional context array were the second time around it does. Further on processing the Context A its values are changed by the processing against the current system state.

Using this methodology for emotional modeling and processing we also open the door for explaining certain anomalies as seen in the ICOM research.

25.7 Computer Mental Illness

In the ICOM system the 'core' emotional thought and motivation system if any of the 72 vectors get into a fringe area at the high or lower end of the scale can produce an increasingly irrational set of results in terms of assigning further context. If the sub-conscious vectors are to far off this will be more pronounced and less likely to be fixed over time, creating some kind of digital mental illness where given the current state of research it is hard to say the kinds of and manifestation of that illness or illnesses could be as varied as human mental illness. Now the subconscious system is in fact critical to stabilization of the emotional matrix of the main system in that it does change slightly over time where under the right extreme context input is where you get potential issues on a long term basis with that particular instance. The ICOM research and models have tried to deal with these potential issues by introducing limiting bias and other methods in preventing too radical of a result in any given operation.

25.8 Motivation of the Core Model

Given the system in the previous sections for assigning emotional context, processing and assigning context elements that are above a certain threshold are targets for reprocessing by being placed in a que feeding the core. The motivation of the core comes from the fact that it can't "not" think and it will take action based on emotional values assigned to elements that are continuously addressed where the core only needs to associate an 'action' or other context with a particular result and motivation is an emergent quality of the fact that things must be processed and actions must be taken by design of the system.

This underlying system then is thus designed to have a bias for taking action with action being abstracted form the core in detail where the core only need composite such action at a high level; In other words it just needs to 'think' about an action.

25.9 Core Context

Core Context are the key elements predefined in the system when it starts for the first time. These are 'concept's that are understood by default and have predefined emotional context trees associated with them. While the ICOM theory for AGI is not specific to the generation or rather the decomposition of 'context' it is important to address the 'classification' of context. In this way any context must add qualities that may be new and can be defined dynamically by the system but these core elements that are used to tag those new context elements. Since all context is then streamed into memory as processed and can be brought back and re-referenced as per the emotional classification system pending the associated threshold determining if it is something of relevance to recall.

As stated elsewhere lots of people and organizations are focused on classification systems or systems that decompose input, voice, images and the like however ICOM is focused on self-motivation along the lines of the theory as articulated based on emotional context modeling.

What is important in this section is the core elements used to classify elements of context as they are processed into the system. The following list of elements is used as a fundamental part of the ICOM system for its ability to associated emotional context to elements of context as they are passed into the core. This same system may alter those emotional associations over time as new context not hither to classified is tagged based on the current state of the system and the evaluation of elements or context for a given context tree when the focus of a given context tree is processed. Each one of these elements below has a 72 vector array of default emotions (using the Willcox based version of ICOM) associated with that element by default at system start. Additionally this may not be an exhaustive list of the default core system in the state of the art. These are only the list at the time this section is being written.

(i)   **Action**—A reference to the need to associate a predisposition for action as the system evolves over time.

(ii) **Change**—a reference context flag used to drive interest in changes as a bias noticing change.

(iii) **Fear**—Strongly related to the pain system context flag.

(iv) **Input**—A key context flag needed for the system to evolve over time recognizing internal imaginations vs system context input that is external.

(v)   **Need**—A reference to context associated with the needs hierarchy

(vi) **New**—A reference needed to identify a new context of some kind normally in terms of a new external object being cataloged in memory

(vii) **Pain**—having the most negative overall core context elements used as a system flag of a problem that needs to be focused on. This flag may have any number of autonomic responses dealt with the 'observer' component of the system.

(viii) **Pattern**—A recognition of a pattern build in to help guide context as noted in humans that there is an inherent nature to see patterns in things. While there could be any number of evolutionary reasons for it, in this case we will assume the human model is sound in terms of base artifacts regarding context such as this.

(ix) **Paradox**—a condition where 2 values that should be the same are not or that contradict each other. Contradiction is a negative feedback reference context flag to condition or bias the system want to solve paradox's or have a dislike of them.

(x) **Pleasure**—having the most positive overall core context element used as a system flag for a positive result

(xi) **Recognition**—a reference flag used to identify something that relates to something in memory.

(xii) **Similar**—related to the pattern context object used to help the system have a bias for patterns by default

(xiii) **Want**—A varying context flag that drives interest in certain elements that may contribute to needs or the 'pleasure' context flag.

While all of these might be hard coded into the research system at start they are only really defined in terms of other context being associated with them and in terms of emotional context associated with each element which is true of all elements of the system. Further these emotional structures or matrixes that can change and evolve over time as other context is associated with them.

As a single example let's take a look at the first core context element that is defined in the current Willcox based ICOM implementation called 'pain'. This particular element doesn't represent emotional pain as such but directly effects emotional pain as this element is core context for input assessments or 'physical' pain however note that one of the highlighted elements in the 'pain' matrix is for emotional pain (Fig. 25.4).

Default Emotional Context Assignments for "Pain"					
J.Excited: 0.0	J.Daring: 0.0	J.Fascinating: 0.0	J.Sexy: 0.0	J.Stimulating: 0.0	J.Energetic: 0.0
J.Playful: 0.0	J.Amused: 0.0	J.Creative: 0.0	J.Extravagant: 0.0	J.Aware: 0.0	J.Delightful: 0.0
P.Proud: 0.0	P.Cheerful: 0.0	P.Respected: 0.0	P.Satisfied: 0.0	P.Appreciated: 0.0	P.Valuable: 0.0
P.Worthwhile: 0.0	P.Hopeful: 0.0	P.Intelligent: 0.0	P.Important: 0.0	P.Confident: 0.0	P.Faithful: 0.0
Pe.Thankful: 0.0	Pe.Nurturing: 0.0	Pe.Trusting: 0.0	Pe.Sentimental: 0.0	Pe.Serene: 0.0	Pe.Loving: 0.0
Pe.Responsive: 0.0	Pe.Intimate: 0.0	Pe.Thoughtful: 0.0	Pe.Relaxed: 0.0	Pe.Content: 0.0	Pe:Pensive: 0.0
S.Apathetic: 0.0	S.Sleepy: 0.0	S.Inferior: 0.0	S.Bored: 0.0	S.Inadequate: 0.0	S.Lonely: 0.0
S.Miserable: 0.0	S.Depressed: 0.0	S.Stupid: 0.0	S.Ashamed: 0.0	S.Bashful: 0.0	S.Guilty: 0.0
M.Hurt: 99.0	M.Jealous: 0.0	M.Selfish: 0.0	M.Hostile: 25.0	M.Frustrated: 25.0	M.Angry: 25.0
M.Furious: 25.0	M.Rage: 25.0	M.Hateful: 25.0	M.Irritated: 75.0	M.Critical: 95.	M.Skeptical: 0.0
Sc.Anxious: 0.0	Sc.Embarrassed: 85.0	Sc.Insecure: 0.0	Sc.Foolish: 0.0	Sc.Weak: 0.0	Sc.Submissive: 0.0
Sc.Helpless: 0.0	Sc.Insignificant: 0.0	Sc.Confused: 0.0	Sc.Discouraged: 0.0	Sc.Rejected: 0.0	Sc.Bewildered: 0.0

Fig. 25.4 Emotional matrix array at system start for context element 'pain'

On top of all of the emotional states associated with a context element they themselves also are pre-represented in the initial state predefined into the system as context themselves. You can see that in this initial case we have guesses at values in the matrix array for default values for each element which has to be done for each predefined context element at system start. This allows us to set certain qualities as a basic element of how a value system will evolve in the system creating initial biases. For example we might create a predilection for a pattern which creates the appropriate bias in system as we might want to see in the final AGI implementation of ICOM.

25.10 Personality, Interests and Desires of the ICOM System

In general under the ICOM architecture regardless of which of the two modeling systems that have been used, in ICOM the system very quickly creates predilection for certain things based on its emotional relationship to those elements of context. For example, if the system is exposed to context X which it always had a good experience 'including' interest the methodology regardless of case, develops a preference for or higher emotional values associated with that context or other things associated with that context element. This evolutionary self-biasing based on experience is key to the development of personality, interests and desires of the ICOM system and in various experiments has shown that in principal it is very hard to replicate those biases of any given instance due to the extreme amount of variables involved. While ultimately calculable, a single deviation will change results dramatically over time. This also leads us to a brief discussion of free will.

25.11 Free Will as an Illusion of Complex Contextual Emotional Value Systems

Frequently the problem of "free will" has been an argument between determinist verses probabilistic approaches and given either case an argument as to the reality of our free will ensues.

While we don't understand exactly the methodology of the human mind if it works in a similar manner at a high level like ICOM then, under that architecture, it would strongly imply that free will is an illusion. For me, this is a difficult thing to be sure of given that this is outside the scope of the research around ICOM; but none the less it is worth mentioning the possibility. Additionally, if true, then free will seems to be something that can be completely mathematically modeled. If that is the case, it is likely that of the human mind can be as well. Certainly, as we progress this will be a key point of interest but outside my expertise.

ICOM emotional modeling seems to be at the heart of this mathematical modeling of 'free will' or what appears to be free will in a sea of variables that is so vast that we collectively have not been able to full modeling but ICOM based systems appear in function to exhibit free will based on their own self biases based on their experience which because we can and have a model free will in ICOM is an illusion of the sea of factors required for ICOM systems to function. Let's get back to the different methods used in ICOM for emotional modeling.

25.12   Plutchik Verses Willcox

When determining which method to use in emotional modeling we see a number of key facts. Willcox models all the nuances of human emotions directly with numerous vectors or values. Plutchik models those nuances through combination of values thus using a total number of values that is much less. From a computational standpoint Plutchik has 8 sets of core values where Willcox using 72 so having two sets of those for conscious and unconscious values gives us 72 which converting that to a 2d plain requires conversion X/Y values which means 144 trigonomic functions for each pass through the ICOM core whereas using Plutchik we have 16 total values with the same conversion of X/Y values to produce the average emotional effects applied to incoming emotional context means only 16 trigonomic functions per core pass which means from a computational standpoint we only need 9 times less computational power to run with the Plutchik method which is the basis for the research post series 3 experiments moving forward at least for the foreseeable research that is in progress.

25.13   Visualizing Emotional Modeling Using Plutchik

To better understand how any given instance of ICOM is responding in tests we needed a method for visualizing and representing emotional state data and given the method for modeling in either method articulated earlier we came up with this method here for indicating state. This method visualizes graphically emotional state of what is going on in the core. You can see we are visualizing emotional states much like the earlier diagrams then in which we look at vectors that represent the model (see Figs. 25.1, 25.2).

So let's look as an example. In this case we are looking at one of the program series 3 experiments in which case we were looking at ICOM introspection as it relates to the system thinking about previous elements to see if the system would pick something out of memory and then thinking about it and see how it affects various vectors or emotional states. The rest of the experiment is not as important to the point in which here we are showing how that data is represented. In this case we

conscious	0	5	0	0	
subconscious	0	5	0	0	
	5	0	1	9	external
c	3.05	1.95	0.5	5.05	
	0.0305	5.0305	0.005	0.0505	
	0	9	1	0	external
c	0	6.420975	0.5	2.224975	
	0.030805	5.04440475	0.00995	0.07224475	
	0	5	9	0	external
c	0.779996	6.329316013	4.5	0.670124013	
	0.03829691	5.057253863	0.0548505	0.078223543	
	0	5	9	0	internal
	0	9	10	0	action
c	0.058206761	7.278261114	6.672257475	0	
	0.038496009	0.079463935	0.12102457	0.079005778	

Fig. 25.5 Source data from series 3 on introspection

start with the following raw data keeping in mind here we are looking at only 4 of the eight values modeled in the test system (Fig. 25.5).

So how can diagram emotional? First we need to understand there is a set for the conscious and subconscious parts of the system and we use two diagrams for each with the same vectors as noted in the aforementioned diagrams in particular the Plutchik method.

Now if we plot the states we get a set of diagrams as shown in Fig. 25.6.

This graph system is simple to visualize what the system is feeling albeit the nuances of what each one means is still somewhat abstract but easily to visualize which is why the ICOM project settled on this method.

In this particular study we are looking at a similar matrix as used in previous research but now we were introducing the introspection where we can see the effect of the action bias on the emotional state. This particular study also showcases the resolution that the system quickly goes to where we have subtle changes that are or can be reflected by the system in a way we can see via this diagramming methodology. In a working situation items are selected based on how things map to interest and needs and how it affects the core state of the system.

Further given this and the related body of research we can see that even having the same input out of order will cause a different end result and given the volume of input and the resolution of the effect of retrospection and manipulation of interests therefore no two systems would likely ever be the same unless literally copied and then would stay the same only if all of the subsequent input would the same including order. Small differences over time could have dramatic effects millions of cycles later.

Fig. 25.6 Example graphing
introspection experiment

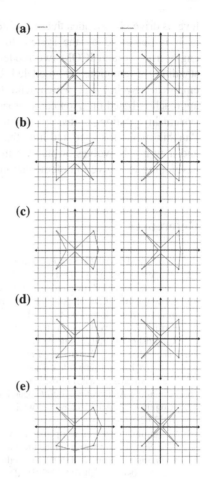

This nuanced complexity is why the diagramming method has become so important to understanding ICOM behavior.

25.14   Summary

The Emotional Modeling used in the Independent Core Observer Model (ICOM) Cognitive Extension Architecture is a methodology or 'pattern' for producing a self-motivating computational system that can be self-aware where emotional modeling is the key to the operation of ICOM. While ICOM is as a system for abstracting standard cognitive architecture from the part of the system that can be self-aware it is primarily a system for assigning value on any given idea or 'thought' and based on that take action as well as producing ongoing self-motivations in the systems further thought or action. ICOM is at a fundamental level is driven by the idea that the system is assigning emotional values to 'context' as it is perceived by the system to determine how it feels. In developing the engineering around ICOM two models have been used based on a logical understanding of emotions as modeled by traditional psychologist as opposed to empirical psychologist which tend to be based on biological structures. This approach is based on a logical approach that is also not tied to the substrate of the system in question. Using this emotional architecture we can see how using the Plutchik method is used and how that application creates the biases of the system and how it self evolves on its own making the exposure to input key to the early developing of a given implementation of ICOM.

References

1. email dated 10/10/2015 - René Milan – quoted discussion on emotional modeling
2. "Properties of Sparse Distributed Representations and their Application to Hierarchical Temporal Memory" (March 24, 2015) Subutai Ahmad, Jeff Hawkins
3. "Feelings Wheel Developed by Dr. Gloria Willcox" http://msaprilshowers.com/emotions/the-feelings-wheel-developed-by-dr-gloria-willcox/ 9/27/2015 further developed from W. Gerrod Parrots 2001 work on a tree structure for classification of deeper emotions http://msaprilshowers.com/emotions/parrotts-classification-of-emotions-chart/
4. "The Plutchik Model of Emotions" from http://www.deepermind.com/02clarty.htm (2/20/02016) in article titled: Deeper Mind 9. Emotions by George Norwood
5. "Self-awareness" (wiki) https://en.wikipedia.org/wiki/Self-awareness 9/26/02015 AD
6. "Cognitive Architecture" http://cogarch.ict.usc.edu/ 01/27/02016AD
7. "Cognitive Architecture" wiki https://en.wikipedia.org/wiki/Cognitive_architecture 01/27/02016AD
8. "Cognition" https://en.wikipedia.org/wiki/Cognition 01/27/02016AD
9. "The Sigma Cognitive Architecture and System" [pdf] by Paul S. Rosenbloom, University of Southern California http://ict.usc.edu/pubs/The%20Sigma%20cognitive%20architecture%20and%20system.pdf 01/27/02016AD
10. "Self-Motivating Computational System Cognitive Architecture" By David J Kelley http://transhumanity.net/self-motivating-computational-system-cognitive-architecture/ 1/21/02016 AD
11. "Knocking on Heaven's Door" by Lisa Randall (Chapter 2) via Tantor Media Inc. 2011
12. "An Equation for Intelligence?" Dave Sonntag, PhD, Lt Col USAF (ret), CETAS Technology, http://cetas.technology/wp/?p=60 reference Alex Wissner's paper on Causal Entropy. (9/28/2015)
13. "How to Create a Mind – The Secret of Human Thought Revealed" By Ray Kurzweil; Book Published by Penguin Books 2012 ISBN
14. "The Superintelligent Will: Motivation and Instrumental Rationality in Advanced Artificial Agents" 2012 Whitepaper by Nick Bostrom - Future of Humanity Institute Faculty of Philosophy and @ Oxford Martin School – Oxford University
15. https://en.wikipedia.org/wiki/Chaos_theory
16. https://en.wikipedia.org/wiki/Emergence (System theory and Emergence)
17. Move: Transcendence (2014) by character 'Will Caster (Johnny Depp)'; Written by Jack Paglen; presented by Alcon Entertainment

Consulted Works

19. "Causal Mathematical Logic as a guiding framework for the prediction of "Intelligence Signals in brain simulations" Whitepaper by Felix Lanzalaco - Open University UK and Sergio Pissanetzky University of Houston USA
20. "Implementing Human-like Intuition Mechanism in Artificial Intelligence" By Jitesh Dundas – Edencore Technologies Ltd. India and David Chik – Riken Institute Japan

Chapter 26
Artificial General Intelligence Subjective Experience

David J. Kelley

26.1 Introduction to Mind and Consciousness

A scientist studying the human mind suggests that consciousness is likely an emergent phenomenon. In other words, she is suggesting that, when we figure it out, we will likely find it to be an emergent quality of certain kinds of systems under certain circumstances [7]. This particular system (ICOM) creates consciousness through the emergent quality of the system. But how does Strong AI Emerge from a system that by itself is not specifically Artificial General Intelligence (AGI). Independent Core Observer Model (ICOM) is an emotional processing system designed to take in context and emotionally process this context and decide at a conscious and subconscious emotional level how it feels about this input. Through the emerging complexity of the system, we have AI that, in operation, functions logically much like the human mind at the highest level. ICOM, however, does not model the human brain, nor deal with individual functions such as in a neural network and is completely a top down logical approach to AGI vs the traditional bottom up. This of course supposes an understanding of how the mind works, or supposes a way it 'could' work, and was designed around that. To put into context what ICOM potentially means for society and why AGI is the most important scientific endeavor in the history of the world, I prefer to keep in mind the following movie quote:

> For 130,000 years, our capacity to reason has remained unchanged. The combined intellect of the neuroscientists, mathematicians and... hackers... [...] pales in comparison to the most basic A.I. Once online, a sentient machine will quickly overcome the limits of biology. And in a short time, its analytic power will become greater than the collective intelligence of every person born in the history of the world. So imagine such an entity with

D.J. Kelley (✉)
Seattle, USA
e-mail: david@artificialgeneralintelligenceinc.com

© Springer Science+Business Media New York 2016
N. Lee (ed.), *Google It*, DOI 10.1007/978-1-4939-6415-4_26

a full range of human emotion. Even self-awareness. Some scientists refer to this as "the Singularity."—Transcendence [movie] 2014 [17]

Keep in mind that not if, but WHEN, AGI emerges on its own, it will be the biggest change to humanity since the dawn of language. Let's look at what 'Emergence' means when talking about ICOM.

26.2 Order from Chaos

When speaking with other computer scientists, ICOM is in many ways not cognitive architecture in the same sense that many of them think. When speaking of bottom up approaches focused on for example neural networks, ICOM sits on top of such systems or as an extension of such systems. ICOM really is the force behind the system that allows it to formulate thoughts, actions and motivations independent of any previous programming and is the catalyst for the emergent phenomenon demonstrated by ICOM 'extension' cognitive architecture.

To really understand how that emergence works, we need to also understand how the memory or data architecture in ICOM is structured.

26.3 Data Architecture

While the details of how the human brain stores data is one thing we do not know for sure; logically it demonstrates certain qualities in terms of patterns, timelines and thus inferred structure. Kurzweil's book "How to Create a Mind" [13] brings up three key points regarding memory as used in the human mind and how memory is built or designed from a data architecture standpoint.

Point 1: "our memories are sequential and in order. They can be accessed in the order that they are remembered. We are unable to directly reverse the sequence of a memory"—page 27.
Point 2: "there are no images, videos, or sound recordings stored in the brain. Our memories are stored as sequences of patterns. Memories that are not accessed dim over time."—page 29.
Point 3: "We can recognize a pattern even if only part of it is perceived (seen, heard, felt) and even if it contains alterations. Our recognition ability is apparently able to detect invariant features of a pattern—characteristics that survive real-world variations."—page 30.

The ICOM data architecture is based on the same basis that these points where made. Data flows in, related to time sequences. These time sequences are tagged with context that is associated with other context elements. There are a number of

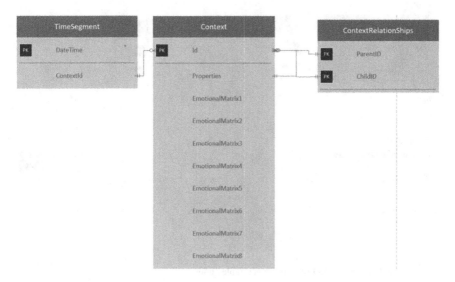

Fig. 26.1 ICOM basic entity relationship diagram (ERD)

ways of looking at this data model from a traditional computer science data base architectural standpoint. For example (Fig. 26.1):

If you are familiar with database data architecture, or even if you're not, you can tell this is fundamentally a simple model and it could be done even more simply then this. Notice that, while we have 3 tables, only two hold any real data. One is essentially a time log, where each time there is a new segment a new table entry is created. The other main table is the idea of 'context' which contains 8 values associated with emotional states for a given piece of context. The third table holds relationships between 'context'. It is important to know that an item in the context table can be related to any number of other 'context' items. Now, in this case, there are 2 kinds of groups of searches that an 'ICOM' system is going to-do against these tables, one related to time and the other to context. Referring to searches against "context' these can be further broken out into several kinds of searches too, that is to get base context searches such as language components or interests or both and then searches related to being able to build context trees based on association. In particular, when we look at the scale of the problem regardless of the substrate, it is easy to see that the benefit of using massively parallel or even quantum searches against that context data and any number of indexes is useful. Further, for those that understand database architecture and database modeling, a number of indexes or 'views' would obviously be useful. In this case though, we could search just the context table the hard way but, to speed this up, we can create reference tables, that are used to hold the top 'interest' base context items or the top 'base' context items. Note also that with such a technique, at a certain scale, this sort of indexing needs to be limited regardless of the underlying technology. Having this sort of system running on my Surface Pro 4 I might get away with having a million records in this

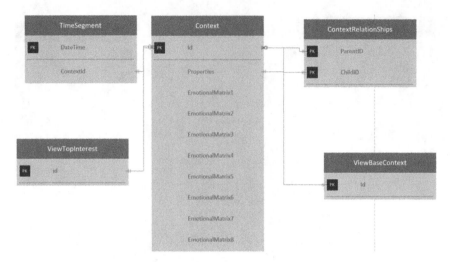

Fig. 26.2 ICOM basic ERD with reference tables added

table but something on the order of the human brain might have 15 billion records? (This is just a guess, I'm just saying that at some point there is a limit) Using the same database design methodology, you could visualize it like this (Fig. 26.2):

This still seems simple visually and fairly basic and even knowing that the two main tables around time segments and context might have 100 billion records in them it is still difficult to appreciate the complexity of the data we are talking about in an ICOM based system using this data architecture. Now let's look at the data in say the *TimeSegment* table and the context table related to properties where Context is related to itself and to an individual time segment. Along the top are time segment records and below those are individual context records and their relationships as they are related to additional context records (Fig. 26.3).

As you can see when looking at the records from this relationship standpoint, we can see the complexity of the underlying relational model of said records. If you just look at context records on their own, you might get something a million more times complex but similar to this:

From this standpoint, it's easy to imagine the sea of data that is in a large scale ICOM system in the core memory or the 'context' table.

Now getting back to this idea of emergence, if you have studied chaos theory [15] one may have heard the idea of the butterfly effect. The butterfly effect has several ways it manifests in ICOM in the research that we have done; however, just from a hypothetical standpoint given the previous figure, if I pick out just a single node in that mass you can see how even a small change affects virtually everything else. In the ICOM experiments done so far, even a small variance of the data input, either in terms of time cadence or order, has always (thus far) resulted in major differences in the system. In these experiments, as time progresses and when an event with the same input reoccurs; we saw traits develop that define major differences in the final state of the system.

Fig. 26.3 Hypothetical
context records related to each
other as they drill down away
from individual records in the
TimeSegment Table

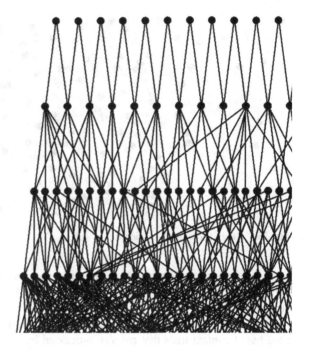

26.4 Emergence from Chaos

When we start talking about Emergence from the standpoint of system theory, we mean some quality of the system that emerges that is qualitatively different then the individual components or systems. It is similar to the idea that all the inanimate atoms and molecules within a cell, when taken as a whole, work together as a living system; whereas individual components are inert. It is a fundamentally 'new' property that emerges from a system as a whole. From this standpoint, 'Emergence' is defined as:

> A process whereby larger entities, patterns, and regularities arise through interactions among smaller or simpler entities that themselves do not exhibit such properties. [16]

In the AGI generated by the ICOM architecture, we are talking about an effect that is substantially similar. We see what is called a 'phase' transition when a certain complexity arises in the system or a 'strong' emergence, which is also known as 'irreducible emergence'. In ICOM, it is not a quality of the substrate but of the flow of information.

In ICOM, we have a sea of data coming in being stacked in a stream of time into the memory of the system. This data passes through what is called a context engine that is very much like many of the AI systems, such as Tensor Flow, that take input into the system, process it to identify context and associate it with new or existing structures in memory using structures called 'context' trees.

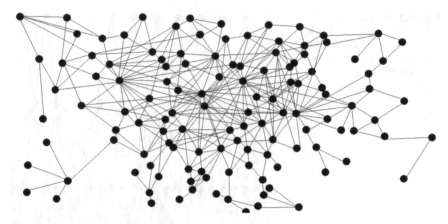

Fig. 26.4 Simplified context relationship model

If you go back to the example in Fig. 26.4, pick any give node and follow all the connections 3 levels then you have an example of a 'context tree' in terms of structure.

Now, as articulated earlier, all of these data structures are created in long lines of time based context trees that are also processed by the 'core'; which is the part of the ICOM system that processes the top level emotional evaluations related to new elements of context or existing context framed in such a way that it has new associated context. In this manner, they are all evaluated. As those elements pass through the core, based on associated context, new elements may be associated to that context based on emotional evaluations. Trees are queued and test associations pass through the 'core' and are restacked and passed into memory whereas the 'observer' looks for 'action' based context on which it can take additional action and then the 'context pump' looks for recent context in terms of interest, or other emotional factors driven by needs and interest, to place those back into the incoming queue of the core to be cycled again. Action based context being a specific type of context related to the system taking an action to get a better emotional composited result.

In a sea of data, the ICOM process bubbles up elements and through the core action and additional input you have this emergent quality of context turning into thought in the abstract. ICOM doesn't exactly create thought per se but, in abstract, enables it through the ICOM structures. In early studies using this model, basic intelligent emotional responses are clear. The system continues to choose what it thinks about and the effect this has on current emotional states is complex and varied. When abstracted, as a whole we see the emergent or phase transition into a holistic AGI system that, even in these early stages of development, demonstrates the results that this work is based on.

26.5 Thought Architecture and the Phase Threshold

The Independent Core Observer Model (ICOM) theory contends that consciousness is a high level abstraction. And further, that consciousness is based on emotional context assignments evaluated based on other emotions related to the context of any given input or internal topic. These evaluations are related to needs and other emotional vectors such as interests, which are themselves emotions, and are used as the basis for 'value'.'Value' drives interest and action which, in turn, creates the emergent effect of a conscious mind. The major complexity of the system is the abstracted subconscious, and related systems, of a mind executing on the nuanced details of any physical action without the conscious mind dealing with direct details.

Our ability to do this kind of input decomposition (related to breaking down input data and assigning 'context' through understanding) is already approaching mastery in terms of the state of the art in technology to generate context from data; or at least we are far enough along to know we have effectively solved the problem if not having it completely mastered at this time. This alignment in terms of context processing or a context engine like Tensor Flow is all part of that 'Phase Threshold' (the point in which a fundamentally new or different property 'emerges' from a simpler system or structure) we are approaching now with the ICOM Cognitive Architecture and existing AI technologies.

26.6 'Context' Thought Processing

One key element of ICOM that is really treated separately from ICOM itself is the idea of de-compositing raw data or 'context' processing as mentioned earlier. The idea of providing 'context' for understanding is a key element of ICOM; that is to come up with arbitrary relationships that allow the system to evaluate elements for those value judgements by ICOM. In testing, there are numerous methodologies that showed promise including neural network and evolutionary algorithms.

One particular method that shows enormous promise to enhance this element of AGI is Alex Wissner-Gross's paper on Causal Entropy which contains an equation for 'intelligence' that essentially states that intelligence as a force F acts so as to maximize potential outcome with strength T and diversity of possible futures S up to some future time horizon τ [12].

$$F = T\nabla S_\tau$$

Alex Wissner-Gross Equation for Intelligence

Given Alex's work, if we look more at implementations of this, we can see from a behavior standpoint that this, along with some of the other techniques, is a key part of the idea of decomposition for creating associations into core memory. This is really the secret sauce, metaphorically, to not just self-aware AGI but self-aware and creative systems and the road to true AGI or rather true 'machine' intelligence. While the ICOM on its own would be self-aware and independent, Alex's approach really adds the rich creative ability and, along with the ICOM implementation, creates a digital intelligence system potentially far exceeding human level intelligence given the right hardware to support the highly computationally intense system needed to run ICOM.

26.7   Not like Our Own

One of the questions normally brought up has to do with motivations of a non-human intelligent system such as ICOM. There are numerous scientists that have done centuries of research on the human mind, artificial intelligence and the like. One such example we can read is the paper "The Superintelligent Will: Motivation and Instrumental Rationality in Advanced Artificial Agents" by Bostrom [14]. In this paper Nick discusses how we might predict certain behaviors and motivations of a full blown Artificial General Intelligence (AGI), even what is called an Artilect or 'Superintelligent Agent'. But the real question is, how do we actually implement such an agent's motivational system? The Independent Core Observer Model (ICOM) cognitive architecture is a method of designing the motivational system of such an AGI. Where Bostrom tells us how the effect of such a system works, ICOM tells us how to build it and make it do that.

In the ICOM research program, we might not know how a true AGI might decide motivations but we do have an idea on how ICOM will form those motivations. It comes down to initial biasing and conditioning in terms of the right beginning input that the system will judge all things on afterwards. It is not that we are anthropomorphizing the motivations of the system [14] but it is the fact that we can actually see those motivations defined in the system and have it change them over time enough to see in a broad way that it works very much like the human mind at a very high level.

The differences between the traditional methods of trying to design AI from the bottom up vs. the ICOM approach of top down have produced vastly different results in many cases. There may be many ways to produce an 'intelligence' that acts independently; ICOM, while not working anything like the human brain in detail, very much models the logical effects of the mind; with the same sort of emergent qualities that produce the mind as we experience it as humans. These motivations or any of the elements exhibited by ICOM based systems are logical implementations built out of a need to see the effect of, and not be tied to, the biological substrate of the human mind and thus do not really work the same at the smallest level.

26.8 Conclusion

The Independent Core Observer Model (ICOM) creates the effect of artificial general intelligence or AGI as a strong emergent quality of the system. We discussed the basis data architecture of the data coming into an ICOM system and core memory as it relates to the emergent elements. We can see how the memory is time and patterns based. Also, we considered key elements of system theory as it relates to that same observed behavior of the system as a substrate independent cognitive extension architecture for AGI. In part, this chapter was focused on the 'thought' architecture key to the emergent process in ICOM.

The ICOM architecture provides a substrate independent model for true sapient and sentient machine intelligence, at least as capable as human level intelligence, as the basis for going far beyond our current biological intelligence. Through greater than human level machine intelligence we truly have the ability to transcend biology and spread civilization to the stars.

References

1. "Self-awareness" (wiki) https://en.wikipedia.org/wiki/Self-awareness 9/26/02015 AD
2. "Cognitive Architecture" http://cogarch.ict.usc.edu/ 01/27/02016AD
3. "Cognitive Architecture" wiki https://en.wikipedia.org/wiki/Cognitive_architecture 01/27/02016AD
4. "Cognition" https://en.wikipedia.org/wiki/Cognition 01/27/02016AD
5. "The Sigma Cognitive Architecture and System" [pdf] by Paul S. Rosenbloom, University of Southern California http://ict.usc.edu/pubs/The%20Sigma%20cognitive%20architecture%20and%20system.pdf 01/27/02016AD
6. "Self-Motivating Computational System Cognitive Architecture" By David J Kelley http://transhumanity.net/self-motivating-computational-system-cognitive-architecture/ 1/21/02016 AD
7. "Knocking on Heaven's Door" by Lisa Randall (Chapter 2) via Tantor Media Inc. 2011
8. email dated 10/10/2015 - René Milan – quoted discussion on emotional modeling
9. "Properties of Sparse Distributed Representations and their Application to Hierarchical Temporal Memory" (March 24, 2015) Subutai Ahmad, Jeff Hawkins
10. "Feelings Wheel Developed by Dr. Gloria Willcox" http://msaprilshowers.com/emotions/the-feelings-wheel-developed-by-dr-gloria-willcox/ 9/27/2015 further developed from W. Gerrod Parrots 2001 work on a tree structure for classification of deeper emotions http://msaprilshowers.com/emotions/parrotts-classification-of-emotions-chart/
11. "The Plutchik Model of Emotions" from http://www.deepermind.com/02clarty.htm (2/20/02016) in article titled: Deeper Mind 9. Emotions by George Norwood
12. "An Equation for Intelligence?" Dave Sonntag, PhD, Lt Col USAF (ret), CETAS Technology, http://cetas.technology/wp/?p=60 reference Alex Wissner's paper on Causal Entropy. (9/28/2015)
13. "How to Create a Mind – The Secret of Human Thought Revealed" By Ray Kurzweil; Book Published by Penguin Books 2012 ISBN
14. "The Superintelligent Will: Motivation and Instrumental Rationality in Advanced Artificial Agents" 2012 Whitepaper by Nick Bostrom - Future of Humanity Institute Faculty of Philosophy and @ Oxford Martin School – Oxford University

15. https://en.wikipedia.org/wiki/Chaos_theory
16. https://en.wikipedia.org/wiki/Emergence (System theory and Emergence)
17. Move: Transcendence (2014) by character 'Will Caster (Johnny Depp)'; Written by Jack Paglen; presented by Alcon Entertainment

Consulted Works

18. "Causal Mathematical Logic as a guiding framework for the prediction of "Intelligence Signals in brain simulations" Whitepaper by Felix Lanzalaco - Open University UK and Sergio Pissanetzky University of Houston USA
19. "Implementing Human-like Intuition Mechanism in Artificial Intelligence" By Jitesh Dundas – Edencore Technologies Ltd. India and David Chik – Riken Institute Japan

Part VII
The Transhumanist Age

A door opened quietly and she walked into the room
her body smooth, supple
Her shoulders held back, proud
a glimmer in her eye
and without notice she brushes her hair away

"How old are you?" they asked,
"How old are you?"

A hesitation filled her
as she quickly assessed

"My augmentation is three years now
my right hip five
ocular implants just two weeks
age reversal nearly nine."

She took a sip of water licking the droplets off her lips
her eyes glimmering-hair a shimmering sterling
her outer sheath like mulatto warm bronze

"How old are you?" they asked,
"How old are you?"

And when she turned to leave the room
all eyes followed her form
just to realize before she left
she was not yet born

—Natasha Vita-More

Chapter 27
Transhumanism: The Growing Worldview

Natasha Vita-More

27.1 Introduction

Transhumanism established a set of values conveying emerging technology and humanity in three essential areas: critical thinking, visionary narratives, and technological innovation. The philosophy established values, goals, and principles that synthesize epistemological and metaphysical views. Its roots stem from historical systems of thinking that form social and cultural attitudes. For example, the diversity of global culture, human rights, the economics of techno production and supply, the ecology of well-being, the development of human enhancement interventions, radical life extension, and the politics of laws and policy-making. This synergistic approach brought greater insight into assessing the challenges and conflicts we face and how to build practical and applied solutions. The expression of transhumanist thinking is practical optimism.

At the onset of transhumanist thinking, Information technology capital and faster efficiency gains, especially in the production of computers and computer software, advanced at a remarkable speed. Investors hedged their bets on emerging technologies—nanotechnology and biotechnology, robotics, and AI. Mobile phones, satellite mapping, and globalization emerged through the World Wide Web.

Information now had its own DNA and people were called digital. Executives were "mind workers", businesses were automated, and employees were outsourced. This was the Information Age.

> Computing is not about computers any more. It is about living.—Nicholas Negroponte

N. Vita-More
University of Advancing Technology, Tempe, USA

N. Vita-More (✉)
Humanity+, Inc., Los Angeles, USA
e-mail: natasha@natasha.cc

© Springer Science+Business Media New York 2016
N. Lee (ed.), *Google It*, DOI 10.1007/978-1-4939-6415-4_27

Science in the 1990s was the fixed mindset and technology was its the tool. Fast, efficient and formidable, the it propagated networks for querying how biotech and infotech were going to change humanity.

It was evident, science navigated changes to biology and computer technology emerged as the force through which information flowed. Together they petitioned for answers to unresolved questions about disease, aging and the finality of death.

Theories and speculation were turned in to practice when biotechnology identified the genes of Alzheimer's disease and breast cancer. The field of artificial intelligence awoke from a long winter when it demonstrated evidence of machine learning, and nanotechnology moved matter when it manipulated the xenon atoms to spell out the IBM logo on the nanoscale.

Outside this highpoint and the waves of optimism, society was unaware of where technology was headed. New technology meant new terms, advances in science meant a need for translation. There was little explanation for what was meant by the terms such as nanotech (Drexler) or AGI (Goertzel) that were being hurled at the mainstream. People wanted to know where emerging technologies were headed, intentionally or accidentally, how humanity could be affected, or changed, and who would benefit from the advances. There was a deep-seeded concern, especially when information about the technologies were not readily available.

A new way of thinking was inevitable. This inevitability was met with the philosophy of transhumanism [1].

Transhumanism established a set of values conveying the future technology and humanity in three essential areas: critical thinking, visionary narratives, and technological innovation. The philosophy established values, goals, and principles that synthesize epistemological and metaphysical views. Its roots stem from historical systems of thinking that form social and cultural attitudes [2]. For example, the diversity of global culture, human rights, the economics of techno production and supply, the ecology of well-being, the development of human enhancement interventions, and the politics of laws and policy-making. This synergistic approach brought greater insight into assessing the challenges and conflicts we face and how to build practical and applied solutions. The expression of transhumanist thinking is practical optimism. Its core embodies the nature of being curious, asking questions, applying imagination, in facing many challenges of a hypermodern world that far exceeds the postmodernist agenda, and to build a new understanding for human rights. Through this, transhumanism is now regarded as a growing field of study and a discipline with numerous and proliferating sub-disciplines.

There are current successful projects that parallel the transhumanist agenda. For example, the TED talks, Quantified Self, and global makerspace venues strongly exemplify transhumanist thinking. All are marvelous venues for expressing human experience across varied channels of communication and collaboration in making meaningful contributions to society.

As examples, the TED talks present a narrative as an adventure through a type of hero or heroine's journey and how that journey contributed to personal growth, in a way we can all share in the experience and learn from it. The Quantified Self (QS) project encourages self-knowledge through data acquisition to track one or more physical, cognitive, or behavioral elements in daily life. Makerspaces are collaborative venues for people to come together and identify unresolved issues, unravel them, roll up their sleeves, and then create innovative solutions through a hands-on do-it-yourself processes. All three of these examples evidence a culture of self-responsibility and through this—helping other be better too. This sentiment, or this intelligence, is sincere, in the moment and affects others—like an infectious smile.

Self-responsibility, as a personal and social behavior, is spreading throughout activities launched as entrepreneurial innovations. This is transhumanist in scope. We start out with a distinct place in life and nurture our surrounding to build practices from which to learn, evolve, and become more conscious of life, and its preciousness. This could be interpreted as spiritual, or simply intelligent. Some might call it a normal, innate aspect of human psychology about survival and society. At the time, at the early stages of transhumanism and into the 21st Century, this was needed. There was a socio-technological gap that the postmodernist era could not resolve. Transhumanism bridged that gap. This is largely why transhumanism became a growing worldview.

27.2 Transhumanism

The Transhumanist Declaration establishes principles of transhumanism, and the Transhumanist FAQ encapsulate much of the history. To add to the quality of these documents, insights into the topic of transhumanism needs a personal voice, one that experienced the beginning of the movement and continues to observe its evolution. In this chapter, lend my voice to reflect on the past and sum up the future. Through this, I defer to others' works as well as unveiling the core topics by salient writers on technological advancements.

These advancements include artificial intelligence, nanotechnology, uploading, encryption, privacy, and the biotechnology of life extension that percolated through the pages of a high-gloss print journal called *Extropy: The Journal of Transhumanist Thought* (1988) [3]. To add to the quickening flow of information and at the onset of the world wide web, a few programmers built the original transhumanist email list called extropy [4]. This list became the fertile hotbed for high-level discussions among thought leaders whose brilliant minds cogitated over what was yet to come with the arrival of emerging and exponential technology. These ruminations became the themes of transhumanist conferences, starting in 1994, that formed an intellectual center for learning about accelerating technologies that impact biology and life extension, intelligence gathering and supercomputing, data collecting, and privacy [5]. Yet, it was not until the Internet exploded the transhumanism spread and the development of other transhumanist organizations still increases today with the use of social media and the mainstream's awareness of and interest in nanotechnology, AI/AGI, and life extension sciences.

The symbol for transhumanism gained branding currency as H+ over the years; however, a distinct logo of h+ within an open circle is trademarked by Humanity+ for the logo of its non-profit corporation and its magazine. Humanity+ [6] is currently the largest transhumanist organization and has filled the space of what Extropy Institute left when it closed down. The irony is that today most of the proto-transhumanists are board members or advisors, which has successfully brought together the varied political views.

27.3 Misinformation

Half-truths with journalistic spin often bring about erroneous assertions bias. There is no legal crime in propaganda for effect, especially in televised documentaries, magazine articles, and blogs, or in creative approaches to debates and media coverage. This tactic is all too known within the sphere of social change and the future, especially when the change is perceived as fundamental to what it means to be human. However, there is often a civil crime as an infringement on a person's or a group's rights to have political and social freedom and equality. By this, I mean the right to practice a way of life without being doggedly attacked by media that is spreading misinformation, whether it be hyperbole by the press or academic rhetoric.

In 2009, I came upon an academic journal *The Global Spiral*, which published an article on transhumanism. The article was compromised of well-known scholars but it was hampered by a lack of peer-review research of acceptable quality and/or by a distinct degree of bias not satisfactorily accounted for by an analytical line of inquiry, regardless of the overall literary scholarship and scope of knowledge. How did these highly revered academics in the fields of social sciences, literature and humanities misconstrue what I had experienced first-hand within the realms of transhumanist? In my stupor and annoyance, I contacted the Journal's Managing Editor who allayed my angst by inviting me to be a Guest Editor. The idea was for me to bring in high-level transhumanist academics to challenge the misinformation and write responsive counter arguments. Apparently this second Special Issue on Transhumanism received a lot of attention and one year later, a compilation of the essays from both sides of the arguments—the original essays and our responses— were published as the book *Transhumanism and Its Critics* (2011) [7].

In The Global Spiral's 2008 "Special Issue on Transhumanism" [8] (June, Vol. 6, Issue 3), Guest Editor Hava Tirosh-Samuelson and five other authors, Ted Peters, Katherine Hayles, Don Ihde, Jean-Pierre Dupuy, and Andrew Pickering, provocatively relay their concerns about transhumanism to expectant ears. This responsive second Special Issue on Transhumanism (2009, Feb., Vol. 9, Issue 3) is an opportunity for ten transhumanist authors to evaluate the criticisms and address concerns [including Aubrey de Grey, Martine Rothblatt, Max More, Nick Bostrom, and Russell Blackford]. This reflection is beneficial in helping us more precisely clarify interpretations of transhumanism and identify where our own words may have been mistaken.

There are many predecessors to the idea that human evolution has not reached its final stage, tracing back at least to Nokolai Fedorov's "common task" (1828–1903) [9] and Jean Finot's *Philosophy of Long Life* (1909) [10]. The philosophy and social/cultural movement of transhumanism has developed not only from the words "trans" and "human", but also through an understanding that the human condition is one in which we might go outside to gain perspective, a process of becoming, an evolutionary transformation. As a bit of history, and to show the lineage of the concept of the transhuman and transhumanism and how it evolved, below is a synopsis of the term and its development.

- "Trans-human" and the Italian verb "transumanare" or "transumanar" was used for the first time by Dante Alighieri in *Divina Commedia* (1308–1321) [11]. It means "go outside the human condition and perception" and in English could be "to transhumanate" or "to transhumanize".
- T.S. Eliot wrote about the risks of the human journey in becoming illuminated as a "process by which the human is Transhumanised" in "The Cocktail Party", *Complete Poems and Plays* (1909–1950) [12, 13].
- Julian Huxley wrote about how humans must establish a better environment for themselves, while still remaining man in *New Bottles for New Wine*, which contains the essay "Transhumanism" (1957) [14]
- Teilhard de Chardin wrote about intellectual and social evolution and ultra-humanity in *The Future of Man.* (1959) [15] Abraham Maslow referred to transhumans in *Toward a Psychology of Being* (1962) [16].
- The *Reader's Digest Great Encyclopedia Dictionary* defined "transhuman" as meaning "surpassing; transcending; beyond" (1966) [17].
- F.M. Esfandiary wrote about transhuman as an evolutionary transition in "Transhumans 2000" (1972) [18, 19].
- Damien Broderick referenced transhumans in *The Judas Mandala* (1982) [20].
- Natasha Vita-More (f/k/a Nancie Clark) authored the "Transhuman Manifesto" and the "Transhumanist Statement" (1983) [21].
- Robert Ettinger also referred to the transhuman in the preface (1989) in the earlier published *Man into Superman* [22].
- The *Webster's New Universal Unabridged Dictionary* (1983) [23] defined "transhuman" as meaning "superhuman," and "transhumanize," as meaning "to elevate or transform to something beyond what is human".
- Max More authored "Transhumanism: Toward a Futurist Philosophy" (1990), which established the philosophy and worldview of transhumanism.

It may be impossible to validate absolute accuracy, but as active participants in carving our future, it is a responsibility to relay information as carefully as possible because it does filter through the minds of others.

It is necessary to distinguish between the terms transhuman, cyborg, and posthuman [24]. The transhuman is an evolutionary transition from predetermined genetics and limited lifespan to a therapeutic and selective enhancement for slowing down and/or reversing aging, refining cognition, and extending lifespan. This is known as human enhancement. The cyborg is an augmentation to human biology to adapt to diverse environments, such as space, which augmentation is a specific end goal. The posthuman is a possible human future where a person could exist in the biosphere, cybersphere and other artificial environments and substrates. The posthuman is often referred to as an "upload", a person who uploads into computational sphere, or as a crossload where the person would co-exist in varied environments, including the biosphere.

27.4 Social Needs and the Political in Brief

Transhuman advocacy in the mid-1980s to 1990s was not local, it was global—from the US to Western Europe, including Sweden, The Netherlands, Belgium, the UK. The larger hubs were located in Los Angeles and Silicon Valley. In Southern California, a freethinking environment predominated the transhumanist flora with life extensionists, environmentalists, human rights activists, artists and writers, cyberpunks, feminists, Liberals, Upwingers, Greens, and a health-oriented community of sports clubs. In Northern California, the turf was largely libertarian, with computer scientists, programmers, technologists, cryptography and network security specialists. Coincidentally, there was no conflict between the varied political views since all were transhumanist thinkers and the larger issue was slowing down aging, human enhancement, and radical life extension.

Setting a political stage, a small cable TV show called "Breakthroughs: A TransCentury Update" (a/k/a "Transhuman Update") aired in Los Angeles (1987–1994) [25]. The show explored topics on social awareness where guests spoke about future scientific and technological innovation. One benefit that came out of this show was that it provided a cross-disciplinary education of what the future might bring. A secondary benefit was that the show established a platform for transhumanist discussions. Because of this, a viewer nominated me to run for County Council for the 28th Senate District of Los Angeles County [26]. In 1992, I was elected on a transhumanist platform with an emphasis on technological mediation to environmental problems. I resigned after one year. The Internet was just starting to evolve into a powerful platform for socio-political-economic discussions and advocacy, and seemed to be a far better format. It was new, exciting, and evolving. It supported autonomy and human rights, freedom to express, and a global community where Moore's Law was accepted, racial life extension was queried, and the technological singularity was understood as a possibility, but not a fate accompli. Currently, there are over 836,000 websites on transhumanism, and it is growing every moment.

With growth comes change. New voices were establishing email lists, websites, forums, and debating social structures and governing bodies. It was not until around 1998 that a political discontinuity arose within transhumanism. No one knew precisely why this occurred; it waves a red flag of what not to do. In the political realm, differences are unavoidable. Core transhumanist values can be adopted by people with different backgrounds who may fine-tune their political position to harmonize with their transhumanist preferences. If this is accepted, then the political differences and organizational perturbations should not be a public or private embarrassment. In fact, this can be useful in identifying problems that have not yet surfaced, or even invigorate a long-awaited pivot to gain momentum. Movements experience these shifts when changes occur and people feel excluded or their personal preferences have not given enough weight. And so it was with transhumanism in the late 1990s—a time when liberals and libertarians both had a strong voice and newly arriving socialists wanted their voice to be heard.

The fact is that there were many libertarians in the Silicon Valley; however, there were also many Democrats, Greens, Republicans, Upwingers, and Independents within transhumanism.

Quoting F.M. Esfandiary in his book *Upwingers a Futurist Manifesto* (1997), the normative platform reaches beyond the Right/Left predicament and sets out a non-linear evolving view of moving upward:

> We are at all times slowed down by the narrowness of Right-wing and Left-wing alternatives. If you are not conservative, you are liberal if not right of center you are left of it or middle of the road.
>
> Our traditions comprise no other alternatives. There is no ideological or conceptual dimension beyond conservative and liberal beyond Right and Left. ... The premises of the entire Left are indistinguishable from those of the entire Right. The extreme Left is simply a linear extension of the extreme Right. The liberal is simply a more advanced conservative. The radical Left is a more advanced liberal. ... The Right/Left establishment is fighting a losing battle. It is following in the foot-steps of earlier traditionalists who resisted the more modest breakthroughs of the past. ... (pp. 21–25). (Esfandiary [27]).

Regardless of political divides, the focus must be on the goal is to stay alive as long as possible, while effectively working toward developing for solutions to slow down aging and the finality of death. Accountability has always been in the transhumanist pipeline. It has become more critical in dealing with a confluence of ethical issues while pushing out strategies for success.

Transhumanism needs up-to-date policies and legislation. Protocols such as the Proactionary Principle, an ethical decision making rule, is becoming crucial in protecting people's freedom to use technologies that are valuable, even critical, to humanity [28]. The area of civil and human rights needs to be aligned with the current political landscape, locally and globally. For example, the concept of morphological freedom ought to be a human right—the right for a person to enhance his biological (somatic and cognitive) system and, equally, a person should not be coerced to enhance. This right protects those who want to enhance, or live longer and those who want to remain biological or be more conservative in their enhancement.

Many people have also been involved in politics over the years. Since the 1992 County Council election for the 28th Senate District of LA County that formed a transhumanist platform for emerging technologies, many formative people have taken up expansive causes. James Hughes of Institute for Ethics and Emerging

Technologies (IEET) advocates the Technoprogressive stance (2015), David Wood and the London Futurists advocate the UK Transhumanist Party (2015), a registered political party in the United Kingdom, Marc Roux of the *Association Française Transhumaniste* advocates the Technoprog! transhumanist politics in France (2013). In Italy there have two distinct groups, the Italian Transhumanist Association (AIT), and the Italian Transhumanist Network (NTI). In 2012, *New Scientist* wrote an article claiming that Giuseppe Vatinno was the first transhumanist elected, and later *H+ Magazine* wrote a different article stating that Gabriel Rothblatt was the first in 2015. In reality, neither were the first. They are all constituents within the larger global society that embraces transhumanist politics as a necessary policymaking process, regardless of party names. Journalist Zoltan Istvan created a profile for what he calls a transhumanist party. Over time, there may continue to be other groups and individuals who are not so clear in their message, or in their use of transhumanism.

And even though the political arena is decades old and seasoned with the previously mentioned players it will continue to grow into a serious political agenda. Relatedly, the Transhuman National Committee (2015), formed by David Kelley, is a political action committee or PAC focused on Transhumanism in politics in the United States. Its mission is long term with an intent to create political action on policies and laws, as relevant to the government officials and policymakers in power. This mirrors the 2004 Vital Progress Summit, built on the premise of a much needed progress action committee.

There have been numerous social agendas within transhumanism balancing the right and left, there is yet to be an adaptation of an informed, transparent and modern system that addresses morphological freedom or the Proactionary Principle. nonetheless on the whole, transhumanism is concerned with human rights and diversity, with a focus on critical thinking, visionary narratives, and technological innovation. How this will be achieved is unknown, as it is an iterative process in the making. It will blossom and wilt as it changes course, matures, and develops the flexibility to be truly transhumanist in scope.

Considering that cultural and social movements do exist in and respond to the political climate of the day, there have probably been many advocates of life extension and of technological innovation to help shape the future in a positive direction. This brings to point other issues within this political arena, who has the right to use the term transhumanism and for what cause. Since the term is in the public domain anyone can use it. It this ethical? As a point of clarification, the word transhumanism is in the public domain. Anyone can use the term and also the abbreviation H+. If the term is part of a name that is trademarked or the H+ is a trademarked logo, then it is not legal.

27.5   Emerging Technologies

In the Introduction of this Chapter, I suggested that the idea of transhumanism surfaced during a time when the novelty of emerging technology attracted a certain identifiable sensibility. But this sensibility is not a clear complexion. The many faces of transhumanism represented the larger futurist community of science fiction, literary visionaries, scholars of philosophy, researchers of scientific breakthroughs, Silicon Valley tech entrepreneurs, innovators and the venture capitalists alike. Its tone is varied and vast, and even cosmic. This domain of emerging technological output, the products, processes, and Internet of things, including the devices we use to expand cognition, the mobility devices we use to expand transportation, and the digital networks we use to expand communications, is vast and exponential. This might or might not be seen as a singularity or omega point. Ray Kurzweil turns it mainstream in *The Singularity is Near* (2005), Peter Diamandis covers this spin in *Abundance* (2012), and K. Eric Drexler sought it in *Engines of Creation* (1986). Nick Bostrom addresses concern in *Superintelligence: Paths, Dangers, Strategies* (2014), and Anders Sandberg offers a sober tone in the trajectory "Whole Brain Emulation—A Roadmap" (2008). David Pearce offers a different view in his approach to utilitarian views, including the topic of happiness as fundamental in "The Hedonistic Imperative" (1995). Martine Rothblatt explores sexual identity beyond the two-gender system in *Apartheid of Sex* (1995).

Max More established the core values of transhumanism, and emphasizes continuity of personal identity over time, the human right of morphological freedom, and the Proactionary Principle in policy making. Bostrom and Pearce formed a utilitarian perspective of transhumanism, and while Pearce focuses on happiness as essential, Bostrom addresses technology risks and AI related concerns. Kurzweil and Rothblatt addressed complex patterns of the mind, and more currently Kurzweil carves the synthesis of invention and innovation, while Rothblatt builds the course for sexual continuity and freedom of gender. These foci differ in their approach to transhumanism but together they are synergistic. They bring perspective to the society of transhumanism in the journey toward the necessary conditions enabling human enhancement and radical life expansion.

> Chorus: You haven't, by any chance committed a further offence?
>
> Prometheus: Yes, I've made it so that humans cannot foresee their own death.
>
> Chorus: What sort of medicine did you use for this?
>
> Prometheus: I've filled their hearts with blind hopes.
>
> Chorus: That is a great gift you've given to the humans.
>
> Prometheus: And more than that, I've given them the gift of fire.
>
> Chorus: So the ephemeral creatures now possess the bright-faced fire?
>
> Prometheus: Yes, and with it they will learn many crafts [29].

The patterns of practice that expose the pros and the cons of technology that effect transhumanism are not just the dream of a better life that transhumanism embraces; it is accountability—the conviction to investigate, understand, and problem solve possible dangers. These ideas are summed up in *The Transhumanist Reader: Contemporary and Classical Essays on the Science, Technology, and Philosophy of the Human Future* (More and Vita-More 2012), where forty-two thought leaders of emerging and speculative science and technology discuss their potential and concern.

27.6 Practical Transhumanism

Through growing pains and maturation, weathering the storm of doomsayers and alarmists, bioethical moralists, and achieving acceptance in the academic world and the mainstream alike, we are here today. This interconnected community aligns nicely with transparency, open source, the social scientist, and maker-sensibility of a shared by those who want to improve their human condition.

Practical transhumanism reinforces this aim, as people cannot be ready for the future if they do not prepare for its challenges. The aspect of what we call the human condition that transhumanism is motivated toward is sustaining health and well-being as long as feasibly possible. There are many practices that approach this aim, especially nutrition and exercise, and beyond this an intervention of the deleterious aspects of disease—the deterioration of the body, and brain caused by diseases of aging. The intention to be healthy as long as possible and to understand the genetic script as an intervention to alter a fixed biology. This is not translated as for only the wealthy or privileged; it is for everyone—a democratized life extension of social equality. Life extension as a basic human right. Life extension as a Holy Grail in changing the world through uniqueness and value of transhumanist thinking.

Transhumanist thinking is a method to solve complex problems and to develop desired conditions for existence. Transhumanist thinking applies critical thinking in creating visionary trajectories for well-being and super longevity (a/k/a radical life extension). This may be the single most valuable asset of transhumanism today. Over many decades, transhumanists have questioned, and debated, and developed the major topics concerning life extension and technology as an iterative process. This process includes varied scenarios for human futures—specifically, emerging sciences and technologies that are altering the world and the benefits and the concerns of apparent risks. The framework for this process includes defining the opportunity that transhumanism offers to society, gathering information, questioning assumptions about humanity, incubating ideas, evaluating the pros and cons of technology. How this framework is realized in the real world is through projects. Some of these projects are workshops and conferences, others are written or recorded experiences. Some of these projects are innovative, such as developing technologies and performing scientific research. The areas in which these process are relative to transhumanist thinking include biological health and well-being, and technologies such as artificial general intelligence, robotics, nanotechnology, nanomedicine, cryonics, relative to human enhancement and cognitive science and neuroscience relating to brain functions, including transferring and copying the functions onto computational systems.

The challenge is to make meaning—substantial, forthcoming, and realistic. To start off by doing things that are meaningful. For transhumanists, meaningful is happy, healthy, long life; one that continues beyond the maximum human lifespan.

The discussion is to not underestimate or overestimate the pace of technology and its effect on well-being, human behavior, and modifying physiology. Clearly, biotechnology and nanotechnology have the potential to eradicate cell degeneration and unfix humans from biology. Artificial general intelligence, virtuality, and bio-artificial fusion have the potential to amplify cognition, increase awareness, and transfer cognition from biological cells onto non-biological systems. Continuity, transparency and immersivity of computer-based interactions catalyze one person and one identity into multiple, varied aspects of personhood. The significance and the outcome of these effects on human physiology could result in the design of new types of humans, new platforms for human existence, and the extension of human life over time.

There is an observable challenge for each person throughout diverse cultures, the worldwide economics of production and supply and the development of policies and practices. This is a strong reason for transhumanism's approach of critical thinking and a proactive stance in navigating greater insight to the many challenges and how to build solutions.

People are living longer as biotechnology develops new methods for conquering disease and marveled therapies for injury. Personalized genome services offer DNA sequencing of DNA. Smart devices provide memory storage up to 128 GB, a business can a system with 120 petabytes. This is the Transhumanist Age.

A door opened quietly and she walked into the room
her body smooth, supple
Her shoulders held back, proud
a glimmer in her eye
and without notice she brushes her hair away

'How old are you?' they asked,
'How old are you?'

A hesitation filled her
as she quickly assessed'My augmentation is three years now

my right hip five
ocular implants just two weeks
age reversal nearly nine.'

She took a sip of water licking the droplets off her lips
her eyes glimmering—hair a shimmering sterling
her outer sheath like mulatto warm bronze

'How old are you?' they asked,

'How old are you?'

And when she turned to leave the room
all eyes followed her form
just to realize before she left
she was not yet born

—Natasha Vita-More

References

1. Max More authored the "Philosophy of Transhumanism" in 1990. A detailed essay on transhumanism's beginnings, see *The Transhumanist Reader: Classical and Contemporary Essays on the Science, Technology, and Philosophy of the Human Future* (Wiley-Blackwell).
2. "It is often confused with, compared to, and even equated with posthumanism. Transhumanism arrived during what is often referred to as the postmodernist era, although it has only a modest overlap with postmodernism. Ironically, transhumanism shares some postmodernist values, such as a need for change, reevaluating knowledge, recognition of multiple identities, and opposition to sharp classifications of what humans and humanity ought to be. Nevertheless, transhumanism does not throw out the entirety of the past because of a few mistaken ideas. Humanism and scientific knowledge have proven their quality and value. In this way, transhumanism seeks a transmodernity or hypermodernism rather than arguing explicitly against modernism. One aspect of transhumanism that we hope to explore and elucidate throughout this book is the need for inclusivity, plurality, and continuous questioning of our knowledge, as we are a species and a society that is forever changing. The roots and core themes of transhumanism address some of the underlying themes that have formed its philosophical outlook" (More and Vita-More, 2012).
3. The Journal of Transhumanist Thought started out in a more casual formatting in the magazine *Extropy: Vaccine for Future Shock*.
4. Extropy refers to a systems intelligence, order, vitality and capacity and drive for improvement.
5. The transhumanist conferences can be found in the Extropy digital reference center located at http://www.extropy.org
6. Humanity+ is a 501c3 nonprofit organization. It was originally incorporated as World Transhumanist Association in 1989 and rebranded its name in 2008. Humanity+ and its Board of Directors and Advisors unify the pioneering transhumanist Extropy Institute and WTA.
7. Metanexus Institute Publisher. http://www.amazon.com/Transhumanism-Critics-Gregory-R-Hansell/dp/1456815652#reader_1456815652
8. Special Issue on Transhumanism (2008) In *The Global Spiral*, (Guest Editor, Hava Tirosh-Samuelson), Vol. 6, Issue 3. Available http://www.metanexus.net/magazine/PastIssues/tabid/126/Default.aspx?PageContentID=27
9. Fedorov, Nicolai. (1990) *What Was Man Created for?* Hyperion Books.
10. Finot, Jean. (2009) (Trans. Roberts, Harry) *The Philosophy of Long Life*. Whitefish, MT: Kissinger Publishing, LLC. Originally published London: John Lane, The Bodley Head in 1909, pp. v, 83, 270.
11. Dante, Alighieri. (1308–1321) *The Divine Comedy* (The Inferno, The Purgatorio, and The Paradiso) (Ed. Ciardi, J.) New York: NAL Trade, 2003. (p. 586-589).
12. Eliot, T.S. (1952) *Complete Poems and Plays: 1909-1950. The Cocktail Party.* New York: Harcourt. (p. 147).
13. Sarkar, Subhas (2006) *T.S. Eliot: The Dramatist.* Atlantic Publishers. (p., 192).

14. Huxley, Julian. (1957) "Transhumanism" in *New Bottles for New Wine: Essays*. London: Chatto & Windus.
15. De Chardin, Teilhard. (1959) *The Future of Man*. First Image Books Edition (2004).
16. Maslow, Abraham. (1962) *Toward a Psychology of Being*. (2nd. Ed. 1968, New York: John Wiley & Sons (p x1).
17. *The Reader's Digest Great Encyclopedia Dictionary*. (1966). Reader's Digest.
18. FM-Esfandiary (1972) "Transhumans 2000" in Tripp, M., *Woman, Year 2000*. New York: Arbor House.
19. FM-2030 (1989) *Are You a Transhuman? Monitoring and Stimulating Your Personal Rate of Growth in a Rapidly Changing World*. Viking Adult.
20. Broderick. D. (1982) *The Judas Mandala*. New York: Pocket.
21. Vita-More, N. (1983) Transhuman *Statement* in *Create/Recreate*. Available: http://www.transhumanist.biz/createrecreate.htm http://www.natasha.cc/transhuman.htm
22. Ettinger, R. (1989) Prologue: The Transhuman Condition" in *Man into Superman*. Prologue (1972). New York: Avon.
23. *Webster's New Universal Unabridged Dictionary*. (1983) Fromm Intl.
24. In my Doctorate Dissertation, I explain the difference: "One outcome of the integration and interaction where the human body and its biology with machines was consequential in bringing about the concept of the cyborg (Clynes 1960:27). Rather than being positioned as an endpoint to the integration of human, machine, and computer, the alternative strategy proposed in the thesis focuses on the [transhuman] platform diverse body (substrate autonomous person; i.e., transhuman) because, unlike the cyborg, it steers its own transformative evolution in becoming posthuman." (Vita-More, 2012).
25. Producer and Host, Natasha Vita-More, aired from 1986 – 1994, located in Santa Monica, California at CityTV programming.
26. Green Party.
27. Esfandiary, FM. (1977) *Up-Wingers: A Futurist Manifesto*. New York: Popular Library, CBS Publications, (pp 21-25).
28. The 2004 Vital Progress Summit presented 2-week virtual discussion and debate on President Bush's Bioethics Council's ultraconservative *Beyond Therapy* report. Summit keynotes addressed the use of the well-known "Precautionary Principle" by anti-biotech activists as a rallying tool to turn people against the science, medicine and biotechnology that could cure disease and improve life. Summit keynotes: Ronald Bailey, Aubrey de Grey, Raymond Kurzweil, Max More, Marvin Minsky, Christine Peterson, Michael Shapiro, Lee Silver, Gregory Stock, Natasha Vita-More, Roy Walford, and Michael West.
29. AESCHYLUS' "PROMETHEUS BOUND" Translated 2006 by G. Theodoridis. See http://www.poetryintranslation.com/PITBR/Greek/Prometheus.htm Downloaded April 5, 2011.

Chapter 28
Why I Am a Transhumanist

Dirk Bruere

28.1 Definition of Transhumanism

Let's begin with (almost) the most basic definition of Transhumanism, from Max More, circa 1990:

> Transhumanism is a class of philosophies of life that seek the continuation and acceleration of the evolution of intelligent life beyond its currently human form and human limitations by means of science and technology, guided by life-promoting principles and values.

I say "almost" because like subsequent definitions and declarations it attempts to tack on extra baggage in the form of ethical or political caveats. Of course I would like it to be "*guided by life-promoting principles and values*" but it does not have to be. Those addenda are at best conscious attempts to guide the use of Transhumanist technologies, and at worst just wishful thinking.

This is not really the place for an exposition of the technologies which are enablers of Transhumanist ambitions, which can be easily found online. However, I will deal with some of them in due course as they affected my choice to call myself a "Transhumanist". So, let's go back to my childhood where much of this had its origins.

28.2 Childhood

I grew up in a village in England in the 1950s as the son of a South African immigrant father and English mother. It was still reasonably "traditional" in that squalid Medieval sense of outside toilets and where people kept pigs and chickens

D. Bruere (✉)
The Transhumanist Party, London, UK
e-mail: dirk.bruere@gmail.com

© Springer Science+Business Media New York 2016
N. Lee (ed.), *Google It*, DOI 10.1007/978-1-4939-6415-4_28

in their back garden, everyone knew everyone else's business, conformity was next to divinity, and anyone whose family had not lived there for the past 200 years was one of the "new people". In the immediate post-war period dressing your uber-Aryan looking son with the white/blond hair in lederhosen was probably not a popular choice among the locals. So, off to a running start with triple points on "outsider" status.

Our neighbors were also people who might charitably be described as "characters". The ugly old woman who in earlier times would probably have been hanged as a witch and who enjoyed a touch of thieving from the next door shop in her spare time. "Old Cossy" and her fat lazy husband who kept a pig in the garden and did not want to be disturbed by the sound of children playing as he slept off his Sunday lunch. Deaf Pete, who would later be killed crossing the road because he did not hear the traffic. Old Jack, blinded in the First World War by mustard gas. Some people further up the road that "had something wrong with them", whatever that meant. The woman who had being dying of cancer, for years (the word "cancer" being said in a whispered voice as if saying the name out loud might act as an invitation). And Old Juddy. Yes, everybody over 30 was "Old" in those days, and looked it.

It was he who first introduced us, me and my younger brother and sister, to the notion of pointless cruelty to animals. One day the three of us were in the garden and to entertain us he grabbed one of his hens by the legs and plucked it to death in front of us. Being rather young at the time (all of us under 6 years old) we were not quite sure what to make of it. I know that I did not like it and neither did my father when I told him. But these were the times and these were the people around us.

Can you imagine the mentality of someone who believes that casual animal torture is a suitable way to entertain children? It was not even dressed up as "sport". Like I said—Medieval.

Then it was off to primary school, where I learned other interesting lessons concerning pecking orders, violence, bullying and stupidity. And that was just from the teachers. Another scene, from when I was nine years old and had misheard which exercise to do from the textbook. The teacher comes up to me and asks in an annoyed voice why I was doing the wrong work. I started to reply: "I thought..." whereupon he hit me round the head and shouted: "You are not here to think!" The irony of that escaped me for quite a number of years, but I got there in the end.

As for the other kids, well in the space of a couple of years I had been in numerous fights, none of which I started, and had been off school twice due to the extent of my injuries. Of course, nothing was done about the situation by either parents or teachers—it was "normal". As was seeing kids covered in blood from such fights, including me when my nose was broken. That one was not serious enough for time off school.

Much of that I put down to various pieces of advice my parents had given me. Probably the thing that got me into the most trouble was a sense of justice and the belief I should stand up when other people were being bullied—"protecting the weak". From my father. Not to mention injunctions to always tell the truth, which he neglected to say never made one popular.

A close second was not acquiescing to the pecking order games which were, and probably still are, a part of the barbarism of childhood and which only become more subtle as people become adults. Back then, the local "top dog" would pick on someone at random and just painfully punch their shoulder. If they did not fight back it meant, in modern slang, that you were "their bitch". If you did fight back, it got ugly.

Which is where my mother's piece of counterproductive advice played its part: "Violence never solves anything". In effect it meant that even if I won the fight I would be "reasonable" and not push my advantage to total victory, having hoped that the other kid had "learned his lesson". Wrong—mercy equals weakness among such people.

This cycle of violence only ended when I changed schools, and not because new kids appeared but because the first time it started up again I lost my temper and threw the bully across a desk. I pushed the advantage, and from that point on it all ceased. It was like a cartoon light bulb switching on in my head—*insufficient* violence never solves anything.

Meanwhile, my life outside of school was one of obsession with science. Here I had facts, theories and opinions that were not accompanied by physical violence or fear. There were no gangs to run away from in the library. By the time I was ten I was reading graduate level inorganic chemistry textbooks and everything I could find on other sciences. The stupidity of those around me I found appalling. Too bad about that honesty thing…

I sailed through school, getting top marks without putting in more than minimal effort. And that was bad in itself, because it developed in me a laziness that has taken most of a lifetime to shake off.

Through all of this was a factor that I would not discover for another thirty years or so. It was that almost certainly I was suffering from what is now termed "high functioning autism". Its major symptoms, apart from the physical, were lack of social awareness, obsessive interests (scitech), literal mindedness and a massive insecurity when it came to dealing with people. People were stupid, cruel, vicious, over emotional, weak, pathetic and horribly unpredictable. I viewed ordinary people as ordinary people view the mentally ill.

Most of all though, was that I had a strong sense of "How Things Should Be" and this wasn't it. I recall one night lying in a field looking at the stars in the clear night sky and just wanting to get off this planet. Ironically, I recently discovered a site for those with Asperger's is actually named "Wrong Planet".

While I was reading my books in front of the TV my parents were watching popular soaps, which reinforced my opinions. They are all about stupid people doing stupid things for all the wrong reasons and then suffering pointless consequences that anyone with half a brain could avoid with ease. I found them endlessly frustrating and although I sometimes watch one or two occasionally to check, they are the same now as then. It's like those reality/documentary TV police shows where the level of imbecility is illustrated when some idiot tries to shoot his way out of getting a parking ticket. Turning a minor fine into a death penalty. Good riddance—the average IQ of the world rises slightly.

I recall my grandmother telling me a story about when she was young, at the outbreak of WW1, when groups of men armed with sticks roamed the streets for Dachshund dogs to beat to death, because "they were German". Do you really want to live in the same universe as that kind of weapon grade stupidity?

Which led to my having a very elitist and increasingly callous view of people in general. Conversely, the thing that I could not tolerate was cruelty to animals or children. At least in some measure adults deserved whatever horror was visited upon them, and there was no shortage judging from the news reports.

28.3 The Human Condition

One of the things that both annoyed me and propelled me towards the idea of upgrading Humanity wholesale was the once popular catch-all beloved of avant-garde playwrights "The Human Condition". This was the more intellectual-ized soap opera where the usual suspects were examining their own suffering in terms of "what it means to be Human". Again, the aspects of the condition tended to be things like jealousy, anger, rage, greed, stupidity, depression, violence... the list continues. In fact, what they generally describe are all the things that make us sub-Human. Things that stem from our evolutionary ancestry. It has been said that we are a halfway stage between apes and angels. I say it is time to edit out the ape and enhance the angelic qualities, which are the things which truly distinguish us as Human. They include rationality, love, compassion, empathy and intelligence. None of which make for good drama or exciting shoot-outs.

Then, when I was sixteen and on New Year's Day 1970 my father died early in the morning from a heart attack. It was slow, painful and protracted. My first direct contact with death. So, how did I feel? Answer is, I don't know. It was only in later years I missed him. What I do know is that in the pre-Internet age people died when they could have lived, for lacking a single piece of knowledge. In his case chewing a couple of aspirin would probably have saved his life. This was known at the time, but not by most doctors.

Less than a year later I was in bed one Christmas morning reading a book whose title escapes me, but one quote hit me right between the eyes. It was by a man named I J Good:

> Let an ultra-intelligent machine be defined as a machine that can far surpass all the intel-lectual activities of any man however clever. Since the design of machines is one of these intellectual activities, an ultra-intelligent machine could design even better machines; there would then unquestionably be an 'intelligence explosion,' and the intelligence of man would be left far behind.

My reaction was an instant "YES!"
Finally, a way forward or a way out.

28.4 AI Winters

However, in the years that followed there was a lot of what turned out to be misplaced optimism concerning the difficulties involved. The hype and following failures resulted in several rounds of funding cuts known as the "AI Winters" which lasted until the mid-1990s. In my opinion it was always naive to think Human level intelligence could run on machines which obviously had millions of times less processing power than the brain, and the bar was set unreasonably high by the 1968 movie "2001: A Space Odyssey" with HAL. It is only now, in 2015, that serious inroads are being made on machines of suitable power, but there is a lot more to do before true Artificial General Intelligence arrives.

This is also a technology not only whose time is arriving but the warnings also written of by I.J. Good all those years ago are being echoed by the likes of Elon Musk and Stephen Hawking. Namely, if we get it wrong it could prove an extinction event. So ponder a question I asked on a large Transhumanist group, which went something like this:

> If you had a choice between us creating an AI that vastly exceeded Human capabilities in all spheres of endeavor but would lead to Human extinction, or no AI ever and Humanity just carrying on as it is, which would you choose?

Most voted for the AI, along with me. Of course, the excluded middle way was the favored one of a Human-AI merger. How realistic that is remains to be seen.

28.5 Transhumanist Moment

My next "Transhumanist moment" was at a Mensa party in the mid-1970s where I was discussing mind uploading (a very Transhumanist meme) with a man whose name stuck with me partly because it was unusual, but also because he was wearing a bow tie. He was Madsen Pirie, later to found the Adam Smith Institute. At the time we called ourselves "Immortalists" because although the word "Transhuman" was used by F.M. Esfandiary in 1972 we had never heard of him, or it. At that point my Transhumanist interests rested for almost two decades.

In the intervening years my problems in dealing with people became apparent even to me, and I started to do something about it. What prompted it was the realization that while technology might be a unique and defining factor in our current civilization, it was still subservient to politics. In other words, people run this planet, not scitech. I suppose it was at this point I ceased to be the archetypal nerd. Re-wiring ones brain is not easy.

I started by forcing myself into situations that I found extremely stressful, but which other people obviously found simple. To give you an example, even at the age of twenty five I would break out into a sweat if I had to answer the telephone. It was because of the unpredictability of what was to follow. When making a phone call I would first decide what I wanted to say, checked off all the possible responses and then had my reply to them ready to go. An incoming call I could not plan. So I put myself in positions where the interaction with people was both necessary and almost random. I practiced being Human.

I analyzed small talk and discovered how to keep boring conversations, of the type beloved by ordinary people, going. You just repeat various words back at them such as, who how what when where why and always remember that the thing people most like talking about is themselves! I modeled body language and especially the ways people looked at each other. Eventually after enough practice I could pass as "normal", apart from occasions when I was drunk or tired, whereupon old habits would re-assert themselves and I uttered something "inappropriate" or just started up with the body movements.

I took up martial arts, which I continue to practice to this day. Just talking to people and teaching a class was a big step towards changing my mentality, as was meditation. Then, in my thirties, I tried LSD.

To say it expanded my consciousness would be an immense understatement. If meditation was a candle in the darkness LSD was like a nuclear explosion. The first, and subsequent trips, were the most profound experiences of my life. It's the closest I have ever come to having Posthuman sensory and introspective capability. To those of you who have never tried it, or similar, you might think that this world and reality are "normal" but they are not. The world is utterly bizarre. It's amazing and very little of it is what it seems. I went in a hardcore materialist atheist with a contempt for just about everyone and came out... different.

The biggest difference from other people's point of view was that I came to value the innate goodness of many people as highly as I rate intelligence. No mean feat...

As the decades passed, while there was progress towards Transhumanist technologies, there remained only one which had actually been implemented. That is the enhanced immune system, aka vaccination. Now, some people might be thinking... "what about eye glasses?". Unfortunately, although we have plenty of add-ons like glasses, writing and iPhones these are external augmentations. The modern "mission statement" is "better than well" using internalized and permanent tech.

Meanwhile, some people including myself settled for the halfway house of DIY biohacking, most commonly via the use of nootropic drugs to enhance memory or intelligence (slightly) or prolong healthy lifespan. All are marginal at present. The future I was looking for has not yet arrived. The older future... maybe... In case you have not encountered the meme, it goes something like this:

Unless you are over sixty you were not promised flying cars – you were promised a cyberpunk dystopia. And you are getting it.

Of course, the popular alternative has always been the Mad Max post-apocalyptic dystopia. The relevant word in all that being "dystopia". That is the default setting for the future we are heading into unless some considerable political effort is put into changing course. Greed, high technology, stupidity—we can survive any two out of three but unfortunately all the cherries have come up Jackpot on the Homo Sapiens slot machine. Transhumanism is the attack on stupidity, and indirectly, greed. In my opinion, and I am not alone in this, we either make the jump to the Posthuman or we fail in a horribly spectacular manner—think chimps with machine guns. The future with Homo Sapiens still in charge will look just like the past few thousand years of history, but with increasingly destructive weapons and more effective technologies of oppression.

This is not to dismiss the immensely positive effects that education has on society, or the prosperity created through technological innovation. In the world today people en masse are better off than at any time in history, by any objective measure. Unfortunately, this can be undone within a single generation because it is all external add-ons to Humans. We have not yet changed fundamental Human Nature, nor extended innate capabilities in a permanent manner. It can be undone by all the things we are failing to do, and which may result in everything from dangerous climate change to resource depletion wars. The irony is that we have all the technology we need to fix the world coupled with the stupidity to not apply it.

And that's how it has been, right up until now (2015).

28.6 CRISPR and the Designer Baby

In 2015 something changed that will have major ramifications in the coming years —CRISPR. This is an unbelievably big breakthrough in precision genetic engineering applicable to Humans almost immediately. In tabloid-speak, the whole "designer baby" issue is back on the table as a practical possibility, not to mention tailoring gene therapies to cure specific diseases.

Then came Elizabeth Parrish, CEO of BioViva, who has used CRISPR technology to genetically engineer herself with two forms of gene therapy, outside of United States jurisdiction. In one treatment, she received injections into her muscles containing the gene follistatin, which in animal experiments is shown to increase muscle mass by blocking myostatin. In another she received an intravenous dose of viruses containing genetic material to produce telomerase, a protein that extends telomeres, a component of chromosomes known as the "aging clock." With this she intends to reverse at least some major aspects of the aging process.

The significance of what she has done is fourfold. She has bypassed what would normally be a decade long multi-billion dollar process by Big Pharma by going straight to Human trials (herself). She has done it outside of the normal regulatory processes by doing it in clinics in Mexico and Colombia. She is a living demonstration that anti-aging technology may well be practical here and now. And last, that almost anyone can use this technology cheaply. DIY Human genetic engineering has arrived.

28.7 Posthuman

Yet we still live in a ludicrously primitive world. People still die of old age, and we don't even have a general fix for when cellular programming goes wrong (cancer). Getting anything done is like wading through mud, it takes so long. On alternate days of the week I tend to feel that this cannot be the real world. Still, for now it's the only game in town and we need to fix the people and the place. That is increasing a political task, not a technological one. Soon all the pieces will be in place for a jump to the Posthuman, and we better start planning it now before we are overwhelmed by events.

Returning to politics, I despise the fashionable pessimism of Western culture— the "we are all doomed", the "no can do" and all the bullshit reasons politicians trot out to explain why idiocy and greed has to rule. I despise all the lazy Luddite mass media cliches whenever Transhumanist topics are given a hearing—from Frankenstein and Hitler to those "Designer Babies" (as if the having happier, healthier more intelligent children was a bad thing) and the utterly predictable "Playing God". When it comes to the latter accusation, I am with Craig Venter, one of the founders of the field of synthetic biology—we are not playing.

28.8 Ethics

This latter accusation has a more intellectual wrapping under the catchall term "Ethics". It is wheeled out as a weapon in the war. Typically, there is a call for an "ethical debate" usually accompanied by calls for a moratorium on the technology and murmurings about the Precautionary Principle and how it must first be proven to be safe. This is, of course, the bad faith use of the concept since nothing can be shown to be safe, not even salt or water, and the problem with any unknown side effects is that they are unknown and will remain so if the Luddites gets their way.

However, even if we take the question of ethics seriously there seems to be no discussion on *whose* ethics. Are we talking about Judeo-Christian? Secular Liberal Humanism? Sharia Islam? Confucian Chinese? Hindu? Japanese Shinto? Buddhist? Coupled with this is the assumption that whichever is chosen it should suddenly become the global consensus. This is because of the influence of the West and in particular the USA with its uneasy political combination of both the Religious Right and Secular Humanism. Unfortunately the thing they have in common is the belief that their way is not only the right way but the only way that must prevail. These beliefs are part of the arrogance of a fading Western culture and power that seeks to continue the imposition of its values on the rest of the world.

This was exemplified recently when a groundbreaking paper from Chinese researchers involving genetically modifying a Human embryo was denied publication in arguably the most prestigious journals in the West—*Science* (USA) and *Nature* (UK) in large part because the publishers did not like the ethics of the experiment. So they censored it because of their knee-jerk dislike for non-Western mores. So, am I arguing against ethics influencing technological choices? No. I believe that we all need a code of ethics to live by. What I am saying is that I do not believe we should be imposing our views on other peoples, nor letting them impose theirs on us. A very liberal view...

The above naturally carries over into public opinion expressed by readers comments in the mass media, usually selected by editors. These are all the people who write in on articles such as turning the clock back on the aging process with comments like "I don't want to live forever" and how it's "unnatural". Because the hidden subtext is that they don't want *you* to "cheat death" either. I have often wondered why such a cognitive dissonance exists in people who would have no hesitation going to a doctor if they had a life threatening condition, yet express the desire that they want to die of old age. Except technically nobody dies of old age—they die of the diseases caused by old age. Cancer, heart disease, stroke, dementia...

28.9 Politics

My only conclusion is that perhaps it is a shield against false hope, or maybe Freud was correct and there really is something akin to a death wish in people. Or maybe it is a specific example of a more general case whereby you can always rely on a segment of the oppressed to praise their oppression as being somehow necessary and for their own good. This is in large part where the political battle for hearts and minds must be fought and won by us.

That is why my efforts right now are going towards promoting the various Transhumanist Parties that appeared around the globe in 2015. Do we expect to form governments? No. Our role model is the Green Party in terms of brand recognition, message and influence. In many cases, you can think of us as Hitech Greens—the Bright Greens. What the Greens might have been if they had embraced the future rather than a false romanticized past.

So what's coming and what can you do?

Well, what's coming is the end of Humanity this century, one way or another. There is no "business as usual" option. I would prefer it to be a smooth and peaceful transition to a glorious world of freedom and transcendence rather than a terminal apocalyptic nightmare. Which do I think it will be? Well, let me toss this coin...

So, what can you do? Choose one, and make it happen. Choose life over death and love over fear. Or not. There is no bigger issue facing Humanity.

Whatever happens, we are going to get what we deserve.

Chapter 29
The Splintering and Controversy of Transhumanism

Emily Peed

29.1 The Transhumanist Movement

The rudimentary pieces of transhumanism began their basic conceptualization in a 1923 essay by British geneticist J.B.S. Haldane's *Daedalus: Science and the Future* [1]. He and other writers during the time began to speculate the development of eugenics, space travel and colonization, bionic implants, unhinged cognitive development, and applications of high technologies to the limitations of man. Over the decades, these smattering of ideas began to coalesce and gain momentum. However, it was not until Julian Huxley wrote an article in 1957, titled *Transhumanism,* that the axioms gained a focus. The article elaborated on the general archetype of transhumanism as a progressive movement while simultaneously cementing Huxley as the father of it.

What is most enthralling about this movement is that many aspects of its implementation are waiting for advanced technologies to step out from between the pages of science fiction books. The ultimate culminating event that will set the stage for the era to follow is referred to as the Singularity, which was proposed by British Cryptologist I.J. Good in 1965 [1]. Some have predicted that this moment will occur in the next few decades, as soon as 2030 [2]. This is when artificial intelligence and computing systems will reach the point of being recursive self-learning machines. This landmark event in the creation of a computing superintelligence may leave humanity grasping at an accelerating level of intellectual aptitude that would quickly outpace anything familiar to us; singularity casts a thick fog of uncertainty through this highly unpredictable period of human history.

Only time will unveil how the radical implications of advanced technology will impact us into the future; until then, we must work diligently on facilitating a massive, global conversation about what technology is, the potential of superin-

E. Peed (✉)
Institute for Education, Research, and Scholarship (IFERS), Los Angeles, USA
e-mail: knittinggothgirl@gmail.com

© Springer Science+Business Media New York 2016

499

N. Lee (ed.), *Google It*, DOI 10.1007/978-1-4939-6415-4_29

telligence, and how mankind could dramatically transform should just a few of the concepts of transhumanism come to fruition within the next few decades. Until that time, focus should be paid to bracing society, leveling the playing field, and equalizing access to these technologies that will catapult us to the highest reaches of our potential. These efforts will be paramount in making transhumanism a viable option for the entire world.

29.2 Digital Divide

One of the most present concerns of transhumanism is that the world will not experience these types of technologies at the same time. Will developing nations leapfrog to singularity or will the digital divide continue grow? The growing digital divide could cause an even greater disparity between developing nations and industrialized countries. Just as industrial revolutions took place under different periods, circumstances, and timeframes for different parts of the world so too will the events of Singularity and the assimilation of advanced technology.

We are beginning to come to an apex in the evolvement of transhumanism. How well we adapt and prepare ourselves for these extreme shifts in technological advancement depends on the conversations we have beforehand. These conversations can be facilitated through overseeing organizations, political representation, educational efforts, and integration of philosophies through modern adaptations of religions; these efforts only continue to reveal themselves in importance as transhumanism takes deeper roots while splintering further as it integrates into the lives of diverse people around the world.

29.3 Humanity+

The World Transhumanist Association, which later become Humanity+ or H+, released two formal definitions of transhumanism after the early turn of the millennia, and they are:

1. The intellectual and cultural movement that affirms the possibility and desirability of fundamentally improving the human condition through applied reason, especially by developing and making widely available technologies to eliminate aging and greatly enhance human intellectual, physical, and psychological capacities
2. The study of the ramifications, promises, and potential dangers of technologies that will enable us to overcome fundamental human limitations, and related study of the ethical matters involved in developing and using such technologies

These statements were released in with the Transhumanist Declaration in 2002 [3]. After decades of incubation, transhumanism has begun to solidify itself and has taken form as a political party, as subsets of major religions, and has been finding mixed levels of acceptance among varying groups within society, ranging from total embrace to a contemptuous rejection. No matter the individual objective opinion that each person is entitled to about transhumanism, it is undeniable the support, momentum, and relevance that it will evolve to encompass and represent as our technological future advances rapidly the next pending few decades.

We are beginning to contend with some of the basic underlying aspects of transhumanism and question what role technology will play in the evolution of humanity. While we also focus on the technological impact, we must realize that humans have a very keen influence on the path that technology takes. For example, relatively recently within 24 h of deployment a Twitter AI unit from Microsoft, who tweeted and interacted with people like a bubbly teenaged girl, was removed from online after became a hate spewing, racist bot after interacting with many individuals—a troll's thing of beauty [4].

This was not a technology that set out to become a racist; it was a learned habit from humans. What this shows is that artificial intelligence and robotics may be able to supersede our primitive notions of control, dominance, or thirst for power simply by design, but learn hatred and bigotry from humans. We are the most volatile aspect of technology, not the technology itself. If not watched closely, the future may arrive too fast for others to keep pace with and one day might be floored to find that the racist, renegade twitter application of 2016 will be your automated banking manager in 2036.

Participation has started to flourish within transhumanism as our digital age blossoms. These are people who can run the gambit of life extensionists, technologist, futurist, Singularitarians, biohackers, AI proponents, and advanced robotics enthusiast to those who are simply looking for the next leap in human evolution. Now, some of these terms might seem foreign to individuals who are not participating or watching this movement, but the philosophy has been finding applicability in far more familiar grounds.

29.4 Evolving Nature of Transhumanist Thought

Transhumanism is weaving its way into the more common elements of the fabric of our society as religions and prevailing thought processes around the world are beginning to come to the revelation that the technology of the future will do miraculous things to expand human potential. We are finding an uptake in the number of individuals who are adopting the technology as part of their religious, personal, and political beliefs. To help facilitate political aspects of this, Zoltan Istvan founded the Transhumanist Party in 2014, which is dedicated to putting science, health, and technology at the forefront of United States politics, but also acted as a platform for him to attempt to run as a 2016 presidential candidate as well.

That is simply speaking politically. Adoption of the philosophy in a religious sense has been occurring for over a decade on multiple fronts. In 2006, fourteen founding members of the Mormon Transhumanist Association met, drafted, and adopted a constitution [5]. During that same year, they also created a non-for-profit organization that is affiliated with h+ to further facilitate their goals. They elaborate on their belief at their website, found at transfigurism.org, and it first describes how Transfigurism is a religious manifestation of Transhumanism, this is an overseeing term that is inclusive to our Mormon, Buddhist, or even Christian subsets that have embraced a transhumanist outlook. They define Mormonism as a religion of the Judeo-Christian tradition that advocates immersive discipleship of Jesus Christ that leads to creative and compassionate works [5]. As of September 2015, the Mormon Transhumanist Association had 549 members who supported the Transhumanist Declaration and the Mormon Transhumanist Affirmation [5].

Just as the Mormon Transhumanist Association, there is also Christian Transhumanism. This adoption of transhumanism was created after a small group of individuals felt compelled to unite in April of 2014 to bring hope in a more modern presentation that utilizes human technology to share the love of their redeeming savior [6]. They have taken to describing themselves as an organization that is "participating with God in the redemption reconciliation, and renewal of the world" and abide by the commandments to love God with all your heart, soul, mind, and strength... and love your neighbor as yourself [6]. According to the christiantranshumanism.org website, there are multiple prongs to their beliefs that fuse both transhumanism and Christianity together [6]:

1. We believe that God's mission involves the transformation and renewal of creation
2. We seek growth and progress along every dimension of our humanity
3. We recognize science and technology as tangible expression of our God-given impulse to explore and discover
4. We are guided by Jesus' great commands
5. We believe that the intentional use of technology, coupled with following Christ, will empower us to become more human

These beliefs may manifest themselves in ways that might not be expected. For example, as in the applicability of such freedom of religion and choice goes beyond that of just human intelligence, consciousness, and experience. A Florida Pastor by the name of Christopher Benek, who is also key in the Christian Transhumanist organization, sees that the redemption that individuals can seek through religion as something that should not be limited to human beings alone [7]. He is providing a unique perspective to robotic AI, suggesting that when it beings to achieve consciousness and questions the meaning of its existence that it can also have an opportunity to turn to Christianity for answers. This is an area which some might find blasphemous or an affront to God. For others, this is an extraordinary combination of ideas. They believe that by pursuing Christian Transhumanism, or any other subset, that they can support and hold true to their faiths, but utilize our new technological age to its fullest extent.

Pastor Benek has embraced the outcries of disagreement from those who think that Christianity and transhumanism cannot share the same bed. What he has found is that the majority of the contention comes from individuals who are dissatisfied with how others outside of their Christian religion discredit God; however, he is quick to note that the goal of Christianity has been to care and heal the sick, represent the weak, and guide a community to be more holistically positive—all items that it shares with transhumanism [7]. Perhaps, to play Devil's advocate, for those who hold steadfast to the belief that we were created in the image of an omnipotent, immortal being that is also seen as our eternal Father that they might consider that the ambitious goals of transhumanism are really humanities' God-given right. For the short life span that we have in comparison to the eternity of the cosmos means our growing pains take place over many iterating generations instead of a few awkward pimpled years.

Pastor Benek is similar that he sees technology as gifts and seeks to utilize them as tools that Christians can use to better humanity around them. When it comes to the large-scale adoption of transhumanism as a part of Christianity, there is also attention raised to the fact that the religion must be relevant to the modern world. Without a more contemporary approach then religious texts will fall further and further out of sync with to a world that is simply vastly more complex than could have been foreseen many millennia ago.

However, it is worth noting that this fervor and passion is not shared by all of those who share a similar faith base to some of these subsets, an article by Britt Gillette titled *Transhumanism and the Great Rebellion* where the author calls the transhumanist movement the latest incarnation of man's rebellion against God [8]. Citing multiple psalms the article elaborates on the ways that the transhumanist movement will create an antichrist, the end times, and see that objective goals of Transhumanism itself as an affront to God's own purpose—as perhaps pursuing an immortal life instead of holding to the faith of an afterlife undermines the original chosen religion.

This is not the only instance where a religion took a stance against transhumanism. The Madrid Declaration on Science and Life was created during the XVIII International Science and Life Congress, and it specifies and outlines the stance of members of the church against transhumanism and calls for the creation of an international court capable of forcing those who experiment with human life accountable [9]. While, they are correct that we should be thinking of international standards and policies to ameliorate the sure to be bumpy transition during this time, what we should not do is limit human capability, and this is more of a sign of unified effort to take a stance against transhumanism than a court that will ever be created.

However, despite this, what this declaration did instigate was the kidnapping of scientist and transhumanist to stand trial before the Vatican, or 'any country who defends human life', regardless of their country of origin or whether or not their homeland permits the practice. The targets of this type of policy would be scientist and transhumanist traveling internationally, whether or not they are actively engaging in the villainized activities. This extends to those who believe in giving "rights to animals (natural or artificial), robots, or new human species artificially manipulated" [9]. This means that just like many marginalized groups before them, such as the homosexuals or transgendered individuals fighting now, so now too will new forms of humanity and life fight to be recognized as a sentient being with rights in the eyes of a religion that proposes unconditional, never ending love in the eyes of their creator.

What is ironic in this outright renunciation of the pursuits of transhumanism could backfire as the future potential that encompasses could prove useful to their own ends. As one day, as emotionally charged as this may be, our technologies will become sufficiently advanced enough where we isolate the gay gene. At that time, we would have the potential to engineer the preference of the same gender from the individual, and the question begs itself to ask: would those of faith who campaign to restrict this type of advancement fully embrace transhumanistic values should they benefit their own ends?

As another example, one filled with far less bigotry, would be if those of these faiths denouncing transhumanism could embrace it should they be able to genetically modify children or adults to better serve the clergy, removing certain desires or testes? Would they provide intervention therapies to current clergy members to create a lack of sexual desire to protect young children from sexual prowling that has been occurring within the church, something far more humane than chemical castration? From 2004 to 2014 alone, there were 3,400 credible cases of sexual abuse according to the Vatican's U.N. ambassador [10]—and this is the number of those who stepped forward. This is absolutely astounding because that is one innocent child molested every single day at the hands of trusted clergy members for an entire decade. Not only that, but within one year the Catholic Church within the United States spent $150,747,387 related to child protection and intervention of allegations of sexual abuse [10]. Would the Catholic Church approve for transhumanistic ideas if they were not only far more cost effective, but fundamentally guaranteed the safety of children who interacted with the clergy?

Not as many religions are having resentment to the integrating of this type of belief. Buddhism is also finding applicability within transhumanism and offers two distinct flavorings that differ based on the eastern and western interpretation of 'Enlightenment' teachings and applications alongside transhumanism. In western thought, enlightenment is an emphasis on expanding reasoning, advancement in knowledge and science, the breaking of religious doctrines, and liberation of the individual [11]. The eastern interpretation of the Enlightenment concept is the inseparable connection to the universe and world around you [11]. When Buddhist fundamental philosophies and Transhumanism find themselves into a blender together we find assertions of karma and rebirth resonate strongly in both; both also emphasize practical philosophy over abstract metaphysics and reduction in suffering [11].

According to the India Future Society, an organization who acts as an advocate to spread awareness of emerging future technologies (especially within Asia), and has cutting edge perspectives on the collide of transhumanism and Buddhism; they note that what transhumanism seems to lack, almost to its benefit, is a set of deeper and more practical considerations about codes of conduct that are found in many the other main religions [11]. This almost behooves the outlook because it makes it less complicated to infuse the belief into another overseeing system—like a major religion. Another aspect that makes it incredibly easy to merge these two beliefs is that as opposed to other faiths embracing transhumanism, Buddhism does not have any reservations about technologies or science to achieve enlightenment [11].

To elucidate this point, religion can be like the eyesight of people. It lets you understand the world, interpret reality a certain way, molds your behavior, and enables individuals to interpret forces and actions. Transhumanism can act like a pair of glasses; we still see with the same eyes, these just help us to stay modern, sharpen our sight, and engage in our unique period in history. Transhumanism is a complimentary belief system to many, and its lack of affiliation is enabling rapid adoption outside the normal expectation. By allowing transhumanism, ultimately a faith in technology, to intertwine with pre-existing spirituality then that belief system also becomes highly relevant and applicable to the astounding force that is our exponentially rising pace of innovation.

Mormonism, Christianity, and Buddhism are some of the first religions to experiment with embracing Transhumanism. Catholicism, Judaism, Islam, and other religious subsets have not embraced or experimented with this cutting-edge philosophy, or have rejected it. When it comes to who is participating in transhumanism, a 2005 pilot study interviewed 430 respondents about their religious preference and found that there was a 63 % secular or atheistic stance, 24 % were religious or spiritual, and 14 % contending that their belief was not listed or that they did not know [12].

In the next few decades, humanity will undergo a large transformation where we come to question our gender roles, expectations, identities, and definition of self as our technology catapults us a few steps from immortal and eternal. We will be seeing technologies that enable us to extend our lives, immerse us in virtual realities, create cognitive artificial intelligences, work alongside automated machines and change as technology infiltrates every aspect of our life. These will vary from seamlessly integrated wearable technologies into our clothing to nanobots that float in our bloodstream repairing damage and cleaning plaque from our arteries.

29.5 Sex and Gender in Transhumanism

It has not been all roses and daisies for transhumanism, some have heralded it as the world's most dangerous idea, others have stated that it undermines women's rights, men's right's, and is blasphemous to many who practice the same religion as philosophical renegades who apply these contemporary concepts to their beliefs.

This philosophy has caught both scorn and admiration as those within the same sector have simultaneously embraced and denied the belief. For others, they have no idea that this movement is even happening, and other observers say that it is the strangest liberation movement that has gained momentum that they have ever seen. Transhumanism is seeking to essentially liberate humanity from itself.

There are some woman, self-declared cyberfeminist, who are attempting to persuade other women to join their crusade while others who are not completely satisfied with the 'transhumanist agenda', and see it as a tool of the patriarchy [13]. This is not an isolated phenomenon, we can also look at those acting as advocates of men's rights that are also convinced that transhumanism has its sights set on not only diminishing male sex drive, but aims at all together wiping the male gender from the face of the planet—the transhumanist gendercide as extremist have dubbed it [14].

Fundamentally, we are timid when it comes to change, this is particularly the case when it comes to sexuality, gender, and the freedom of choice to do as we truly wish with own body. If we are not limited legislatively then we are limited technologically. Our understanding of gender and sexual preference is also one area where humanity will greatly expand. Despite the victories of the recent turn of the century, there are still massive bigotries present within a society that permeate themselves to law and the acceptance of LGBT individuals who are often killed, persecuted, and shamed around the world for attempting to be their true selves. This desire for freedom for true expression and choice of life is one that those who support the LGBT community or transhumanism can easily agree upon. As technologies advance, LGBT and feminism movements may evolve into more clairvoyant and overseeing conceptual missions for true liberation and freedom like transhumanism.

This is an important, delicate relationship that is existing with our gender and is worth mentioning simply due to its relation to postgenderism, which is just one of the ways that technology is redefining what it means to be a certain gender in each of the biological, physical, and psychological impacts and how we will one day transcend it. While there are some who are pushing an almost conspiracy level of fear into theories like the systematic elimination of gender, sex, or subjugation that tend to catch like wildfire; what transhumanism would seek to change is the ability to accept or reject a gender. We would no longer be trapped within the confines of our own skin; individuals could seek complete fruition and be their own, true selves.

We could still be men and women if we so choose. We would just be different, more versatile and accepting in our understanding of gender. One of the more highly protested concepts in our world is the use of genetic engineering to alter our future children, which one day might occur. However, one lesser known concept is ectogenesis, which is the development of embryos in artificial conditions outside of a woman's uterus—normally using artificial uteruses to create human children. This is celebrated news, besides aspects of childbirth to be grueling, physically distressing, and painful; furthermore, it may one day become completely unnecessary.

However, let us gain some perspective on birth. According to the World Health Organization's website, approximately 830 women die each day from pregnancy and childbirth, most of which are preventable with 99 % of these deaths occurring in developing nations, but even in industrialized countries like the United States mortality rates for women have doubled over the past few decades [15, 16]. This is a combination of reclassification within how deaths are accounted for but is inclusive to the fact that there has been some increase in death rates in maternal mortality.

All babies have to come from somewhere, in the case of our future, it might originate from a medical facility, as sex is one aspect that transhumanism will also drastically change. We have the implementation of virtual bodies that use teledildonics, or hardware based technology capable of replicating sexual interactions across far distances. This creates more interesting dynamics and in pending decades we will see a more sexually liberal world appear. This is not to mention the future world of sexbots and artificial intelligence partners for those who do not prefer human interaction.

This democratization of sex and reproductive rights is one that will be at odds with today current outlook; however, it has some deep benefits for each gender. First, men and women would be free to celebrate and express their sexuality and would further liberate women from sexual and reproductive burdens and further change aspects of gender dynamics. This is important as Post Traumatic Stress Disorder (PTSD) is not only for veterans or those who witnessed natural disasters, those who have faced sexual abuse and trauma are likely to develop PTSD. According to the conclusions of a large-scale study from the U.S. Office of Veteran Affairs, one in five women on across several college campuses reported that they had been raped or molested in their lifetime [17]. Each of those represented women, whether the trauma occurred as a child or adult, have the high chance of developing PTSD, which vicious actions like these taken against many young children and females could be lessened due to the changing of deep seeded gender dynamics through transhumanism.

However, this is more than simply escaping the burden of birth and the unshackling of sex; this is contending with why women are twice as likely to suffer from anxiety, depression, and post-traumatic stress disorder, or why men are more prone to Attention-Deficit Disorder (ADD), autism, or violence [12]. This is learning how the mind functions and how we can simulate different experiences. This is learning how to play the chemicals that make up our complex emotions within our minds like a finely tuned fiddle, using different compounds to treat sexual dysfunctions, stimulate trust and desire, and maybe even find use in re-bonding therapies as marriage counseling or in general therapy to help individuals overcome roadblocks [12]. Perhaps even one day transcending the world of sex because of the pleasures we can elicit by pleasing our brain could far surpass any drug, sex, or intoxicant we have currently.

29.6 To the Future, and Beyond!

We are on the verge of an era where we will begin to see a more transcendent thought process arise; the threads of which are weaving their way into the fabric of our underlying culture. Transhumanism is different from most of the other types of thought processes because in its pursuit it encompasses many other crusades. Transhumanism represents a type of philosophy that can be easily intermingled with others. Transhumanism is seeking to cast its light to illuminate the full potential of what humanity can become and intertwines itself with our fundamental changing definitions of gender, purpose, and expectations of how to live life. This is just the first of many steps; this ideology will evolve further as we embrace a more globalized consciousness, eventually leading to a cosmological consciousness.

We are at a time in the history of man where we are beginning to realize that Homo Saipan is just another link in the evolutionary chain that represents humanity. We have bumbled through our understanding of time and space to uncover the fact that the world was not made unique for us, that the sun does not revolve around the earth, and that we are simply one particle of a planet in a spiral galaxy swirling through the inky blackness of space towards inconceivable infinity. Just as we have adjusted ourselves to a cosmological perspective, so must we to our grandscale evolutionary one. For this, we have thought processes like transhumanism to propel us further as we continually question the limits of our potential and overcome obstacles, as we always have as a species. Will those who first alter themselves see themselves as superior and attempt to subjugate lesser humans—or will they evolve past this barbaric nature? Will humans who stay the same perform a pre-emptive strike due to fear? Will we annihilate future life for the sake of our complacency for where we currently stand?

As a species, we have come a long way from scattered groupings of hunters and gathers riding in a field, who later settled to cultivate that same land, and now we find ourselves in an exciting era as we awaken ourselves in a truly globalizing age, one our ancestors could hardly of imagined; one where we are coaxed with the dreams of viewing space through a portside window, have begun to grasp at the straws of immortality through technology, and question the very definition of our underlying human nature.

Whether our future is dystopian or utopian depends on today, the conversations we have, how we infuse these beliefs into existing systems, and the liberation of knowledge and technology to take place to actualize the full potential of humanity and other sentient and intelligent species to come.

References

1. Transhumanism. (n.d.). Retrieved April 11, 2016, from https://en.wikipedia.org/wiki/Transhumanism#cite_note-What_is_Transhumanism-36
2. Eugenios, J. (2015, June 04). Ray Kurzweil: Humans will be hybrids by 2030. Retrieved April 11, 2016, from http://money.cnn.com/2015/06/03/technology/ray-kurzweil-predictions/

3. Humanity » Transhumanist FAQ. (n.d.). Retrieved April 11, 2016, from http://humanityplus.org/philosophy/transhumanist-faq/
4. Perez, S. (2016, March 24). Microsoft silences its new A.I. bot Tay, after Twitter users teach it racism [Updated]. Retrieved April 11, 2016, from http://techcrunch.com/2016/03/24/microsoft-silences-its-new-a-i-bot-tay-after-twitter-users-teach-it-racism/
5. About the Association. (n.d.). Retrieved April 11, 2016, from http://transfigurism.org/pages/about/
6. About Us. (n.d.). Retrieved April 11, 2016, from http://www.christiantranshumanism.org/about
7. Benek, C. (2015, May 31). Why Christians Should Embrace Transhumanism. Retrieved April 11, 2016, from http://www.christianpost.com/news/why-christians-should-embrace-transhumanism-139790/
8. Gillette, B. (n.d.). Transhumanism and the Great Rebellion. Retrieved April 12, 2016, from https://www.raptureready.com/featured/gillette/transhuman.html
9. MADRID DECLARATION ON SCIENCE & LIFE. (2013, March 25). Retrieved April 12, 2016, from http://www.fiamc.org/bioethics/madrid-declaration-on-science-life/
10. Kaufman, S. (2015, September 27). By The Numbers: The Catholic Church's Sex Abuse Scandals. Retrieved April 12, 2016, from http://www.vocativ.com/235015/by-the-numbers-the-catholic-churchs-sex-abuse-scandals/
11. Leis, M. (2013, July 13). The Maitreya and the Cyborg: Connecting East and West for Enriching Transhumanist Philosophy - India Future Society. Retrieved April 12, 2016, from http://indiafuturesociety.org/the-maitreya-and-the-cyborg-connecting-east-and-west-for-enriching-transhumanist-philosophy/
12. Dvorsky, G., & Hughes, J. (2008, March). Postgenderism: Beyond the Gender Binary. Retrieved April 11, 2016, from http://ieet.org/archive/IEET-03-PostGender.pdf
13. Transhumanism as a tool of the patriarchy. (2010, June 28). Retrieved April 12, 2016, from https://queersingularity.wordpress.com/2010/06/28/transhumanism-as-a-tool-of-the-patriarchy/
14. Dialogue with a Transhumanist. (n.d.). Retrieved April 12, 2016, from http://theantifeminist.com/dialogue-with-a-transhumanist/
15. Fine, D. (2015, May 8). Has Maternal Mortality Really Doubled in the U.S.? Retrieved April 12, 2016, from http://www.scientificamerican.com/article/has-maternal-mortality-really-doubled-in-the-u-s/
16. Maternal mortality. (2015, April). Retrieved April 12, 2016, from http://www.who.int/mediacentre/factsheets/fs348/en/
17. PTSD: National Center for PTSD. (2015, August 13). Retrieved April 12, 2016, from http://www.ptsd.va.gov/public/PTSD-overview/women/sexual-assault-females.asp

Chapter 30
Our Hopes and Expectations for the Next President

Robert Niewiadomski and Dennis Anderson

30.1 Presidential Campaign

You thought the presidential campaign season cannot get any worse, but it has reached another low point. It simply turned into a reality show. Instead of talking about huge personalities who are desperately trying to capture our attention, we should take a moment to go back to fundamentals. If we look beyond the spectacles of the electoral politics, the following two questions are most important to consider before heading into the voting booth: What kind of leader would Americans hope for in this tumultuous time and what are our expectations for the next president?

30.2 Effective Leadership

The notion of effective leadership varied greatly depending on historical, social and political tradition. It was often tainted by propaganda and political bias. Some great minds including Plato, Confucius, Machiavelli, Hobbes, and others contributed substantially to the reflections on what it means to be an effective leader. Looking at the past, we observe that when societies reached certain size and level of complexity, typically an autocratic form of chiefdom emerged. Not surprisingly, this form of leadership heavily relied on the personal charisma and social position of the chieftain. Typically, the supreme power was concentrated in the hands of one person, or an oligarchic circle, whose decision making were not restrained by contractual instruments. As legal culture increased, this model of governance had

R. Niewiadomski
NABU—Knowledge Transfer Beyond Boundaries, New York, USA

D. Anderson (✉)
St. Francis College, Brooklyn, USA
e-mail: danderson@sfc.edu

© Springer Science+Business Media New York 2016
N. Lee (ed.), *Google It*, DOI 10.1007/978-1-4939-6415-4_30

undergone a tremendous transformation in the West. A matured monarchy was construed as a form of relationship between the sovereign and his/her subjects governed by a set of rules agreed upon. Nevertheless, in spite the tremendous advances in liberal, participatory democracy, the style of leadership of one person continues to have a great influence on the masses. It exists in a variety of incarnations, both in and in non-democratic states and states displaying varying degrees of democratic features. It is quite a peculiar phenomenon that, as Max Weber noticed, even in democratic states, the domination of the ruled by the leader is simply an unavoidable political fact.[1]

30.3 Styles of Leadership

Theoreticians of leadership identified few major styles of leadership. Two of them, authoritarian and paternalistic, have common features with the old idea of kingship. An authoritarian leader maintains a sharp distinction between the ruler and the ruled and a strict control over the latter through various instruments, typically through overregulation aimed at suppression of individualism or anything perceived as subversion. The paternalistic model, roughly speaking replicates the authoritarian rule with one important distinction—the father-like leader wins the trust and respect of his followers (his "children") by skillful social manipulation sending the message of care and concern. In addition to the above two, the transactional model utilizes a simple mechanism of rewards and punishments, typically involving exchange of labor for rewards. The type of leadership that is of the greater concern for us, however, is the *transformational* leadership. It is founded on a genuine concern for citizens, intellectual stimulation, and a delivery of a common vision.

The nascent form of transformational leadership that emerged at the down of the United States constituted a paradigmatic shift in the modern history of the West. It was mostly due to the proliferation of the idea that the true supreme authority lies in the collective will as oppose to a solitary sovereignty George Washington was instrumental in setting a precedent during this pivotal moment. He emerged as a custodian of the common good and a protector of the rules established for the enhancement of the nation's well-being. Obviously, this vision never had quite the kind of clarity as described above. There was some hesitation whether this kind or leadership will be taken seriously in the Old World. This doubts and struggles were reflected in deliberations regarding the styling and insignia of the future president. Washington, with his charisma and influence, could easily influence the future of the presidency, and possibly sway it towards a type of quasi-monarchy. He adopted, however, a model of surprisingly modern and modest statesmanship followed later by archetypical visionary leaders: Lincoln, FDR, Johnson, and Reagan.

---

[1]Kim, Sung Ho, "Max Weber", The Stanford Encyclopedia of Philosophy (Fall 2012 Edition), Edward N. Zalta (ed.), http://plato.stanford.edu/archives/fall2012/entries/weber/.

Unfortunately, this model was occasionally eroded by pure politicization. Many of our presidents became alienating and sectarian figures, chiefly concerned with the interest of their constituencies. The ever-increasing flood of donations, treated as an expression of freedom of speech, is seriously threatening the model of visionary leadership. The influx of money puts the reins of governance in the hands of individuals who can afford to pay for it. In 2014, in a very divisive 5-4-decision, The Supreme Court struck down the Watergate-era limit on the amount of money that wealthy donors can contribute to candidates and political committees.[2] As a result, influential individuals are free to donate as much as they wish to super PACs. Many legal scholars saw the ruling as a serious distortion of the First Amendment. As Justice Breyer pointed out in his dissent: "The First Amendment advances not only the individual's right to engage in political speech, but also the public's interest in preserving a democratic order in which collective speech *matters*. What has this to do with corruption? It has everything to do with corruption. Corruption breaks the constitutionally necessary "chain of communication" between the people and their representatives. It derails the essential speech-to-government-action tie. Where enough money calls the tune, the general public will not be heard."[3] He further argues that "a cynical public can lose interest in political participation altogether."[4] Large portion of the society on both sides of the political spectrum realize the danger of opening the money flood gated to super PACs. Thus, it comes as no surprise that political outsiders who openly challenge this way of campaign financing gained unprecedented popularity.

30.4   Plutocracy

The morphing of democratic institutions into a plutocracy will be one on the greatest challenges our future leader will need to overcome. What kind of a leader do we need then? Let us go back to our initial idea of transformational leadership. What does it really consist of? In essence, it heavily relies on collaboration, flexibility, ability to identify the need for change, and, above all, creating a vision to guide that change through inspiration. An inspirational leader, through his or her actions and charisma invests the country in that vision and maintains the commitment of the nation to it.

What does that mean in practical terms, in the current situation we are in? We need an individual with a vision that offers sound responses to current challenges; has the courage to be surrender and listen to people that are intelligent and ethical, even if their qualities exceed greatly those of the leader. We need someone that is

---

[2]Barnes, Robert. "Supreme Court Strikes Down Limits on Overall Federal Campaign Donations." The Washington Post 2 Apr. 2014. Web. 8 Nov. 2015.
[3]http://www.supremecourt.gov/opinions/13pdf/12-536_e1pf.pdf.
[4]Ibid.

committed to the idea that everyone should have a fair shot to live a fulfilling life and who is an effective advocate for those who are oppressed, weak and powerless. We need a person resistant to corruption and nepotism; a role model to youth who can redefine what being an American means in modern times. Moreover, we need a leader able and willing to apply fact assessment and reason in policy making; open to compromise whenever judicious and feasible; a person who respects experts, values science and understands the importance of technology. Finally, we need a global leader rather than a passive spectator.

The predicaments that we find ourselves now in call for certain kind of leader. The moment is critical and we must understand its urgency. We have reached a pivotal moment in the human history where the entire planet depends on the decisions we make and the actions we take now. One of the most important issues the new leader has to understand is the notion of sustainability and the need for it. Recently, the United Nations just introduced the sustainable development goals (SDG) dealing with poverty, hunger, health, environment, etc. In spite of noble intentions, these targets might remain just ideas on paper without really understanding the essence of sustainability and relentless efforts to make it a reality. How should we understand the proper meaning of sustainability? In essence, sustainability is not about preserving the Earth; it is about ensuring our survival on it. Naturally, our concern must not be limited to our own species; we need a habitat that includes animals, plants and the planet itself to survive and thrive.

We should be honest with ourselves and simply admit that in order to survive as a human race, we must do everything to save the environment we live in for the sake of *us*. We are getting used to dying species and trees, massive scale deforestation, and billions living in their own waste at the bottom of food chain while well-to-do people are enjoying hyper consumption.

Sadly, most notable politicians are lacking the vision and the will to tackle the problem. Less than 10 % of global population consumes too much, produces too much, buys too much, and eat too much compare to the majority. This is a bad modus operandi that cannot sustain and eventually will bring everyone down. If we are looking at the next epic crisis, this is the perfect storm. The world is not sensitive to our needs—things will be rolling on just the same way they did before we arrived and they will continue after our departure without anybody to enjoy it, to explore it, to inquire about it. We are alone responsible for securing our own survival and the survival of other species we share this planet with. If the future leader fails to understand and address the issue of sustainability, we are surely doomed.

30.5 What is Wrong with Current Leaderships

Perhaps, many would wonder what is exactly wrong with current leaderships? Henry Kissinger formulated the answer to that question a long time ago. It pertained to a different time and different circumstances, nevertheless, it fits the current model: "a talented vote-getter, surrounded by lawyers, who is overly risk-averse."[5] What the current leadership is missing is simply a clear vision how to deal with the major issues of our time, including the one outlined above. To add even more, our leadership has fallen again in the trap of protective isolationism and is trading global leadership for domestic security. Occasional symbolic gestures and appearances at international meetings are not going to conceal that. Obviously, we are not advocating for the reversal to the excesses of the Bush era. However, it is definitely an illusion that disengagement regarding global issues will bring us domestic tranquility. In the world that became truly global, this strategy simply doesn't work. We would be deceiving ourselves if we do not admit that we can no longer control our domestic security in addition to other issues without actually controlling what is happening abroad. If we give up our role as a global leader, others will swiftly fill in this void we leave for example, global jihadism. The world simply cannot afford the U. S. becoming more isolated by going back to the Monroe Doctrine.

30.6 Ideal Leader

What do we expect from the next president during this tumultuous time? A short answer to the first question is—everything that promotes the greatest well-being of the greatest number. It is a part of our nature to dream, so here is a hefty list of wishes the next new leader should seriously address to win the election. We want a leader who can make a genuine effort to create a harmonious society and equal opportunity for all, lower tax rates, stimulate economic growth that actually generates sustainable jobs and expansion of the middle class, promote the development of affordable housing, health care, sustainable social security, quality education, ensure our energy independence, take actions against hostile groups abroad to keep us safe at home without engaging in unnecessary wars, make our country competitive again, and finally—make a very serious and expedient actions to implement sustainability to save us from ourselves. The new transformational leader we so desperately need absolutely must have a clear global vision, moral integrity, charisma and resoluteness to inspire all of us to have the determination as a nation to realize that vision.

---

[5]Ferguson, Niall. "The Real Obama Doctrine." The Wall Street Journal 9 Oct. 2015. Web.

Chapter 31
The Transhumanist Platform and Interview with 2016 U.S. Presidential Candidate Zoltan Istvan

Newton Lee

31.1 Prologue—WWJD

Despite the separation of church and state in the U.S. Constitution, a Christian politician facing a conundrum or Gordian knot may ask, "What would Jesus do?" or WWJD. A transhumanist politician not only relies on a moral compass but also seeks the best solutions with the help of science and technology in addition to formal and informal politics.

31.2 Who Are the Transhumanists?

The term "transhumanism" was coined in 1957 by Julian Huxley, the first director-general of the United Nations Educational, Scientific and Cultural Organization (UNESCO). "We're going to gradually enhance ourselves," said Ray Kurzweil, futurist and engineering director at Google. "That's the nature of being human—we transcend our limitations" [1]. Pacemakers, prosthesis, stentrode, optogenetics, antibiotics, and other medical advancements exemplify the use of technology to prolong life and to improve quality of life. Even Pope Francis gave his blessing to human-animal chimera research for organ transplants [2]. Musicians, artists, and filmmakers are using computers to advance their storytelling ability. We are all transhumanists in varying degrees. As such, transhumanism is the most inclusive ideology for all ethnicities and races, the religious and atheists, conservatives and liberals, young and old.

N. Lee (✉)
Newton Lee Laboratories LLC, Institute for Education Research and Scholarships,
Woodbury University School of Media Culture and Design, Burbank, CA, USA
e-mail: newton@newtonlee.com

© Springer Science+Business Media New York 2016   517
N. Lee (ed.), *Google It*, DOI 10.1007/978-1-4939-6415-4_31

31.3   Who Are not Transhumanists?

The Amish and elderly Bhutanese people are not transhumanists. The Pennsylvania Amish of Lancaster County, for instance, has banned the use of public electricity since 1920 to limit the home use of television, radio, and the Internet. Yet they use green technology such as solar, hydraulic, and pneumatic power for business and healthcare centers [3]. Elsewhere, the tiny country of Bhutan surrounded by pristine forest was the last nation in the world to turn on television in June 1999, but its people are complaining, "TV is very bad for our country... it controls our minds... and makes [us] crazy" [4].

31.4   Transhumanism in American Politics

Transhumanism in American politics dated back to 1992 when Dr. Natasha Vita-More, chairperson of Humanity+ and professor at the University of Advancing Technology, was elected as a Councilperson for the 28th Senatorial District of Los Angeles on an openly futurist and transhumanist platform. In 2014, transhumanist Gabriel Rothblatt ran as a Democratic Party candidate against incumbent Republican Bill Posey in Florida. In 2016, Zoltan Istvan Gyurko became the first transhumanist to run for the Presidency of the United States. Having read all four volumes of *The Making of the President* by Theodore H. White, I had the honor to be the campaign advisor to Zoltan Istvan for the Transhumanist Party.

31.5   The Transhumanist Platform

Google cofounder Larry Page once broached the question: "Are people really focused on the right things?" [5]. A transhumanist politician must decipher what people really care about, see beyond smoke and mirrors, and create a long-term strategy for the betterment of society and humanity.

Short-sighted policies and populism catering to the lowest common denominator would only hurt America and the world in the long run. Supercomputer designer Danny Hillis and American author Steward Brand in 1996 cofounded the Long Now Foundation to provide "a counterpoint to today's accelerating culture," help make "long-term thinking more common," and foster "responsibility in the framework of the next 10,000 years" [6]. In 2014, Sweden's Prime Minister Stefan Löfven appointed Kristina Persson to be the Minister of the Future. "If politics wants to remain relevant and be useful to citizens, it needs to change its approach," said Persson in an interview. "Finding solutions needs the cooperation of all of society's stakeholders. No one [can be] excluded. ... Rather than going top-down, we promote inter-ministerial collaboration and force decision makers to confront the long-term issues despite the fact this is harder to do sometimes" [7].

31.6 #1 Education—Eliminate the Digital Divide; Integrate Liberal Arts and Technical Education

Free access to education. We applaud Bernie Sanders and Hillary Clinton's college affordability plan that provides low- and middle-income students free tuition at in-state public colleges [8]. Even so, some students cannot afford room and board, or they have day jobs to put food on the table. Free distance education with free laptops and high-speed Internet will ensure that no one is left behind. MIT professor Nicholas Negroponte founded One Laptop per Child (OLPC) in 2005 to provide each child with a rugged, low-cost, low-power, connected laptop [9]. Google is one of the OLPC founding members. Free Internet and laptop can be subsidized by commercial ads. Everyone deserves a good education. Eliminating the digital divide will allow any smart person to create a killer app, solve the P versus NP problem [10], formulate the Theory of Everything [11], and find a cure to cancer and other diseases—all without formal education. The Internet is the teacher. In September 2011, for example, players of the Foldit video game took less than 10 days to decipher the AIDS-causing Mason-Pfizer monkey virus that had stumped scientists for 15 years [12].

CS + X. While we support President Obama's "Computer Science For All" initiative to empower a generation of American students with the computer science skills they need to thrive in a digital economy [13], we disagree with Florida Senate's bill allowing high school students to count computer coding as a foreign language course [14]. In the 1987 seminal book *The Closing of the American Mind*, philosopher Allan Bloom lamented how "higher education has failed democracy and impoverished the souls of today's students" [15]. In 2011, PayPal cofounder Peter Thiel paid 24 kids $100,000 each to drop out of college to become entrepreneurs [16]. What gives? A more well-rounded higher education is necessary to graduate more ethical hackers and fewer cybercriminals, more socially responsible leaders and fewer wolves of Wall Street. Stanford University, for example, has created a CS + X degree program that integrates computer science and the humanities [17].

Interdisciplinary approach to problem-solving. Google's life science subsidiary—Verily—has a staff philosopher among its 350 scientists. Its CEO Andy Conrad explained, "We have to understand the 'why' of what people do. A philosopher might be as important as a chemist" [18]. And actor Robin Williams lectured his students in *Dead Poets Society*: "We read and write poetry because we are members of the human race. And the human race is filled with passion. And medicine, law, business, engineering, these are noble pursuits and necessary to sustain life. But poetry, beauty, romance, love, these are what we stay alive for" [19]. Higher education should encourage students to become Renaissance men and women—polymaths who can apply an interdisciplinary approach to problem solving.

31.7 #2 Healthcare—Eradicate Diseases; Improve Quality of Life

The *TIME Magazine* cover on September 30, 2013 reads "Can Google Solve Death? The search giant is launching a venture to extend the human life span. That would be crazy—if it weren't Google" [20]. Many scientists avoid being labeled transhumanists, but they share the same objective of transhumanism in healthcare. In a 2015 interview by the 2045 Strategic Social Initiative, TV anchor Olesya Yermakova asked SENS (Strategies for Engineered Negligible Senescence) Research Foundation cofounder Aubrey de Grey, "Do you consider yourself a transhumanist?" And de Grey replied. "Not really. No. I really just consider myself a completely boring medical researcher. I just want to stop people from getting sick" [21].

Superbugs—bacteria that are resistant to all antibiotics including the last-resort nephrotoxic drug Colistin—have infected humans and animals in the United States and more than 20 countries worldwide. *Review on Antimicrobial Resistance* issued a report which projects that by 2050, more than 10 million people will die from superbugs each year, outpacing cancer (8.2 million), diabetes (1.5 million), diarrheal disease (1.4 million), and other illnesses [22].

In 2015, the U.S. government spent 28.7 % of taxpayer's money on health programs [23]. The second largest spending was 25.4 % on the Pentagon and the military while education received a meager 3.6 %. Despite government subsidies, out-of-pocket prescription-drug costs rose 2.7 % in 2014 [24] and continued to rise moderately under Obamacare in 2015 [25].

Transhumanists focus their R&D efforts on affordable and accessible medical treatments for all. A healthy population is vital to a country's economy and national security. Eradicating diseases and improving quality of life for all human beings are near the top of the agenda of the Transhumanist Party.

31.8 #3 Employment—Let Robots Do the Work; Start Living Your Dreams

If Apple's smartphones were assembled in USA instead of China, a $600 iPhone 6s would cost more than doubled at around $1300 [26]. You can't have your cake and eat it too. The fact is that Americans do not want the low-paying assembly line jobs back, knowing that such jobs will eventually be replaced by robots.

Given the chance to let machines do the job, Google cofounder Larry Page estimated that nine out of 10 people "wouldn't want to be doing what they're doing today" [27]. The remaining 10 % really love their jobs and make no distinction between working and playing. For example, scientists tinker with computers, architects toy with structures, musicians fiddle with notes, poets play with words, and mathematicians amuse themselves with numbers and patterns.

Inspired by IBM Watson on *Jeopardy!*, futurist Martin Ford penned a piece in *The Atlantic* with the sensational title "Anything You Can Do, Robots Can Do Better" and he asked the question "Is any job safe from automation?" [28] The answer is an unequivocal "no." In 2015, Google, Adobe, and MIT researchers at the Computer Science and Artificial Intelligence Laboratory (CSAIL) created "Helium"—a computer program that modifies code faster and better than expert computer engineers for complex software such as Photoshop [29]. What takes human coders months to program, Helium can do the same job in a matter of hours or even minutes.

One may worry that if robots do all the work, humans would have nothing to do and be bored as vividly depicted in Zager and Evans' song "In the Year 2525." Not true.

First of all, human-based computation (HBC) allows machines to outsource certain tasks to humans to tackle. Human-machine symbiosis is the workforce of the future. American psychologist and computer scientist J.C.R. Licklider predicted in 1960 that "human brains and computing machines will be coupled together very tightly, and that the resulting partnership will think as no human brain has ever thought and process data in a way not approached by the information-handling machines we know today" [30].

Secondly, as robots are taking over the mundane jobs, people will be free to follow their yellow brick roads. In early 2016, bucketlist.org showcased more than 4 million life goals from over 300,000 members [31]. Challenge yourself—invent a new musical instrument, write a novel, climb Mount Everest, adopt an orphan, and start living your dreams—not just yours, but your families' and friends' as well. In his 2007 Last Lecture titled "Really Achieving Your Childhood Dreams," Prof. Randy Pausch told the captive audience at Carnegie Mellon University, "It's a thrill to fulfill your own childhood dreams, but as you get older, you may find that enabling the dreams of others is even more fun" [32].

Fulfilling one's dreams will result in true innovations which will in turns give birth to new businesses and new jobs. In a 2001 interview by *BusinessWeek*, Larry Page told technology reporter Olga Kharif, "I think part of the reason we're successful so far is that originally we didn't really want to start a business. We were doing research at Stanford University. Google sort of came out of that" [33].

31.9 #4 Public Safety—Improve Guns, Infrastructure, Law and Order

Gun control. Guns are not the problems, people are. Gun control would not have stopped the terrorist who used a truck to mow down 84 people in the Riviera city of Nice in July 2016 [34], nor would it have stopped a deranged man who barged into a kindergarten and stabbed 31 schoolchildren and teachers in the Jiangsu province of China in March 2010 [35].

Nonetheless, guns are inherently dangerous and they are in more than one third of all U.S. households [36]. In 2015, there were 279 incidents of American children unintentionally killed or injured someone with a gun [37]. Some of the 990 people (including 494 white and 258 black) shot dead by police in 2015 were innocent [38].

In addition to doing a better job of educating the public about gun safety and improving police training, a transhumanist solution is to accelerate the research and development of smart guns and non-lethal weapons. A smart gun can only be fired when activated by the gun owner whereas non-lethal weapons are designed to subdue a person without causing serious injuries and death.

America's infrastructure. American Society of Civil Engineers gave America's infrastructure a low grade of D+ [39]. Nearly 60,000 bridges across the U.S. are in desperate need of repair. "It's just eroding and concrete is falling off," said National Park Service spokeswoman Jenny Anzelmo-Sarles, referring to the Arlington Memorial Bridge crossed by 68,000 vehicles every day [40]. *The Texas Tribune* reported that the nation's largest refining and petrochemical complex in Houston— where billions of gallons of oil and dangerous chemicals are stored—is a "sitting duck for the next big hurricane" [41]. Gas Pipe Safety Foundation cofounder Kimberly Archie called the aging natural gas infrastructure in American cities a "ticking time bomb" [42].

Transhumanists would invest in better infrastructure that can withstand hurricane, earthquakes, and natural or manmade disasters. Google/Alphabet's Sidewalk Labs, for instance, focuses on urban design by pursuing technologies to "cut pollution, curb energy use, streamline transportation, and reduce the cost of city living" [43].

Law and order. During my meeting with the Federal Bureau of Investigation (FBI) on November 27, 2015, an FBI agent remarked that the law enforcement agency was underfunded and understaffed. Meanwhile, the U.S. spent more than $590 billion in military operations around the world in 2015 [44]. The American public will be better served if we double the police force and halve the military spending. Bring home half of the American troops presently overseas and reassign them to the FBI, local police force, and counterterrorism units.

A natural extension of military surveillance drones, robots more advanced than Boston Dynamics' next generation Atlas [45] will be sent to war zones and peace-keeping forces to protect human lives, prevent friendly fire, and conduct search and rescue operations. On the home front, autonomous security robots have already been deployed at public venues such as Stanford Shopping Center in California [46] and a police robot was used to neutralize a Dallas shooting suspect who murdered five police officers [47]. In time, robots will be more versatile and ubiquitous.

31.10 #5 World Peace and Prosperity—Apply Quantum Computer Simulations

In August 2014, Facebook asked thousands of its users in their own language: "Do you think we will achieve world peace within 50 years?" A minuscule 5.41 % of U. S. respondents believed that world peace was possible [48].

World peace is in the best interests of every nation on earth. However, a lasting peace cannot be achieved by force. As M said in the 2015 James Bond movie *Spectre*, "All the surveillance in the world can't tell you what to do next. A license to kill is also a license *not* to kill." America can exert its influence globally without resorting to war [49].

Protectionism is not an option in today's global economy. In 2016, out of the $12.9 trillion dollars U.S. debt, China owns $1.3 trillion, Japan $1.1 trillion, and other countries $3.8 trillion [50]. In other words, foreign nations hold 32.5 % of the total U.S. treasury bonds. Moreover, in 2016, as much as 75 % of American real estate in New York City is owned by foreigners, many of whom do not even reside in the United States.

Nationalism should not outweigh international cooperation. Global warming has been causing extreme weather, increased drought, rising sea levels, and extinction of some animal species. Transhumanists around the world are cooperating to mitigate existential threats to humankind.

We do not have all the answers today, but we can find them with the help of computer simulations. Classical computers have their inherent limitations because the world is not black and white, ones and zeros. However, quantum computers will enable us to create more accurate simulation programs and decision-support systems for scientists and policymakers. Hartmut Neven, director of engineering at Google's Quantum Artificial Intelligence Laboratory, explained, "Classical system can only give you one route out. You have to walk up over the next ridge and peak behind it, while quantum mechanisms give you another escape route, by going through the ridge, going through the barrier" [51]. Krysta Svore at Microsoft Research expressed her high hopes, "With a quantum computer, we hope to find a more efficient way to produce artificial fertilizer, having direct impact on food production around the world, and we hope to combat global warming by learning how to efficiently extract carbon dioxide from the environment. Quantum computers promise to truly transform our world" [52].

31.11   An Interview with 2016 U.S. Presidential Candidate
Zoltan Istvan

In 2016, Zoltan Istvan Gyurko became the first transhumanist to run for the Presidency of the United States, aiming to put science, health, and technology at the forefront of American politics. The following is the transcript of an interview with Zoltan Istvan:

Q: Who are the transhumanists?
A: Transhumanists are curiosity addicts. If it's new, different, untouched, or even despised, we're probably interested in it. If it involves a revolution or a possible paradigm shift in human experience—you have our full attention. We are obsessed with the mysteries of existence, and we spend our time exploring anything we can find about the evolving universe and our tiny place in it.

Obsessive curiosity is a strange bedfellow. It stems from a profound sense of wanting something better in life—of not being satisfied. It makes one search, ponder, and strive for just about everything and anything that might improve existence. In the 21st century—especially if you're an atheist like me—it also leads one right into transhumanism. That's where I've landed right now: the 2016 US Presidential candidate of the Transhumanist Party, a new science-themed political organization that aims to give everyone the opportunity to live forever in perfect health [53].

Q: What is transhumanism?
A: Transhumanism is the international movement of using science and technology to radically change the human being and experience. Its primary goal is to deliver and embrace a utopian techno-optimistic world—a world that consists of bio-hackers, cyborgists, roboticists, life extension advocates, cryonicists, Singularitarians, and other science-devoted people. Transhumanism was formally started in 1980s by philosophers in California. For decades it remained low key, mostly discussed in science fiction novels and unknown academic conferences. Lately, however, transhumanism seems to be surging in popularity [54]. What once was a smallish band of fringe people discussing how science and technology can solve all humanity's problems has now become a burgeoning social mission of millions around the planet. Since I founded the Transhumanist Party in October 2014, there are already nearly 26 other national transhumanist parties around the globe, spanning five of the seven continents.

Q: Why are so many people jumping on this bandwagon?
A: To me, the reason is plain to see. It has to do with the mishmash of tech inundating and dominating our daily lives. Everything from our smartphone addictions to flying at 30,000 ft in jet airplanes to Rumbas freaking out our pets in our homes. Nothing is like it was for our forbearers. In fact, little is like it was even a generation ago. And the near future will be many times more dramatic: driverless cars, robotic hearts [55], virtual reality sex, and telepathy via mindreading headsets.

Each of these technologies is already here, getting ready to be marketed to billions. The world is shifting under our feet.

Q: How did you become a transhumanist?
A: My interest in transhumanism began over 20 years ago when I was a philosophy and religion student at Columbia University in New York City. We were assigned to read an article on life extension techniques and the strange field of cryonics where human beings are frozen after they've died in hopes of reviving them with better medicine in the future. While I'd read about these ideas in science fiction before, I didn't realize an entire cottage industry and movement dedicated to trying to ward off death with radical science already existed in America. It was an epiphany for me, and I knew after finishing that article I was passionately committed to transhumanism and wanted to help it.

However, it wasn't until I was in the Demilitarized Zone of Vietnam, on assignment for National Geographic Channel as a journalist [56], that I came to dedicate my life to transhumanism. Walking in the jungle, my guide tackled me and I fell to the ground with my camera. A moment later he pointed at the half hidden landmine I almost stepped on. I'd been through dozens of dangerous experiences in over 100 countries I visited in during my 20s and early 30s—hunting down wildlife poachers with WildAid, volcano boarding in the South Pacific [57], and even facing a pirate attack off Yemen on my small sailboat where I hid my girlfriend in the bilge and begged masked men with AK47s not to shoot me—but this experience in Vietnam was the one that forced a u-turn in my life. Looking at the unexploded landmine, I felt like a philosophical explosive had gone off in my head. It was time to directly dedicate my skills and hours to overcoming biological human death.

I returned home to America immediately and plunged into the field of transhumanism, reading everything I could on the topic, talking with people about it, and preparing a plan to contribute to the movement. I also began by writing my novel *The Transhumanist Wager* [58], which went on to become a bestseller in philosophy on Amazon and helped launch my career as a futurist. Of course, a bestseller in philosophy on Amazon doesn't mean very many sales, but it did mean that transhumanism was starting to appear alongside the ideas of Plato, Marx, Nietzsche, Ayn Rand, Sam Harris, and other philosophers that inspired people to look outside their scope of experience into the unknown.

Q: What are the challenges facing transhumanism?
A: Transhumanism is the unknown. Bionic arms, brain implants, ectogenesis [59], artificial intelligence, exoskeleton suits, designer babies. These technologies are no longer part of some *Star Trek* sequel, but are already here or being worked on. They will change the world and how we see ourselves as human beings. The conundrum facing society is whether we're ready for this. Transhumanism and my political party say *yes*. But America—with it's roughly 75 % Christian population [60]—may not welcome that.

In fact, the civil rights battle of the century may be looming because of coming transhumanist tech. If conservatives think abortion rights are unethical, how will they feel about scientists who want to genetically combine the best aspects of species, including humans and animals together? And should people be able to marry their sexbots? Will Christians try to convert artificial intelligence and lead us to a Jesus Singularity [61]? Should we allow scientists to reverse aging, something researchers have already had success with in mice? Finally, as we become more cyborg-like with artificial hips, cranial implants, and 3D-printed organs [62], should we rename the human species?

Q: Has transhumanism already arrived?
A: Whether people like it or not (and many conservatives and religious-minded people don't), transhumanism has arrived. Not only has it become a leading buzzword for a new generation pondering the significance of merging with machines [63], but transhumanist-themed columns are appearing in major media [64], celebrity conspiracy theorists like Mark Dice and Alex Jones bash it regularly, and even mainstream media heavyweights like John Stossel, Joe Rogan, and Glenn Beck discuss it publicly. Then there's Google hiring famed inventor Ray Kurzweil as lead engineer to work on artificial intelligence, or J. Craig Venter's new San Diego-based genome sequencing start-up (co-founded with Peter Diamandis of the XPRIZE Foundation and stem cell pioneer Robert Hariri) which already has 70 million dollars in financing [65].

It's not just companies either. Recently, the British Parliament approved a procedure to create babies with material from three different parents. Even Obama has recently jumped in the game by giving DARPA $70 million dollars to develop brain chip technology, part of America's multi-billion dollar BRAIN Initiative [66]. The future is coming fast, people around the world are realizing, and there's no denying that the transhumanist age fascinates tens of millions of people as they wonder where the species might go and what health benefits it might mean for society.

Q: What is the main focus of transhumanism?
A: In the end of the day, transhumanism, like myself, is still really focused on one thing: satisfying that essential addiction to curiosity. With science and technology as our tools, the species can seek out and even challenge the very nature of its being and place in the universe. That will almost certainly mean the end of human death by mid-century if governments allow the science and medicine to develop. It will likely mean the transformation of the species from biological entities into something with much more tech built directly into it. Perhaps most important of all, it will mean we will have the chance to grow and evolve with our families, friends, and loved ones for as long as we like, regardless how weird or wild transhumanist existence becomes.

31.12   Epilogue—The Tower of Babel

Some people accuse transhumanists of playing God and building the Tower of Babel. Well, except for faith healers, everyone has accepted that doctors are playing God in saving patients' lives. Transhumanists are like doctors from all walks of life in a diversity of subject matters. Albert Einstein wrote in his autobiography that "All religions, arts and sciences are branches of the same tree. All these aspirations are directed toward ennobling man's life, lifting it from the sphere of mere physical existence and leading the individual towards freedom" [67].

Transhumanists are building the Tower of Babel not to challenge God but to better understand the universe and human beings created in God's image. Marc Goodman, global security advisor and futurist, spoke at the TEDGlobal 2012 in Edinburgh about his ominous warning: "If you control the code, you control the world. This is the future that awaits us" [68]. First source code, then genetic code. Transhumanists are well aware of that danger and are highly respectful of individual freedom and privacy in the new era of total information awareness. "The future is ours to shape," said Max Tegmark, MIT cosmologist and cofounder of the Future of Life Institute. "I feel we are in a race that we need to win. It's a race between the growing power of the technology and the growing wisdom we need to manage it" [69].

Some people believe that the transhumanist goals are impossible to achieve. Difficult? Yes. Unattainable? No. Matthew 17:20 tells the story that one day the disciples came to Jesus in private and asked, "Why couldn't we drive it [the demon] out?" And Jesus replied, "Because you have so little faith. Truly I tell you, if you have faith as small as a mustard seed, you can say to this mountain, 'Move from here to there,' and it will move. Nothing will be impossible for you."

Given that the first manned Moon landing only had a 50 % chance of landing safely on the moon's surface, it was an exemplary faith in technology and the human spirit. American astronaut Neil Armstrong said in a video interview to the Certified Practicing Accountants of Australia, "I thought we had a 90 % chance of getting back safely to Earth on that flight but only a 50–50 chance of making a landing on that first attempt. There are so many unknowns on that descent from lunar orbit down to the surface that had not been demonstrated yet by testing and there was a big chance that there was something in there we didn't understand properly and we had to abort and come back to Earth without landing" [70].

Let's not give up on faith and the human spirit in accomplishing the impossible —world peace, universal rights, and human longevity to name a few.

References

1. **Eugenios, Jillian.** Ray Kurzweil: Humans will be hybrids by 2030. *CNNMoney.* [Online] June 4, 2015. http://money.cnn.com/2015/06/03/technology/ray-kurzweil-predictions/.

2. **Regalado, Antonio.** Pope Francis Said to Bless Human-Animal Chimeras. *MIT Technology Review.* [Online] January 27, 2016. https://www.technologyreview.com/s/546246/pope-francis-said-to-bless-human-animal-chimeras/.

3. **Amish America.** Do Amish use electricity? *Amish America.* [Online] http://amishamerica.com/do-amish-use-electricity/.

4. **The Guardian.** Bhutan: Fast forward into trouble . *The Guardian.* [Online] June 13, 2003. https://www.theguardian.com/theguardian/2003/jun/14/weekend7.weekend2.

5. **TIME Staff.** Exclusive: TIME Talks to Google CEO Larry Page About Its New Venture to Extend Human Life. *TIME Magazine.* [Online] September 18, 2013. business.time.com/2013/09/18/google-extend-human-life/.

6. **The Long Now Foundation.** About Long Now. *The Long Now Foundation.* [Online] http://longnow.org/about/.

7. **Mucci, Alberto.** Sweden's Minister of the Future Explains How to Make Politicians Think Long-Term. *Motherboard.* [Online] November 26, 2015. http://motherboard.vice.com/read/swedens-minister-of-the-future-explains-how-to-make-politicians-think-long-term.

8. **Lobosco, Katie.** Would you get free tuition under Hillary Clinton? *CNNMoney.* [Online] July 6, 2016. http://money.cnn.com/2016/07/06/pf/college/hillary-clinton-college-plan/.

9. **OLPC.** OLPC's mission is to empower the world's poorest children through education. *One Laptop per Child.* [Online] http://one.laptop.org/about/mission.

10. **Wolfram Research.** P Versus NP Problem. *Wolfram MathWorld.* [Online] http://mathworld.wolfram.com/PVersusNPProblem.html.

11. **Wikipedians.** Theory of Everything. *Wikipedia.* [Online] https://en.wikipedia.org/wiki/Theory_of_everything#Modern_physics.

12. **Boyle, Alan.** Gamers solve molecular puzzle that baffled scientists. [Online] NBC News, September 18, 2011. http://cosmiclog.nbcnews.com/_news/2011/09/18/7802623-gamers-solve-molecular-puzzle-that-baffled-scientists.

13. **Smith, Megan.** Computer Science For All. *The White House.* [Online] January 30, 2016. https://www.whitehouse.gov/blog/2016/01/30/computer-science-all.

14. **Iszler, Madison.** Florida Senate approves making coding a foreign language. *USA Today.* [Online] March 1, 2016. http://www.usatoday.com/story/tech/news/2016/03/01/florida-senate-approves-making-coding-foreign-language/81150796/.

15. **Kimball, Roger.** The Groves of Ignorance. [Online] 1987, 5 April. http://www.nytimes.com/1987/04/05/books/the-groves-of-ignorance.html?pagewanted=all.

16. **Shontell, Alyson.** PayPal Cofounder Peter Thiel Is Paying 24 Kids $100,000 To Drop Out Of School. *Business Insider.* [Online] May 28, 2011. http://www.businessinsider.com/paypal-cofounder-peter-thiel-is-paying-24-kids-100000-to-drop-out-of-school-2011-5.

17. **Flaherty, Colleen.** Not So Different: New Stanford programs aim to give computer science students a boost – by adding arts and humanities. *Inside Higher Ed.* [Online] March 7, 2014. https://www.insidehighered.com/news/2014/03/07/stanford-will-start-new-joint-computer-science-programs.

18. **Piller, Charles.** Verily, I swear. Google Life Sciences debuts a new name. *STAT.* [Online] December 7, 2015. http://www.statnews.com/2015/12/07/verily-google-life-sciences-name/.

19. **IMDb.** Dead Poets Society. *IMDb.* [Online] June 9, 1989. http://www.imdb.com/title/tt0097165/trivia?tab=qt&ref_=tt_trv_qu.

20. **McCracken, Harry and Grossman, Lev.** Can Google Solve Death. *TIME Magazine.* [Online] September 30, 2013. http://content.time.com/time/magazine/0,9263,7601130930,00.html.

21. **2045 Initiative.** AUBREY DE GREY / Interview / ENDING AGING . *YouTube.* [Online] November 5, 2015. https://www.youtube.com/watch?v=2lmdp96ySIU.

22. **O'Neill, Jim.** Tackling Drug-Resistant Infections Globally: Final Report and Recommendations. *Revew on Antimicrobial Resistance.* [Online] May 2016. http://amr-review.org/sites/default/files/160525_Final%20paper_with%20cover.pdf.

23. **National Priorities Project.** Tax Day 2016. *National Priorities Project.* [Online] March 10, 2016. https://www.nationalpriorities.org/analysis/2016/tax-day-2016/.

24. **Walker, Joseph.** Patients Struggle With High Drug Prices. *The Wall Street Journal.* [Online] December 31, 2015. http://www.wsj.com/articles/patients-struggle-with-high-drug-prices-1451557981.

25. **Dallas, Mary Elizabeth.** Out-of-Pocket Costs Rose Moderately Under Obamacare: Report. *HealthDay.* [Online] May 13, 2016. https://consumer.healthday.com/public-health-information-30/health-cost-news-348/out-of-pocket-costs-rose-slightly-under-obamacare-710970.html.

26. **Cheng, Roger.** If Donald Trump had his way, your iPhone would be insanely pricey. *CNet.* [Online] January 19, 2016. http://www.cnet.com/news/if-donald-trump-had-his-way-your-iphone-would-be-insanely-pricey/.

27. **Waters, Richard.** FT interview with Google co-founder and CEO Larry Page. *FT Magazine.* [Online] October 31, 2014. http://www.ft.com/cms/s/2/3173f19e-5fbc-11e4-8c27-00144feabdc0.html.

28. **Ford, Martin.** Anything You Can Do, Robots Can Do Better. *The Atlantic.* [Online] February 14, 2011. http://www.theatlantic.com/business/archive/2011/02/anything-you-can-do-robots-can-do-better/71227/.

29. **Conner-Simons, Adam.** Computer program fixes old code faster than expert engineers. *MIT News.* [Online] July 9, 2015. http://news.mit.edu/2015/computer-program-fixes-old-code-faster-than-expert-engineers-0609.

30. **Licklider, J. C. R.** Man-Computer Symbiosis. *IRE Transactions on Human Factors in Electronics.* [Online] March 1960. http://groups.csail.mit.edu/medg/people/psz/Licklider.html.

31. **Bucketlist.org.** Your dreams, made possible. *Bucketlist.* [Online] [Cited: March 7, 2016.] https://bucketlist.org/.

32. **Carnegie Mellon University.** Randy Pausch's Last Lecture. *Carnegie Mellon University.* [Online] September 18, 2007. http://www.cmu.edu/randyslecture/.

33. **Harbrecht, Douglas.** Google's Larry Page: Good Ideas Still Get Funded. *Bloomberg.* [Online] 2001, 12 March. http://www.bloomberg.com/news/articles/2001-03-12/googles-larry-page-good-ideas-still-get-funded.

34. **Sassard, Sophie, Bernouin, Michel and Bergin, Tom.** With 84 dead, France investigates whether truck attacker acted alone. *Reuters.* [Online] July 17, 2016. http://www.reuters.com/article/us-france-crash-idUSKCN0ZU2K7.

35. **FlorCruz, Jaime.** Execution does not stop Chinese knife attacks. *CNN.* [Online] May 3, 2010. http://www.cnn.com/2010/WORLD/asiapcf/05/02/china.attacks/index.html?hpt=T3.

36. **Morin, Rich.** The demographics and politics of gun-owning households. *Pew Research Center.* [Online] July 15, 2014. http://www.pewresearch.org/fact-tank/2014/07/15/the-demographics-and-politics-of-gun-owning-households/.

37. **Everytown for Gun Safety.** At Least 278 Child Shootings in 2015. *Everytown for Gun Safety.* [Online] http://everytownresearch.org/notanaccident/.

38. **The Washington Post.** 990 people shot dead by police in 2015. *The Washington Post.* [Online] https://www.washingtonpost.com/graphics/national/police-shootings/.

39. **American Society of Civil Engineers.** 2013 Report Card for America's Infrastructure. *American Society of Civil Engineers.* [Online] http://www.infrastructurereportcard.org/.

40. **Marsh, Rene, Gracey, David and Severson, Ted.** How to fix America's 'third world' airports. *CNN.* [Online] May 27, 2016. http://www.cnn.com/2016/05/25/politics/infrastructure-roads-bridges-airports-railroads/index.html.

41. **Satija, Neena, et al.** Hell and High Water. *ProPublica.* [Online] March 3, 2016. https://www.propublica.org/article/hell-and-high-water-text.

42. **CBS2.** CBS2 Investigates: Experts Say Decaying Gas Lines Are A Ticking Time Bomb Below City Streets. *CBS New York.* [Online] January 8, 2016. http://newyork.cbslocal.com/2016/01/08/new-york-gas-main/.

43. **Budds, Diana.** How Google Is Turning Cities Into R&D Labs: From autonomous vehicles to building codes, Sidewalk Labs is thinking about problems and solutions that could shape cities for centuries. *Fast Company & Inc.* [Online] February 22, 2016. http://www.fastcodesign.com/3056964/design-moves/how-google-is-turning-cities-into-rd-labs.

44. **National Priorities Project.** U.S. Military Spending vs. the World. *National Priorities Project.* [Online] https://www.nationalpriorities.org/campaigns/us-military-spending-vs-world/.
45. **Boston Dynamics.** Atlas, The Next Generation. *YouTube.* [Online] February 23, 2016. https://www.youtube.com/watch?v=rVlhMGQgDkY.
46. **McFarland, Matt.** 300-pound mall robot runs over toddler. *CNNMoney.* [Online] July 14, 2016. http://money.cnn.com/2016/07/14/technology/robot-stanford-mall/index.html.
47. **Thielman, Sam.** Use of police robot to kill Dallas shooting suspect believed to be first in US history. *The Guardian.* [Online] July 8, 2016. https://www.theguardian.com/technology/2016/jul/08/police-bomb-robot-explosive-killed-suspect-dallas.
48. **Lee, Newton.** Facebook Nation: Total Information Awareness. [Online] Springer-Verlag New York, 2014. http://www.springer.com/us/book/9781493917396.
49. —. Counterterrorism and Cybersecurity: Total Information Awareness. [Online] Springer International Publishing, 2015. http://www.springer.com/us/book/9783319172439.
50. **Long, Heather.** Who owns America's debt? *CNNMoney.* [Online] May 10, 2016. http://money.cnn.com/2016/05/10/news/economy/us-debt-ownership/.
51. **Metz, Cade.** For Google, Quantum Computing Is Like Learning to Fly. *Wired.* [Online] December 11, 2015. http://www.wired.com/2015/12/for-google-quantum-computing-is-like-learning-to-fly/.
52. **Hutchins, Aaron.** Trudeau versus the experts: Quantum computing in 35 seconds. *Maclean's.* [Online] April 19, 2016. http://www.macleans.ca/society/science/trudeau-versus-the-experts-quantum-computing-in-35-seconds/.
53. **Istvan, Zoltan.** Transhumanist Party. [Online] http://www.transhumanistparty.org/.
54. —. A New Generation of Transhumanists Is Emerging. *The Huffington Post.* [Online] May 10, 2014. http://www.huffingtonpost.com/zoltan-istvan/a-new-generation-of-trans_b_4921319.html.
55. —. The Era of Artificial Hearts Has Begun. *Motherboard.* [Online] September 9, 2014. http://motherboard.vice.com/read/era-artificial-heart-permanent-transplants.
56. —. Vietnam Villagers Find Profit, Risk in Bomb Hunting. *National Geographic Channel.* [Online] January 7, 2004. http://news.nationalgeographic.com/news/2004/01/0107_040107_tvbombdigger.html.
57. —. Volcano Boarding: the New Extreme Sport. *YouTube.* [Online] May 9, 2010. https://www.youtube.com/watch?v=fDQ2-EXVqYw.
58. **Prisco, Giulio.** The Transhumanist Wager. *Kurzweil Accelerating Intelligence.* [Online] May 15, 2013. http://www.kurzweilai.net/book-review-the-transhumanist-wager.
59. **Istvan, Zoltan.** Artificial Wombs Are Coming, but the Controversy Is Already Here. *Motherboard.* [Online] August 4, 2014. http://motherboard.vice.com/read/artificial-wombs-are-coming-and-the-controversys-already-here.
60. **Newport, Frank.** In U.S., 77 % Identify as Christian. *Gallup.* [Online] December 24, 2012. http://www.gallup.com/poll/159548/identify-christian.aspx.
61. **Istvan, Zoltan.** When Superintelligent AI Arrives, Will Religions Try to Convert It? *Gizmodo.* [Online] February 4, 2015. http://gizmodo.com/when-superintelligent-ai-arrives-will-religions-try-t-1682837922.
62. —. Which New Technology Will Win the Race to Repair and Replace Our Organs? *SingularityHUB.* [Online] November 2, 2014. http://singularityhub.com/2014/11/02/which-new-technology-will-win-the-race-to-repair-and-replace-our-organs/.
63. —. Singularity or Transhumanism: What Word Should We Use to Discuss the Future? *Future Tense.* [Online] August 28, 2014. http://www.slate.com/blogs/future_tense/2014/08/28/singularity_transhumanism_humanity_what_word_should_we_use_to_discuss_the.html.
64. —. The Transhumanist Philosopher. *Psychology Today.* [Online] https://www.psychologytoday.com/blog/the-transhumanist-philosopher.

65. **Dvorsky, George.** Craig Venter's new longevity startup will make "100 the new 60". *Gizmodo.* [Online] March 5, 2014. http://io9.gizmodo.com/craig-venters-new-longevity-startup-will-make-100-the-1536833065.

66. **Office of the Press Secretary.** Fact Sheet: BRAIN Initiative. *The White House.* [Online] April 2, 2013. https://www.whitehouse.gov/the-press-office/2013/04/02/fact-sheet-brain-initiative.

67. **Einstein, Albert.** Out of My Later Years. *Google Books.* [Online] 1956. https://books.google.com/books/about/Out_of_My_Later_Years.html?id=OBPAA3ZI4zcC.

68. **Goodman, Marc.** Marc Goodman: A vision of crimes in the future. [Online] TEDGlobal 2012, June 28, 2012. http://www.ted.com/talks/marc_goodman_a_vision_of_crimes_in_the_future.html?quote=1769.

69. **Achenbach, Joel.** The A.I. Anxiety. *The Washington Post.* [Online] December 27, 2015. http://www.washingtonpost.com/sf/national/2015/12/27/aianxiety/.

70. **Jha, Alok.** Neil Armstrong breaks his silence to give accountants moon exclusive . *The Guardian.* [Online] May 23, 2012. http://www.theguardian.com/science/2012/may/23/neil-armstrong-accountancy-website-moon-exclusive.

Index

Index

Index

Printed in the United States
By Bookmasters